Nanomagnetism

Nanomagnetism

An Interdisciplinary Approach

Georgia C. Papaefthymiou

CRC Press is an imprint of the
Taylor & Francis Group, an **informa** business
A CHAPMAN & HALL BOOK

First Edition published 2022
by CRC Press
6000 Broken Sound Parkway NW, Suite 300, Boca Raton, FL 33487-2742

and by CRC Press
4 Park Square, Milton Park, Abingdon, Oxon, OX14 4RN

CRC Press is an imprint of Taylor & Francis Group, LLC

ISBN: 978-1-439-81846-6 (hbk)
ISBN: 978-1-032-14090-2 (pbk)
ISBN: 978-1-315-15701-6 (ebk)

DOI: 10.1201/9781315157016

Typeset in Palatino
by SPi Technologies India Pvt Ltd (Straive)

eResources: Support materials are available for this title at https://www.routledge.com/9781439818466

To the memory of my parents, the love of my children, and my precious granddaughter, Georgia Sue.

Contents

Preface

Nanomagnetism is an area of interest to scientists and engineers across scientific disciplines. The field has experienced immense growth in the last two decades due to advancements in sample preparation, nanopatterning techniques and magnetic measurement instrumentation. New discoveries are scattered throughout the scientific literature and conference proceedings. Presentation of these novel results is often given without an explanation of the underlying fundamental principles. Yet, understanding these principles is essential for completed appreciation of the impact of the innovation for a broad range of investigators.

Colleges and universities across the United States, and across the World, are struggling to meet the needs of the new interdisciplinary age of nanoscience. They increasingly appropriate funds for the development and introduction of new interdisciplinary science courses. Scientists working in industrial laboratories—in nanotechnology and biotechnology—need to master concepts across disciplinary boundaries. Pharmaceutical and medical researchers and physicians need an education grounded in the physics of the magnetic probes introduced to patients for imaging enhancement, drug delivery, or hyperthermia therapy.

As a result, the 21st century has witnessed unprecedented activity in interdisciplinary science. The boundaries between disciplines have become diffuse, opening vast new areas of research and development at the interface of traditional science domains. Students and research practitioners address interdisciplinary problems early on in their scientific careers, which requires a background in more than one fundamental field. Many collaborative research initiatives among academic, national and industrial laboratories demand the establishment of a new interdisciplinary scientific culture. It has become imperative that scientists trained in one discipline understand the language of another if they are to participate in fruitful interdisciplinary collaborations.

There is no other area of science where the blurring of scientific disciplines is truer than nanoscience. Nanoscience lies at the molecular/solid interface where quantum mechanics meets classical physics. The ability to manipulate matter at the atomic scale has brought materials science and engineering to the crossroads of physics and chemistry. The continuous unraveling of the molecular underpinnings of biological processes has made possible the new area of biomimetic materials science and the synthesis of bio-inspired functional nanomaterials, further marrying materials science to biology and medicine.

This book is conceived as a response to the perceived need for textbooks in new interdisciplinary courses in nanomagnetism and magnetic nanomaterials and their many applications across the disciplines. The first part of the book introduces the fundamentals of magnetism assuming only basic familiarity with calculus, classical physics and quantum chemistry. Latter parts deal with the synthesis, characterization and applications of magnetic nanoparticles and nanocomposites of various dimensionalities in science, medicine and technology.

Preface

Acknowledgment

I would like to acknowledge my PhD mentor, C. S. Wu at Columbia University, for inspiring me to life-long scholarship and my post-doctoral mentor, R. B. Frankel at MIT, for teaching me how to think as a physicist. Without their influence, this book could have never been written.

Introduction

Manifestations of magnetism abound in nature. Our planet is a giant magnetic dipole. The geomagnetic field is responsible for natural phenomena such as rock magnetism, polar auroras, magnetic compass navigation, bird migration and bacterial magneto taxis. The scientific discipline of magnetism is perhaps the oldest known to mankind. Magnetite was known to the ancient Greeks and was put to use by the ancient Chinese in the development of the magnetic compass. However, magnetism still presents formidable challenges to science and technology in the 21st century, within the new area of nanomagnetism.

The field of magnetism—from the experimental observations of Faraday and Ampère on current-carrying wires, to the grand synthesis of Maxwell's theory of classical electrodynamics—is established firmly within the realm of classical physics. But magnetism in materials is inherently a quantum mechanical phenomenon anchored in the Exchange Interaction and the Pauli Exclusion Principle. It should therefore come as no surprise that magnetic materials of nanosize dimensions—a length scale where the classical properties of solids are modified by quantum mechanical behavior associated with atoms and molecules—exhibit a wealth of novel magnetic behavior, or quantum-size effects, to be understood in its fundamental underlying principles and harnessed for practical technological applications.

The fundamental entity of magnetism is the magnetic dipole, which can be defined both classically, through the magnetic field of a current-carrying wire loop, and quantum mechanically, through the orbital and spin angular momenta of electrons circulating the nucleus. Thus, all matter exhibit magnetic properties, classified by its response to an external magnetic field into diamagnetism, paramagnetism, superparamagnetism, or more complex magnetic behaviors due to the long-range magnetic order of individual atomic moments.

Magnetic ordering in solids is a collective or cooperative phenomenon not associated with single atoms or molecules. It leads to different types of ordering: ferro-magnetism, antiferromagnetism, ferri-magnetism, heli-magnetism, spin glass and other types of magnetism. Thus, a fundamental question in nanomagnetism is: How many magnetically interacting or spin-exchange-coupled atoms are needed for the onset of cooperative magnetic ordering? This question has been addressed both theoretically and experimentally in atomic cluster beam experiments and in supramolecular cluster chemistry investigations. These studies attempt to delineate the cluster/particle boundary in terms of magnetic behavior and thus define the smallest possible size of magnetic nanoparticles.

The upper limit in nanoparticle size range arguably could be defined by the characteristic length scale of a single magnetic domain, of the order of 100 nm, within which all atomic moments theoretically are oriented along a single preferred crystallographic axis. In the bulk, magnetostatic energy is minimized by the spontaneous formation of magnetic domain walls and random orientation of individual domains, rendering the overall material *macro*scopically non-magnetic. Thus, the study and the technological applications of nanomagnetism require the preparation of well-defined, monodispersed particle assemblies of single- or sub-single-magnetic domain size. The nanoparticle magnetic parameters to be characterized and tailored to specific applications are the uniaxial magnetic anisotropy density, K_u, and the magnetic coercivity, H_c. K_u imparts bistability to the nanoparticle and determines its superparamagnetic relaxation properties; H_c relates to spin reversal mechanisms. The former is of interest in biomedical applications while the latter, in the design of magnetic recording media.

Life has evolved in the presence of the Earth's magnetic field. Iron—the paramount magnetic element—is plentiful on the surface of the Earth. In living organisms, we encounter examples of magnetic nanoparticle formation by the integration of inorganic magnetic nanoparticles within organic tissue *via* the process of biomineralization. Magnetic biomineral cluster nucleation and growth is invariably promoted within the confined spaces of preformed membrane vesicles or protein cages that limit particle growth to a single- or sub-single magnetic domain size, as encountered in magnetotactic bacteria and the iron storage protein ferritin, respectively. This paradigm of biology has inspired today's nanoengineers to use bottom-up synthetic approaches of nanoparticle formation and stabilization within preformed porous, polymeric, or inorganic templates or to coat the nanoparticles by forming elaborate nanoarchitectures through sol-gel or microemulsion processes. The resulting nanoparticles—smaller than the biological cell which is of micrometer dimensions—can be functionalized further to interact with specific cell receptors and even enter the cell, of interest in targeted drug delivery and gene transfection studies. Furthermore, bifunctional magnetic nanoparticles have been prepared successfully to possess both magnetic and fluorescent properties. These advances have major implications for disease diagnosis and treatment modalities.

Our newfound ability to manipulate matter at nanoscale dimensions has ushered the nano-age of the 21st century and the new fields of nanoscience, nanotechnology and nanomedicine. It has been realized that the novel magnetic behavior of nanomagnets, and therefore their engineering to specific applications, originates from the large number of atoms lying at the surface or interface sites where lattice strain and spin canting exist due to the abrupt interruption of crystallographic and magnetic order at the surface. Nanocomposites, comprised of magnetic nanoparticles embedded within a non-magnetic matrix, add further complexity and versatility to magnetic behavior. The magnetic nanoparticles or the host matrix may be conducting or insulating in nature. In conductive matrices, novel electronic transport properties have been observed *via* electronic spin scattering at interfaces which gives rise to giant magnetoresistance (GMR) in magnetic granular media and magnetic multilayers. The scientists, Albert Fert and Peter Grünberg, credited for the discovery of the GMR effect, were awarded the 2007 Nobel Prize in Physics. Their work has enabled the introduction of new read-head technology to be used with high-density magnetic recording media. Current research in spintronics and molecular electronics promise to further revolutionize information processing in the future.

This book presents the fundamentals of magnetism pertaining to magnetic nanostructures, that is, magnetic finite-size effects at the nanoscale and their applications to nanotechnology and biomedicine. It also discusses the physical, chemical and nanotemplating synthesis techniques for the production of magnetic nanoparticles and reviews experimental techniques that have been critical to the determination of the *macro*scopic and *micro*scopic magnetization of nanoparticles. Studies of the onset of magnetic ordering with increasing cluster size and the transition from the molecular to the solid state are presented along with explanations of the synthesis and characterization of functionalized core/shell nanoparticles for applications in biotechnology and nanomedicine. The book explores the role of nanomagnetism in high-density magnetic recording media, nanostructured permanent magnets, biotechnology and nanomedicine.

Part I

Fundamental Concepts in Magnetism and Magnetic Materials

1

The Magnetic Field

Faraday, in his mind's eyes, saw lines of force traversing all space.

James Clerk Maxwell A Treatise on Electricity and Magnetism (1873)

1.1 Overview and Historical Background

Magnetic and electrical forces of magnetite and rubbed amber (ελέκτρον in Greek), respectively, were known to the ancient Greeks. Magnetite, a magnetic oxide of iron mined in Magnesia (Μαγνησία), an ancient Greek province in Asia Minor, is mentioned in Greek texts as early as the 8th century BC. Thales of Miletus (Θαλής ο Μιλήσιος), the first natural philosopher according to Aristotle, was the first to study magnetic forces in the 6th century BC. The oldest practical application of magnetite is in the construction of the magnetic compass that some historians attribute to the Chinese as far back as the 26th century BC. The modern use of the magnetic compass for navigation purposes dates to *ca.* 1100 AD. Since antiquity, interest in magnetism and electrification has never left science and remains a major focus today.

William Gilbert is considered the father of the Modern Science of Magnetism and Electricity. He spent many years experimenting with magnetism and assembled his results and all that was known about the subject in his treatise "On the Magnet, Magnetic Bodies, and the Great Magnet of the Earth" (*De Magnete, Magneticisque Corporibus, et de Magno Magnete Tellure*), published in 1600. He introduced the term electric force as the force of interaction between two bodies electrified by rubbing and magnetic force as the force of interaction between the poles of two magnetized bodies. Similar poles repelled each other, just like similar charges do, and unlike poles attracted each other, as opposite charges do. He noted, however, that there was a fundamental difference between electric and magnetic interactions; two electrified objects attracted or repelled each other, while two magnetized objects tended to align relative to each other. He also alluded to the non-existence of the magnetic monopole as he noted that a bar magnet cut in half resulted in two smaller bar magnets rather than two magnetic monopoles.

The formulation of the quantitative laws of electrostatics, magnetostatics (steady magnetic fields) and quasistatics (slowly varying magnetic fields) dates back to the 18th and 19th centuries with the work of Charles-Augustin de Coulomb, Hans Christian Oersted and Michael Faraday. In 1820, Oersted discovered that a current-carrying wire produces a magnetic field when he accidentally put the wire close to a compass and noticed that the compass turned in a direction perpendicular to the wire. His finding, together with the

DOI: 10.1201/9781315157016-2

subsequent discovery by Faraday that a changing magnetic field induces an electric current in a nearby wire circuit, established the inextricable interconnection between electric and magnetic phenomena. Theoretical work by James Clerk Maxwell, inspired by Faraday's observations, completed this interconnection by showing that a changing electric field produces a magnetic field, culminating with the formulation of the famous Maxwell's equations of Electrodynamics that are the basis of most modern electro-technology.

In "A Dynamical Theory of the Electromagnetic Field" published in 1865, Maxwell predicted the existence of electromagnetic waves that propagate through space with the speed of light. In other words, Maxwell deduced that light is an electromagnetic wave, unifying optics with electricity and magnetism. The existence of electromagnetic waves was confirmed experimentally in 1888, almost ten years after Maxwell's death, by the generation and detection of radio waves by Heinrich Hertz. Maxwell's work was what propelled Albert Einstein to develop his theory of special relativity and establish magnetism as a relativistic phenomenon of electrostatics. In this way, electrodynamics has played a fundamental role in the development of modern science as we know it today.

Within magnetic materials, magnetic ordering is governed by quantum mechanisms making magnetism inextricably connected with yet another fundamental scientific theory, quantum mechanics. Nanoscale magnetic materials, or nanomagnets, thus exhibit properties associated with both classical and quantum physics. These properties, and their technological applications, are the subject of our explorations in this book. Before proceeding, however, we must introduce fundamental concepts in classical and quantum magnetism, which we must master in order to be able to investigate and appreciate the new field of nanomagnetism.

1.2 The Magnetic Field of a Current-Carrying Wire

The region of space around a current-carrying wire is modified, in the sense that it possesses properties non-existent in the absence of the current. This idea can be most simply described by the observation that a magnetic compass placed in the vicinity of the current-carrying wire experiences a torque that rotates it away from the direction of the Earth's magnetic field and orients it into a new direction, as originally observed by Oersted. Upon reducing the current in the wire to zero, the compass returns to its original orientation, along the direction of the Earth's magnetic field, as depicted in Figure 1.1. We ascribe this modification of space to the existence of a magnetic field, $\vec{B}$, which we wish to depict with magnetic field lines as we do with the Earth's magnetic field or the magnetic field around a bar magnet, shown in Figure 1.2.

Field lines form closed loops emanating from the magnet's North Pole (N) and retuning through the magnet's South Pole (S). The density of field lines is high inside the bar magnet and near the poles, where the magnetic field is strong, and low at points away from the poles, where the magnetic field is weak. Thus, a quantitative depiction of the magnetic field around the current-carrying wire would require not only determination of the direction of orientation of the compass, which determines the direction of the magnetic field at that point in space, but also the magnitude of the torque experienced by the compass, which determines the strength of the magnetic field at that point.

Alternatively, this property of space ascribed to the presence of the magnetic field $\vec{B}$ can be described by a second observation, namely that a charged particle moving in a region of

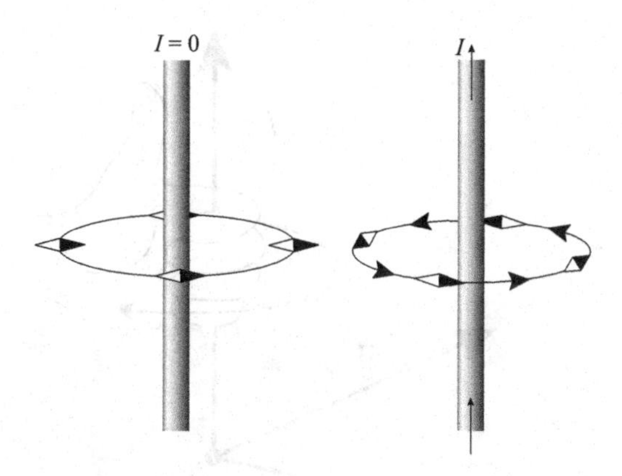

FIGURE 1.1
Orientation of magnetic compass in the region of space around a wire. On the left, in the absence and on the right, in the presence of a current I.

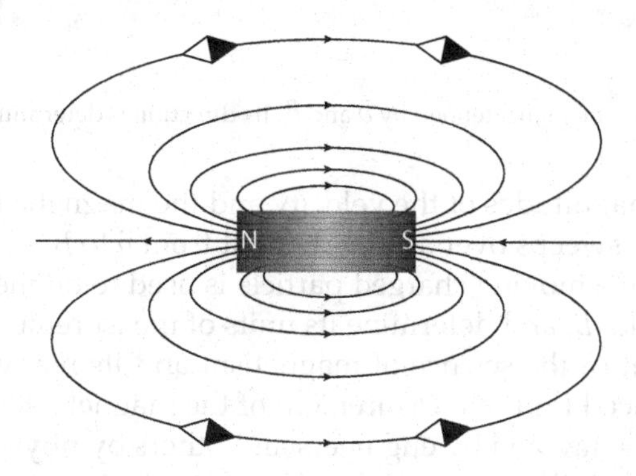

FIGURE 1.2
Depiction of the magnetic field around a bar magnet and orientation of a magnetic compass.

space where a magnetic field exists is deflected from its path. The direction of deflection depends on the sign of the charge, positive or negative, and on the direction of its instantaneous velocity, $\vec{v}$. Thus, according to Newtonian mechanics, the moving charged particle must experience a force. This magnetic force is given by Eq. (1.1)

$$\vec{F}_B = q\vec{v} \times \vec{B} \tag{1.1}$$

It is defined in terms of the cross (or vector) product of two vectors: the instantaneous velocity, $\vec{v}$, of the particle and the magnetic field vector, $\vec{B}$, at the location of the particle. The direction of the force is determined by the Right-Hand Rule (RHR); when your palm is aligned along the direction of the velocity and then rotated into the direction of the magnetic field, your thumb points along the direction of the force, as shown in Figure 1.3, provided the charge q is positive. If q is negative, simply flip your thumb over into the opposite direction to obtain the direction of the force.

Using the definition of the cross product, we recognize that the magnetic force will vanish if $\vec{v}$ and $\vec{B}$ are collinear vectors, that $\vec{F}_B$ is experienced in a direction normal to both $\vec{v}$ and $\vec{B}$, that the magnitude of the force is given by

$$F_B = qvB\sin\theta \tag{1.2}$$

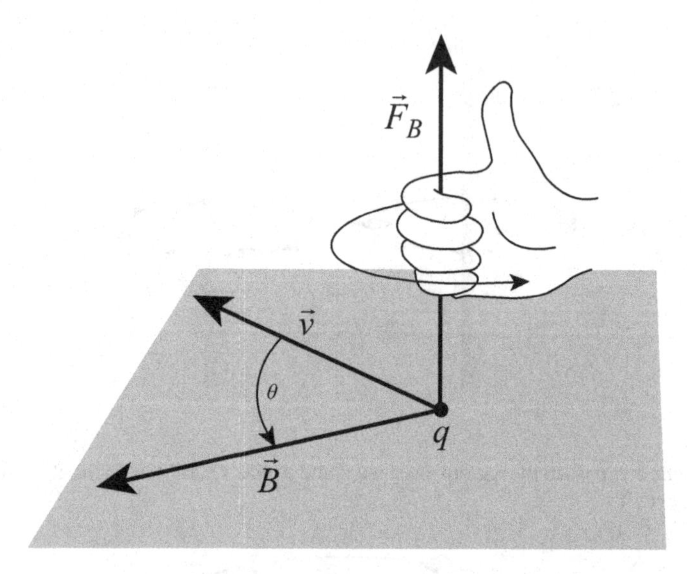

FIGURE 1.3
The magnetic force is normal to the plane defined by $\vec{v}$ and $\vec{B}$. Its direction is determined by the RHR, as depicted.

where v and B are the magnitudes of the velocity and the magnetic field, respectively, and θ is the angle your palm sweeps over as you rotate it from $\vec{v}$ to $\vec{B}$.

The magnetic force on a moving charged particle is used to define the physical quantity we call the magnetic field, $\vec{B}$, and determine its units of measurement. Due to the long history of the development of the science of magnetism and its many applications, various units have been introduced for the measurement of the magnetic field and other magnetic physical quantities, some favored by engineers and others by physicists. The SI system of units (from the French *Le Système International d' Unités*), which was internationally adopted in 1960, is the system we use primarily in this book. In SI units, the force is measured in newtons (N), the charge in coulombs (C), the velocity in meters per second (m/s) and the magnetic field in Tesla (T). Thus, 1 T is the strength of the magnetic field that exerts a force of 1 N on a particle carrying a charge of 1 C and moving with a velocity of 1 m/s perpendicular to $\vec{B}$. Thus,

$$\mathrm{T} = 1\mathrm{N}\cdot\mathrm{s}/1\mathrm{C}\cdot\mathrm{m}$$

Other commonly used units are the Gauss (G) or kilo Gauss (kG) and the Oersted (Oe) or kilo Oersted (kOe); 1 T = 10 kOe = 10 kG.

From detailed investigations of the torque exerted by a current-carrying conductor on a magnet, Jean-Batiste Biôt and Félix Savart arrived at an expression, known as the Biôt–Savart Law, for calculating the magnetic field $\vec{B}$ around a conductor as given in Eq. (1.3) and depicted in Figure 1.4.

In Eq. (1.3), $d\vec{s}$, measured in meters, is an infinitesimal displacement vector in the direction of the current I, measured in amperes (A), r, measured in meters, is the distance from $d\vec{s}$ to point P, the point where the magnetic field is being calculated and $\hat{r}$ is the unit vector pointing from $d\vec{s}$ to P. The magnetic field is then given in Tesla.

$$d\vec{B} = \frac{\mu_0}{4\pi}\frac{Id\vec{s}\times\hat{r}}{r^2} \tag{1.3}$$

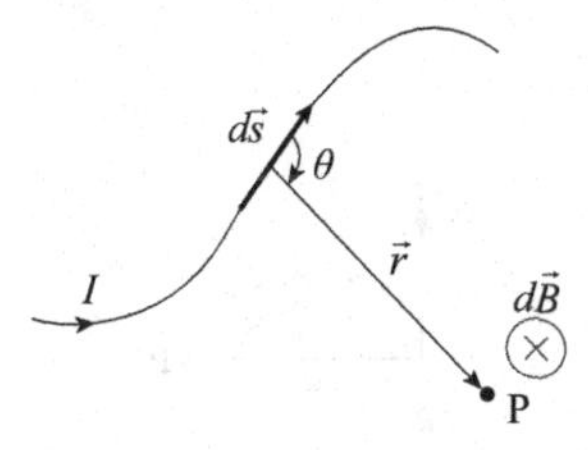

FIGURE 1.4
According to the Biôt–Savart Law, the current element $Id\vec{s}$ contributes a field $d\vec{B}$ at point P, directed into the page.

The constant μ_0 is called the permeability of free space and is given by

$$\mu_0 = 4\pi \times 10^{-7}\, T \cdot \text{m/A}$$

The total magnetic field $\vec{B}$ at P is in the direction of $d\vec{s} \times \hat{r}$ and may be calculated by integrating the expression of Eq. (1.3) over the length of the conductor.

$$\vec{B} = \frac{\mu_0}{4\pi} \int \frac{Id\vec{s} \times \hat{r}}{r^2} \tag{1.4}$$

The above equation is ensured by the fact that the vector field $\vec{B}$ obeys the superposition principle; that is, the contribution $d\vec{B}$ to the field from a certain current element, $Id\vec{s}$, is independent from those of other parts of the conductor, leading to the total $\vec{B}$ being the direct vector sum of each independent contribution $d\vec{B}$, given by Eq. (1.4) above.

This expression for $\vec{B}$ is the most general and may be applied to any current distribution. Carrying out this vector integration, however, may be extremely challenging depending on the geometry of the problem. Analytical expressions may be derived for some geometries of high symmetry; however, for very complicated geometries, it may only be amenable to numerical integration.

Example 1.1: The Magnetic Field of a Current-Carrying Wire

Let us use the Biôt–Savart Law to calculate the magnetic field $\vec{B}$ at point P, a distance a away from a straight, infinitely long wire carrying a current I in the z-direction, as shown in Figure 1.5.

The cross product $d\vec{s} \times \hat{r}$ determines the direction of the magnetic field $d\vec{B}$ contributed at P by each current element $Id\vec{s}$. It is clear that the contributions from all current elements are perpendicular to the page and pointing into the page. Thus, contributions from different sections of the wire are collinear, leading to the reduction of the vector integral of Eq. (1.4) to a simple scalar integral over the entire length of the wire

$$B = \frac{\mu_0}{4\pi} \int_{-\infty}^{\infty} \frac{Ids \sin \beta}{r^2} \tag{1.5}$$

where β is the angle between $d\vec{s}$ and $\hat{r}$ and $\vec{r}$ is the displacement vector from $d\vec{s}$ to P. Using trigonometric identities and relationships pertaining to the triangle formed by line

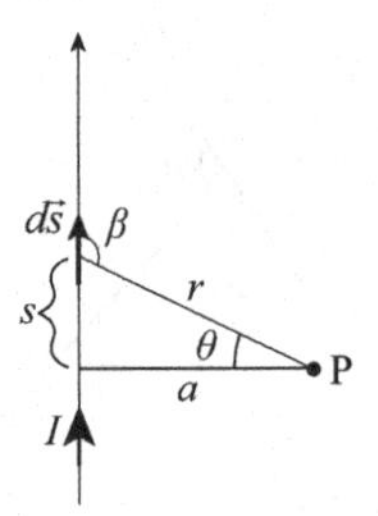

FIGURE 1.5
Calculation of the magnetic field around an infinitely long, straight current-carrying wire in the z-direction.

segments a, s and r in Figure 1.5, we recognize that $\sin\beta = \cos\theta$, $r^2 = a^2 + s^2$, $s = a\tan\theta$ and $\frac{ds}{d\theta} = a\frac{d\tan\theta}{d\theta}$ or $ds = a\,(1+\tan^2\theta)d\theta$. The integral in Eq. (1.5) can be simplified by changing the integration variable from ds to $d\theta$, to yield

$$B = \frac{\mu_0}{4\pi} I \int_{-\pi/2}^{\pi/2} \frac{\cos\theta a\left(1+\tan^2\theta\right)d\theta}{a^2\left(1+\tan^2\theta\right)} = \frac{\mu_0}{4\pi a} I \int_{-\pi/2}^{\pi/2} \cos\theta d\theta = \frac{\mu_o I}{2\pi a} \tag{1.6}$$

The strength of the magnetic field is proportional to the magnitude of the current, I, and inversely proportional to the distance, a, of point P from the wire.

In highly symmetrical situations, however, one may use the symmetry in the geometry of the problem to arrive at the value of the magnetic field without having to compute complicated integrals. This is achieved by using an alternate law that relates the magnetic field $\vec{B}$ to its source, the current I, in highly symmetric situations. Ampère's Circuital Law of Eq. (1.7) states that the line integral of the magnetic field around a closed loop, called Ampèrian loop, is equal to μ_0 multiplied by the current enclosed within the Ampèrian loop. The integrand, $\vec{B}\cdot d\vec{l}$, in Eq. (1.7) indicates the dot (or scalar) product, between vectors $\vec{B}$ and $d\vec{l}$, $\vec{B}\cdot d\vec{l} = B(dl)\cos\theta$, where θ is the angle between $\vec{B}$ and $d\vec{l}$.

$$\oint \vec{B}\cdot d\vec{l} = \mu_0 I_{enclosed} \tag{1.7}$$

Consider the special, highly symmetric cylindrical geometry of the straight, infinitely long, current-carrying wire of Example 1.1, as depicted in Figure 1.6. For each infinitesimal $d\vec{s}$, the Biôt–Savart Law gives a magnetic field contribution $d\vec{B}$, which (a) lies in the plane normal to the wire that contains point P, (b) is circumferential around the wire and (c) has the same magnitude for all points lying on the locus of the circumference of the circle of radius a, centered at the wire.

Applying Ampère's Law to this problem, we choose an Ampèrian loop to be a circle in the plane perpendicular to the wire, centered at the wire, and of radius a, the distance at which we wish to evaluate the magnetic field. $\vec{B}$ and $d\vec{l}$ are circumferential, collinear vectors pointing in the same direction and thus, $\vec{B}\cdot d\vec{l} = Bdl$. The integral on the left side of Eq. (1.7) simplifies to the product of B, constant over the Ampèrian loop and the length of the Ampèrian loop, $2\pi a$. The strength of the magnetic field, B, as a function of a, is then readily derived, given by Eq. (1.8). B decreases inversely proportional with respect to a.

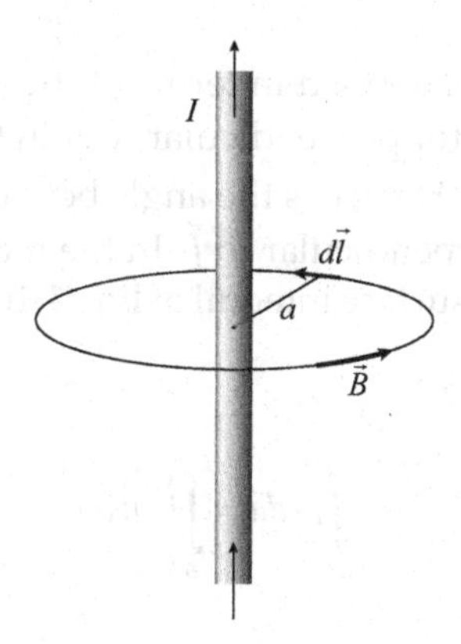

FIGURE 1.6
Ampère's Circuital Law. Use a circular loop concentric to the current as the Ampèrian loop. $\vec{B}$ is tangential to the Ampèrian loop, as shown.

The direction of $\vec{B}$ can be obtained by the RHR; point the thumb of your right hand along the direction of the current, then your fingers rotate in the direction of $\vec{B}$, as shown in Figure 1.6.

$$B = \frac{\mu_0 I}{2\pi a} \tag{1.8}$$

Note that the result obtained is identical to that of Eq. (1.6), which was derived using the Biôt–Savart Law. The magnetic field in the vicinity of a long, straight, current-carrying wire can thus be approximated to the field derived above for the idealized symmetry of the infinitely long wire, as long as we stay away from the edges of the wire. Its direction is circumferential, or in cylindrical coordinates, along the ϕ-direction. Thus,

$$\vec{B} = \frac{\mu_0 I}{2\pi a}\hat{\phi} \tag{1.9}$$

The field is non-uniform, its direction and strength vary as you move around in the region of space in the vicinity of the wire.

We can apply Ampère's Law to determine the magnetic field inside the wire as well, if we know the current density $\vec{J}$ inside the conductor. The current density is defined as the amount of current per unit cross-sectional area, normal to the direction of the current, flowing in the conductor. For a general cross section of area A, shown in Figure 1.7, and uniform

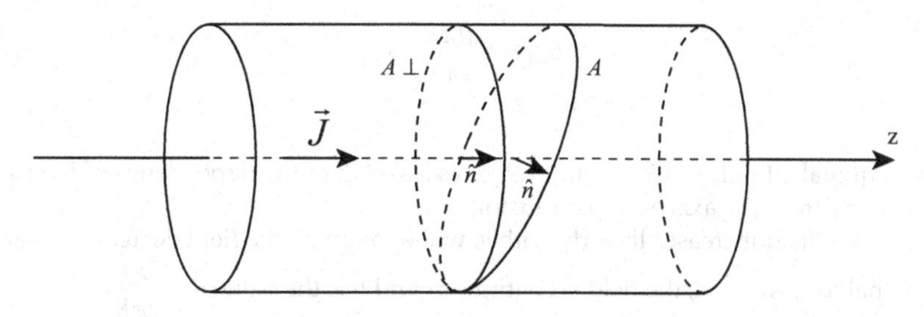

FIGURE 1.7
Depiction of current density $\vec{J}$ and cross-sectional area A in a current-carrying conductor.

current density $\vec{J}$, $I = \vec{J} \cdot \vec{A} = \vec{J} \cdot \hat{n} A$, defined in terms of the dot or scalar product of vectors $\vec{J}$ and $\hat{n}$, with $\hat{n}$ being the unit vector perpendicular to A in the general direction of the current, or, $I = \vec{J} \cdot \hat{n} A = JA\cos\theta = JA_{\perp}$. Here, θ is the angle between $\vec{J}$ and $\hat{n}$, while the dot product picks the component of A perpendicular to $\vec{J}$. In the most general case of non-uniform $\vec{J}$, the total current is given by the surface integral of Eq. (1.10) taken over any cross-sectional area A of the conductor.

$$I = \int_A \vec{J} \cdot d\vec{a} = \int_A \vec{J} \cdot \hat{n}\, da \tag{1.10}$$

In Example 1.2, we make use of the current density to derive expressions for the magnetic field inside and outside an infinitely long, straight, conducting wire.

Example 1.2: The Magnetic Field of an Infinitely Long, Straight Conductor of Radius *R*

Consider an infinitely long, straight conductor with radius R, shown in Figure 1.8. It carries uniform current density $\vec{J}$ in the z-direction. Determine the magnetic field, magnitude and direction, inside and outside the conductor as a function of a, the distance from the z-axis.

The problem has axial symmetry. We can apply Ampère's Circuital Law of Eq. (1.7) to determine $\vec{B}$. Due to symmetry, you expect the magnetic field to be circumferential around the z-axis, its direction being along $\hat{\varphi}$ according to the RHR. Choose a circular Ampèrian loop of radius a, where B is to be evaluated, centered on the axis, as shown in Figure 1.8. Inside the conductor, Ampèrian loop 1, the amount of current enclosed within the Ampèrian loop is not the total current flowing in the conductor. Therefore, $\oint \vec{B}_{in} \cdot d\vec{l} = \mu_0 I_{enclosed} = \mu_0 \int_A \vec{J} \cdot d\vec{a} = \mu_0 J \pi a^2 = B_{in} 2\pi a$ which gives

$$\vec{B}_{in} = \frac{\mu_0 J a}{2}\hat{\varphi} \tag{1.11}$$

Outside the conductor, the Ampèrian loop 2 encloses the total current $I = J\pi R^2$, so

$$\vec{B}_{out} = \frac{\mu_0 J \pi R^2}{2\pi a}\hat{\varphi} \tag{1.12}$$

This is equivalent to Eq. (1.9). Figure 1.9 gives a sketch of the dependence of B on a, the distance away from the axis of the conductor.

For $a < R$ the field increases linearly with a, while for $a > R$, the field decreases inversely proportional to a. At $a = R$, the field is continuous and has the value $\vec{B} = \frac{\mu_0 I}{2\pi R}\hat{\varphi}$.

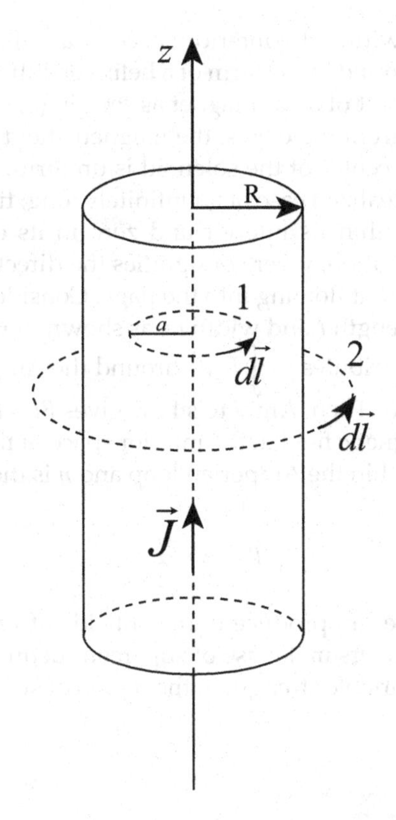

FIGURE 1.8
Depiction of Ampèrian loops, inside and outside the conductor, used in determining $\vec{B}$.

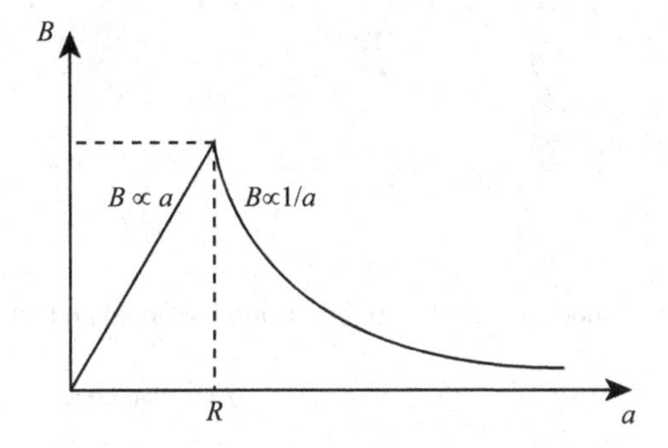

FIGURE 1.9
Dependence of the strength of the magnetic field with distance from the axis.

1.3 Solenoids and Uniform Magnetic Fields

Many applications in the study and characterization of magnetic materials, including nanomaterials, require the use of homogeneous or uniform magnetic fields. Such homogeneous magnetic fields can be produced in the gap between the poles of a horseshoe magnet shown in Figure 1.10(a), or within the interior space of a tightly wound solenoid, (b). A long current-carrying wire wound in the form of a helix constitutes a solenoid. Incidentally, its magnetic field resembles that of a bar magnet as seen in (c).

As long as we stay away from the edges, the magnetic field at the center of the gap of the horseshoe magnet or the center of the solenoid is uniform, that is, constant in magnitude and direction. For the idealized case of an infinitely long, tightly wound solenoid, the magnetic field is uniform within its interior and zero in its exterior space. Figure 1.11 depicts such an idealized situation, where ⊙ signifies the direction of current flowing out of the page and ⊗ that of current flowing into the page. Consider an Ampèrian loop in the shape of a parallelogram of length ℓ and width w, as shown in Figure 1.11.

The only segment that contributes to $\oint \vec{B} \cdot d\vec{l}$ around the Ampèrian loop is the segment of length ℓ inside the solenoid. Then, Ampère's Law gives $Bl = \mu_0 NI$, or $B = \mu_0(N/l)I = \mu_0 nI$ for the magnitude of the magnetic field in the interior space of the solenoid, where N is the number of loops enclosed within the Ampèrian loop and n is the number of loops per unit length of the solenoid. Thus,

$$\vec{B} = \mu_0 nI\,\hat{z} \tag{1.13}$$

By varying the current, we can produce uniform fields of variable strength within the solenoid. Modern magnetometers make use of superconducting solenoid coils to produce uniform magnetic fields of variable strength in the region of space close to the center of the solenoid.

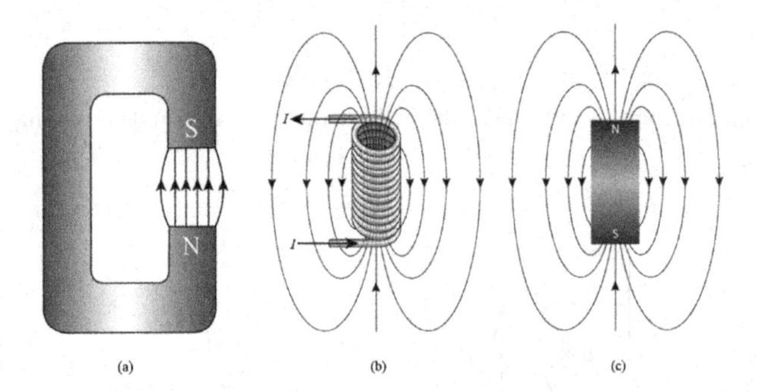

FIGURE 1.10
Magnetic field lines for (a) a horseshoe magnet, (b) a tightly wound solenoid and (c) a bar magnet.

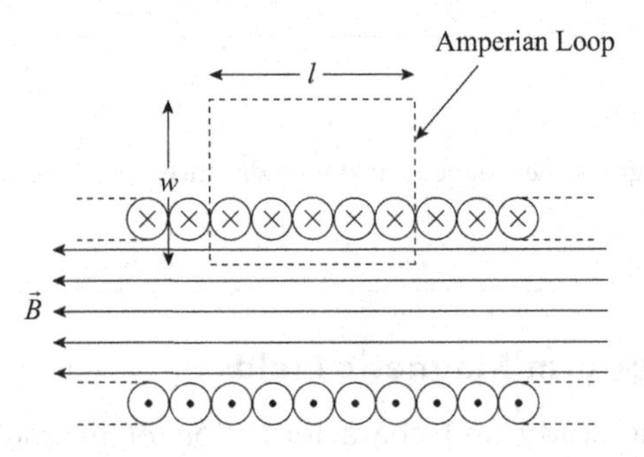

FIGURE 1.11
Calculating the magnetic field in the interior of an infinitely long solenoid oriented along the z-direction of a cylindrical coordinate axis system.

1.4 The Lorentz Force and Cyclotron Motion

Let us consider such a region of space where a homogeneous magnetic field exists. Figure 1.12 shows a cross section perpendicular to the direction of the magnetic field, which is directed into the page.

A positively charged particle, of mass m and charge q moving with instantaneous velocity $\vec{v}$, say in the y-direction, experiences a magnetic force according to Eq. (1.1). The force is normal to both $\vec{v}$ and $\vec{B}$ and, therefore, lies in the plane of the page, as shown. It deflects the positively charged particle from its path (to the left), but it does not speed it up or slow it down, as it acts normal to its velocity and therefore its displacement, $d\vec{l} = \vec{v}dt$. Therefore, the amount of work done over an infinitesimal displacement $d\vec{l}$, is identically equal to zero ($dW = \vec{F}_B \cdot d\vec{l} = q\left(\vec{v} \times \vec{B}\right) \cdot \vec{v}dt \equiv 0$). This is a fundamental observation about magnetic forces and is often stated as "magnetic forces do no work". Thus, the magnitude of $\vec{v}$ remains equal to its original value, but its direction changes continuously. The trajectory of the particle is circular, with $\vec{F}_B$ acting as the centripetal force and imparting a centripetal acceleration $a_c = v^2/r$ to the particle, r being the radius of the particle's circular trajectory. Applying Newton's second law of motion, $\vec{F} = m\vec{a}$, we obtain Eq. (1.14), which relates the magnitudes of the magnetic field and the velocity of the particle to the radius of its circular trajectory.

$$qvB = \frac{mv^2}{r} \tag{1.14}$$

The particle executes uniform circular motion with $r = mv/qB$ and rotational period $T = 2\pi r/v = 2\pi m/qB$, or equivalently, angular frequency, ω, given by Eq. 1.15.

$$\omega = \frac{qB}{m} \tag{1.15}$$

We note that the angular frequency is independent of the radius r or the velocity v. This is known as the "cyclotron frequency". It is this property of the circular trajectories of charged particles confined within homogeneous magnetic fields on which the design of

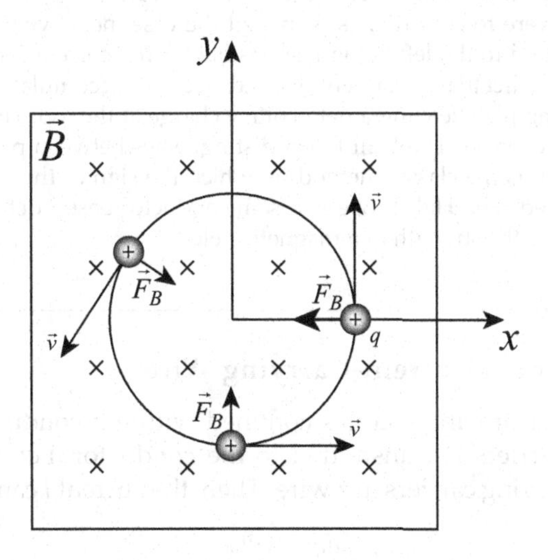

FIGURE 1.12
Positive charge q with velocity $\vec{v}$ perpendicular to a magnetic field $\vec{B}$ moves on a circular trajectory.

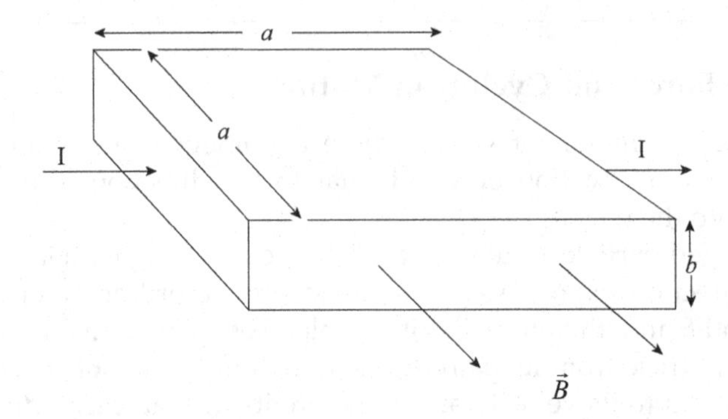

FIGURE 1.13
For current flowing along the y-direction in the presence of a magnetic field in the x-direction, a potential difference develops across the bar along the z-direction.

the cyclotron is based that is used to accelerate charged particles by applying alternating electric fields synchronous to the motion over limited parts of their trajectories. (See Exercise 1.12)

Example 1.3: The Hall Effect

Consider a conducting square bar of length a on a side and thickness b, as shown in Figure 1.13. A current I flows to the right, while a uniform magnetic field $\vec{B}$ is applied in the plane of the bar perpendicular to the direction of the current.

Assuming that the charge carriers carry positive charge q, they experience a magnetic force $\vec{F}_B = q\vec{v} \times \vec{B}$ that deflects them downward. Thus, the positive charge accumulates at the bottom surface of the conducting plate, leaving an equivalent negative charge on the top surface. This charge separation continues until the downward magnetic force is counterbalanced by the upward electrical force on the charge carriers, $q\vec{E} = q\vec{v} \times \vec{B}$, where $\vec{E}$ is the electric field within the plate due to the charge separation. At equilibrium, a constant potential difference, ΔV, exists between the top and bottom surface of the plate $\Delta V = Eb$, with the bottom surface being at the higher potential. This is known as the Hall voltage, named after Edwin Hall, who discovered the effect in 1879.

Note that if one were to consider, as is indeed the case, negatively charged electrons moving with velocity $\vec{v}$ to the left being responsible for the current I, the magnetic force $\vec{F}_B = -e\vec{v} \times \vec{B}$ would predict that negatively charged electrons accumulate to the bottom surface of the conducting plate leaving a net positive charge at the top. Thus, the polarity of the Hall voltage is reversed. The Hall Effect distinguishes between positive *vs.* negative charge carriers and it is the classic method by which the sign of the charge carriers in a material is established. The Hall Probe is a semiconductor-based detector that uses the Hall Effect to measure the strength of a magnetic field.

1.5 Magnetic Force on a Current-Carrying Wire

Since current consists of moving charges confined within a conductor, the force experienced by the charge carriers is transmitted to the conductor. Let λ represent the linear charge density of the moving carriers in a wire. Then, the current I can be expressed according to Eq. (1.16)

$$I = \frac{dq}{dt} = \frac{\lambda dl}{dt} = \lambda v \tag{1.16}$$

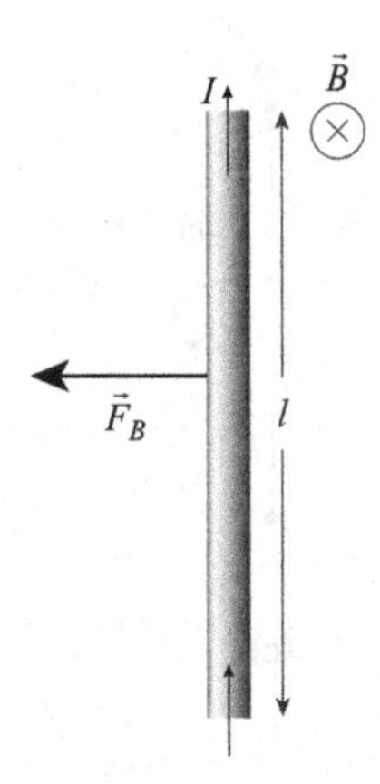

FIGURE 1.14
A wire carrying a current upward in the presence of a magnetic field pointing into the page experiences a magnetic force to the left.

Thus, the magnetic force on a segment of the wire is

$$\vec{F}_B = \int\left(\vec{v}\times\vec{B}\right)dq = \int\left(\vec{v}\times\vec{B}\right)\lambda dl = \int I\left(d\vec{l}\times\vec{B}\right) = I\int\left(d\vec{l}\times\vec{B}\right) \tag{1.17}$$

where the integral runs over the length of the segment. Consider a long straight wire carrying a current I upward in the presence of a magnetic field pointing into the page, Figure 1.14.

Since $\vec{B}$ and $d\vec{l}$ are perpendicular to each other, the magnetic force experienced by a wire of length l is $F_B = IlB$. The force points to the left according to the RHR applied to Eq. (1.17). Extend your palm in the direction of the current (upward) and rotate your fingers into the direction of the magnetic field (into the page), your thumb points in the direction of the magnetic force on the wire (to the left).

1.6 The Magnetic Dipole Moment

1.6.1 Torque Considerations in a Uniform Magnetic Field

Figures 1.15(a–d) depict an electric dipole, a magnetic dipole according to Gilbert, a magnetic dipole according to Ampère and the forces producing a torque on an electric dipole in a homogeneous electric field, respectively. We mentioned earlier that a magnetic compass, or small bar magnet, orients in the direction of the local magnetic field. The rotation of the compass away from its original direction implies the application of a torque. We can express this torque using the Gilbert model of a magnetic dipole, $\vec{m}$, which was introduced in analogy to the physical electric dipole, $\vec{p}$, as shown in Figure 1.15(a) and (b). The physical electric dipole, $\vec{p} = q\vec{d}$, defined in terms of two equal but opposite point charges, $+q$ and $-q$, displaced by $\vec{d}$, with the displacement vector $\vec{d}$ pointing from the negative to the positive charge, leads directly to a torque, $\vec{\tau} = \vec{p}\times\vec{E}$, when $\vec{p}$ is placed in a homogeneous electric field $\vec{E}$, by considering the pair of forces, $+q\vec{E}$ and $-q\vec{E}$, exerted on the dipole that lead to the

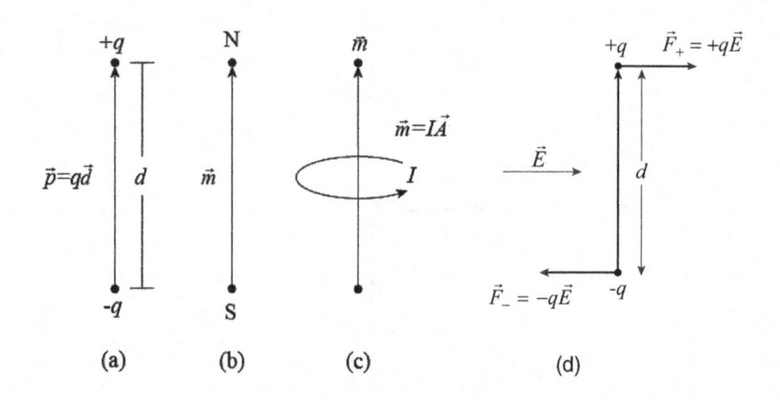

FIGURE 1.15
Models of the electric dipole (a) and the magnetic dipole (b) and (c). In Gilbert's model of the magnetic dipole (b), the north and south poles behave in analogy to the positive and negative point charges of an electric dipole (a). In Ampère's model (c), there are no poles. (d) Torque on an electric dipole due to a uniform electric field $\vec{E}$.

rotation of the dipole around its center, Figure 1.15(d). The dipole rotates until it lies parallel to $\vec{E}$. At this orientation, the torque vanishes.

By analogy to the electrical case, the Gilbert model leads to a torque given by Eq. (1.18)

$$\vec{\tau} = \vec{m} \times \vec{B} \tag{1.18}$$

when the magnetic dipole, $\vec{m}$, is placed in a uniform magnetic field, $\vec{B}$. This analogy, however, treats magnetic poles as separate individual entities, $+q_m$ (north pole) and $-q_m$ (south pole) similar to the $+q$ and $-q$ charges of the electric dipole that can be isolated and can exist separate from each other. No such separation (of the north and south poles of a magnetic dipole) has ever been achieved, however. Thus, in this book, we adopt Ampère's model of a magnetic dipole, Figure 1.15(c), that of a "current loop", which is physically feasible and, as we will see, constitutes the elemental entity of magnetism.

Consider a small current loop carrying a current I as shown in Figure 1.16(a). Painstaking application of the Biôt–Savart Law leads to the mapping of the magnetic field around such a current-carrying loop, as shown. The field is reminiscent of that of a small bar magnet, Figure 1.16(b). Field lines are continuous; the strength of the magnetic field is strong close to the dipole where the field lines are dense and weak away from the dipole where the density of field lines diminishes. By definition, the magnetic dipole of the current loop is given by Eq. (1.19)

$$\vec{m} = I\vec{A} \tag{1.19}$$

where $\vec{A}$ is a vector, whose magnitude equals the area of the current loop measured in m^2 and whose direction is determined by the RHR; rotating your fingers in the direction of the current I, your thumb points in the direction of $\vec{A}$ and thus of $\vec{m}$. The magnitude of the magnetic moment for a circular current loop of radius r is thus given by Eq. (1.20).

$$m = I\pi r^2 \tag{1.20}$$

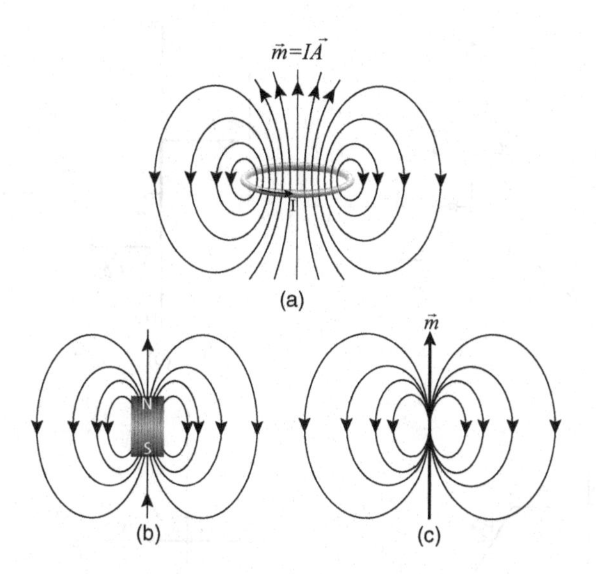

FIGURE 1.16
(a) Magnetic field lines around a current loop carrying current I. (b) Magnetic field lines around a bar magnet. (c) Magnetic field lines of a point magnetic dipole.

In the limit of $r \to 0$, $I \to \infty$, m remains finite. This defines the point magnetic dipole $\vec{m}$, depicted in Figure 1.16(c).

Example 1.4: Torque on a Magnetic Dipole in a Uniform Magnetic Field

A square loop of side length a, carrying current I, constitutes a magnetic dipole moment $\vec{m} = Ia^2\,\hat{n}$, where $\hat{n}$ is the unit vector perpendicular to the area of the loop according to the RHR. The loop shown in Figure 1.17(a) is centered at the origin of a Cartesian coordinate axis system and lies in the xz-plane; for current in the counterclockwise direction $\hat{n} = \hat{y}$. What is the response of the loop to the application of a uniform magnetic field $\vec{B}$ along the positive z-axis?

Let us consider the magnetic force $\vec{F}_B = \int I d\vec{l} \times \vec{B}$ (Eq. (1.17)) exerted on each segment of the wire loop due to $\vec{B}$. We take $d\vec{l}$ in the direction of the current, which flows in a counter-clockwise direction around the square loop. For the back-side segment of the loop, $d\vec{l}$ and $\vec{B}$ are parallel vectors, so their cross product vanishes; for the front-side segment, they are antiparallel and thus the cross product is also zero. No magnetic force is exerted on these two segments.

The situation for the top and bottom segments is entirely different, however, as $d\vec{l}$ and $\vec{B}$ are perpendicular to each other, leading to maximum possible magnetic force. The top and bottom segments feel exactly the same magnitude of magnetic force, but in opposite directions.

$$F_B = IaB$$

At the bottom, $\vec{F}_B$ points to the right, along $+\hat{y}$. At the top, $\vec{F}_B$ points to the left, or the $-\hat{y}$ -direction. Each force exerts a maximum possible torque on the square loop around the x-axis. Each torque is given by $\vec{\tau} = \vec{r} \times \vec{F}_B$, where $\vec{r}$ is the lever arm of the force from the origin. The torques reinforce each other, both point along $+\hat{x}$-direction, to give a total torque twice

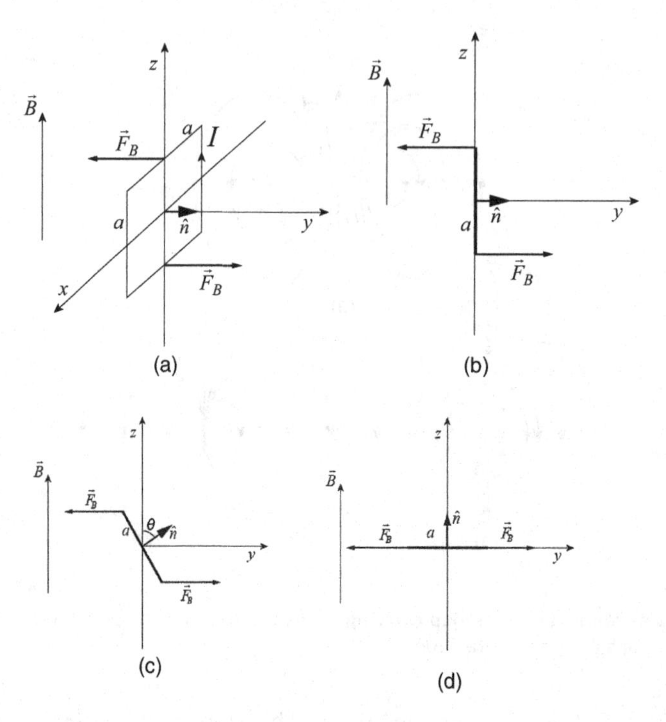

FIGURE 1.17
Calculating the torque on a square loop (see text).

as large. The magnitude of the lever arm is $|\vec{r}| = \frac{a}{2}$ and we obtain the maximum total torque on the square loop to be

$$\tau_{\max} = 2\frac{a}{2}F_B = a(IaB) = Ia^2B$$

Figure 1.17(b) shows an end view of the loop along the x-axis. The torque is in a sense as to rotate $\hat{n}$ toward the direction of $\vec{B}$. Now, suppose that the plane of the loop is tilted to the left of the xz-plane, so that $\hat{n}$ is no longer perpendicular to $\vec{B}$, but makes an angle θ with respect to $\vec{B}$, as shown in Figure 1.17(c). The lever arm is reduced to $\frac{a}{2}\sin\theta$ and the magnitude of the torque is accordingly reduced from its maximum value to

$$\tau = 2\frac{a}{2}\sin\theta F_B = a(IaB)\sin\theta = Ia^2B\sin\theta$$

The front and back sides of the loop now feel equal and opposite forces, which however do not contribute to the torque. We can write the above equation in vector form as $\vec{\tau} = Ia^2\,\hat{n}\times\vec{B} = \vec{m}\times\vec{B}$, recapturing Eq. (1.18). Assuming that the loop is pivoted onto the x-axis in a manner that allows rotation around it, the loop will rotate till the unit vector $\hat{n}$ becomes parallel to $\vec{B}$, the plane of the loop being perpendicular to $\vec{B}$ (Figure 3.17(d)). At this orientation, the torque vanishes. Depending on whether there are frictional losses on the pivot, the current loop will either come to static equilibrium with its magnetic moment $\vec{m}$ parallel to $\vec{B}$, or (in the case of frictionless pivot) $\vec{m}$ will oscillate about $\vec{B}$. In real experiments with compass needles, friction is always present and thus the magnetic dipole comes to equilibrium when $\vec{m}$ is parallel to $\vec{B}$. At the equilibrium position, the total force and torque on the dipole is zero if the magnetic field is uniform, as considered above. If $\vec{B}$ were not uniform, however, the dipole would feel a net force (see Section 1.6.3 and Exercise 1.13).

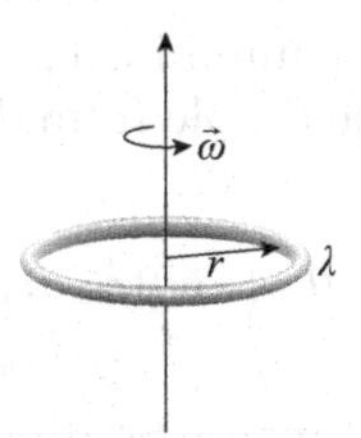

FIGURE 1.18
A rotating ring, with constant angular velocity, carrying linear charged density λ constitutes a magnetic dipole.

Parenthetically, let me mention that you do not necessarily need a conducting, current-carrying wire to produce a magnetic dipole. A ring of radius r made of non-conducting material but carrying a uniform linear charge density λ and rotating around its axis of symmetry with uniform angular velocity $\vec{\omega}$ also constitutes a magnetic dipole moment (Figure 1.18). The rotating charged ring is equivalent to a circular current

$$I = \frac{dq}{dt} = \lambda\frac{dl}{dt} = \lambda v = \lambda r\omega$$

and therefore, the magnetic dipole moment

$$\vec{m} = I\vec{A} = \lambda r\vec{\omega}\left(\pi r^2\right) = \lambda\pi r^3\vec{\omega} \tag{1.21}$$

Incidentally, a spinning charged disk carrying uniform surface charge density σ also constitutes a magnetic dipole, as it can be considered to consist of a superposition of infinitely thin rotating charged rings. Similarly, a spinning uniformly charged sphere also constitutes a magnetic dipole, as it can be decomposed into an infinite number of stacked-up rotating disks. Thus, magnetic moments can be associated with current loops or with spinning charged spheres. As we elaborate further in Chapter 2, both of these occur in nature.

1.6.2 Energy Considerations in a Uniform Magnetic Field

As we saw in Example 1.4, in the presence of a uniform magnetic field $\vec{B}$, a magnetic dipole moment $\vec{m}$ rotates under the influence of the torque exerted by $\vec{B}$, until $\vec{m}$ lies along the direction of $\vec{B}$. The dipole comes to equilibrium at this orientation. To rotate the dipole at a constant angular speed back to an angle θ away from the direction of $\vec{B}$, rotational work must be performed by applying an external torque on the dipole equal and opposite to that of the magnetic field

$$\Delta W = \int_0^\theta \tau\, d\theta = \int_0^\theta mB\sin\theta\, d\theta = mB\int_0^\theta \sin\theta\, d\theta = mB - mB\cos\theta$$

One can then associate an orientational potential energy U to the magnetic dipole in a uniform magnetic field given by

$$U = -\vec{m}\cdot\vec{B} = -mB\cos\theta \tag{1.22}$$

The orientational potential energy is minimized at $\theta = 0$. According to the work–energy theorem, the work done in rotating the dipole from $\theta = 0$ to θ must be equal to the difference in potential energy

$$\Delta U = U_f - U_i = -mB\cos\theta - (-mB\cos 0) = mB - mB\cos\theta = \Delta W$$

We can then discuss the rotation of magnetic dipoles into the direction of an external magnetic field in terms of energy considerations. At its equilibrium position, the orientational potential energy of the dipole with respect to the external field is minimized. U is minimized when vectors $\vec{m}$ and $\vec{B}$ are parallel to each other, as we concluded on torque considerations. The same is true in electrostatics, for the case of an electric dipole $\vec{p}$ in a uniform electric field and $\vec{E}$. The orientational potential energy, $U = -\vec{p}\cdot\vec{E}$, is minimized when vectors $\vec{p}$ and $\vec{E}$ are parallel to each other.

We will see in Chapter 2 that circulating electrons in their atomic orbitals constitute microscopic electric current loops, and thus, microscopic magnetic moments, which contribute to the atomic moments responsible for the magnetization of magnetic materials. It is the torques exerted on such elemental current loops that are collectively responsible for the rotation of the compass needle along the direction of the local magnetic field, where its orientational potential energy is minimized.

1.6.3 Force Considerations in a Non-Uniform Magnetic Field

If the magnetic field is inhomogeneous, the magnetic dipole will experience a net magnetic force in addition to the torque discussed in Section 1.6.1. Consider again Figure 1.17(a) of a square current-carrying wire loop constituting a magnetic dipole moment $\vec{m} = m_y\,\hat{y} = Ia^2\,\hat{y}$ placed in a magnetic field $B(y)\hat{z}$, whose strength increases along the y-direction, $\frac{dB}{dy} > 0$. Once the dipole has been rotated due to the torque (Figure 3.17(d)), the magnetic force on the wire pointing to the right, call it F_{y+}, would be of greater magnitude than that pointing to the left, F_{y-}. Thus, the wire loop will experience a net force pulling it into the region of the stronger magnetic field. The net force on the dipole is given by the difference

$$\vec{F}_B = Ia\frac{dB}{dy}\Delta y\,\hat{y}$$

Using $\Delta y = a$, the width of the loop, we obtain

$$\vec{F}_B = Ia^2\frac{dB}{dy}\hat{y} = m_y\frac{dB}{dy}\hat{y}$$

One can prove that in the general case of a dipole $\vec{m} = m_x\,\hat{x} + m_y\,\hat{y} + m_z\,\hat{z}$ in a non-uniform magnetic field $\vec{B}(x,y,z)$, the magnetic force is given by

$$\vec{F}_B = \vec{\nabla}\left(\vec{m}\cdot\vec{B}\right) \tag{1.23}$$

where $\vec{\nabla}$, pronounced del, is the vector differential operator $\vec{\nabla} = \frac{\partial}{\partial x}\hat{x} + \frac{\partial}{\partial y}\hat{y} + \frac{\partial}{\partial z}\hat{z}$ (see Appendix).

1.7 Time-Varying Currents and Maxwell's Equations

If we were to consider only magnetostatics, that is, steady-state currents, we would be done with the fundamental concepts of magnetism and magnetic interactions in free space. There are, however, many areas of magnetism where time-varying currents and, therefore, time-varying magnetic fields, must be considered. Such a consideration, however, does not simply extend the concepts introduced so far; on the contrary, it forces us to consider new phenomena which unravel the whole complexity that arises from the fundamental interconnection between magnetism and electricity. To proceed, we must introduce another physical quantity associated with vector fields, that of the flux of the magnetic field, Φ_B. Consider a cross-sectional area A, whose orientation is defined by the unit vector $\hat{n}$ normal to A, in a region of space with a homogeneous magnetic field $\vec{B}$, as shown in Figure 1.19.

We define the flux of the magnetic field through A in terms of the dot product of Eq. (1.24),

$$\Phi_B = \vec{B} \cdot \vec{A} = \vec{B} \cdot A\hat{n} \tag{1.24}$$

which is generalized to the surface integral of $\vec{B}$, in the case of inhomogeneous fields or curved surfaces, according to Eq. (1.25).

$$\Phi_B = \int \vec{B} \cdot d\vec{a} \tag{1.25}$$

If the flux of the magnetic field through a conducting wire loop is changing, a current is observed to flow in the wire loop. An electromotive force, $\mathcal{E}$, is induced within the conducting wire loop that is responsible for pushing the current around the loop. Through experimental observations, Faraday concluded that the induced electromotive force, or "emf", is directly related to the time rate of change of the magnetic flux through the wire loop according to Eq. (1.26), known as Faraday's Law.

$$\mathcal{E} = \oint \vec{E} \cdot d\vec{l} = -\frac{d\Phi_B}{dt} \tag{1.26}$$

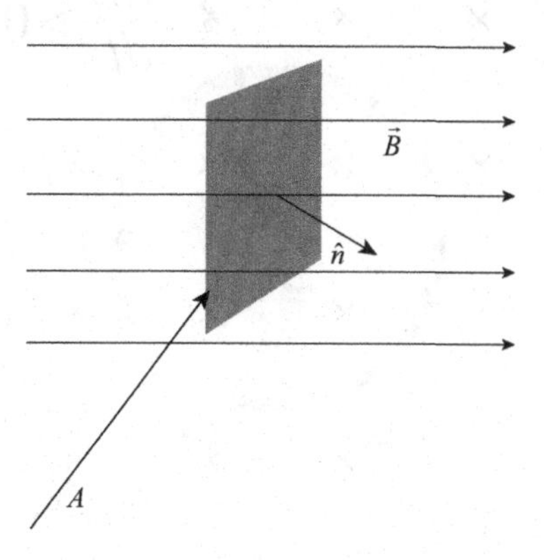

FIGURE 1.19
Definition of the flux of $\vec{B}$ over an area A. $\hat{n}$ depicts the unit vector in a direction perpendicular to A.

The minus sign, also known as Lenz's Law, indicates that the sign of the induced electromotive force, and therefore the sense of the induced current, opposes the change in magnetic flux. For example, if the magnetic flux is decreasing, the induced current flows in a direction so that the magnetic field due to the induced current itself adds to the external flux as if to prevent the flux from decreasing. If, on the other hand, the magnetic field flux through the wire loop is increasing, the induced $\mathcal{E}$ and correspondingly sense of the induced current is in the reverse direction, so that the induced flux subtracts from the external flux, consistent with the notion that the induced current attempts to prevent the increase in flux. We conclude that $\mathcal{E}$ is always induced in a direction as to oppose any change in magnetic flux.

Consider the region of space, such as the interior of a long solenoid, depicted in Figure 1.20. A uniform magnetic field points into the page. The field strength is increasing with time at a constant rate, $\frac{dB}{dt} > 0$. A circular wire loop of radius r exists within this region. According to Faraday's Law, a counterclockwise steady-state current is induced in the wire loop due to the induced electromotive force of Eq. (1.26), where $\vec{E}$ is a circumferential electric field induced within the wire, and which drives the current, Figure 1.20. The question then arises: In the absence of a wire, is there any electric field induced?

The answer is yes. An electric field is indeed induced in the region of space where there exists a changing magnetic field, irrespective of the presence of a wire loop. The properties of this induced electric field can be summarized as follows: The field lies in the plane normal to the direction of the magnetic field, it is circumferential around the direction of the magnetic field, and its sense of direction is determined by the RHR. For decreasing magnetic field strength, place your thumb along the direction of the magnetic field, then your fingers curl in the direction of the induced electric field, $\vec{E}$. For increasing magnetic fields, place your thumb in the direction opposite to the magnetic field. The direction of the induced $\vec{E}$ is indicated by the direction in which your figures curl.

A charged particle moving in such a region of space will feel a generalized force known as the Lorentz force, given by Eq. (1.27), where q and $\vec{v}$ are the charge and the velocity of the particle, respectively.

$$\vec{F} = q\vec{E} + q\vec{v} \times \vec{B} \tag{1.27}$$

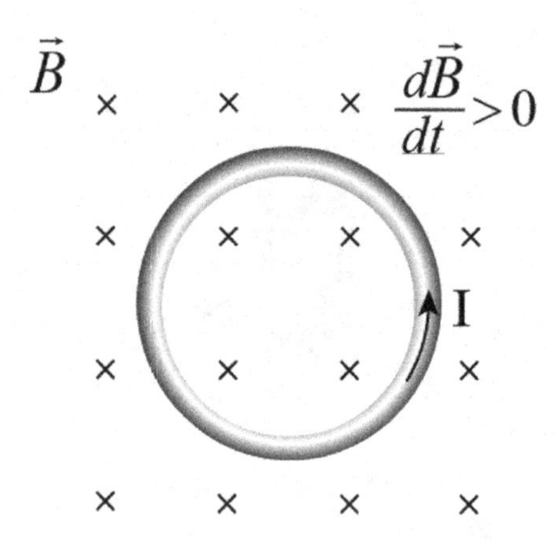

FIGURE 1.20
Faraday's Law of Induction. A steady current I is induced in the counterclockwise direction when the magnetic field $\vec{B}$, pointing into the page, increases with time at a constant rate.

The induced electric field, $\vec{E}$, is fundamentally different than the electrostatic field due to a distribution of stationary charges in the vicinity. For the case of the electrostatic field, the inverse square law dependence of the strength of the electric field with distance away from the localized charge distribution, as dictated by Coulomb's Law, guarantees vanishing line integral of the electrostatic field around a closed loop, in contradiction to Eq. (1.26). That is, electrostatic fields are "conservative fields", while induced Faraday electric fields are not. However, both electrostatic and induced electric fields interact similarly with a charged particle and thus, no distinction needs to be made in Eq. (1.27) as to the origin of $\vec{E}$.

Maxwell, inspired by Faraday's observations, considered this interconnection between $\vec{B}$ and $\vec{E}$ further and suggested, as to complete the symmetry in the problem, that a changing electric field must produce a magnetic field. For a concrete example, let us consider charging a capacitor, as depicted in Figure 1.21. While the electric field is increasing in the region of space between the plates of the capacitor, Maxwell suggested that a circumferential magnetic field is induced in the plane perpendicular to the electric field, in the sense determined by the RHR. Placing your thumb in the direction of the increasing electric field, your palm rotates in the direction of the induced magnetic field, as indicated in Figure 1.21.

If the capacitor were being discharged, that is if $\frac{d\vec{E}}{dt} < 0$, the magnetic field would be induced in the clockwise direction. This led to amending Ampère's Law according to Eq. (1.28).

$$\oint \vec{B} \cdot d\vec{l} = \mu_0 \left(I_{enclosed} + \varepsilon_0 \frac{d\Phi_E}{dt} \right) \tag{1.28}$$

Here, $\varepsilon_0 = 8.854 \times 10^{-12}\ C^2/N{\cdot}m^2$ is the permittivity of free space and $\varepsilon_0 \frac{d\Phi_E}{dt}$ is Maxwell's displacement current. Equations (1.26) and (1.28) constitute two of the celebrated four Maxwell's equations of electrodynamics involving the magnetic field $\vec{B}$. The third is a direct corollary of the absence of the magnetic monopole leading to magnetic field lines closing upon themselves, and thus, vanishing magnetic field flux over a closed surface. If you were to consider the flux of a magnetic field over a closed surface of arbitrary shape,

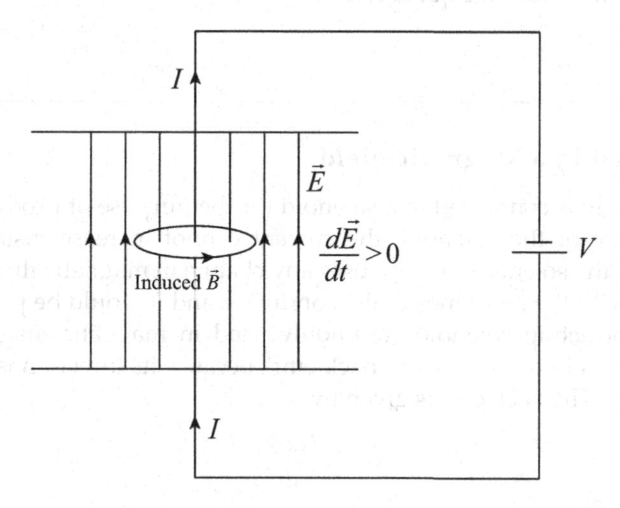

FIGURE 1.21
Depiction of the induced magnetic field $\vec{B}$ in the region of space between the plates of a parallel plate capacitor while charging.

TABLE 1.1
Maxwell's Equations

$\oint \vec{E}\cdot d\vec{a} = \frac{q_{enclosed}}{\varepsilon_0}$	$\vec{\nabla}\cdot\vec{E} = \frac{\rho}{\varepsilon_0}$	Gauss's Law
$\oint \vec{E}\cdot d\vec{l} = -\frac{d\Phi_B}{dt}$	$\vec{\nabla}\times\vec{E} = -\frac{\partial\vec{B}}{\partial t}$	Faraday's Law
$\oint \vec{B}\cdot d\vec{a} \equiv 0$	$\vec{\nabla}\cdot\vec{B} \equiv 0$	No magnetic monopoles
$\oint \vec{B}\cdot d\vec{l} = \mu_0\left(I_{enclosed} + \varepsilon_0\frac{d\Phi_E}{dt}\right)$	$\vec{\nabla}\times\vec{B} = \mu_0\left(\vec{J} + \varepsilon_0\frac{\partial\vec{E}}{\partial t}\right)$	Ampère's Law with Maxwell's correction

you will always find vanishing flux, as all magnetic field lines that enter (exit) the interior space of the closed surface must exit (enter) it.

$$\oint \vec{B}\cdot d\vec{a} = 0 \tag{1.29}$$

The fourth of Maxwell's equations is Coulomb's Law of electrostatics, usually expressed in the equivalent alternate expression of the inverse square law, known as Gauss's Law

$$\oint \vec{E}\cdot d\vec{a} = \frac{q_{enclosed}}{\varepsilon_0} \tag{1.30}$$

Gauss's Law relates the flux of the electric field over a closed surface to the charge enclosed within the volume bounded by the surface. The above equations are Maxwell's equations expressed in integral form. One can arrive at their differential form by applying the fundamental theorems of vector calculus. The Appendix introduces vector differential calculus and the fundamental theorems. Table 1.1 summarizes Maxwell's Equations in integral and differential form for quick reference.

1.8 Energy Stored in a Magnetic Field

When a power supply is connected to a solenoid for the purpose of producing a magnetic field within the core of the solenoid, the current cannot increase instantly due to the induced emf across the solenoid that opposes any change in magnetic flux through it. This has nothing to do with the resistance of the conductor and it would be present even in the case of a superconducting solenoid commonly used in magnetic magnetometers. The induced emf is known in this case as a "back emf" because its direction is opposite to that of the power supply. The back emf is given by

$$\mathcal{E} = -\frac{Nd\Phi_B}{dt} \tag{1.31}$$

where N is the total number of turns in the solenoid and Φ_B is the flux of the magnetic field through a single turn. Using Eq. (1.13) for the field of a solenoid, $B = \mu_0 nI$, the total flux of the magnetic field through the solenoid at any time t can be expressed in terms of the instantaneous current flowing in the solenoid

$$N\Phi_B(t) = lA\mu_0 n^2 I(t)$$

where A is the cross-sectional area of the solenoid, n is the number of turns per unit length and l is its length. The rate at which energy is delivered by the power supply against the back emf is given by

$$\frac{dU}{dt} = I\mathcal{E} = lA\mu_0 n^2 I \frac{dI}{dt}$$

The total work delivered by the power supply to bring the current up to a final value I is given by

$$\int_0^U dU = lA\mu_0 n^2 \int_0^I IdI = \frac{lA\mu_0 n^2 I^2}{2} = \frac{lAB^2}{2\mu_0} \quad (1.32)$$

This energy is stored in the field and it is regained when the power supply is disconnected. Due to the induced back emf again the current does not drop to zero instantaneous, the current persists and decays exponentially.

Since the product lA that appears in Eq. (1.32) gives the volume of the solenoid, one concludes that the energy per unit volume associated with a magnetic field is

$$u_B = \frac{1}{2\mu_0} B^2 \quad (1.33)$$

This implies that the magnetic properties of a region of space around a bar magnet can also be described in terms of the local magnetic energy density due to the existence of a magnetic field.

Exercises

1. Consider a circular current loop of radius r lying in the xy-plane and centered at the origin. If it carries a steady current I in the counterclockwise direction, (a) calculate the magnetic field $\vec{B}$, magnitude and direction, for points on the z-axis, (b) sketch qualitatively the magnetic field lines on and off the z-axis and (c) determine the behavior of the magnetic field far from the current loop, that is when $z >> r$. Express this limiting field in terms of the magnetic moment of the current loop, $\vec{m} = I\vec{A}$, where A is the area of the loop.
2. *Helmholtz Coils.* The magnetic field of the current loop in the previous problem is not homogeneous; it falls off sharply with increasing z. You can create a more homogeneous field in the region of space between two such circular coils placed a distance s apart. Consider two circular coils of radius R centered on the z-axis and placed parallel

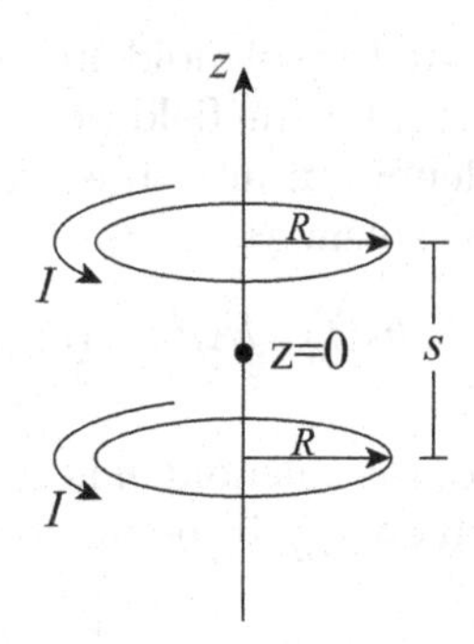

FIGURE 1.22
Schematic representation of a Helmholtz coils arrangement.

to the xy-plane with their centers at (0, 0, $\pm\, s/2$), both carrying steady current I in the counterclockwise direction, as shown in Figure 1.22. (a) Calculate the magnetic field on the z-axis as a function of z and show that $\partial B/\partial z = 0$ at $z = 0$, halfway between the coils. (b) Determine the value of s for which the second derivative $\partial^2 B/\partial^2 z$ is also zero at $z = 0$. (c) Calculate the resulting magnetic field at the center. This arrangement is known as Helmholtz coils and is used in laboratory experiments to produce a fairly uniform field at the center.

3 Calculate the magnetic field at point P for the steady-state current configurations shown in Figure 1.23.

4 A long straight conductor of radius R carries a current I uniformly distributed over its cross-sectional area. (a) Calculate the magnetic field $\vec{B}(r)$ as a function of r, the distance from the center of the wire, for $r < R$ and $r > R$. (b) Plot the magnitude of the magnetic field $vs.$ r. Is the field continuous at $r = R$? (c) Sketch the magnetic field lines inside and outside the wire.

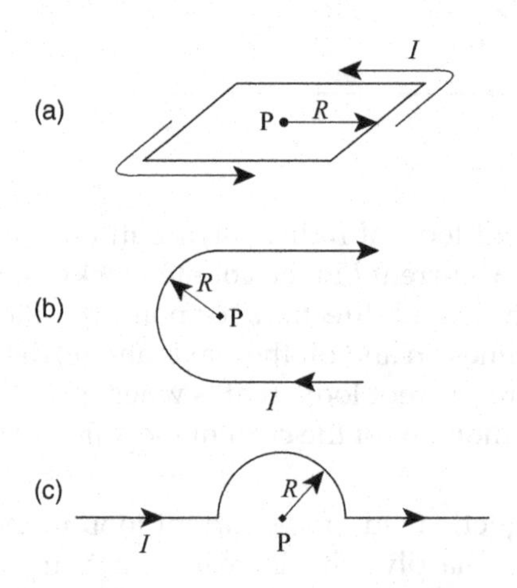

FIGURE 1.23
Various steady-state current configurations.

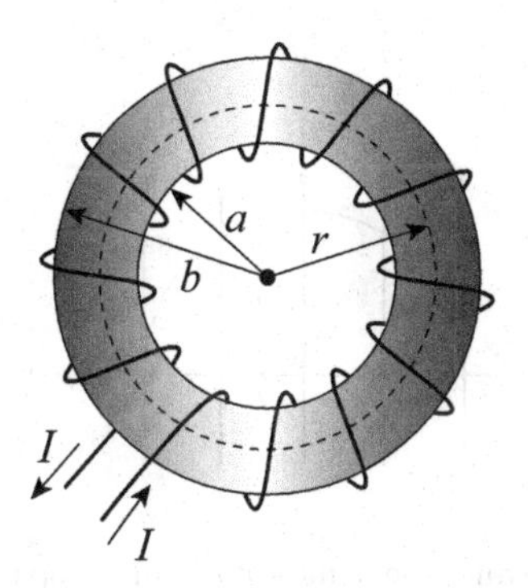

FIGURE 1.24
Calculating the field of a toroid.

5 A wire carrying current I is tightly wound in the shape of a donut or, torus, of inner radius a and outer radius b, as shown in Figure 1.24. The winding is tight enough so that each turn can be considered to be a closed loop. If the total number of wire loops of the toroid is N, use symmetry arguments to derive the magnetic field, both magnitude and direction at a distance r from the center of the toroid, for $a < r < b$.

6 Consider two parallel, long, straight wires placed a distance d apart and carrying a current I each, in the same direction. (a) Draw a pictorial representation showing the directions of the currents, fields and forces to deduce whether the forces between the two currents are attractive or repulsive and derive the magnitude of the force per unit length experienced by either wire. (b) If $d = 1$ m and $I = 1$ A, what is the force per unit length? (Note that a specified force per unit length can be used to define the unit of current. This is in fact how the Ampère is defined.)

7 A rectangular loop carrying current I in the counterclockwise direction is placed close to a very long wire carrying similar current I to the right, as shown in Figure 1.25. Calculate the force, magnitude and direction, exerted on the rectangular loop.

8 A conducting wire shaped into a semicircular loop of radius r lies on the xy-plane with its straight segment along the x-axis, as shown in Figure 1.26. A uniform magnetic field $B\hat{y}$ exists pointing in the y-direction. If the loop carries a current I, calculate the magnetic force on (a) the straight and (b) the curved segments of the wire loop.

9 A rectangular loop of length a and width b carrying current I lies in a uniform magnetic field $\vec{B}$ as shown in Figure 1.27. By considering all forces on the loop, prove that the loop experiences a torque $\vec{\tau} = \vec{m} \times \vec{B}$, where $\vec{m} = I\vec{A}$ is the magnetic moment of the loop.

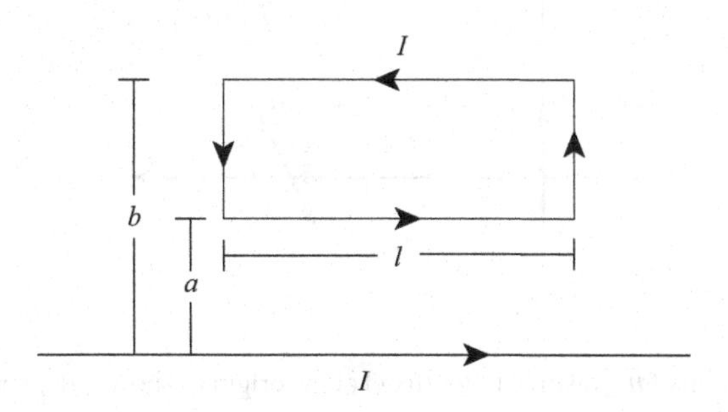

FIGURE 1.25
Force on a current-carrying wire loop.

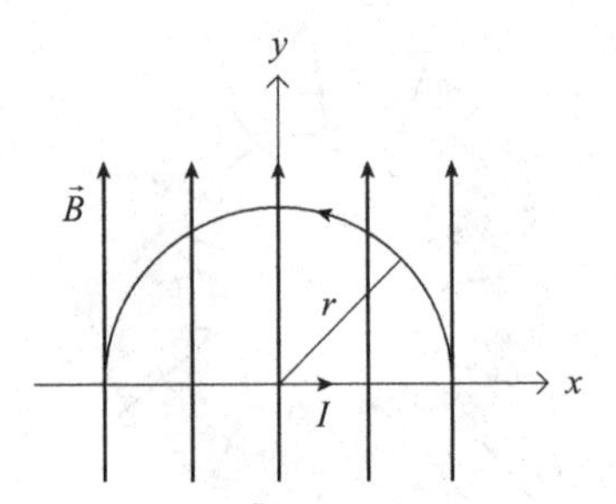

FIGURE 1.26
Current I flows in the counterclockwise direction in the semicircular loop of radius r in a uniform magnetic field.

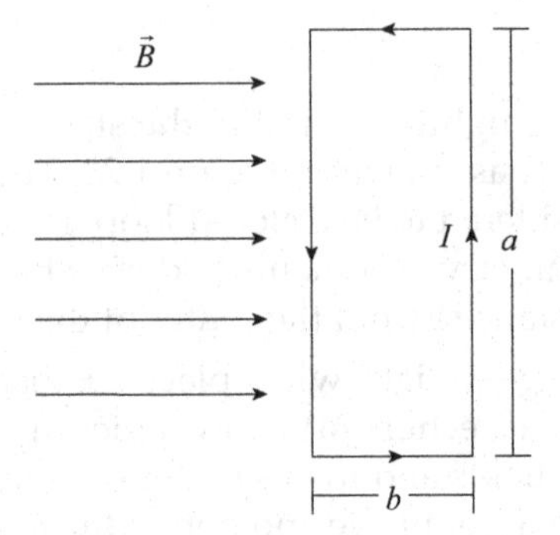

FIGURE 1.27
Calculating the torque on a current loop.

10 Consider a point magnetic dipole $\vec{m}_1$ located at the origin with its magnetic moment fixed in the z-direction. A second identical point dipole $\vec{m}_2$ is placed on the x-axis a distance r away from the origin, at an angle θ relative to the x-axis, as shown in Figure 1.28. If the second dipole is free to rotate, in which direction would $\vec{m}_2$ point at equilibrium? At equilibrium, are the dipoles parallel or antiparallel? Repeat the process for the case of placing $\vec{m}_2$ at a distance r from the first dipole but now placed on the z-axis. What is the relative orientation of the two dipoles at equilibrium in this case?

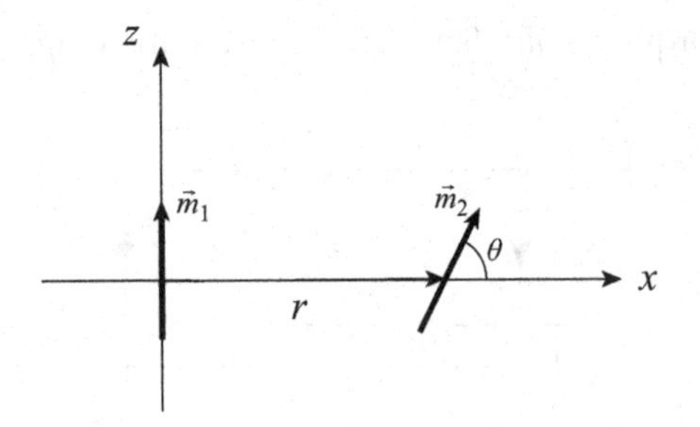

FIGURE 1.28
At equilibrium, the orientation of $\vec{m}_2$ relative to $\vec{m}_1$ (fixed at the origin as shown) depends on its position relative to the origin.

11 In a laboratory experiment, electrons in a glass chamber containing rarefied neon gas are accelerated through a potential difference ΔV until they acquire a speed $|\vec{v}| = 1.11 \times 10^7$ m/s. They then enter a region of space with a homogeneous magnetic field $B = 8.4 \times 10^{-4}$ T, normal to their velocity. The circular trajectory of the electrons is made visible *via* ionizing collisions with neon atoms in the glass chamber. If the radius of the circular orbit is measured to be 7.50 cm, determine the ratio $\frac{e}{m}$, of the electronic charge to the electronic mass.

12 J. J. Thomson was the first to measure the $\frac{e}{m}$ ratio of the electron in 1897. He first passed the beam of electrons through a region of space with uniform crossed electric $\vec{E}$ and magnetic $\vec{B}$ fields (mutually perpendicular and both perpendicular to the direction of the beam). He first adjusted the strength of the electric field so that the electron beam passed undeflected through the crossed $\vec{E}$ and $\vec{B}$ fields. He then turned off the electric field and measured the radius of curvature of the electron beam, R, in the magnetic field. (a) Derive a relationship between the speed of the electrons and the strength of the crossed fields necessary for zero deflection of the electron beam. (b) Derive an expression for the $\frac{e}{m}$ ratio in terms of E, B and R.

13 The cyclotron is a charged particle accelerator used in nuclear scattering experiments and the production of radioactive products for use in science and medicine. It makes use of a vertical magnetic field to confine the particles in horizontal semi-circular orbits in two D-shaped regions or "dees", as shown schematically in Figure 1.29(a). Across the gap between the "dees" an alternating voltage is applied with a frequency of oscillation $f = \omega/2\pi$, where ω is the cyclotron frequency. As the magnetic field performs no work on the particles, it is only when they cross the gaps between the "dees" that they get accelerated. Figure 1.29(b) shows schematically, the particle trajectory before the accelerated particle exits the "dees" to be directed to the target. Consider a proton with a kinetic energy of 12 MeV in a cyclotron with a magnetic field of 2 T. Calculate the cyclotron frequency and the radius of the orbit.

14 A non-uniform magnetic field exerts a force on a magnetic dipole. Consider a horizontal conducting ring currying current I in the clockwise direction. A strong magnet is placed underneath the ring along its axis, as shown in Figure 1.30. If the magnetic field at the ring's location makes an angle θ relative to the vertical, find the magnitude and direction of the net force on the ring.

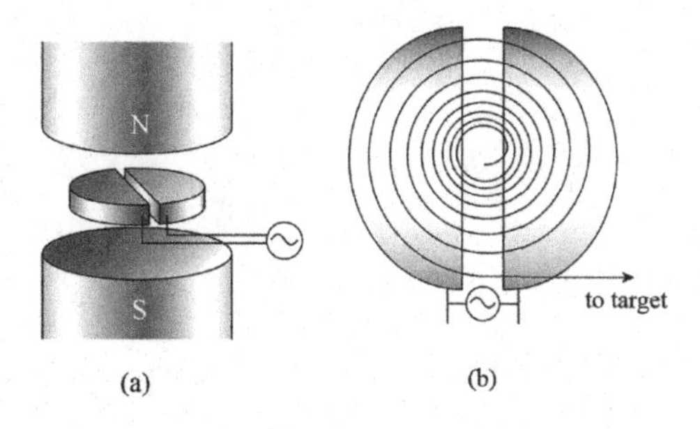

FIGURE 1.29
(a) Schematic of the cyclotron and (b) proton trajectory.

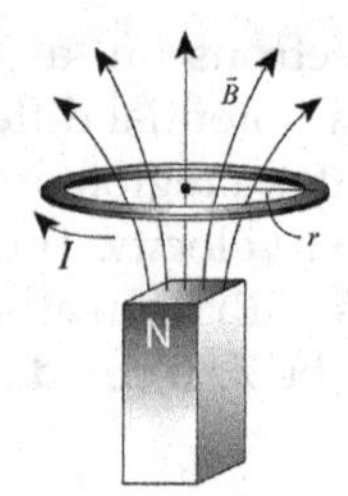

FIGURE 1.30
Calculating the force exerted by an inhomogeneous magnetic field on a magnetic dipole.

2

Classical and Quantum Treatment of Diamagnetism and Paramagnetism

2.1 Langevin Diamagnetism

The elementary interactions between magnetic dipoles, current loops and magnetic fields considered in Chapter 1, constitute the basis of our qualitative understanding of the response of matter to the application of a magnetic field. All matter is composed of atoms that have negatively charged electrons circulating around positively charged nuclei. The circulating electrons form tiny current loops and, therefore, elementary magnetic moments. Let us consider an electron in a uniform circular motion in an orbit of radius r and tangential velocity $\vec{v}$, counterclockwise around the z-axis, as shown in Figure 2.1.

If the time it takes for the electron to complete a full revolution is T, the orbiting electron can be visualized to constitute a current loop of a clockwise steady-state current of magnitude

$$I = \frac{dq}{dt} = \frac{e}{T} = \frac{ev}{2\pi r} \tag{2.1}$$

The corresponding magnetic moment ($\vec{m} = -I\pi r^2 \hat{z}$) is given by Eq. (2.2).

$$\vec{m} = -\frac{1}{2} evr\,\hat{z} \tag{2.2}$$

The minus sign accounts for the negative charge on the electron, making the magnetic moment point in the negative z-direction. In an atom, the centripetal force is provided by the attractive Coulomb force between the nucleus and the electron, which for the simple case of the hydrogen atom is given by

$$F_c = \frac{1}{4\pi\varepsilon_0}\frac{e^2}{r^2} = m_e\frac{v^2}{r} \tag{2.3}$$

Here, m_e is the mass of the electron and $\frac{v^2}{r}$ is the centripetal acceleration. Suppose now that a magnetic field $\vec{B}$ is introduced in the positive z-direction (upward), as shown in Figure 2.2. The circulating electron will feel an additional force $\vec{F}_B = -e\vec{v} \times \vec{B}$ along the radial direction and pointing toward the center of the circle, adding to the centripetal force.

DOI: 10.1201/9781315157016-3

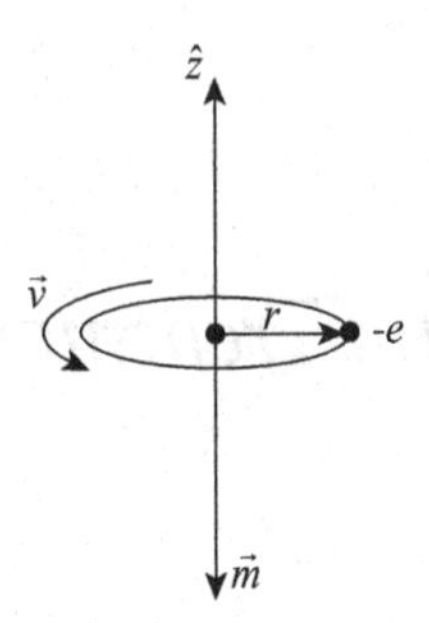

FIGURE 2.1
A circulating electron constitutes a magnetic dipole $\vec{m}$.

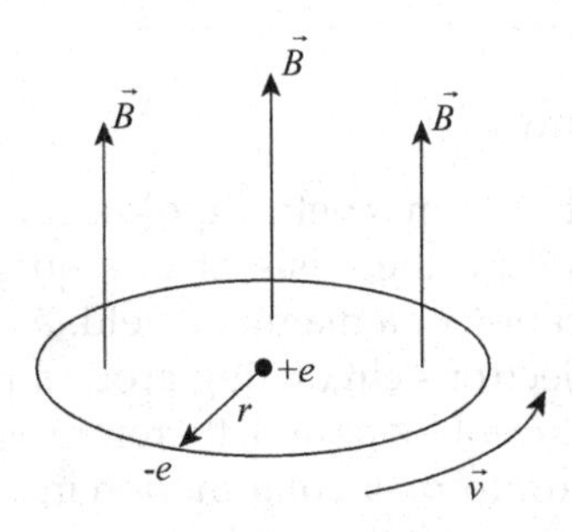

FIGURE 2.2
An electron, such as in the hydrogen atom, circulates around a proton in the horizontal plane. A magnetic field $\vec{B}$ is applied along the vertical direction.

Assuming that the radius remains the same (Bohr orbital quantization will ensure this), the increase in the strength of the centripetal force will result in an increase in the speed of the electron to a new speed v'. Equation (2.3) then becomes

$$F_c' = \frac{1}{4\pi\varepsilon_0}\frac{e^2}{r^2} + ev'B = m_e\frac{v'^2}{r} \tag{2.4}$$

Subtracting Eq. (2.3) from Eq. (2.4) gives an estimate of the increase in the speed of the electron in its circular orbit

$$F_c' - F_c = ev'B = \frac{m_e}{r}\left(v'^2 - v^2\right) = \frac{m_e}{r}(v'+v)(v'-v) \tag{2.5}$$

Assuming that the change in speed, $\Delta v = v' - v$, is small compared to v, then $v' + v \cong 2v'$ leads to

$$\Delta v = \frac{erB}{2m_e} \tag{2.6}$$

A change in speed means a change in the magnetic moment $\vec{m}$ of Eq. (2.2)

$$\Delta\vec{m} = -\frac{1}{2}e(\Delta v)r\hat{z} = -\frac{e^2r^2}{4m_e}\vec{B} \tag{2.7}$$

The introduction of the magnetic field results in a change in $\vec{m}$ that is proportional in magnitude but opposite to $\vec{B}$. Note that if the magnetic field were to be introduced in the minus *z*-direction (downward), the magnetic force would point radially outward, and thus, subtract from the Coulomb centripetal force, resulting in a decrease in the speed of the electron. Thus, the change in magnetic moment will again be in the opposite direction than the direction of the field.

If the concept of speeding up or slowing down the electron by the application of a magnetic field alarms you, as it should, since the magnetic force does no work on the electron, its direction being always perpendicular to the electron's displacement, it is because I have not told you the whole story, yet. The force responsible for changing the speed of the electron is of electrical nature, $\vec{F} = -e\vec{E}$, where $\vec{E}$ is the induced electric field due to the change in the strength of the magnetic field from zero to $\vec{B}$ or, equivalently, the change in magnetic flux from zero to Φ_B through the current loop of the orbiting electron, according to the second of Maxwell's Equations, Faraday's Law of Induction (Table 1.1). For either direction of the applied field $\vec{B}$, the induced electromotive force, $\mathcal{E}$, opposes the change in flux through the current loop. $\mathcal{E}$ exists only during the time $\vec{B}$ is changing, but the change in magnetic moment, $\Delta\vec{m}$, persists as long as $\vec{B}$ is present.

The change in magnetic moment, $\Delta\vec{m}$, in Eq. (2.7), was derived under the assumption that the electronic orbit is in a plane perpendicular to $\vec{B}$, that is in the horizontal plane. If the orbit were confined in a vertical plane containing $\vec{B}$, there would be no change in the magnetic moment, as the magnetic flux through the plane of the orbit would be zero. For the general case of an orbit tilted at an angle θ relative to the *z*-axis, as shown in Figure 2.3, only the projection of the plane of the orbit onto the horizontal plane contributes to the flux (Eq. (1.24)). To calculate the induced change in magnetic moment, the radius r in Eq. (2.7) must be replaced by its component parallel to the *xy*-plane, $r\sin\theta$, as shown in Figure 2.3

For random orientation of the orbital plane, the change in magnetic moment (Eq. 2.7) would depend on the average value of $r^2\sin^2\theta$ over all angles given by Eq. (2.8).

$$\left\langle r^2\sin^2\theta\right\rangle = \frac{\int_0^{2\pi}\int_0^{\pi}\left(r^2\sin^2\theta\right)r^2\sin\theta\, d\theta\, d\phi}{4\pi r^2} = \frac{2}{3}r^2 \tag{2.8}$$

where the bracket indicates the average of the quantity enclosed over all angles. Here, $r^2\sin\theta d\theta d\varphi$ is an infinitesimal surface element on the surface of a sphere of radius r, as shown in Figure 2.3, and $4\pi r^2$ is the area of the sphere. On average, the expected change in magnetic moment is reduced by a factor of 2/3, compared to the result obtained in Eq. (2.7). Thus, atoms with no net magnetic moment show only a diamagnetic response to the application of an external magnetic field given by Eq. (2.9).

$$\Delta\vec{m} = -\frac{e^2r^2}{6m_e}\vec{B} \tag{2.9}$$

This classical approach to calculating the induced diamagnetism was first introduced by Langevin in 1905 (P. Langevin, Annalles de Chimie et Physique 5 (1905) 70), prior to the advent of quantum mechanics, and it is known as Langevin diamagnetism.

Due to this diamagnetic response, a magnetization density $\vec{M}$, magnetic dipole moment per unit volume, is build up within the material, whose magnitude is proportional to the strength of the applied field and its direction opposite to the direction of the field. The

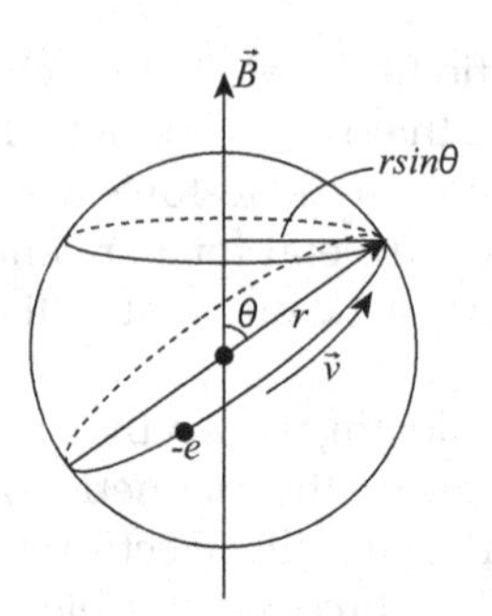

FIGURE 2.3
Depiction of an electron orbit tilted at an angle θ relative to $\vec{B}$ and its projection perpendicular to $\vec{B}$.

observed induced magnetization is known as diamagnetism. Since all matter contains circulating electrons diamagnetism is a universal property of matter. It is the manifestation of Faraday's Law and Lenz's Law, implying a tendency of materials to expel magnetic flux from their interior when an external magnetic field is applied. Materials known as type I superconductors, at temperatures below their critical temperature for superconductivity, develop macroscopic resistance-less currents upon application of an external magnetic field expelling all magnetic flux from their interior, and are thus also known as "perfect diamagnets". However, in most materials, diamagnetic effects are due to microscopic current loops bound to the atoms and their degree of magnetic flux expulsion is small, resulting in diamagnetic susceptibilities of the order of 10^{-4} to 10^{-9}.

Example 2.1: A Perfect Diamagnet

It is observed that for applied magnetic fields smaller than their critical fields for superconductivity, superconductors expel all magnetic flux from their interior, a phenomenon known as the "Meissner Effect", first discovered in 1933 by Meissner and Ochsenfeld. The metal Niobium (Nb) is a superconductor with a critical temperature for superconductivity of $T_C = 9.1$ K and a critical field of $H_C = 0.82$ T. Consider a long rod of Nb with length l and radius R ($l >> R$), its axis lying along the z-direction. It is exposed to an external magnetic field, $\vec{B} = B_0\hat{z}$ (<0.82 T), along its axis. The rod is subsequently cooled below its critical temperature of 9.1 K, upon which it expels all magnetic flux from its interior. We want to find an expression for the current density induced on the surface of the Nb rod upon entering its superconducting state, in order to achieve complete flux expulsion.

Figure 2.4 depicts the magnetic field lines at temperatures above T_C (field lines penetrate the rod) and below T_C (field lines are expelled from the interior of the rod). The expulsion of all magnetic field flux lines from the interior of the superconductor is due to the establishment of macroscopic, superconducting (resistance-less) surface currents on the circumferential surface of the rod, characterized by a surface current density, K. Because these currents do not change in time, they are called "persistent currents". They must produce a field within the superconductor, which is equal and opposite to the externally applied magnetic field. Thus, the rod behaves as a solenoid.

Application of Ampère's Circuital Law (see Figure 1.11) gives $B_0 l = \mu_0 K l$, where K is the surface current density measured in Amp/m and Kl the total current enclosed within the Ampèrian loop, Figure 2.4(b). Thus, the induced surface current is $\vec{K} = -\frac{B_0}{\mu_0}\hat{\phi}$, the minus sign indicating that the current is induced in the clockwise $-\hat{\phi}$ direction (for a right-handed cylindrical coordinate axis system).

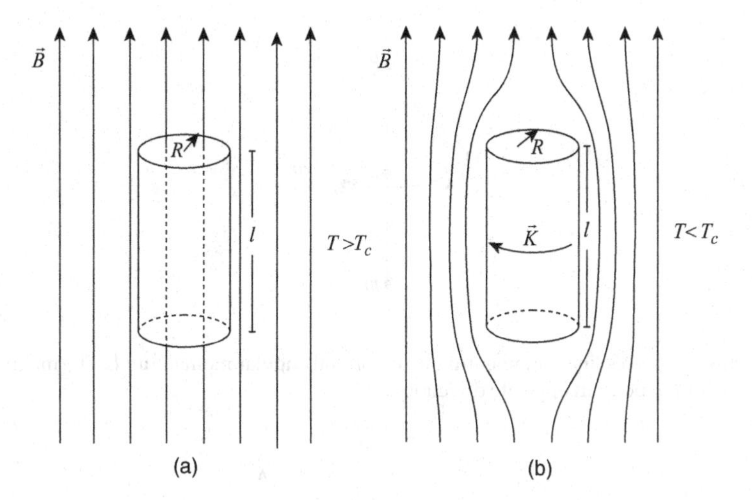

FIGURE 2.4
The Meissner Effect: (a) Magnetic field lines penetrate through the *Nb* rod when in its normal conducting state $T > T_C$. (b) Field lines are entirely expelled from the interior of the *Nb* rod in its superconducting state $T < T_C$.

2.2 Orbital and Spin Angular Momentum

2.2.1 Concepts from Classical Mechanics

According to Newtonian mechanics, the circulating electron of Figure 2.1 has instantaneous linear momentum $\vec{p} = m_e\vec{v}$ and orbital angular momentum

$$\vec{L} = \vec{r} \times \vec{p} = rm_e v\,\hat{z} \tag{2.10}$$

in the positive z-direction, as seen in Figure 2.5. The magnetic moment $\vec{m}$ of the electron given by Eq. (2.2), can thus be expressed in terms of its orbital angular momentum $\vec{L}$ according to Eq. (2.11).

$$\vec{m} = -\frac{e}{2m_e}\vec{L} \tag{2.11}$$

Therefore, the magnetic moment of the electron is proportional to its orbital angular momentum and antiparallel to it. The factor of proportionality is known as the gyromagnetic ratio γ, where

$$\vec{m} = \gamma\vec{L} \tag{2.12}$$

Let us now revisit the case addressed by Langevin, of applying a magnetic field $\vec{B}$ in a direction other than the normal to the plane of the orbit. What modifications must we

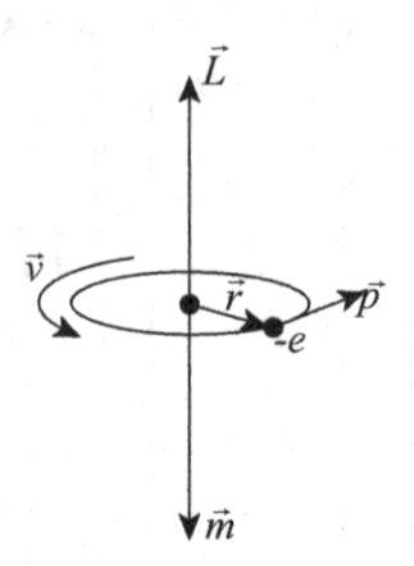

FIGURE 2.5
A circulating electron possesses linear momentum $\vec{p}$ and orbital angular momentum $\vec{L}$. The magnetic moment $\vec{m}$ and angular momentum $\vec{L}$ point in opposite directions.

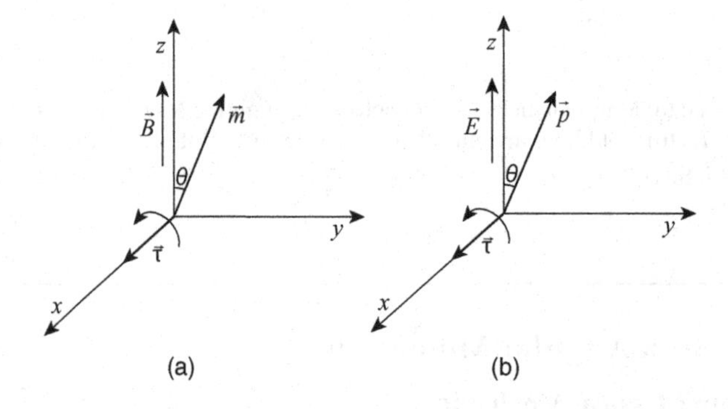

FIGURE 2.6
(a) A magnetic dipole $\vec{m}$ in a magnetic field $\vec{B}$ along the z-direction experiences a torque along the x-direction. (b) An electric dipole $\vec{p}$ in an external electric field along the z-direction experiences a torque along the x-direction.

introduce in calculating $\Delta\vec{m}$, due to the coupling of the magnetic moment to the angular momentum?

According to Eq. (1.22), the magnetic dipole in Figure 2.6(a) has orientational potential energy $U = -\vec{m}\cdot\vec{B} = -mB\cos\theta$, which is minimized when the magnetic moment is parallel to the magnetic field ($\theta = 0$). According to Eq. (1.18), the magnetic dipole also experiences a torque $\vec{\tau} = \vec{m}\times\vec{B}$ that tends to rotate it into the direction of the magnetic field, thus, minimizing its energy. Our experience with compass needles and macroscopic current loops, discussed in Chapter 1, also dictates the rotation of the magnetic dipole into the direction of the field. This rotation is analogous to the rotation of an electric dipole in an electric field due to the torque, Figure 2.6(b). The analogy drawn with the electric dipole needs to be qualified, however, due to the fact that the circulating electron moment is coupled to an angular momentum vector, unlike the electric dipole moment.

Newton's second law of motion for rotational dynamics states that the torque equals the time rate of change of the mechanical angular momentum, $\vec{\tau} = \frac{d\vec{L}}{dt}$. Equation (2.12) then leads to the relationship

$$\frac{d\vec{m}}{dt} = \gamma\frac{d\vec{L}}{dt} = \gamma\vec{m}\times\vec{B} = -\frac{e}{2m_e}\vec{m}\times\vec{B} \tag{2.13}$$

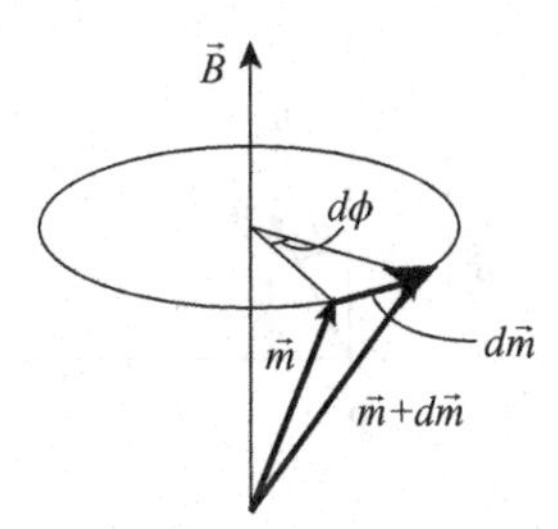

FIGURE 2.7
Larmor precession of a magnetic moment $\vec{m}$ about a magnetic field $\vec{B}$.

According to the definition of the cross product, Eq. (2.13) implies that the change in $\vec{m}$ is in a direction perpendicular to both $\vec{m}$ and $\vec{B}$, causing the magnetic moment to precess about the direction of the magnetic field $\vec{B}$, in the counterclockwise direction, while the magnitude of the magnetic moment, $|\vec{m}|$, remains constant. The magnetic moment vector prescribes the surface of a cone of angle 2θ, as it precesses around the magnetic field, Figure 2.7. The vector sum of $\vec{m}$ and $d\vec{m}$ gives the new position of the precessing magnetic moment. The frequency of precession is known as the Larmor precession frequency

$$\omega_L = \gamma B \tag{2.14}$$

We conclude that when we consider the coupling of the magnetic dipole moment to the angular momentum vector the interaction between the magnetic field and the magnetic dipole moment leads to precession, rather than alignment. This presents a fundamental difference between the electric and magnetic case. An electric dipole is not coupled to an angular momentum vector, and thus the torque on an electric dipole leads readily to the rotation of the dipole into the direction of the electric field. In the isolated system of a magnetic dipole moment $\vec{m}$ in a magnetic field $\vec{B}$, the precession will persist indefinitely. In a real material, however, the circulating electrons are not in isolation. Many atoms are in close proximity. Interatomic interactions and thermal agitation lead to energy exchange among atoms, which allow eventual moment alignment along the direction of the magnetic field, as observed macroscopically. This is analogous to the precession of a toy spinning top around the gravitational field. If it were not for external interactions leading to frictional losses, the top would precess indefinitely.

Example 2.2:

Let us consider a particular case in order to gain further insight into the precessional motion of the magnetic moment about $\vec{B}$. Using spherical coordinates, consider a magnetic field $\vec{B}$ pointing in the z-direction, and a magnetic moment $\vec{m}$ precessing about $\vec{B}$ with angular frequency ω_L, its instantaneous direction specified by spherical coordinates θ and ϕ (Figure 2.8). Assume that the magnetic dipole moment initially lied in the xz-plane, that is $\phi = 0$ at $t = 0$. As time evolves, $\vec{m}$ precesses around the z-axis and at some later time t, it points at a direction specified by angles θ and ϕ.

Carrying out the cross product, the vector differential equation of Eq. (2.15) is reduced to three scalar differential equations for the components of $\vec{m}$ along the x-, y- and z-directions, Eq. (2.16). The solutions of these three differential equations are given in Eq. (2.17), with $\omega_L = -\gamma B$.

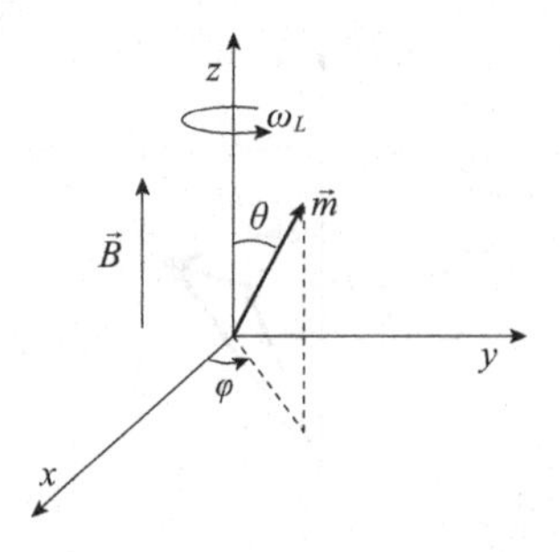

FIGURE 2.8
Coordinate axis system for the depiction of the Larmor precession of magnetic moment $\vec{m}$ about the magnetic field $\vec{B}$ in the z-direction as a function of time.

$$\frac{d\vec{m}}{dt} = \gamma \vec{m} \times \vec{B} \tag{2.15}$$

$$\frac{dm_x}{dt} = \gamma m_y B \qquad \frac{dm_y}{dt} = -\gamma m_x B \qquad \frac{dm_z}{dt} = 0 \tag{2.16}$$

$$m_x(t) = m \sin\theta \cos(\omega_L t) \quad m_y(t) = m \sin\theta \sin(\omega_L t) \quad m_z = m \cos\theta \tag{2.17}$$

Thus, the z-component of the magnetic moment, m_z, remains constant with time, while the x- and y-components, m_x and m_y, vary harmonically with time, as the projection of $\vec{m}$ on the xy-plane rotates in the counterclockwise direction about the z-axis, with angular frequency ω_L during the precession.

The gyromagnetic ratio γ is a constant of proportionality that connects the magnetic moment with the angular momentum (Eq. 2.12) and the precession frequency with the magnetic field (Eq. 2.14). The precession of the magnetic moment about $\vec{B}$ implies that the magnetic field does not only cause magnetic moments to eventually line up in the direction of B, but it can also induce dynamical magnetic phenomena. One such phenomenon observed in materials, as discussed later, is "magnetic resonance", where electromagnetic radiation incident on the precessing magnetic moment can be absorbed only if the angular frequency of the incident electromagnetic wave is equal to $|\gamma B|$.

2.2.2 Concepts from Quantum Mechanics

According to quantum mechanics, the orbital electronic energy states of the circulating electrons in an atom are quantized, that is, electrons can occupy only a discrete set of quantum states, characterized by the principal quantum number n, orbital quantum number l and azimuthal quantum number m_l. Thus, in atomic physics, the electronic orbital angular momentum depends on the electronic state occupied by the electron and it is measured in units of $\hbar$, Planck's constant h divided by 2π. The total orbital angular momentum has the value of $\sqrt{l(l+1)}\hbar$ and its z-component the value of $m_l\hbar$, where m_l can take one of a total of $(2l+1)$ possible values given by $m_l = (-l, -l+1, \ldots +l-1, +l)$. Quantization of angular momentum restricts the angle θ formed between the direction of the angular momentum and the direction of $\vec{B}$ to a discrete number of values, rather than the continuum expected classically and assumed in the Langevin theory of diamagnetism.

The orbital quantum number l can take the values of $l = 0, 1, 2, 3, 4\ldots$ corresponding to states labeled $s, p, d, f\ldots$ electronic states. Let us consider the ground state of the hydrogen atom. In a semi-classical model, the electron is depicted moving with velocity v in a circular orbit of radius r around a proton placed at the center of the circle. The classical analog of a magnetic moment due to a current loop attributed to the circulating electron was given in Eq. (2.11). According to quantum mechanics, however, in the ground state of the atom, the electron occupies an s state with zero orbital angular momentum. For the ground state of the hydrogen atom, one would predict zero magnetic moment due to the vanishing orbital angular momentum for an s electronic state.

Nevertheless, the hydrogen atom in its ground state does exhibit a magnetic moment, in accord with the classical model of the elemental current loop. It has been determined experimentally to have a value given by

$$\mu = -\frac{e}{2m_e}L = -\frac{e}{2m_e}\hbar \tag{2.18}$$

where the value of the classical angular momentum L of the electron has been replaced by $\hbar$, according to quantum mechanics. Here, we follow convention to refer to quantum mechanical magnetic moments by the Greek letter μ, as opposed to those of classical current loops denoted by m. The magnitude of the magnetic moment of the hydrogen atom in its ground state even serves as the definition of the unit of measurement for atomic magnetic moments, called the Bohr magneton, μ_B. Its value in SI units is $\mu_B = 9.274 \times 10^{-24}\text{Am}^2$.

$$\mu_B = \frac{e\hbar}{2m_e} \tag{2.19}$$

The observation of a magnetic moment in the ground state of the hydrogen atom ($l = 0$), indicates that electrons must possess an intrinsic magnetic dipole moment, not associated with the orbital quantum number and therefore their orbital motion. Two types of experimental observations made in the 1920s suggested that electrons possessed intrinsic angular momentum. One was the closely spaced splitting of the hydrogen spectral lines, known as fine structure splitting, and the other was the Stern–Gerlach experiment on a beam of silver atoms split into two beams by an inhomogeneous magnetic field. In 1925, Goudsmit and Uhlenbeck (S. Goudsmit and G. E. Uhlenbeck, *Physica* 6 (1926) 273) postulated that the electron possessed an intrinsic angular momentum, denoted by $\vec{s}$, or spin angular momentum, independent of its orbital motion. Even free electrons traveling in a straight line, not bound to an atom, possess spin angular momentum and, therefore, magnetic dipole moment.

An electron with an orbital angular momentum $\vec{l}$ in an electronic state characterized by an orbital quantum number l and an azimuthal quantum number m_l has an effective orbital magnetic dipole moment $\mu_{eff} = \sqrt{l(l+1)}\mu_B$, whose projection along the z-axis is given by $m_l\mu_B$.

The electronic spin angular momentum $\vec{s}$ is also characterized by a quantum number $s = \frac{1}{2}$ and an azimuthal quantum number $m_s = \pm\frac{1}{2}$. Thus, the magnitude of the spin angular momentum is given by $\sqrt{s(s+1)}\hbar$ and its z-component is $m_s\hbar$. We attribute an intrinsic spin magnetic dipole moment to the electron in analogy to that attributed to its orbital angular momentum. By convention, however, the spin magnetic moment is defined so that its z-component takes on the values of $\mp\mu_B$ (opposite to the direction of the spin angular momentum) even though the spin is half-integral. This introduces a constant, known as the g-factor, whereby the component of the magnetic moment along a particular axis is given by $-gm_s\mu_B$ and the effective spin magnetic dipole moment given by $\mu_{eff} = g\sqrt{s(s+1)}\mu_B$. For a free electron, the g-factor is equal to 2. In atoms, however, electrons generally possess both orbital (g = 1) and spin (g = 2) angular momenta, and thus, the g-factor can take on various values depending on the value of the orbital angular momentum.

The magnetic moment of the hydrogen atom in its ground state, where the electron occupies an s orbital state with no orbital angular momentum (l = 0), is due entirely to its spin angular momentum. In p, d, f (l = 1, 2, 3) and higher orbital angular momentum electronic states, there are both orbital and spin angular momentum contributions; the total angular momentum of the electron, $\vec{j} = \vec{l} + \vec{s}$, would be characterized by total angular momentum quantum number, j, and azimuthal quantum number, m_j. This time, application of a magnetic field $\vec{B}$, say in the z-direction, will induce precession of the "atomic magnetic moment" around the z-axis, as shown in Figure 2.7, with the z-component of the atomic magnetic moment being a constant of the motion. This is the basis of paramagnetism.

Nuclei also possess intrinsic spin magnetic dipole moments that are associated with the nuclear spin angular momentum $\vec{I}$. The nuclear spin quantum number I can have half-integral and integral values and its z-component is characterized by the azimuthal quantum number m_I, which can take any of the (2I+1) possible values $m_I = I, (I-1),\ldots 0,\ldots, -(I-1), -I$. The total spin angular momentum of the nucleus is then given by $\sqrt{I(I+1)}\hbar$ and its z-component by $m_I\,\hbar$. The nucleus has a finite size, albeit very small, with nuclear radii of the order of 10^{-15} m, allowing the visualization of spin nuclear magnetic dipole moments according to the classical analog of a rotating charged sphere spinning with angular velocity $\vec{\omega}$. The spin magnetic dipole moment of the electron, however, is not amenable to a similar classical analog visualization, due to the fact that for all practical purposes, the electron is considered to be a point charge. Electronic spin, and its associated magnetic dipole moment, is of pure quantum mechanical origin, grounding magnetism firmly into the realm of quantum mechanics.

The unit of measurement of nuclear magnetic moments is the nuclear magneton defined in Eq. (2.20) in analogy to the Bohr magneton, where m_p is the mass of the proton.

$$\mu_N = \frac{e\hbar}{2m_p} \tag{2.20}$$

In SI units, the value of $\mu_N = 5.050 \times 10^{-27}\,\text{Am}^2$, three orders of magnitude smaller than the Bohr magneton. Thus, due to the fact that protons are very massive compared to electrons, 1,836 times heavier, the magnetic moments of nuclei do not contribute considerably to the magnetization of a material. Their effect is, however, measurable in various spectroscopic measurements that probe the intrinsic magnetic properties of materials through the detection of hyperfine interactions, such as in ^{57}Fe-Mössbauer spectroscopy, in nuclear magnetic resonance (NMR) spectroscopy and in magnetic resonance imaging (MRI).

2.3 Atomic Magnetic Moments

Atoms contain many electrons, each spinning around its own axis, as they orbit around the nucleus. Vector addition of all electronic moments can lead to two possibilities: (a) individual electronic magnetic moments may be oriented in such a way that they cancel each other out, leading to zero total atomic moment or (b) the moment cancelation may be only partial or incomplete, leading to a net atomic magnetic moment. The former case leads to diamagnetism while the latter to paramagnetism. Atoms possessing closed electronic shells, such as, the noble gases (argon, helium, krypton, neon, radon and xenon) are diamagnetic. Only atoms that possess open or incomplete electronic shells can be paramagnetic, since closed electron shells have zero total angular momentum.

In most atoms, except the heavier ones, the coupling between the orbital and spin angular momentum, known as "spin–orbit coupling", is relatively weak. In transition metals, for example, the electronic orbital angular momenta add separately to produce total orbital angular momentum $\vec{L}$, and total spin angular momentum $\vec{S}$ for the atom, according to Eq. (2.21), where the summation goes over all electrons in the atom.

$$\vec{L} = \sum_i \vec{l}_i \qquad \vec{S} = \sum_i \vec{s}_i \tag{2.21}$$

This yields a total atomic angular momentum

$$\vec{J} = \vec{L} + \vec{S} \tag{2.22}$$

This type of coupling, whereby L, S and J are good quantum numbers for the atomic states, is known as $L - S$ coupling or Russell–Saunders coupling. Under this coupling scheme, the total orbital and total spin angular momenta of the atom have separately well-defined values. As we will see, within a material, the orbital motion of the electrons is usually strongly bound to intrinsic crystallographic axes, making rotation of the plane of the orbit, upon the application of a magnetic field, difficult. The spin magnetic moments are generally free to precess about the applied magnetic field direction, however, making the electronic spin primarily responsible for paramagnetism.

The magnetic moment of an isolated atom and its z-component can then be expressed as

$$\mu_{eff} = g\mu_B\sqrt{J(J+1)}, \quad \mu_z = -g\mu_B M_J \tag{2.23}$$

where M_J, the azimuthal quantum number, is associated with J, and μ_B is the Bohr magneton. The g-factor for a multi-electron atom is given by the Landé equation

$$g = 1 + \frac{J(J+1) + S(S+1) - L(L+1)}{2J(J+1)} \tag{2.24}$$

In the case of $S = 0$, $J = L$. Then, the above equation gives $g = 1$, as expected for the orbital magnetic moment. On the contrary, if $L = 0$, then $J = S$; the Landé equation yields $g = 2$, as expected for a spin magnetic moment.

Independent experiments performed by Lamb and Retherford (W. E. Lamb Jr. and R. C. Retherford, *Phys. Rev.* 72 (1947) 241) and by Kusch and Foley (P. Kusch and H. M. Foley, *Phys. Rev.* 74 (1948) 250) in the 1940s determined the precise value of the electronic g-factor to be 2.000238, rather than 2. In the first experiment, the spectroscopic shift in the electronic spectrum of the hydrogen atom was used for the precise measurement of g. In the second, an atomic beam experiment was used to make careful measurements of the magnetic moment of atoms containing a single electron beyond a closed shell. The discrepancy from $g = 2$ was interpreted according to, the novel at the time, the Theory of Quantum Electrodynamics, where the electron freely emits and absorbs photons, which carry an angular momentum $\vec{L} = 1$, wielding on the average an angular momentum for the electron larger that ½. Thus, the measurement of the g-factor of the electron provided some of the early confirmations of the validity of the theory of quantum electrodynamics. Lamb and Kusch shared the 1955 Nobel Prize in Physics for their precise measurements of g.

In the case of strong "spin–orbit" coupling encountered in heavier elements, such as the rare earths, an alternate coupling scheme must be introduced, known as $j - j$ coupling, according to which the total angular momentum $\vec{j}_i$ for each electron is first obtained and the total angular momentum $\vec{J}$ of the atom is given by a sum over all $\vec{j}_i$'s.

$$\vec{j}_i = \vec{l}_i + \vec{s}_i \qquad \vec{J} = \sum_i \vec{j}_i \tag{2.25}$$

Under this coupling scheme, both orbital and spin angular momenta contribute to paramagnetism.

The ground state of a multi-electron atom is obtained by a particular combination of electronic angular momentum quantum numbers that minimizes the energy of the multi-electron system. Under Russell–Saunders coupling, it can be determined by following "Hund's Rules" and the "Pauli Exclusion Principle". The latter simply states that no two electrons in an atom can share all four quantum numbers (n, l, m_l and m_s) while the former contains three rules that must be satisfied in the following order: (a) electronic states are populated so as to maximize S, (b) within the constraints of (a) electronic states are populated as to maximize L and (c) $J = |L - S|$ if the shell is less than half-full and $J = |L + S|$ if the shell is more than half-full. The first two of Hund's rules together with the Pauli Exclusion Principle aim at minimizing the electrostatic repulsion energy between electrons, preventing, for example, electrons with parallel spins from sharing the same space, while the third rule aims at minimizing the spin–orbit interaction energy. By convention, having determined the values of S, L and J for the atom, its atomic state is given in spectroscopic notation by using the terminology ${}^{2S+1}L_J$, where J is the total angular momentum number, L the total orbital angular momentum and the number $(2S + 1)$ gives the spin multiplicity. Again by convention, L is given not as a number but as a letter, S, P, D, F, *etc.*, corresponding to $L = 0, 1, 2, 3$, *etc.*

Example 2.3:

Consider the two most common oxidation states of iron, the ions Fe^{3+} (ferric) and Fe^{2+} (ferrous). The outer electrons occupy an incomplete 3d electronic shell. For d electrons, $l = 2$. Thus, there are $2l + 1 = 2 \times 2 + 1 = 5$ orbital electronic states available. For Fe^{3+} there are 5 electrons, while for Fe^{2+} there are 6 electrons that occupy these 3d states.

Let us consider the case of the ferric ion first with five 3d electrons. Following the first of Hund's rules that requires maximizing S, each electron will have $m_s = +\frac{1}{2}$ and each must occupy a different m_l (−2, −1, 0, +1, +2) state. Thus, the five d electrons together contribute zero orbital angular momentum, $\vec{L} = 0$, but maximum spin angular momentum $\vec{S} = 5 \times (1/2) = 5/2$. The spin multiplicity is $2 \times 5/2 + 1 = 6$ and $J = 5/2$. The ionic ground state is then an orbital singlet, or S state, given by ${}^6S_{5/2}$.

In the case of the ferrous ion, Fe^{2+}, a sixth electron also resides in the 3d orbital states. This sixth electron, beyond the half-filled shell, must share an orbital state with one of the five electrons already occupying the d electronic shell. According to Pauli Exclusion Principle, the two electrons sharing the same orbital state must pair up in an antiparallel spin fashion, one "spin-up", $m_s = +1/2$, and the other "spin-down", $m_s = -1/2$. Thus, the spin

TABLE 2.1
3*d* Electron Population for Ferric (Top) and Ferrous (Bottom) Ions

m_l	$m_s = +1/2$	$m_s = -1/2$
2	↑	
1	↑	
0	↑	
−1	↑	
−2	↑	

m_l	$m_s = +1/2$	$m_s = -1/2$
2	↑	↓
1	↑	
0	↑	
−1	↑	
−2	↑	

Vectors indicate "spin-up" and "spin-down" electrons.

angular momentum is now reduced to $\vec{S} = 4 \times \frac{1}{2} = 2$, but in this case, the sixth electron contributes to the orbital angular momentum. Following the second of Hund's rule, the orbital angular momentum for the ion is $\vec{L} = 2$ corresponding to a D state. The spin multiplicity is $2 \times 2 + 1 = 5$ and $J = 2 + 2 = 4$, giving ${}^5D_4$ for the Fe^{2+} ion state. Table 2.1 depicts the electron population of the 3*d* shell for the two cases.

2.4 Bound Currents and the Auxiliary Field H

Having introduced some fundamental concepts to establish the quantum mechanical origin of magnetism, we can return to our development of the subject in the realm of classical physics, as many fundamental observations in magnetism were made prior to the advent of quantum mechanics and the discovery of the spin of the electron. In addition, due to the fact that in a macroscopic specimen, there is a large number (Avogadro's number per mole of material) of atoms and therefore atomic moments, macroscopic magnetic measurements cannot detect the granularity or quantization of magnetization. Thus, classically, the magnetization per unit volume in the material can be assumed to be a continuous vector field, in the same sense as water is depicted as a continuous fluid even though it is composed of discrete water molecules.

Assuming that there is no interaction among neighboring atomic moments, or between atomic moments and their environment, the magnetic moments of the atoms in the material, prior to the application of an external magnetic field, are expected to point in random directions, rendering the material as a whole non-magnetic. Upon application of a magnetic field in the +*z*-direction, the first response of the atomic moments is to precess about the *z*-axis. Each moment, considered classically to arise from a microscopic current loop, acquires an orientational potential energy $U = -\vec{m} \cdot \vec{B} = -mB\cos\theta$, where θ is the angle between the direction of the magnetic moment and the applied magnetic field. Interatomic interactions within the material, however, facilitate moment rotation toward the direction of the magnetic field, minimizing the magnetic orientational potential energy. In the simplest visualization, all atomic moments in the material become collinear with the field and a positive maximum magnetization per unit volume, $\vec{M} = N\vec{m}$, is induced in the material, where N is the number of atoms per unit volume and $\vec{M}$ is measured in amperes per meter (A/m).

Conversely, when a material is introduced in a region of space where there already existed a magnetic field $\vec{B}$, the magnetic field within and around the material is altered, due to the induced magnetization. Classically, the induced magnetization is considered to arise from oriented, microscopic current loops, each constituting a magnetic dipole. Figure 2.9 depicts a chunk of a material of thickness t with such oriented microscopic current loops due to the presence of the magnetic field. For uniform magnetization $\vec{M}$ throughout the material, currents between adjacent current loops cancel each other out, except at the cutaway surface of thickness t. Near the surface, there is incomplete cancelation, resulting in a perceived surface current circulating in the counterclockwise direction. By convention, the term "bound" surface current is used to describe such a current in order to distinguish it from any real or "free" current that may be flowing in a conducting material due to a voltage source. The value and direction of the "bound surface current density", $\vec{K}_b$, is given by the cross product of the magnetization density, $\vec{M}$, at the surface with the normal to the surface, $\hat{n}$, as indicated by Eq. (2.26).

$$\vec{K}_b = \vec{M} \times \hat{n} \tag{2.26}$$

In Figure 2.9, a pictorial representation of the elemental current loops is given, with $\vec{M}$ pointing in the direction of $\vec{B}$, and of the resulting bound surface current, $\vec{K}_b$, circulating in the counterclockwise direction on the surface.

In Eq. (2.26), $\vec{M}$ is the magnetization density evaluated at the surface and $\hat{n}$ is the unit vector normal to the surface, pointing from the interior to the exterior of the material, as depicted in Figure 2.9. The bound surface current is given by Eq. (2.27), where t is the thickness of the material.

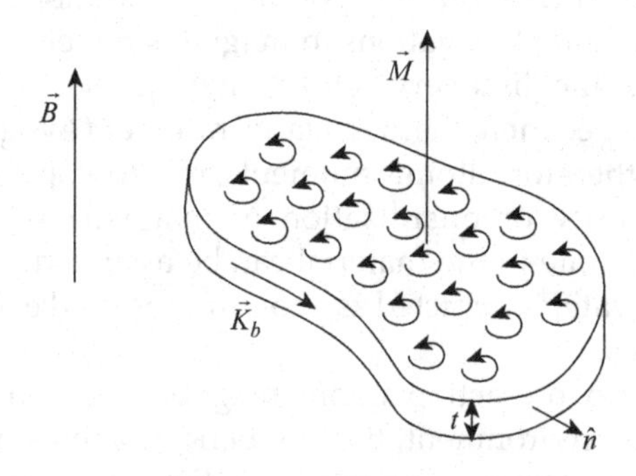

FIGURE 2.9
Cross section of magnetized material due to oriented microscopic current loops in the presence of an applied field $\vec{B}$. For uniform magnetization $\vec{M}$, a bound surface current density, $\vec{K}_b$, is induced in the counterclockwise direction.

$$I_b\left(surface\right)=\int_0^t \vec{K}_b\cdot d\vec{t} \tag{2.27}$$

This bound surface current produces a magnetic field inside and outside the material, which modifies the original magnetic field $\vec{B}$ that existed at that region, prior to the introduction of the material. If the magnetization within the material is non-uniform, then a "bound volume current density", $\vec{J}_b$, will also exist within the material. Non-uniform magnetization may arise from internal structural characteristics of the material or may be induced by a non-uniform external magnetic field. The bound volume current density is given by the curl of the magnetization vector field, Eq. (2.28).

$$\vec{J}_b=\vec{\nabla}\times\vec{M} \tag{2.28}$$

The vector differential operator is defined by $\vec{\nabla}=\frac{\partial}{\partial x}\hat{x}+\frac{\partial}{\partial y}\hat{y}+\frac{\partial}{\partial z}\hat{z}$ (see Appendix). The bound current passing through an infinitesimal cross-sectional area $d\vec{a}$ would then be $\vec{J}_b\cdot d\vec{a}$ and the total current over a finite cross-sectional area A would be

$$I_b\left(volume\right)=\int_A \vec{J}_b\cdot d\vec{a}=\int_A\left(\vec{\nabla}\times\vec{M}\right)\cdot d\vec{a} \tag{2.29}$$

Using the fundamental theorem for curls, also known as Stoke's theorem (see Appendix), the surface integral of the curl of the magnetization in Eq. (2.29) can be written as the line integral of the magnetization $\vec{M}$ over a closed path bounding area A.

$$\int_A\left(\vec{\nabla}\times\vec{M}\right)\cdot d\vec{a}=\oint\vec{M}\cdot d\vec{l} \tag{2.30}$$

In the interior of the material, the fourth of Maxwell's equations must be modified by including the volume bound current of Eq. (2.29) in addition to any free current, I_f, that may be flowing in the material to give Eq. (2.31), where for simplicity, the bound volume current is represented simply by I_b.

$$\oint\vec{B}\cdot d\vec{l}=\mu_0\left(I_f+I_b\right)=\mu_0\left(I_f+\oint\vec{M}\cdot d\vec{l}\right) \tag{2.31}$$

Combining the two line integrals on the left side of the equation, we obtain

$$\oint\left(\frac{1}{\mu_0}\vec{B}-\vec{M}\right)\cdot d\vec{l}=I_f \tag{2.32}$$

The integrand defines a new physical quantity,

$$\vec{H}=\frac{1}{\mu_0}\vec{B}-\vec{M} \tag{2.33}$$

Ampère's Circuital Law inside the material can then be expressed as

$$\oint\vec{H}\cdot d\vec{l}=I_f \tag{2.34}$$

while the magnetic field inside the material is given by

$$\vec{B} = \mu_0\left(\vec{H} + \vec{M}\right) \tag{2.35}$$

Please, note that $\vec{H}$ has the same dimensionality as $\vec{M}$. In free space, where $\vec{M} = 0$, we conclude that $\vec{B} = \mu_0\vec{H}$. That is, $\vec{B}$ is proportional to $\vec{H}$ with the constant of proportionality being the permeability of free space. The quantity $\vec{H}$ is often also referred to as the magnetic field, measured in amperes/meter, even though $\vec{B}$ is the fundamental definition of the magnetic field, measured in Tesla (SI units). To distinguish the two fields, some authors call $\vec{B}$ the "magnetic flux density" or "magnetic induction". We will continue to call $\vec{B}$ the magnetic field and $\vec{H}$ the *H*-field. Generally, it would be clear by the context which field one refers to. In free space, the two fields point in the same direction. In a material, they could point in different directions, including pointing in opposite directions. In experimental studies, it is customary to use $\vec{H}$, rather than $\vec{B}$ to denote the magnetic field. This is so because, *via* the Circuital Law, it is $\vec{H}$ that is directly related to the free current, I_f, rather than $\vec{B}$. It is the free current that you freely control experimentally.

Example 2.4:

An infinitely long conducting cylinder of radius *R* is exposed to an external magnetic field $\vec{B}$ along its axis; a uniform magnetization $\vec{M}$ along the axis of the cylinder is induced. Find (a) the induced bound volume current density $\vec{J}_b$, (b) the induced bound surface current density $\vec{K}_b$ and (c) the $\vec{H}$-field inside ($r < R$) and outside ($r > R$) the conductor.

Let us assume that the axis of the cylinder is along the *z*-direction, as shown in Figure 2.10. (a) According to Eq. (2.28), $\vec{J}_b = \vec{\nabla} \times \vec{M}$. Since $\vec{M}$ is uniform all its spatial derivatives vanish, giving $\vec{\nabla} \times \vec{M} = 0$. Therefore, $\vec{J}_b = 0$. (b) From Eq. (2.26), $\vec{K}_b = \vec{M} \times \hat{n}$, where $\hat{n}$ is the unit vector along the radial direction, perpendicular to the surface and pointing from the interior to the exterior of the conducting cylinder. Since $\vec{M}$ is along the *z*-direction, vectors $\vec{M}$ and $\hat{n}$ are perpendicular to each other, resulting in $\vec{K}_b = \vec{M} \times \hat{n} = M\sin\left(\frac{\pi}{2}\right)\hat{\phi}$. Thus, $\vec{K}_b = M\hat{\phi}$, indicating that a bound surface current density is circulating around the circumference of the conductor in the counterclockwise direction. (c) Inside the conductor $\vec{H} = \frac{1}{\mu_0}\vec{B} - \vec{M}$, while outside the conductor, where $\vec{M} = 0$, $\vec{H} = \frac{1}{\mu_0}\vec{B}$.

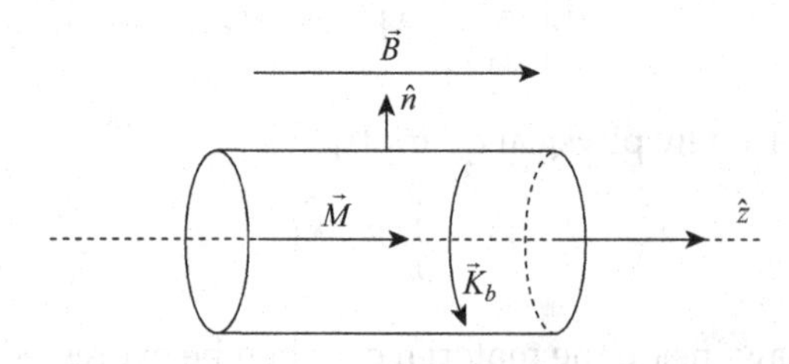

FIGURE 2.10
An infinitely long conducting cylinder with uniform magnetization $\vec{M}$ along the *z*-direction. The equivalent bound surface current $\vec{K}_b$ is shown in the counterclockwise direction.

2.5 Magnetic Susceptibility

In diamagnetic and paramagnetic materials, the magnetization induced by an applied field is readily reduced back to zero upon the removal of the magnetic field. In small applied fields, the induced magnetization is proportional to the applied magnetic field. This proportionality is usually expressed in terms of the $\vec{H}$-field according to Eq. (2.36), where the constant of proportionality, χ, is the magnetic susceptibility of the material (at higher fields saturation effects make the relation between M and H more complex).

$$\vec{M} = \chi \vec{H} \tag{2.36}$$

Such materials where $\vec{M}$ and $\vec{H}$ are collinear are called linear magnetic media. The magnetic field is then given by

$$\vec{B} = \mu_0 \left(\vec{H} + \vec{M} \right) = \mu_0 \left(1 + \chi \right) \vec{H} = \mu \vec{H} \tag{2.37}$$

Thus, inside linear media $\vec{B}$ is also proportional to $\vec{H}$, as is the case of free space. The constant of proportionality, however, is now

$$\mu = \mu_0 \left(1 + \chi \right) \tag{2.38}$$

The constant μ is called the permeability of the material. This is the reason why μ_0 is called the permeability of free space. Figure 2.11 sketches the variation of the magnitude of the induced magnetization, $\vec{M}$, as a function of the strength of the magnetic field $\vec{H}$ for a diamagnetic and a paramagnetic material.

The *M vs. H* curve is linear in both cases, with the slope giving the magnetic susceptibility of the material, negative for diamagnets and positive for paramagnets. Upon reducing the applied field back to zero, the data retrace the original linear curve, as indicated by the arrows.

$$\chi = \frac{M}{H} \tag{2.39}$$

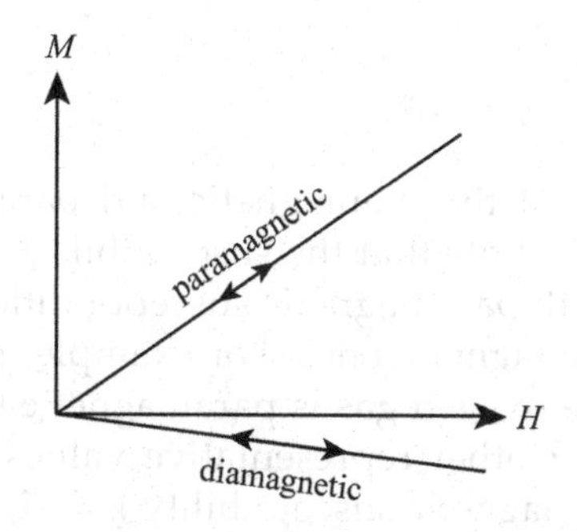

FIGURE 2.11
Variation of induced magnetization *M vs. H* for a diamagnetic and a paramagnetic material.

TABLE 2.2

Magnetic Susceptibilities of Diamagnetic and Paramagnetic Materials at 20°C

Diamagnetic Materials	χ (1 × 10−5)
Ammonia	−0.26
Bismuth	−16.6
Mercury	−2.9
Silver	−2.6
Carbon (diamond)	−2.1
Carbon (graphite)	−1.6
Lead	−1.8
Sodium chloride	−1.4
Copper	−1.0
Water	−0.91
Nitrogen	−0.0005
Hydrogen	−0.00021
Paramagnetic Materials	**χ (1 × 10−5)**
Iron oxide (FeO)	720
Iron ammonium alum	66
Uranium	40
Platinum	26
Tungsten	6.8
Cesium	5.1
Aluminum	2.2
Lithium	1.4
Magnesium	1.2
Sodium	0.72
Oxygen	0.19

Source: Hyperphysics.

Table 2.2 gives the values of the diamagnetic and paramagnetic susceptibility for some representative materials. Note that the susceptibility is a dimensionless constant, of the order of 10^{-3} to 10^{-7}, with paramagnetic susceptibilities being two or three orders of magnitude larger than diamagnetic ones. For example, nitrogen gas is diamagnetic with $\chi = -0.0005 \times 10^{-5}$, while oxygen gas is paramagnetic with $\chi = 0.19 \times 10^{-5}$ at room temperature. See Table 2.2 for other representative values. An exception is noted for superconductors that have a magnetic susceptibility $\chi = -1$, that is they exhibit "perfect diamagnetism" below their critical temperature for superconductivity, as discussed in Example 2.1.

Example 2.5:

An infinitely long, densely coiled solenoid with n turns per unit length carries a current I. Determine the magnetic field $\vec{B}$ inside the solenoid if the core of the solenoid was filled with (a) a paramagnetic material with magnetic susceptibility χ_p, (b) a diamagnetic material with susceptibility χ_d and (c) a superconductor cooled below its critical temperature for superconductivity.

For the case of free space, we applied Ampère's Circuital Law $\oint \vec{B} \cdot d\vec{l} = \mu_0 I_{enc}$ to obtain the magnetic field inside an infinitely long solenoid as $\vec{B} = \mu_0 nI\,\hat{z}$ (Eq. 1.13), along the axis of the solenoid, where I is the free current in the solenoid driven by a voltage source. When the core is filled with a material, induced bound currents will contribute to the magnetic field $\vec{B}$ in addition to the contribution of the free (real) current flowing in the solenoid. As we have no information on the induced bound currents, we can use the symmetry of the problem and apply Ampère's Circuital Law for $\vec{H}$, which involves only the free current. Due to symmetry, we expect the field in the solenoid to point along the axis of the solenoid. A rectangular Ampèrian loop (Figure 1.11) enclosing N turns would then give $\oint \vec{H} \cdot d\vec{l} = I_f(enclosed)$ (Eq. (2.34)). This leads to Eq. (2.40), which is analogous to Eq. (1.13) for the case of an empty core.

$$\vec{H} = nI\,\hat{z} \tag{2.40}$$

From Eq. 2.37, $\vec{B} = \mu_0(1+\chi)\vec{H} = \mu_0(1+\chi)nI\,\hat{z}$. Since $\chi_p > 0$ and $\chi_d < 0$, the magnetic field $\vec{B}$ in the core of the solenoid is slightly enhanced relative to vacuum if the core is filled with a paramagnetic medium and slightly decreased if filled with a diamagnetic medium. Finally, when the core is filled with a superconductor below its critical temperature, the field inside the solenoid is reduced to zero, since $\chi = -1$ in this case.

In Exercise 2.6, you are asked to calculate the change in B, relative to an empty core, when the core is filled with graphite or platinum. Due to the fact that diamagnetic and paramagnetic susceptibilities are of the order of 10^{-5}, you will find that the change in B is very small.

In subsequent chapters, we will see that, unlike paramagnetic and diamagnetic media, ferromagnetic materials can greatly increase the magnetic field within the core. This fact is broadly utilized in technological applications in the design of electronic devices.

2.6 Curie's Law

The first systematic study of the magnetic susceptibility of various substances was undertaken by Pierre Curie and published in 1895 (P. Curie, Ann. Chem. Phys. 5 (1895) 289). He reported that the magnetic susceptibilities were independent of temperature for diamagnets, but inversely proportional to temperature for paramagnets. The lack of temperature dependence of the susceptibility for diamagnets is consistent with Langevin's theory of diamagnetism discussed earlier. The behavior of paramagnets as a function of temperature is known as Curie's Law, given by Eq. (2.41)

$$\chi = \frac{C}{T} \tag{2.41}$$

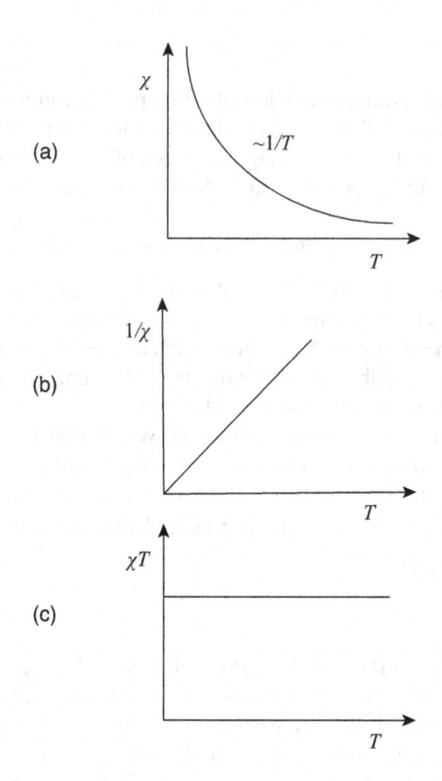

FIGURE 2.12
Curie's Law of paramagnetism. See text.

Here, C is the constant of proportionality, known as the Curie constant and T is the temperature in Kelvin. Figure 2.12(a) gives a sketch of χ *vs.* T for Curie's Law. In experimental studies, one often plots $1/\chi$ *vs.* T, where a linear relationship is expected (Figure 2.12(b)) or χT *vs.* T (Figure 2.12(c)) that is constant, in order to determine the value of C.

These experimental observations of Curie remained without theoretical interpretation for ten years till Langevin's work was published in 1905, where a classical theoretical model for paramagnetism was put forward, in the same paper in which he addressed the case of diamagnetism (P. Langevin, Annalles de Chimie et Physique 5 (1905) 70). Despite the simplicity of his model, which we discuss in the following section, it predicts the susceptibility of many materials to a very good approximation.

2.7 Langevin Paramagnetism

Consider a material made up of atoms, each with a net magnetic moment $\vec{m}$. Assuming that there is no interaction among moments or between moments and their environment, the direction of $\vec{m}$ should be random. As depicted in Figure 2.13, the number of atomic magnetic moments lying within θ and $(\theta + d\theta)$ should be proportional to the area $da = 2\pi(r \sin\theta)(rd\theta)$, where r is the radius of a sphere centered at the origin. For the case of $r = 1$, a sphere of unit radius, this expression simplifies to$2\pi \sin\theta d\theta$.

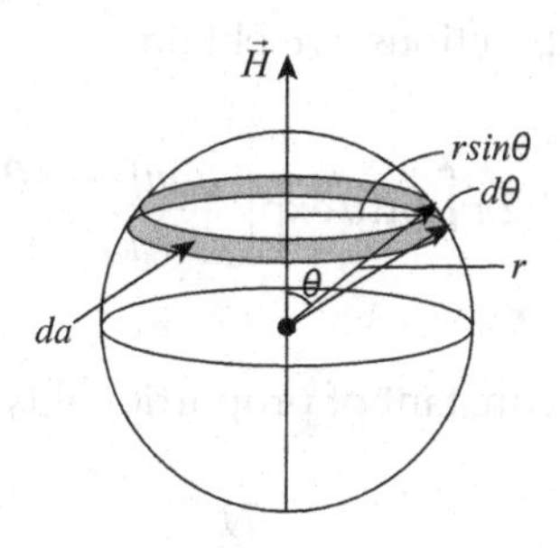

FIGURE 2.13
Unit sphere ($r = 1$) defining the parameters used in calculating the Langevin paramagnetism.

The number of atoms per unit volume with atomic moment pointing within this prespecified angle range is given by

$$dN = K2\pi \sin\theta d\theta \tag{2.42}$$

where K is a proportionality constant. In the absence of a magnetic field, these atomic moments are oriented randomly, with an equal probability of pointing along any angle θ, relative to the z-axis. Then,

$$N = \int_0^N dN = K2\pi \int_0^\pi \sin\theta d\theta = K4\pi$$

giving a value of $K = \frac{N}{4\pi}$, where N is the number of atoms per unit volume and 4π is the total solid angle. Thus, K is a constant independent of angle, as expected.

Upon application of a magnetic field along the z-axis, however, the moments will tend to align along the direction of the magnetic field. At a finite temperature T, thermal agitation will prevent complete alignment of the magnetic moments along $\vec{B}$, resulting in a finite probability that still a number of moments will point within the range of angles from θ to $(\theta + d\theta)$ treated classically as a continuum. However, the proportionality constant K will now be a function of the angle θ. In the presence of the field, each atomic moment acquires an orientational potential energy given by

$$U = -\vec{m} \cdot \vec{B} = -\mu_o \vec{m} \cdot \vec{H} = -\mu_0 mH \cos\theta \tag{2.43}$$

For non-interacting atomic moments, classical Boltzmann statistics allow us to calculate the probability that a moment point along θ

$$p(\theta) = \exp(-U/kT) = \exp(\mu_0 mH \cos\theta / kT) \tag{2.44}$$

where $k = 1.38 \times 10^{-23}$ J/K is the Boltzmann's constant and T is the temperature in Kelvin. Incorporating this probability into Eq. 2.42, we obtain

$$dN = K(2\pi \sin\theta) p(\theta) d\theta = K2\pi \sin\theta \exp\left(\frac{\mu_0 mH \cos\theta}{kT}\right) d\theta \tag{2.45}$$

Integrating over all possible directions, we obtain

$$N = K2\pi \int_0^\pi \sin\theta \exp\left(\frac{\mu_0 mH \cos\theta}{kT}\right) d\theta \tag{2.46}$$

which allows us to evaluate the constant of proportionality K as a function of θ

$$K = \frac{N}{2\pi \int_0^\pi \sin\theta \exp\left(\frac{\mu_0 mH \cos\theta}{kT}\right) d\theta} \tag{2.47}$$

The induced magnetization along the z-axis, the direction of the field, is then given by the expression

$$\vec{M} = \int_0^N \vec{m} \cos\theta dN = \frac{N\vec{m} \int_0^\pi \cos\theta \sin\theta \exp\left(\mu_0 mH \cos\theta / kT\right) d\theta}{\int_0^\pi \sin\theta \exp\left(\mu_0 mH \cos\theta / kT\right) d\theta} \tag{2.48}$$

The integrals appearing in the numerator and denominator of Eq. (2.48) can be evaluated by using change of variables, $u = \cos\theta$ and $du = -\sin\theta d\theta$ to obtain

$$\vec{M} = N\vec{m}\left[\coth\left(\mu_0 mH/kT\right) - \left(kT/\mu_0 mH\right)\right] = N\vec{m}L\left(\mu_0 mH/kT\right) \tag{2.49}$$

Equation (2.49) gives the induced magnetization density in a paramagnetic material when exposed to an external magnetic field, within the realm of classical mechanics. $L(x)$ is known as the Langevin function, where $x = \mu_0 mH/kT$. Since $N\vec{m}$ is the maximum possible magnetization, when the magnetization points along the direction of the applied magnetic field, we can call it the saturation magnetization $\vec{M}_s$. This allows us to introduce the reduced magnetization

$$\frac{M}{M_s} = L(x) = \coth x - \frac{1}{x} \tag{2.50}$$

Note that $-1 < L(x) < 1$, tending to +1 in large positive magnetic fields and −1 in large negative magnetic fields at low temperatures. For the case of $x < 1$, $L(x)$ can be expressed as an infinite power series in x

$$L(x) = \frac{x}{3} - \frac{x^3}{45} + \frac{2x^5}{945} - \ldots \tag{2.51}$$

Figure 2.14 gives a graphical representation of the Langevin equation for the reduced paramagnetic magnetization, which describes the universal behavior of classical paramagnets.

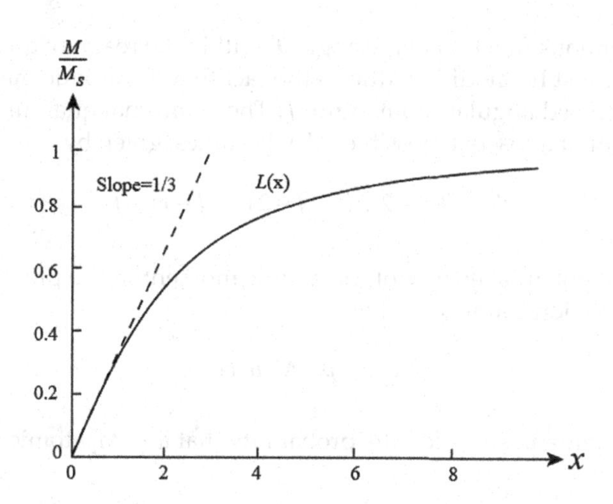

FIGURE 2.14
Reduced paramagnetic magnetization for a classical paramagnet according to Langevin.

In most experimental cases encountered in the laboratory $\mu_0 mH/kT << 1$, that is, magnetic orientational energies are much smaller than thermal energies. Then, Langevin's equation, $L(x)$, can be fairly well approximated with the first term in the series of Eq. (2.51)

$$L(x) = \frac{x}{3} \tag{2.52}$$

giving a magnetization density

$$M = M_s \frac{\mu_0 mH}{3kT} = Nm \frac{\mu_0 mH}{3kT} = N \frac{\mu_0 m^2 H}{3kT} \tag{2.53}$$

and magnetic susceptibility

$$\chi = \frac{M}{H} = \frac{N\mu_0 m^2}{3kT} = \frac{C}{T} \tag{2.54}$$

Thus, Langevin's theory of paramagnetism predicts the experimental observations of Curie, that paramagnetic materials exhibit a susceptibility that varies inversely proportional to temperature, and determines the Curie constant, C, in terms of the properties of the material and fundamental constants of nature, the permeability of free space and Boltzmann's constant.

$$C = \frac{N\mu_0 m^2}{3k} \tag{2.55}$$

2.8 Quantum Theory of Paramagnetism

In his classical theory of paramagnetism, Langevin assumed that the angle θ between the atomic magnetic moment and the magnetic field could vary continuously between 0 and π. This assumption is reflected in the fact that the Boltzmann probability factor, $p(\theta)$ in

Eq. (2.44), is a continuous function of the angle θ. Within the realm of quantum mechanics, Langevin's theory must be modified, due to the fact that the atomic magnetic moment is coupled to the quantized angular momentum $\vec{J}$. The azimuthal quantum number, M_J, can only take a value out of a discrete possible $(2J + 1)$ values, given by

$$J, J-1, J-2, \ldots, -(J-2), -(J-1), -J \tag{2.56}$$

The orientational potential energy of the atomic moment in the presence of a magnetic field along the z-direction is now quantized

$$U = -\mu_0 g M_J \mu_B H \tag{2.57}$$

According to Boltzmann statistics, the probability that any M_J atomic state is populated is given by

$$p(U) = \exp(-U / kT) = \exp(\mu_0 g M_J \mu_B H / kT) \tag{2.58}$$

If there are N atoms per unit volume, the magnetization equals the product of N times the average value of the magnetic moment along z.

$$M = N \frac{\sum_{-J}^{+J} g M_J \mu_B \exp(\mu_0 g M_J \mu_B H / kT)}{\sum_{-J}^{+J} \exp(\mu_0 g M_J \mu_B H / kT)} \tag{2.59}$$

where a summation over the allowed values of M_J has now replaced the integration over θ of Eq. (2.48) of Langevin's classical model. Carrying out the summation, and after considerable manipulation, the above expression for the magnetization leads to

$$\frac{M}{M_s} = B_J(x) \tag{2.60}$$

where $x = \mu_0 g J \mu_B H / kT$, $M_s = N g J \mu_B$ is the saturation magnetization and $B_J(x)$ is the Brillouin function defined as

$$B_J(x) = [(2J+1)/2J] \coth[(2J+1)x/2J] - (1/2J) \coth(x/2J) \tag{2.61}$$

The first experimental confirmation that the magnetization follows the Brillouin function, as predicted by quantum mechanics, rather than the Langevin function, as predicted by classical physics, was presented in the classic experimental measurements by W. Henry in 1952 for Cr^{3+} ($^4F_{3/2}$) in potassium chromium alum, Fe^{3+} ($^6S_{5/2}$) in iron ammonium alum and Gd^{3+} ($^8S_{7/2}$) in gadolinium sulfate octahydrate. Their results are presented in Figure 2.15, which plots the average magnetic moment per ion, $\bar{\mu}$, in Bohr magnetons as a function of H/T, at low temperatures and strong magnetic fields.

The Brillouin function in the quantum mechanical treatment of the atomic magnetic moment of multielectron atoms replaces the Langevin function of Eq. (2.50) encountered in the case of the classical approach. Indeed, in the case where $\vec{J} \to \infty$, when there is no

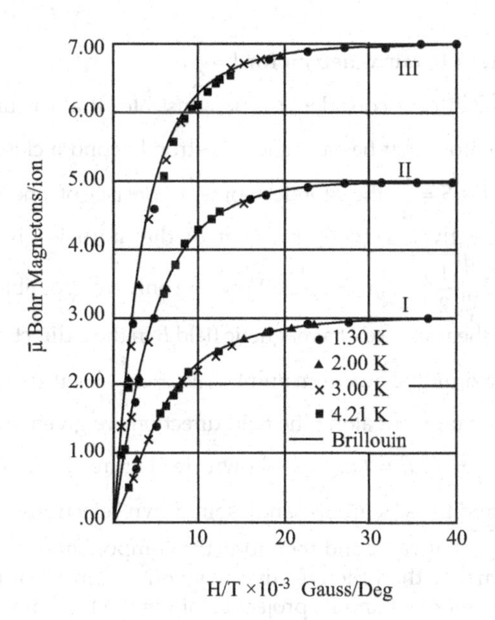

FIGURE 2.15
Plot of average magnetic moment per ion, $\bar{\mu}$ *vs.* H/T for (I) potassium chromium alum ($J = S = 3/2$), (II) iron ammonium alum ($J = S = 5/2$), and (III) gadolinium sulfate octahydrate ($J = S = 7/2$); $g = 2$ in all cases, the normalizing point is at the highest value of H/T. From W. E. Henry, "Spin Paramagnetism of Cr^{3+}, Fe^{3+} and Gd^{3+} at liquid helium temperatures and in strong magnetic fields", *Phys. Rev. B*, 88 (1952) 559 (Copyright the American Physical Society, reprinted with permission).

quantization, no restriction in the value of the *z*-component of the angular momentum, the Brillouin function reduces to the Langevin function, $B_{J\to\infty}(x) = L(x)$. This is an example of the Bohr correspondence principle, which relates quantum mechanical results to the corresponding classical case. For $J = 1/2$, the Brillouin function is reduced to a simple function, namely $B_{1/2}(x) = \tanh(x)$.

In most experimental situations, exempting very low temperatures and extremely high magnetic fields, $x << 1$ and the Brillouin function can be expanded using the Taylor's expansion of coth(x) to obtain

$$B_J(x) = \frac{(J+1)x}{3J} + f(x^3) \tag{2.62}$$

Thus, at magnetic fields commonly available in the laboratory, the susceptibility is given to a first approximation by the first term of Eq. (2.62)

$$\chi = \frac{M}{H} \approx \frac{N\mu_0\mu_{eff}^2}{3kT} \tag{2.63}$$

Equation (2.63) also predicts an inverse temperature dependence of the susceptibility in accord with Curie's Law. In small applied fields, magnetic susceptibility measurements allow the determination of ion effective magnetic moments, $\mu_{eff} = g\sqrt{J(J+1)}\mu_B$, and therefore of the angular momentum. In large magnetic fields, the magnetization saturates and the susceptibility measurements give a measure of the saturated magnetic moment per atom or ion, $\frac{M_s}{N} = Jg\mu_B$.

Example 2.6: Quantum Paramagnetism for $J = \frac{1}{2}$

The simplest case we can consider to demonstrate the calculation of quantum paramagnetism is an atom that has a single *s* electron beyond a closed shell. In this case $\vec{L} = 0$, $\vec{S} = \frac{1}{2}$ and $\vec{J} = \vec{L} + \vec{S} = \frac{1}{2}$; the atomic moment consists of one spin per atom. The magnitude of the effective magnetic moment of the atom is given by Eq. (2.23) as $\mu_{eff} = g\sqrt{J(J+1)}\mu_B = 2\sqrt{\left(\frac{1}{2}\right)\left(\frac{3}{2}\right)}\mu_B = \sqrt{3}\mu_B$. There are only two possible orientations for the magnetic moment in the presence of a magnetic field $\vec{B}$ in the *z*-direction, corresponding to the two values of the azimuthal quantum number, $M_J = \pm\frac{1}{2}$. Then, the allowed components of the atomic magnetic moment along the field direction are given according to Eq. (2.23) by $\mu_z = -gM_J\mu_B = -2\left(\pm\frac{1}{2}\right)\mu_B = \mp\mu_B$, as shown in Figure 2.16. These two allowed orientations are referred to as "spin-up" and "spin-down" electrons.

Note that $J < \sqrt{J(J+1)}$ always, and therefore, the component of the magnetic moment along $\vec{B}$ is always smaller than the effective magnetic moment of the atom. From the value of the magnetic moment and its projection along the field direction, we can obtain the angle θ between the direction of the total magnetic moment and $\vec{B}$. Thus, quantum mechanics predicts that μ_{eff} can point along two allowed directions relative to the direction of the field, namely at 54.7° and 125.3° angles, while it precesses around the magnetic field.

Let us now repeat the calculation of quantum paramagnetism of Eq. (2.59) for this simple case of $J = 1/2$ and $g = 2$, where $gM_J\mu_B$ reduces to $\pm\mu_B$.

$$M = N\frac{-\mu_B \exp(\mu_B\mu_0 H/kT) + \mu_B \exp(-\mu_B\mu_0 H/kT)}{\exp(\mu_B\mu_0 H/kT) + \exp(-\mu_B\mu_0 H/kT)} = N\mu_B \tanh\left(\frac{\mu_B\mu_0 H}{kT}\right) \quad (2.64)$$

Since $N\mu_B$ represents the maximum possible magnetization, that is, the saturation magnetization M_s, we obtain

$$\frac{M}{M_s} = \tanh\left(\frac{\mu_B\mu_0 H}{kT}\right) \quad (2.65)$$

This function is plotted in Figure 2.17. It resembles the Langevin function of classical paramagnetism (Figure 2.14), but exhibits a sharper slope and faster approach to saturation.

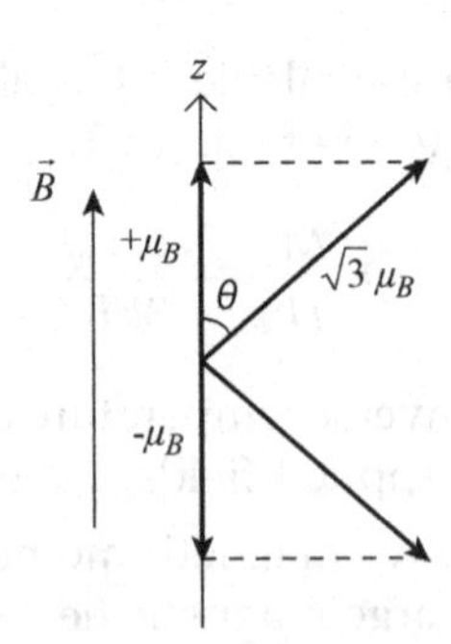

FIGURE 2.16
A spin ½ atomic moment in an applied field $\vec{B}$.

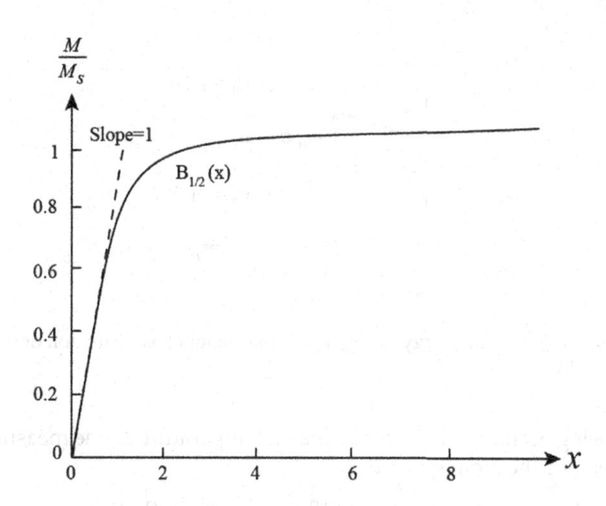

FIGURE 2.17
The reduced magnetization of a spin ½ paramagnet follows a tanh (x) function.

In small fields, Taylor's expansion for tanh(x) gives

$$\tanh(x) = x - \frac{1}{3}x^3 + \frac{2}{15}x^5 - \frac{17}{315}x^7 + \ldots \tag{2.66}$$

Indicating that, in first approximation, $\tanh\left(\frac{\mu_B \mu_0 H}{kT}\right) \approx \frac{\mu_B \mu_0 H}{kT}$ and the susceptibility can be expressed as

$$\chi = \frac{N \mu_0 \mu_B^2}{kT} \tag{2.67}$$

Again, the result is consistent with Curie's Inverse Temperature Law.

Example 2.7: Magnetic Resonance

Figure 2.16 depicts the possible orientation directions of the magnetic moment of a spin 1/2 paramagnet in a magnetic field. There are two allowed orientations corresponding to $M_J = \pm 1/2$, resulting in $\mu_z = \pm \mu_B$; that is, the z-component of the magnetic moment is either parallel or antiparallel to the applied field. Therefore, the classical orientational potential energy $U = -\vec{\mu} \cdot \vec{B}$ is quantized and can take on one of the two possible values, $U = \mu_B B$ or $U = -\mu_B B$. Figure 2.18 plots the magnetic orientational potential energy as a function of the strength of the magnetic field $\vec{B} = \mu_0 \vec{H}$. In the absence of an external magnetic field, the quantum states corresponding to $M_J = \pm\frac{1}{2}$ are degenerate; that is, they have the same energy and are equally populated. The application of a magnetic field raises the degeneracy, splitting the two magnetic states, with the $M_J = -\frac{1}{2}$ state being at lower energy as shown in Figure 2.18. This is known as Zeeman Splitting. The energy difference between the states increases linearly with B and is given by $g\mu_B B$.

At $T \to 0$, only the lower energy electronic state will be populated. At finite temperature, under thermal equilibrium conditions, the number of atoms populating each energy state is proportional to the Boltzmann factor, exp ($-U/kT$), indicating that although a larger number of atoms occupy the lower energy state, the higher energy state has also a finite

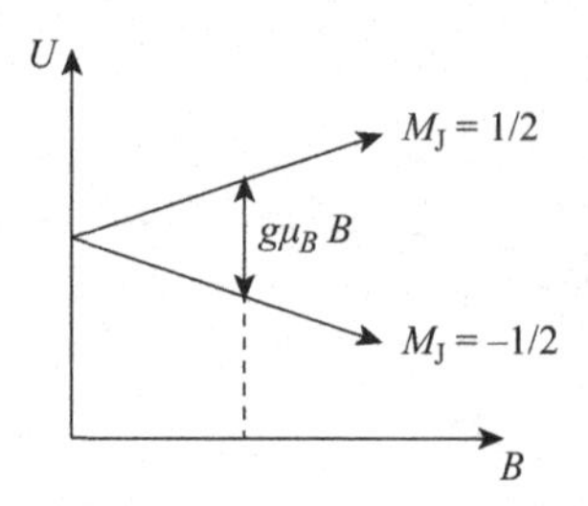

FIGURE 2.18
Magnetic orientational potential energy of a spin 1/2 paramagnet as a function of the strength of an applied magnetic field.

probability for being occupied. This means that at temperature T, one measures an average magnetic moment for the atom given by

$$\langle \mu \rangle = \frac{-\mu_B \exp(\mu_B B / kT) + \mu_B \exp(-\mu_B B / kT)}{\exp(\mu_B B / kT) + \exp(-\mu_B B / kT)} \tag{2.68}$$

Transitions between states differing by $\Delta M_J = \pm 1$ are allowed by quantum mechanics and can take place by absorption or emission of electromagnetic radiation. In this case, one can induce transitions between the lower and higher energy states, by the absorption or emission of electromagnetic radiation of energy $\hbar\omega = g\mu_B B$. This is known as Electron Paramagnetic Resonance (EPR) or Electron Spin Resonance (ESR).

Under usual laboratory conditions, this energy difference falls in the microwave region of the electromagnetic radiation spectrum, with a frequency $\nu = \omega/2\pi$ of the order of 10 GHz. To observe the transition, a sample of the material is placed in a strong uniform magnetic field, B_z, along the z-direction, produced by an electromagnet. It is also subjected to a weak alternating magnetic field B_x acting at right angles. By slowly varying the frequency of the alternating field in the microwave region, or by changing the strength of the static B_z magnetic field, the resonance condition can be achieved, at which the energy absorption rises sharply to a maximum. This occurs when the microwave frequency equals the Larmor precession frequency of the atomic moment in the static field B_z.

2.9 The Effect of Crystalline Fields

In our development of the magnetic properties of materials so far, we have accounted only for the interaction of atomic moments with the applied field $\vec{B}$ and the effect of thermal energies. Thus, our model is truly applicable to an ensemble of free atoms, such as in the case of an ideal gas, where the only interaction between atoms is that of occasional elastic collisions that allow for thermal equilibrium among atoms to be reached. Furthermore, we obtained the magnetic moment of the ground state of the atom or ion by applying Hund's rules and the Pauli Exclusion Principle.

This approach is valid only for free atoms, where the electronic states are determined solely by the spherically symmetric central Coulomb potential of the nucleus. For atoms or ions embedded within a solid, there is no clear rule as to how to arrive at the magnetic moment of the ground state. The difference arises from the fact that within a solid, the electrons are acted upon by the electric fields produced by neighboring ions, in addition to the central field of the nucleus. How do we establish the ground state moment of a magnetic ion within a solid? How can we account for the crystalline electrostatic field at the site of the ion?

Experimentally, one can measure the magnetic susceptibility of the material, which allows the determination of the effective magnetic moment of the ions in the solid, through

Eq. (2.63). By comparing the experimentally determined magnetic moment with that predicted for the free atom or ion, one can get a measure of the strength of the interaction of the magnetic ion with its environment, that is, the strength of the crystalline field.

Most magnetic materials owe their magnetism to the presence of $3d$ or $4f$ ions. The $3d$ elements are the first row of the transition metals (atomic numbers 21–30) on the fourth row of the Periodic Table and the $4f$ elements are two rows below (atomic numbers 57-71), also known as rare earths. These are highlighted in the Periodic Table shown in the Appendix.

In the case of the rare-earth ions, the partially filled $4f$ electronic shell, responsible for the magnetism of the ion, lies deep within the ion below the filled $5s$ and $5p$ electronic states. They are thus shielded from their neighbors within a solid, which leads to a very weak interaction with their environment. With the exception of Eu^{3+} and Sm^{2+}, susceptibility measurements yield values for their effective magnetic moments in solids close to those predicted for the free ions (Table 2.3).

In sharp contrast, the experimental values of the effective magnetic moments of $3d$ transition-metal-ion containing compounds, as seen in Table 2.4, deviate strongly from the free ion moment $\mu_{eff} = g\sqrt{J(J+1)}$, except for the case of the $3d^5$ and $3d^{10}$ electronic states, corresponding to half-filled and full-filled electronic shell configurations where $L = 0$. In the transition metals, the $3d$ electronic state is the outermost electronic shell, as the filled $4s$ shell of the atom is depleted in ionic compounds. The $3d$ electrons are the valence electrons that participate in ionic bonding with neighboring ions and thus interact strongly with their environment. Furthermore, the measured values suggest that the effect of the environment is primarily to neutralize the contribution of the orbital angular momentum, known as quenching of the angular momentum, with the magnetic moment in most cases being close to the spin-only value $\mu_{eff} = g\sqrt{S(S+1)}$.

TABLE 2.3

Ground State Magnetic Moments for Rare-Earth Ions

Rare-Earth Ions	Number of 4f Electrons	Ground State	Theory $\mu_{eff} = g\sqrt{J(J+1)}$	Experiment μ_{eff}
La^{3+}	0	$^1S_0$	0.00	Diamagnetic
Ce^{3+}, Pr^{4+}	1	$^2F_{5/2}$	2.54	2.4
Pr^{3+}	2	$^3H_4$	3.58	3.6
Nd^{3+}	3	$^4I_{5/2}$	3.62	3.6
Pm^{3+}	4	$^5I_4$	2.68	–
Sm^{3+}	5	$^6H_{5/2}$	0.84	1.5
Eu^{3+}, Sm^{2+}	6	$^7F_0$	0.00	3.6
Gd^{3+}, Eu^{2+}	7	$^8S_{7/2}$	7.94	8.0
Tb^{3+}	8	$^7F_6$	9.72	9.6
Dy^{3+}	9	$^6H_{15/2}$	10.63	10.6
Ho^{3+}	10	$^5I_8$	10.60	10.4
Er^{3+}	11	$^4I_{15/2}$	9.59	9.4
Tu^{3+}	12	$^3H_6$	7.57	7.3
Yb^{3+}	13	$^2F_{7/2}$	4.54	4.5
Lu^{3+}, Yb^{2+}	14	$^1S_0$	0.00	Diamagnetic

Source: From A. H. Morris, *The Physical Principles of Magnetism*, IEEE (2001) p. 55.

TABLE 2.4
Ground State Magnetic Moments for the Iron Group Ions

Transition Metal Ion	Number of 3*d* Electrons	Ground State	Theory $\mu_{eff} = g\sqrt{J(J+1)}$	Theory $\mu_{eff} = g\sqrt{S(S+1)}$	Experiment μ_{eff}
K^+, Ca^{3+}	0	$^1S_0$	0.00	0.00	Diamagnetic
Ti^{3+}, V^{4+}	1	$^2D_{3/2}$	1.55	1.73	1.7
V^{3+}	2	$^3F_2$	1.63	2.83	2.8
V^{2+}, Cr^{3+}, Mn^{4+}	3	$^4F_{3/2}$	0.77	3.87	3.8
Cr^{2+}, Mn^{3+}	4	$^5D_0$	0.0	4.90	4.9
Mn^{2+}, Fe^{3+}	5	$^6S_{5/2}$	5.92	5.92	5.9
Fe^{2+}	6	$^5D_4$	6.70	4.90	5.4
Co^{2+}	7	$^4F_{9/2}$	6.64	3.87	4.8
Ni^{2+}	8	$^3F_4$	5.59	2.83	3.2
Cu^{2+}	9	$^2D_{5/2}$	3.55	1.73	1.9
Cu^+, Zn^{2+}	10	$^2S_0$	0.00	0.00	Diamagnetic

Source: From A. H. Morris, *The Physical Principles of Magnetism*, IEEE (2001) p. 67.

Let us consider the effect of the environment for the 3*d* electron case, in greater detail. To understand the effect of the environment on ion magnetic moments, one must start by considering the electric field, referred to as the crystalline electric field (CEF), which is produced by neighboring ions at the crystallographic site of the magnetic ion. Two major theoretical frameworks have been advanced by scientists in order to address this problem. The first, known as Crystal Field Theory, developed by Hans Bethe (H. A. Bethe, *Ann. Phys.* (Leipzig) 3 (1929) 133) and John van Vleck (J. H. Van Vleck, *J. Chem. Phys.* 3 (1935) 807) in the 1930s, treats neighboring ions as negative point charges and uses concepts from group theory to classify and label new valence electronic states that are established in the presence of the crystal field. The second, known as Ligand Field Theory, is an extension of Crystal Field Theory that incorporates molecular orbital overlap between the central ion and the neighboring ions.

We now illustrate the results of Crystal Field Theory by considering the case of ferric ions in magnetic iron oxides, which occur in many magnetic nanomaterials of important practical applications. In Fe^{3+}-ion-containing materials, the magnetism derives from the five valence 3*d* electrons on the outer electronic shell of the ion. As discussed in Example 2.3, the ground state is $^6S_{5/2}$, indicating $L = 0$ and $S = 5/2$ for the free ion. In an environment of spherical symmetry, that is, free space, the five 3*d* orbitals have the same energy. Each is occupied by one electron, making the *d* electron cloud spherically symmetric around the nucleus, with no preferred direction in space. In magnetic iron oxides, iron is octahedrally or tetrahedrally coordinated to oxygen atoms. Thus, the crystallographic symmetry around the iron ion is octahedral or tetrahedral, as depicted in Figure 2.19.

In the case of pure octahedral coordination symmetry, Figure 2.19(a), the Fe^{3+} ion is found at the center of a perfect octahedron, the corners of which are each occupied by an oxygen ion. Electrons occupying *d* states with lobes pointing directly toward the negatively charged oxygen ions feel a strong electrostatic repulsion. In order to minimize electrostatic energy, the electronic charge clouds shift away from the axes to occupy the space in between axes, thus lifting the 5-fold degeneracy of the free ion 3*d* electronic state.

Crystal Field Theory predicts that the new states can be expressed as linear combinations of the original hydrogen-like *d* electron states of the free ion. To illustrate the case, let

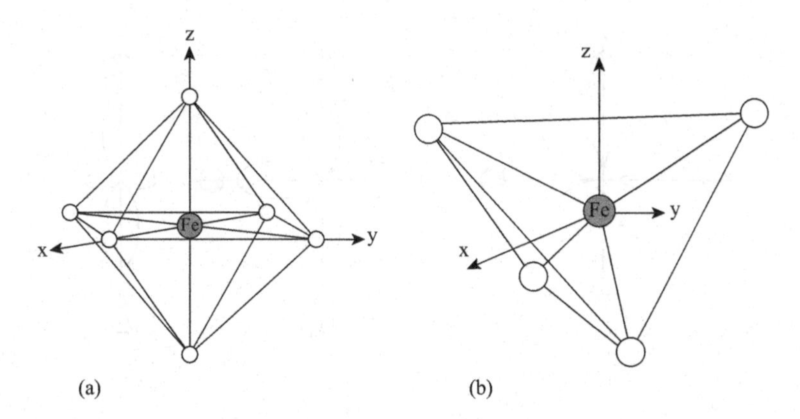

FIGURE 2.19
Metal Fe^{3+} ions in (a) octahedral and (b) tetrahedral coordination complexes.

us consider a single 3*d* electron and form new *d* electron states by considering certain linear combinations of the five, hydrogen-like 3*d* ($l = 2, m_l = -2, -1, 0, 1, 2$) wave functions.

$$d_{xz} = \frac{d_{+1} + d_{-1}}{\sqrt{2}}$$

$$d_{yz} = -i\frac{d_{+1} - d_{-1}}{\sqrt{2}}$$

$$d_{xy} = -i\frac{d_{+2} - d_{-2}}{\sqrt{2}} \tag{2.69}$$

$$d_{x^2-y^2} = \frac{d_{+2} + d_{-2}}{\sqrt{2}}$$

$$d_{3z^2-1} = d_0$$

The spatial orientation of the electronic charge clouds or lobes of these new wave functions is sketched in Figure 2.20.

States d_{xz}, d_{yz}, d_{xy} have lobes pointing between ligands, depicted here as dots on the axes, while sates $d_{x^2-y^2}$ and d_{3z^2-1} have lobes pointing along the axes, toward the ligands. The former have lower energy, while the latter have higher energy due to the stronger electrostatic repulsion of the negatively charged ligands. This splits the originally degenerate five 3*d* states into two groups of states of different energy, as indicated in Figure 2.21(a). A similar, but inverted splitting is observed in the case of tetrahedral coordination symmetry, Figure 2.21(b). The energy splitting Δ is called the crystal field splitting and measures the strength of the crystalline field at the site of the Fe^{3+} ion. The lower energy group in the octahedral case contains the three states d_{xz}, d_{yz}, d_{xy} of equal energy, labeled t_{2g} in group theoretical terminology. Electrons occupying these states spend their time in the space between axes. The higher energy group contains the two electronic states $d_{x^2-y^2}$ and d_{3z^2-1} of the same energy, labeled e_g. Electrons occupying these states spend their time along the *x*-, *y*- or *z*-axes.

Thus, in a perfect octahedral environment, the 5-fold degeneracy of the original 3*d* state is partially lifted to produce 3-fold degenerate (t_{2g}) and 2-fold degenerate (e_g) states. The

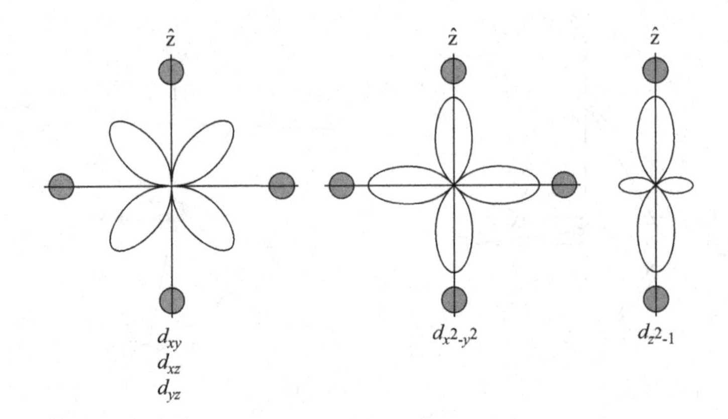

FIGURE 2.20
Depiction of electronic charge clouds of group theoretical wave functions derived from 3*d* single-electron atomic functions according to Eq. (2.69).

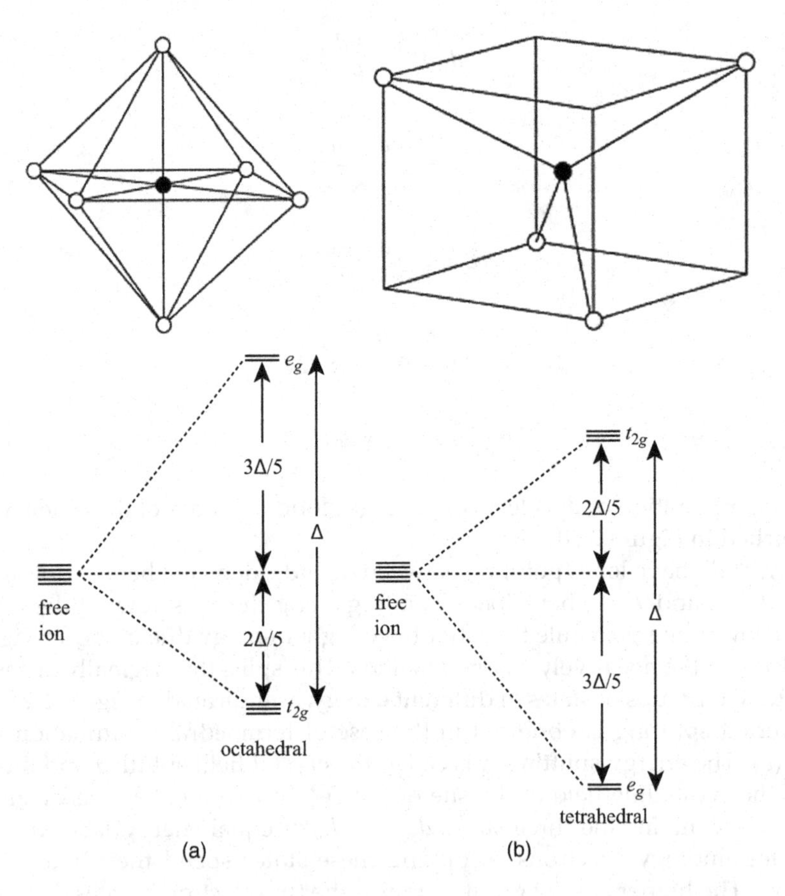

FIGURE 2.21
Crystal field splitting of the 3*d* electronic states of Fe^{3+} in (a) octahedral, and (b) tetrahedral coordination symmetry.

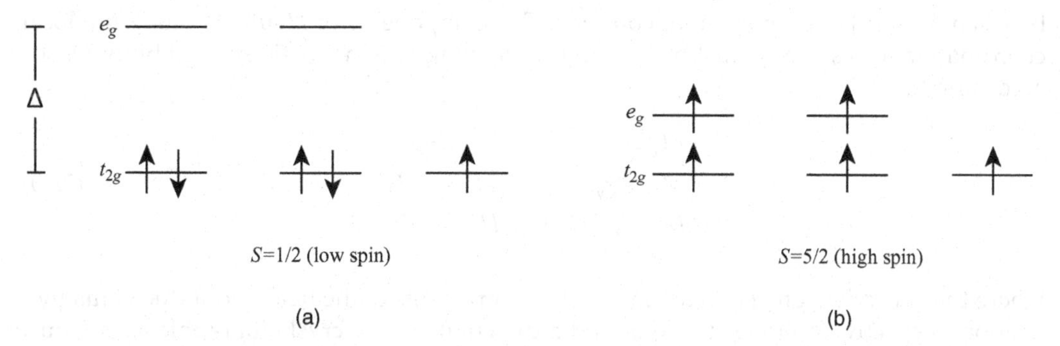

FIGURE 2.22
Spin configurations for Fe^{3+} in (a) strong and (b) weak crystalline field environment.

remaining degeneracy in the orbital states can be further lifted by introducing distortions to the octahedral arrangement of ligands. For example, if the octahedron were to be elongated or compressed along the *z*-axis, the symmetry will be reduced from octahedral to axial. The symmetry can be further reduced to rhombic, if additionally, a similar distortion were to be introduced within the *xy*-plane. Under tetrahedral symmetry, the same groups of electronic states are obtained, but in this case, it is the 2-fold e_g-states that lie lower in energy and the t_{2g} states higher.

The size of the crystal field splitting determines the spin of the ground state of the ion in the solid and thus its magnetic moment. As shown in Figure 2.22, if the crystal field energy splitting Δ is large (strong crystal field) meaning that Δ is larger than the electronic repulsion between the electrons occupying the same spatial wave function, the ground electronic state is obtained by doubly populating each of the t_{2g} states with spin-paired electrons reducing the spin on the ion to $S = 1/2$, as shown in Figure 2.22(a). This is the electronic arrangement that minimizes the energy of the ion, and thus, determines the ground electronic state. If, however, Δ is small (weak crystal field), a high spin of $S = 5/2$ will result from populating each of the states by one electron, as illustrated in Figure 2.22(b). Experimentally, this is what is observed. Ferric ion compounds can be classified as high-spin and low-spin systems, depending on the degree of the electro-negativity of the coordinating ligands.

For the case of Fe^{2+} coordination complexes, there are six 3*d* valence electrons resulting in a high-spin system of $S = 2$ and low-spin system of $S = 0$ ground state, the latter producing a diamagnetic state. Thus, we conclude that the environment can play a decisive role in determining the ion's spin angular momentum. The simple rules we have previously applied to the case of free atoms or ions, namely Hund's rules and the Pauli Exclusion Principle, are still operational but not sufficient to determine the ground electronic state and the effective magnetic moment of an ion embedded in a solid.

Furthermore, due to coupling of the electronic orbital angular momentum to the local crystallographic axes, one would expect the induced magnetization upon application of an external magnetic field to depend on the direction of $\vec{H}$ relative to the local coordination axes of symmetry of the ion, and it may not necessarily be induced in the direction of the field. Most measurements are made on powder or polycrystalline samples, where one detects the average over all directions. When measurements are made on single crystals, however, axes anisotropies can be detected. Thus, while under spherical symmetry, the induced magnetization is related to the applied field by a simple proportionality factor, $\vec{M} = \chi\vec{H}$; when the effect of the environment is taken into consideration, the relation between M and H becomes more complex. The components of H and M along the local coordination axis system must be introduced, leading to a magnetic susceptibility tensor according to

$$\begin{aligned} M_x &= \chi_{xx}H_x + \chi_{xy}H_y + \chi_{xz}H_z \\ M_y &= \chi_{yx}H_x + \chi_{yy}H_y + \chi_{yz}H_z \\ M_z &= \chi_{zx}H_x + \chi_{zy}H_y + \chi_{zz}H_z \end{aligned} \tag{2.70}$$

where the subscripts on H and M signify the components of the field and induced magnetization, respectively, along the x-, y- and z-directions of the crystallographic axes. Thus, one defines the magnetic susceptibility tensor of Eq. (2.71) to fully characterize the magnetic properties of the material.

$$\chi = \begin{pmatrix} \chi_{xx} & \chi_{xy} & \chi_{xz} \\ \chi_{yx} & \chi_{yy} & \chi_{yz} \\ \chi_{zx} & \chi_{zy} & \chi_{zz} \end{pmatrix} \tag{2.71}$$

In free space, prior to the application of the external magnetic field, the direction of the atomic moments in an ensemble of atoms, such as in the case of a rarefied gas, is truly random. There is no preferred direction in space. In real materials, the crystallographic structure lowers the symmetry by introducing preferred crystallographic directions along which the magnetic energy is minimized.

Exercises

1. Estimate the diamagnetic moment induced by an applied field $B = 5$ T for (a) the hydrogen atom in its ground state (Bohr radius $r = 0.53\ (\overset{0}{A})$) and (b) the iron atom (atomic radius $r = 1.42\overset{0}{A}$), assuming that the electrons are circulating around the nucleus at a distance equal to their respective atomic radii ($1\overset{0}{A} = 10^{-10}$ m).
2. A uniform ring of mass M and radius R carries a total charge Q uniformly distributed over its circumference. It is spinning around its axis with uniform angular velocity $\vec{\omega}$. Find the gyromagnetic ratio (magnetic dipole momentum/angular momentum).
3. A uniform sphere of mass M and radius R carries total charge Q uniformly distributed over its volume. It is spinning around its axis with uniform angular velocity $\vec{\omega}$. Find its gyromagnetic ratio.
4. What is the Larmor precession frequency of an atom with $L = 2$, $S = 1$ and $J = 3$ in the presence of a field $B = 5$ T?
5. Use Hund's rules and the Pauli Exclusion Principle to derive the ionic ground state and corresponding magnetic moment of Cu^{2+}, Sm^{3+}, V^{2+}, Co^{2+} and Ho^{3+}.
6. A long, densely coiled solenoid with 300 turns per meter, carries a current $I = 20$ A. Find the magnetic field $\vec{B}$ in the core of the solenoid (a) when the core is empty, (b) filled with graphite, and (c) filled with platinum.

7 Using the Taylor's series expansion for coth(x), prove that $L(x) = \frac{x}{3}$, for $x \to 0$; and $L(x) = 1$ for $x \to \infty$.

8 Prove that the Brillouin function, $B_J(x)$, reduces to the Langevin function, $L(x)$, in the limit of $J \to \infty$; and to $\tanh(x)$ when $J = \frac{1}{2}$.

9 What are the possible orientations that the atomic moment of an atom with $\vec{L} = 2$ and $\vec{S} = 1$ can have in the presence of an applied field $\vec{B}$? Calculate the allowed angles between the direction of $\vec{B}$ and the atomic magnetic moment.

10 $FeCl_3$ is a paramagnetic compound at room temperature with the Fe^{3+} ion in the high spin state, $^6S_{5/2}$. In the presence of an applied magnetic field, the 5-fold degeneracy of the ground state is lifted by Zeeman splitting. For an applied field of $B = 1$ T, (a) at what frequency would magnetic resonance be observed for electronic transitions between the Zeeman levels? (b) What is the average magnetic moment $<\mu>$ measured at T = 20 K?

11 Nuclear magnetic moments contribute insignificantly to the magnetization of materials due to the fact that the nuclear magneton is almost 2,000 times smaller than the Bohr magneton. NMR, however, can contribute equally to EPR in studying the internal magnetism of materials and nanomaterials. In NMR, one observes transitions between Zeeman-split levels of nuclear magnetic moments of nuclei containing an odd number of nucleons. The most commonly used nuclei are the proton, ^1H and ^{13}C. In an applied magnetic field of 10 T, what is the resonant frequency for (a) ^1H-NMR and (b) ^{13}C-NMR?

3

Long-Range Magnetic Order

"Je suppose que chaque molécule éprouve de la part de l' ensemble des molecules environmantes une action égale à celle d' un champ *uniform* proportionnel à l' intensité d' aimantation et de meme direction qu' elle".

***Pierre Weiss*, L'hypothèse du champ moléculaire et la propriété ferromagnétique, presented at the conference of the Société Francaise de Physique on April 4, 1907**

3.1 The Curie–Weiss Law and the Weiss Molecular Field

Diamagnetic and paramagnetic materials have limited, if any, technological applications, as they do not possess permanent magnetization. They play no role in nanomagnetism, the subject of this book. Their study serves, however, an essential role in introducing important concepts in magnetism that form the foundations for further exploration into the physics of cooperative magnetic phenomena associated with magnetic ordering. We have so far considered isolated atomic moments and their response to an applied magnetic field, to thermal agitation and to the local crystalline environment. No interactions between neighboring atomic moments have been considered. In this chapter, interactions between atomic magnetic moments are introduced and their implications examined.

It was observed early on that not all paramagnetic materials behaved as simple paramagnets, in the sense that the temperature dependence of their magnetic susceptibility did not follow the Curie inverse temperature Law of Eq. (2.41). Instead, they obeyed a modified law, known as the Curie–Weiss Law given by Eq. (3.1), where θ is a constant with the dimensionality of temperature. Figure 3.1 gives a graphical representation of Eq. (3.1) in comparison to that of the Curie Law. It is seen that while Curie's Law indicates that $\chi \to \infty$ as $T \to 0$, the Curie–Weiss Law predicts that $\chi \to \infty$ as $T \to \theta$, where the value of θ depends on the material.

$$\chi = \frac{C}{T-\theta} \tag{3.1}$$

This behavior is characteristic of ferromagnets in the temperature region above their critical temperature for ordering. The constant C is known as the Curie–Weiss constant.

In a 1907 paper, Pierre Weiss theorized that such a behavior, incipient to ferromagnetic ordering, could be understood in terms of interacting atomic magnetic moments (P. Weiss, *J. de Physique* 6 (1907) 661). He accounted for such an interaction phenomenologically by

DOI: 10.1201/9781315157016-4

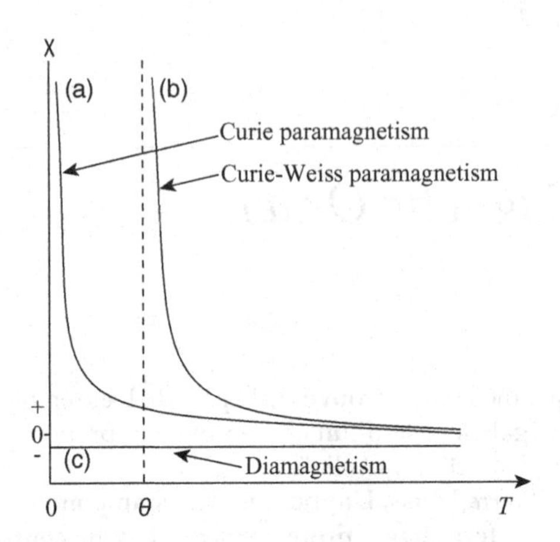

FIGURE 3.1
Variation of paramagnetic susceptibility with absolute temperature according to the Curie Law (a) and the Curie–Weiss Law (b). In contrast, diamagnetic susceptibility is temperature-independent (c).

postulating the existence of an internal magnetic field due to interatomic magnetic interactions. Thus, in addition to the externally applied field, local atomic moments experience this internal magnetic field mediated by neighboring magnetic moments. He called this internal field the "molecular field". The molecular field is not a real field and, thus, it does not enter Maxwell's equations. He assumed it to be proportional to the magnetization $\vec{M}$ according to Eq. (3.2), where λ is the proportionality constant.

$$\vec{H}_{mf} = \lambda \vec{M} \tag{3.2}$$

Furthermore, he assumed that all sites within the material experienced the same molecular field. The name "molecular field" is due to historical reasons; it reflects the belief at the time that all matter consisted of molecules. Today, we know that ferromagnetic metals actually possess atomic, rather than molecular structure.

In Section 2.8, our treatment of quantum paramagnetism for non-interacting magnetic moments deduced that the reduced magnetization, M/M_s, follows the Brillouin function, $B_J(x)$, where x depends on the applied field and temperature, $x = \frac{\mu_0 g J \mu_B H}{kT}$.

$$\frac{M}{M_s} = B_J(x) \tag{3.3}$$

In the presence of the molecular field, variable x in Eq. (3.3) must be modified to read

$$x = \frac{\mu_0 g J \mu_B (H + \lambda M)}{kT} \tag{3.4}$$

where the molecular field has been added to the applied magnetic field H. In order to investigate the inherent magnetic behavior of the system, we set $H = 0$ to obtain

$$x = \frac{\mu_0 g J \mu_B \lambda M}{kT} \tag{3.5}$$

and seek simultaneous solutions for Eqs. (3.3) and (3.5). In Eq. (3.3), M_s stands for the saturation magnetization, which occurs at $T = 0$, when all moments are aligned in the direction of the magnetic field. Using $M_s = NgJ\mu_B$, the maximum possible magnetization according to quantum mechanics, we can recast Eq. (3.5) into

$$\frac{M}{M_s} = \frac{kTx}{N\mu_0 g^2 J^2 \mu_B^2 \lambda} \tag{3.6}$$

Simultaneous solutions to Eqs. (3.3) and (3.6) can be found graphically, as shown in Figure 3.2, where we plot M/M_s as a function of x for $H = 0$, as given by Eqs. (3.3) and (3.6). We observe that Eq. (3.6) gives a linear dependence of the reduced magnetization on x, with a slope that is proportional to T. For slopes larger than the initial slope of the Brillouin function, there are no simultaneous solutions except at the origin, where M vanishes. For less steep slopes compared to the initial slope of the Brillouin function, there is also a solution at a non-zero value of x, where the linear and the Brillouin functions intersect. It can be shown that the non-zero value is the stable solution. The temperature corresponding to a slope equal to the initial slope of the Brillouin function is thus a critical temperature, T_C, which delineates two physical regimes. For $T > T_C$, the material remains paramagnetic, while for $T < T_C$ a non-zero magnetization occurs. When the material is cooled below this critical temperature, it becomes spontaneously magnetized, even in the absence of an applied field. This "spontaneous magnetization" is the hallmark of ferromagnetism.

At T_C, the material undergoes a magnetic phase transition from a paramagnetic to a magnetically ordered state. The relationship between T_C and λ can be obtained by equating the initial slope of the Brillouin function at $x \to 0$ with that of the linear relationship of Eq. (3.6), to obtain

$$T_C = \frac{N\mu_0 g^2 J(J+1)\mu_B^2 \lambda}{3k} = \frac{N\mu_0 \mu_{eff}^2 \lambda}{3k} \tag{3.7}$$

where we have made use of the approximation $B_J(x) = (J + 1)x/3J$ for small x and $\mu_{eff} = g\sqrt{J(J+1)}\mu_B$. Table 3.1 gives pertinent information for some common ferromagnets.

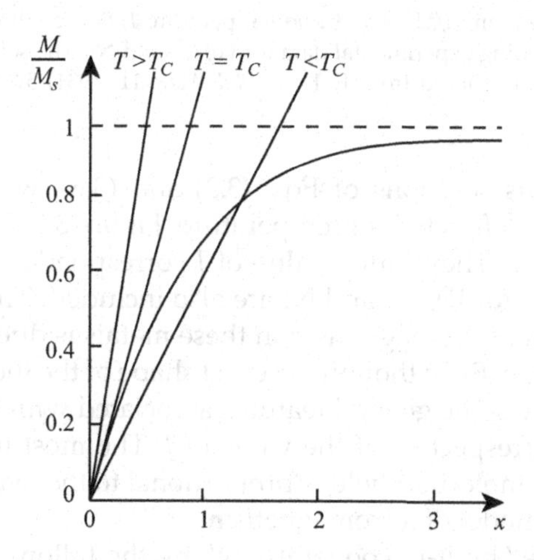

FIGURE 3.2
Plots of reduced magnetization *vs.* x (for $H = 0$) give simultaneous graphical solutions of Eqs. (3.3) and (3.6). The curved line depicts the Brillouin function.

TABLE 3.1

Critical Temperature and Magnetic Moments of Some Common Ferromagnets

Material	T_C (K)	Magnetic Moment (μ_B/Formula Unit)
Fe	1,043	2.22
Co	1,394	1.715
Ni	631	0.605
Gd	289	7.5
MnSb	587	3.5
EuO	70	6.9
EuS	16.5	6.9

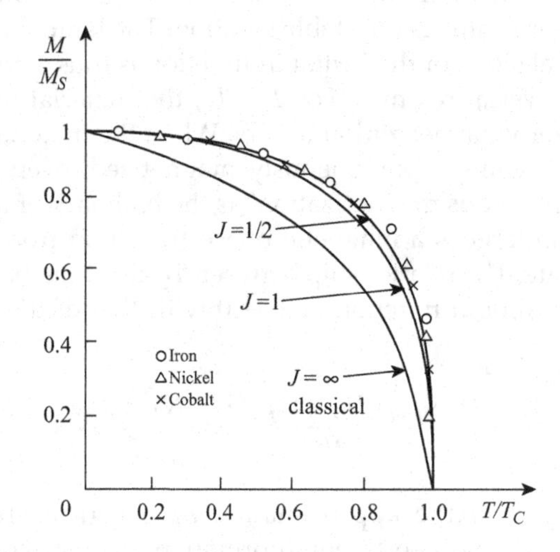

FIGURE 3.3
Reduced spontaneous magnetization M/M_s *vs.* reduced temperature T/T_C according to the Weiss molecular field theory of ferromagnetism, including experimental data for Fe, Co and Ni. The solid lines are theoretical for different values of J, as indicated. (Adapted from F. Tyler, *Phil. Mag.* 11 (1931) 596, Copyright Taylor & Francis, reproduced with permission).

Using the simultaneous solutions of Eqs. (3.3) and (3.6), we can plot the predicted reduced magnetization as a function of temperature. Figure 3.3 shows M/M_s as a function of T/T_C for $J = ½$, 1 and ∞. The infinite value of J corresponds to the classical Langevin model. Experimental data for Fe, Co and Ni are also included. The data points fall on the $J = ½$ curve, indicating that ferromagnetism in these metals is due to electronic spin alone with no orbital contribution. Even though the exact shape of the theoretical curves depends on the value of J, there are some general features associated with the appearance of spontaneous magnetization, irrespective of the value of J. The most important observation is that by assuming that the molecular field is proportional to the magnetization, Weiss built-in cooperativity into his model of ferromagnetism.

You can appreciate the "built-in cooperativity" by the following observations. At low temperatures, the curve flattens out, implying that a small increase in temperature does not have a sizable effect on the reduced magnetization. In this region, it is very hard to misalign the atomic moments by thermal excitation. As the temperature is raised thermal energies misalign more and more moments away from the magnetization direction, thus causing the magnetization to decrease. Finally, as the temperature approaches the critical temperature a precipitous collapse of the magnetization occurs, attaining total atomic moment disorder at T_C, where the magnetization vanishes. The above observations imply that the greater the magnetization, or moment alignment, the greater the force that tends to align any particular moment in the direction of the magnetization, that is, cooperativity. Such cooperative interactions in solids lead to phase transitions at the critical temperature. The magnetization vanishes at $T > T_C$, M is continuous at the critical temperature but its first derivative with respect to temperature, $\frac{\partial M}{\partial T}$, is discontinuous. Thus, within the molecular field model, the transition from a non-magnetic to a ferromagnetic phase is classified as a "second-order phase transition".

Let us now consider the implications of the Weiss theory in the presence of a small applied field H at $T \geq T_C$. For high temperatures and small fields ($x << 1$) the reduced magnetization of Eq. (3.3) can be approximated according to Eq. (3.8)

$$\frac{M}{M_s} \approx \frac{J+1}{3J} \cdot \frac{\mu_0 g J \mu_B (H + \lambda M)}{kT} \tag{3.8}$$

Using Eq. (3.7), the reduced magnetization can be related to the critical temperature according to Eq. (3.9).

$$\frac{M}{M_s} \approx \frac{T_C}{\lambda M_s} \left(\frac{H + \lambda M}{T} \right) \tag{3.9}$$

This can be rearranged to read

$$\frac{M}{M_s} \left(1 - \frac{T_C}{T} \right) \approx \frac{H T_C}{\lambda M_s T} \tag{3.10}$$

We then obtain for the magnetic susceptibility

$$\chi = \frac{M}{H} \approx \frac{T_C}{\lambda (T - T_C)} \tag{3.11}$$

Since λ and T_C are constants characteristic of the material, we conclude that the magnetic susceptibility for small H is inversely proportional to $(T - T_C)$,

$$\chi \propto \frac{1}{T - T_C} \tag{3.12}$$

implying that the experimental constant θ of Eq. (3.1) must be identified with the critical temperature for spontaneous magnetization, T_C.

Experimental studies on various magnetic materials have established that the constant θ in the Curie–Weiss Law of Eq. (3.1) can assume both positive and negative values. Positive values of θ are associated with ferromagnetic ordering, where neighboring moments orient parallel to each other, while negative values of θ are associated with antiferromagnetic ordering, where neighboring moments orient antiparallel to each other.

The critical temperature of ferromagnets is known as the Curie temperature, T_C, while that of antiferromagnets as the Néel temperature, T_N. We will discuss antiferromagnetism in more detail later.

We can get an estimate of the size of the molecular field by inserting $H_{mf} = \lambda M_s = \lambda N g J \mu_B$ into Eq. (3.7) to obtain

$$H_{mf} = \frac{3kT_C}{\mu_0 g(J+1)\mu_B}, \text{ or equivalently } B_{mf} = \frac{3kT_C}{g(J+1)\mu_B} \tag{3.13}$$

As seen from Table 3.1, the values of T_C for the elemental room-temperature ferromagnets Fe, Co and Ni are of the order of 10^3 K. Thus, for a ferromagnet with $J = ½$, $g = 2$ and $T_C = 1{,}000$ K, $B_{mf} = \frac{kT_C}{\mu_B} = 1500\text{T}$. This is an enormous magnetic field, much larger than any field produced in the laboratory, including pulsed fields. The National High Field Laboratory in Tallahassee, Florida, the world's largest and highest-powered facility in the production of high magnetic fields, reports on its website (www.magnet.fsu.edu) that the highest DC magnetic field experimental facility available has $B = 45$ T, while its pulsed-field facility, operated at Los Alamos National Laboratory in New Mexico, affords non-destructive pulsed-field magnets up to $B = 60$ T. Pulsed magnets create very high magnetic fields but only for a fraction of a second. The question then arises: what is the nature of the interatomic magnetic interactions within materials that produce such excessively high, albeit "phenomenological", magnetic fields? It is the origins and implications of these interactions that we address now.

3.2 The Exchange Interaction

The origin of the Weiss molecular field remained unexplained till the advent of quantum mechanics. Heisenberg, in a 1928 paper, showed that cooperative magnetic ordering in solids is due to the quantum mechanical "exchange interaction", which has no classical analog (W. Heisenberg, *Z. Physik*, 49 (1928) 619). Understanding interatomic exchange is the key to understanding magnetic ordering in materials. The necessity to invoke quantum mechanics becomes clear after considering a failed attempt to explain the phenomenon of magnetic ordering classically.

As we discussed in Section 1.5, a magnetic dipole produces a magnetic field in the region of space around it. A second dipole brought into the vicinity will orient in the direction of the local magnetic field due to the first dipole. This interaction between two magnetic dipole moments is known as the "dipole–dipole" interaction. Thus, classically two magnetic moments $\vec{\mu}_1$, $\vec{\mu}_2$ interacting through the dipole–dipole interaction have mutual magnetic orientational potential energy $U_{d\text{-}d}$, which is given by Eq. (3.14). The value of $U_{d\text{-}d}$ depends on the distance r between the dipole moments and their degree of mutual alignment. Here, $\hat{r} = \frac{\vec{r}}{r}$ is the unit vector along the displacement vector from $\vec{\mu}_1$ to $\vec{\mu}_2$.

$$U_{d-d} = \frac{\mu_0}{4\pi r^3}\left[\vec{\mu}_1 \cdot \vec{\mu}_2 - 3\left(\vec{\mu}_1 \cdot \hat{r}\right)\left(\vec{\mu}_2 \cdot \hat{r}\right)\right] \tag{3.14}$$

$U_{d\text{-}d}$ is minimized when the moments are aligned parallel to each other along the line joining them. One can estimate the order of magnitude of the strength of the dipole–dipole interaction by allowing $\mu_1 \approx \mu_2 \approx 1\mu_B$ and $r \approx 0.1$ nm, yielding a value of $U_{d-d} \approx \frac{\mu_0 \mu_B{}^2}{4\pi r^3} \sim 10^{-23}$ J. Such an interaction could induce "spontaneous magnetization" due to moment alignment when thermal energies, kT, are of the order of ~10^{-23} J, or when T~1 K. Many paramagnetic systems do exhibit magnetic ordering at the milli-Kelvin temperature range due to this interaction. However, it cannot account for ferromagnetic ordering in solids, as the ordering temperature of common ferromagnets is $T_C \sim 10^3$ K, three orders of magnitude higher.

It is only within the realm of quantum mechanics and the "exchange interaction" between electrons that the phenomenon of long-range magnetic order in materials can be understood, reminding us that magnetism is firmly grounded in quantum mechanics, since the electronic exchange interaction has no classical analog. We have already come across the exchange interaction when we discussed the Pauli Exclusion Principle and Hund's rules in the formation of atomic moments for incompletely filled orbitals of isolated atoms and ions. The process of applying Hund's rules and the Pauli Exclusion Principle ensured: (a) minimization of the electrostatic repulsion energy between electrons and (b) prevention of any two electrons from sharing the same orbital and spin quantum numbers. These rules are a manifestation of "*intra*-atomic exchange". Magnetic ordering in solids is the result of "*inter*-atomic exchange".

The atomic wave functions we considered in Chapter 2 were single electron wave functions localized on the magnetic atom or ion. In materials, the electronic wave functions of neighboring ions may overlap, forming bonding or antibonding molecular orbitals and in metals, the outer electrons are delocalized into extended conduction electron band states. Thus, one must address the more complex subject of the wave function of a multi-electron system and account for electron–electron interactions explicitly. The simplest example one may consider is the two-electron wave function, for the case of two electrons moving in the same potential field. Common examples are the two electrons in the helium atom or the two electrons in the hydrogen molecule. For the purpose of demonstration of principle, we will consider the case of the hydrogen molecule.

If the electrostatic Coulomb repulsion between the electrons is at first ignored, the two electrons in the hydrogen molecule move in a common electrostatic potential V due to the hydrogen nuclei; their quantum state must be represented by wave function $\psi(r_1, r_2)$ that satisfies the Schrödinger equation of Eq. (3.15). Here, $\mathcal{H}$ is the Hamiltonian operator for the two electron system, r_1 and r_2 are the spatial coordinates of the electrons from a common origin, ∇^2 is the Laplacian operator and E is the energy or eigenvalue of the two non-interacting electron state.

$$\mathcal{H}\psi = \left[-\frac{\hbar^2}{2m}\left(\nabla_1^2 + \nabla_2^2\right) + V(r_1) + V(r_2)\right]\psi = E\psi \tag{3.15}$$

In seeking a solution to the above equation, Heitler and London (W. Heitler and F. London Z. Phys. 44 (1927) 455) assumed that the two-electron wave function $\psi(r_1, r_2)$ of the hydrogen molecule accepts separable solutions of the form $\psi_a(r_1)\psi_b(r_2)$, where $\psi_a(r_1)$ and $\psi_b(r_2)$ are atomic wave functions localized at each atomic site, with subscripts a, b representing the two hydrogen atoms in the molecule. This approach is known as the Heitler–London approximation. It is easy to prove by substitution that $\psi(r_1, r_2) = \psi_a(r_1)\psi_b(r_2)$ is a solution of Eq. (3.15) with $E = E_a + E_b$, where E_a and E_b are the energies of the atomic single-electron states, $\psi_a(r_1)$ and $\psi_b(r_2)$, satisfying the "one electron" Schrödinger equations Eqs. (3.16) and (3.17), respectively.

$$\mathcal{H}_1\psi = \left[-\frac{\hbar^2}{2m}\nabla_1^2 + V(r_1)\right]\psi = E_a\psi \tag{3.16}$$

$$\mathcal{H}_2\psi\left[-\frac{\hbar^2}{2m}\nabla_2^2 + V(r_2)\right]\psi = E_b\psi \tag{3.17}$$

So far, we have assumed that the electrons are distinguishable. In reality, the electrons are identical, indistinguishable particles. In other words, once the hydrogen molecule has been formed from the original two atoms, you can no longer tell which electron was contributed from which atom. Rapid interexchange of electrons occurs at a frequency of the order of 10^{18} times per second in the hydrogen molecule. This means that interchanging them should not alter any observable property of the two-electron system; that is, the value of the probability density $|\psi(r_1, r_2)|^2$ should be unaltered. In Eq. (3.18), we list additional separable solutions to Eq. (3.15) that incorporate the concepts of the interexchange and indistinguishability of the electrons. The second equation is derived from the first by interchanging the electrons, where electron 1 is now associated with atom *b* and electron 2 with atom *a*. The third and fourth equations are linear combinations of the first two, where $1/\sqrt{2}$ is a normalizing factor, assuming that $\psi_a(r_1)$ and $\psi_b(r_2)$ are already normalized.

$$\begin{aligned}
&(i)\ \psi(r_1,r_2) = \psi_a(r_1)\psi_b(r_2)\\
&(ii)\ \psi(r_1,r_2) = \psi_a(r_2)\psi_b(r_1)\\
&(iii)\ \psi(r_1,r_2) = \frac{1}{\sqrt{2}}\left[\psi_a(r_1)\psi_b(r_2) + \psi_a(r_2)\psi_b(r_1)\right]\\
&(iv)\ \psi(r_1,r_2) = \frac{1}{\sqrt{2}}\left[\psi_a(r_1)\psi_b(r_2) - \psi_a(r_2)\psi_b(r_1)\right]
\end{aligned} \tag{3.18}$$

The above wave functions represent mathematical solutions to Eq. (3.15). We must determine which one of the above four possibilities constitutes an acceptable physical solution to our problem. Upon exchange, the wave function must remain unaltered except for a possible change in sign. For linear combination (*iii*), the wave function remains symmetric upon exchange ($\psi(1,2) = \psi(2,1)$), while for linear combination (*iv*), it is antisymmetric ($\psi(1,2) = -\psi(2,1)$). There is experimental evidence that electrons, and all spin ½ elementary particles, or fermions, have antisymmetric wave functions, which change sign upon particle exchange. Linear combination (*iv*) preserves the value of $|\psi(r_1, r_2)|^2$ upon electronic exchange and it is antisymmetric ($\psi(1,2) = -\psi(2,1)$). Thus, only the last linear combination (*iv*) satisfies all physical conditions to be considered an acceptable physical solution to Eq. (3.15) for two non-interacting, indistinguishable electrons. The requirement for an antisymmetric wave function is equivalent to the Pauli Exclusion Principle, because if the wave function were to remain unaltered upon electronic exchange, the two electrons would be occupying the same quantum state, which is forbidden by the Pauli Exclusion Principle.

According to quantum mechanics, the energy of the non-interacting two-electron system is then given by

$$E = E_a + E_b = \iint \psi^*(r_1, r_2)\ \mathcal{H}\psi(r_1, r_2)\, d\tau_1 d\tau_2 \tag{3.19}$$

where the integration is over the volumes (τ_1, τ_2) occupied by the electrons. If we were to assume that the Coulomb repulsion energy between the two electrons is small compared to their potential energy due to their attraction to the hydrogen nuclei, we can use perturbation theory to obtain an estimate of the average electron–electron interaction energy by evaluating the integral

$$E_{\text{int}} = \iint \psi^*(r_1, r_2) \ \mathcal{H}_{12}\psi(r_1, r_2) d\tau_1 d\tau_2 \tag{3.20}$$

In the above equation $\mathcal{H}_{12}$ is the Hamiltonian for the Coulomb interaction between the two electrons, $\mathcal{H}_{12} = e^2/(4\pi\varepsilon_0 r_{12})$, with r_{12} being the distance between the two electrons. Using the antisymmetric wave function (*iv*) of Eq. (3.18), we find that the expression of Eq. (3.20) contains terms of the form

$$C_{12} = \iint \psi_a^*(r_1)\psi_b^*(r_2) \ \mathcal{H}_{12} \ \psi_a(r_1)\psi_b(r_2) d\tau_1 d\tau_2 \tag{3.21}$$

and

$$J_{12} = \iint \psi_a^*(r_2)\psi_b^*(r_1) \ \mathcal{H}_{12} \ \psi_a(r_1)\psi_b(r_2) d\tau_1 d\tau_2 \tag{3.22}$$

The term C_{12} corresponds to the average value of the classical electrostatic Coulomb repulsion energy between the two electrons. The term J_{12} has no classical analog; it is due to the indistinguishability of the two electrons and is called the "exchange interaction". It is purely quantum mechanical in nature, even though it is a function of the electrostatic Coulomb interaction $U_{e\text{-}e} = e^2/(4\pi\varepsilon_0 r_{12})$. As we will see shortly, this exchange interaction brings about spin correlation between electrons and it is ultimately responsible for cooperative spin-alignment and long-range order in solids.

In order to fully appreciate how the exchange interaction produces cooperative spin alignment, we must first note that the two-electron wave function $\psi(r_1, r_2)$ is actually a function of both spatial and spin coordinates and should more appropriately be designated as $\psi(r_1, r_2, s_1, s_2)$, where r_1 and r_2 give the positions and s_1 and s_2 give the spins of the electrons. If there is no interaction between the orbital and spin angular momenta, as is the case when the orbital angular momentum is quenched (*i.e.*, zero), a frequent occurrence in transition metal compounds, the spatial and spin coordinates are separable, and the atomic wave functions can be expressed according to Eq. (3.23).

$$\psi_a(r_1) = \varphi_a(r_1)\chi_a(r_1) \text{ and } \psi_b(r_2) = \varphi_b(r_2)\chi_b(r_2) \tag{3.23}$$

In the above equations, χ and φ stand for the spin and orbital or spatial part of the atomic wave function, respectively. Introducing a quantization axis along the z-direction, say by applying a magnetic field, the electron azimuthal spin quantum numbers can assume values $s_m = +1/2$ or $s_m = -1/2$, "spin-up" or "spin-down" respectively relative to the direction of the magnetic field. Indistinguishability in interchanging the spin orientations between the two electrons can be built into the wave function by considering linear combinations between "spin-up" and "spin-down" electron states according to Eq. (3.24), in analogy with the linear combinations (*iii*) and (*iv*) of Eq. (3.18)

$$\begin{aligned} \chi &= \chi_+(1)\chi_-(2) + \chi_+(2)\chi_-(1) \\ \chi &= \chi_+(1)\chi_-(2) - \chi_+(2)\chi_-(1) \end{aligned} \tag{3.24}$$

Here, $\chi_+(1)$ means that electron 1 is in the spin-up state and so on. The first linear combination above describes a symmetric spin wave function or a spin triplet state, of total spin $S = 1$, and spin multiplicity $2S+1 = 3$. The second linear combination gives an antisymmetric spin state with total spin $S = 0$ with spin multiplicity $2S+1 = 1$, or spin singlet. There are two other combinations that give rise to spin triplet states, namely

$$\chi = \chi_+(1)\chi_+(2) \text{ and } \chi = \chi_-(1)\chi_-(2) \tag{3.25}$$

in analogy with combinations (*i*) and (*ii*) in Eq. (3.18). These triplet states represent the three azimuthal spin states $m_s = 0, +1, -1$. The singlet, antisymmetric state is denoted by χ_S, while the three symmetric triplet spin states are denoted as χ_T.

The total wave function is therefore a separable function of the spatial and spin coordinates of the two electrons and must be antisymmetric when both space and spin coordinates are interchanged between the two electrons. We can build-in the antisymmetric property by requiring that either the spin or the spatial part of the wave function is antisymmetric to obtain the two acceptable linear combinations of Eq. (3.26).

$$\begin{aligned} \psi_{singlet} &= \frac{1}{\sqrt{2}}\left[\varphi_a(r_1)\varphi_b(r_2)+\varphi_a(r_2)\varphi_b(r_1)\right]\chi_S \\ \psi_{triplet} &= \frac{1}{\sqrt{2}}\left[\varphi_a(r_1)\varphi_b(r_2)-\varphi_a(r_2)\varphi_b(r_1)\right]\chi_T \end{aligned} \tag{3.26}$$

Here, $1/\sqrt{2}$ is the normalization factor, assuming that the atomic electron functions are already normalized. In the first of the above equations, the spatial part of the wave function is symmetric while the spin part is antisymmetric, rendering the total wave function antisymmetric, as required for a fermion wave function. In the second equation, the spatial part is antisymmetric while the spin part is symmetric making the total wave function again antisymmetric, as required.

Calculating the electron–electron interaction energy, E_{int}, according to perturbation theory (Eq. 3.20), using the singlet and triplet wave functions of Eq. (3.26), one obtains for the total energy of the two interacting electrons

$$E_{singlet} = E_a + E_b + C_{12} + J_{12} \tag{3.27}$$

and

$$E_{triplet} = E_a + E_b + C_{12} - J_{12} \tag{3.28}$$

Here,

$$C_{12} = \iint \varphi_a^*(r_1)\varphi_b^*(r_2)\ \mathcal{H}_{12}\ \varphi_a(r_1)\varphi_b(r_2)d\tau_1 d\tau_2 \tag{3.29}$$

and

$$J_{12} = \iint \varphi_a^*(r_2)\varphi_b^*(r_1)\ \mathcal{H}_{12}\ \varphi_a(r_1)\varphi_b(r_2)d\tau_1 d\tau_2 \tag{3.30}$$

The Coulomb interaction between the two electrons raises the energy compared to the non-interacting electron system. Depending on the sign of J_{12}, either the spin singlet or the spin triplet is the lower-energy state. Thus, the Coulomb interaction, together with the indistinguishability of the electrons, brings about parallel or antiparallel spin coupling between electrons.

In covalent bonding, as is the case of the hydrogen molecule, we have been considering, the singlet and triplet states correspond to bonding and antibonding molecular orbitals. For the ground state of the hydrogen molecule, the value of J_{12} is negative making the singlet state, with antiparallel spin alignment, the ground state of the hydrogen molecule. Figure 3.4(a) shows the energy, in units of Rydbergs, of two hydrogen atoms as a function of their interatomic separation. A Rydberg equals 13.6 eV and corresponds to the energy required to ionize a hydrogen atom from its ground state. Negative energy corresponds to binding. The curve labeled VW is the result of a classical calculation, essentially the Van der Waals energy, of two neutral atoms. Curves corresponding to bonding and antibonding states according to quantum mechanics are also shown. It is observed that the exchange interaction, responsible for the formation of the singlet and triplet states, lowers the ground energy of the hydrogen molecule by 2.72 eV compared with the classical Van der Waals state. Figure 3.4(b) gives schematically the electronic energy splitting due to the electronic exchange interaction, J_{ex}.

The sum $E = E_a + E_b$ is the total energy of the two isolated hydrogen atoms. When the inter-electron interaction, (J_{ex}), is turned-on bonding and antibonding molecular orbitals are formed with the bonding orbital lying lower in energy, making this the ground state of the hydrogen molecule. Figure 3.4(c) shows the symmetry of the spatial part of the bonding and antibonding wave functions. The spatial part of the bonding wave function is symmetric under the exchange of electrons associated with atoms "a" and "b", while that of the antibonding wave function is antisymmetric.

We conclude that the exchange interaction is responsible for the diamagnetic ground state of the hydrogen molecule with a singlet, $S = 0$, ground state. Similar reasoning applied to the molecular orbitals of the oxygen molecule leads to a paramagnetic, $S = 1$, ground state, as seen in Figure 3.5, which compares the electronic energy levels in the hydrogen and oxygen molecules. Atomic s states split into two molecular states, while p states split into six. This is so because each p state in the oxygen atom contains three orbitals ($l = 1$, $m_l = 0, \pm 1$) forming three bonding and three antibonding orbitals. Thus, two hydrogen atoms, each with electronic configuration $1s^1$, brought together, result in the formation of two molecular orbitals, Figure 3.5(a). Two oxygen atoms, each with the electronic configuration $1s^2$, $2s^2$, $2p^4$, brought together, leads to the formation of ten molecular states, Figure 3.5(b). Electrons populate these states in a way that minimizes the overall energy, while obeying the Pauli Exclusion Principle. The proximity of the two highest populated states in energy favors parallel spin alignment of the last two electrons, thus producing a paramagnetic ground state.

In an extended solid, one has to consider a large number of atoms in close proximity, and thus, a large number of interacting electrons residing in overlapping atomic orbitals localized on neighboring atoms. The exchange interaction applied to this multi-electron system can lead to cooperative parallel or antiparallel spin alignment. Thus, magnetic ordering in solids becomes feasible through this quantum mechanical process.

What we have been saying is that the quantum mechanical exchange interaction modifies the ordinary Coulomb interaction making it dependent on the relative spin orientation of the electrons. If two electrons have antiparallel spins, they can share the same space, which increases their Coulomb repulsion due to their great proximity. On the other hand, if they have parallel spins, they must keep spatially apart, which decreases their Coulomb repulsion energy. This implies that the quantum mechanical exchange interaction is fundamentally of electrostatic origin.

Let us inquire into whether the strength of the exchange interaction can actually explain the occurrence of spontaneous magnetization in ferromagnets. A comparison of the

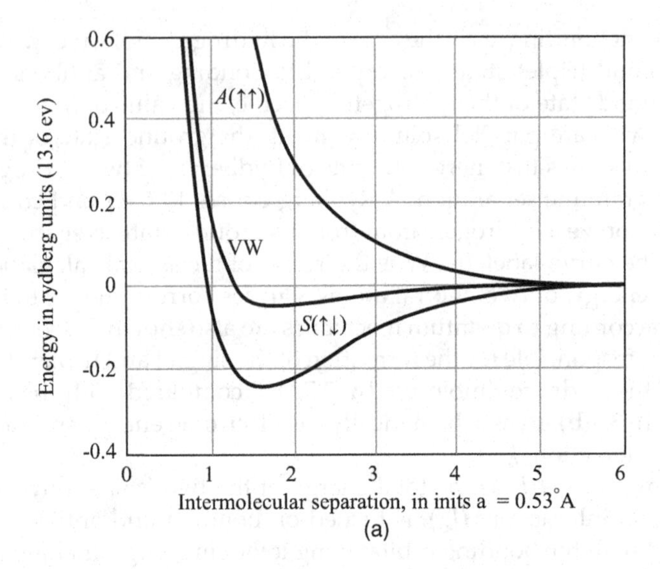

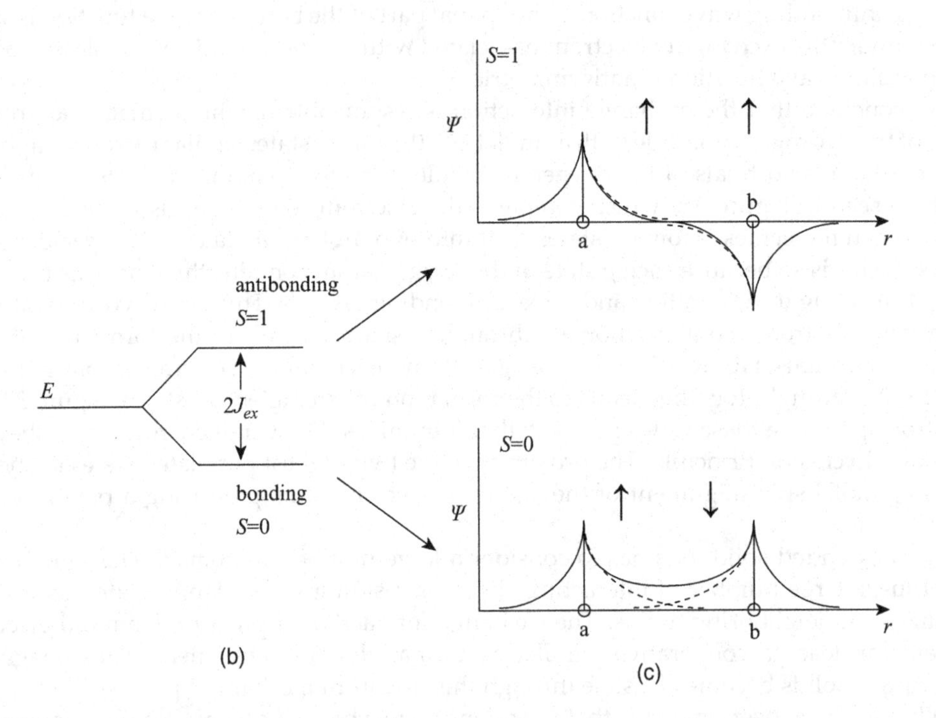

FIGURE 3.4
(a) Representation of the energy of two hydrogen atoms as a function of interatomic separation in units of a, the Bohr radius. The curve labeled VW corresponds to the Van der Walls interaction. The curves labeled $S(\uparrow\downarrow)$ and $A(\uparrow\uparrow)$ correspond to symmetric (bonding) and antisymmetric (antibonding) states. (b) Schematic energy level diagram for the ground state of the hydrogen molecule. E corresponds to the energy of two isolated hydrogen atoms and J_{ex} is the strength of the electron exchange interaction. (c) Representation of the spatial symmetry of the molecular orbitals. The spatial part of the bonding (ground state) wave function is symmetric under exchange, while that of the antibonding is antisymmetric.

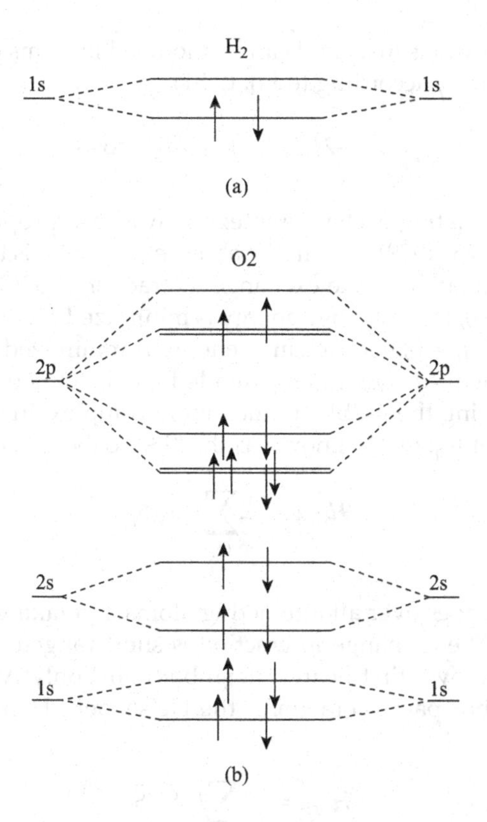

FIGURE 3.5
Depiction of molecular orbital energy states in H_2 (a) and O_2 (b) in relation to atomic orbital energy states. Atomic *s* states split into two molecular states while *p* states split into six.

strength of the Coulomb interaction between two electrons with that of the dipolar interaction between their magnetic moments is in order. For two electrons a distance $r = 0.1$ nm apart, we obtain $U_{e\text{-}e} = \dfrac{e^2}{4\pi\varepsilon_0 r} \approx 2.1\times10^{-18}$ J, which in thermal energies (kT) corresponds to a temperature $T = 1.4 \times 10^5$ K. This is five orders of magnitude stronger than the dipole–dipole interaction energy, $U_{d\text{-}d} \approx \dfrac{\mu_0 \mu_B^{\;2}}{4\pi r^3} \sim 10^{-23}$ J, which would correspond to $T = 1$ K, as calculated earlier. Thus, the strength of the exchange interaction can explain both the high critical temperature for spontaneous magnetization and the large value of the molecular field strength observed in ferromagnetic materials.

3.3 Direct Exchange

3.3.1 Localized Moments and the Heisenberg Exchange Hamiltonian

Calculating exchange integrals of the form of Eq. (3.30) is extremely difficult. Fortunately, in his 1928 paper, Heisenberg showed that the interatomic exchange interaction between two neighboring atoms or ions in a solid can be modeled in terms of the spins, $\vec{S}_i, \vec{S}_j$, localized on interacting atoms i, j according to Eq. (3.31)

$$E_{ex} = -2J_{ij}\vec{S}_i \cdot \vec{S}_j = -2J_{ij}S_iS_j \cos\theta \tag{3.31}$$

This approach to calculating exchange integrals was also proposed by Dirac (P. A. M. Dirac, Proc. Roy. Soc., A123 (1929) 714). In the above equation, J_{ij} is the "exchange integral", which occurs in the calculation of the exchange interaction (Eq. (3.30)), and θ is the angle between the spins. If $J_{ij} > 0$, the exchange energy is minimized for $\cos\theta = 1$, inducing parallel spin alignment ($\theta = 0$). If $J_{ij} < 0$, the exchange energy is minimized for $\cos\theta = -1$, inducing antiparallel spin alignment ($\theta = \pi$). In an extended solid, sums over all interacting spins must be considered making the problem much more complex. In general, one defines the exchange Hamiltonian of Eq. (3.32), known as the "Heisenberg Exchange Hamiltonian"

$$\mathcal{H}_{\text{Heis}} = -2\sum_{i>j} J_{ij}\vec{S}_i \cdot \vec{S}_j \tag{3.32}$$

Here, the summation goes over all interacting atoms and taking $i > j$ ensures that pairs are not counted twice. The exchange interaction is short-ranged, however, sufficing that the summations be taken over first-nearest neighbor pairs, only. Assuming that all J_{ij}'s are equal for nearest neighbor pair interactions, the Heisenberg Hamiltonian can be greatly simplified to

$$\mathcal{H}_{\text{Heis}} = -2\sum_{i>j} J_{ex}\vec{S}_i \cdot \vec{S}_j \tag{3.33}$$

This Hamiltonian has been extensively used in the study of magnetic ordering in solids since J_{ex} can be treated as an adjustable parameter to interpret the results of physical measurements. It will serve as the fundamental model for interatomic spin interactions in our development of the field of nanomagnetism.

For the transition metals, it is the spin alignment of the 3*d* shell electrons that are responsible for magnetic ordering. According to the Heisenberg Hamiltonian, the greater the 3*d* orbital overlap between adjacent atoms, the stronger the exchange integral J_{ex}. Figure 3.6(a) shows the 3*d* and 4*s* electron densities for iron as a function of distance from the nucleus. Half the inter-nuclear distance ($\frac{1}{2}R$~1.3 Å) in metallic iron, indicated by the arrow, falls at the tail of the 3*d* electron density distribution. Thus, in metallic iron, there is a partial overlap of 3*d* electrons between neighboring atoms and, therefore, the direct Heisenberg exchange interaction, which is usually depicted in terms of the 3*d* electron lobes as in Figure 3.6(b), can explain the occurrence of spontaneous magnetization in iron.

Assuming that in a solid, all atoms are (a) in contact with each other and (b) the same distance $R = 2r_a$ apart (where r_a is the atomic radius), the strength of the exchange integral, J_{ex}, for the transition metals has been calculated to follow the universal curve shown schematically in Figure 3.7. This curve, known as the Bethe–Slater curve, gives the variation of J_{ex} as a function of R/r_{3d}, the interatomic distance normalized to the radius of the 3*d* shell.

This model correctly predicts that for Fe, Co and Ni, the exchange integral is positive, producing ferromagnetism (Table 3.1), while for Mn and Cr the exchange integral is negative, producing antiferromagnetism (Table 3.2). Let us visualize the situation where two atoms of the same transition metal element are brought closer and closer together without

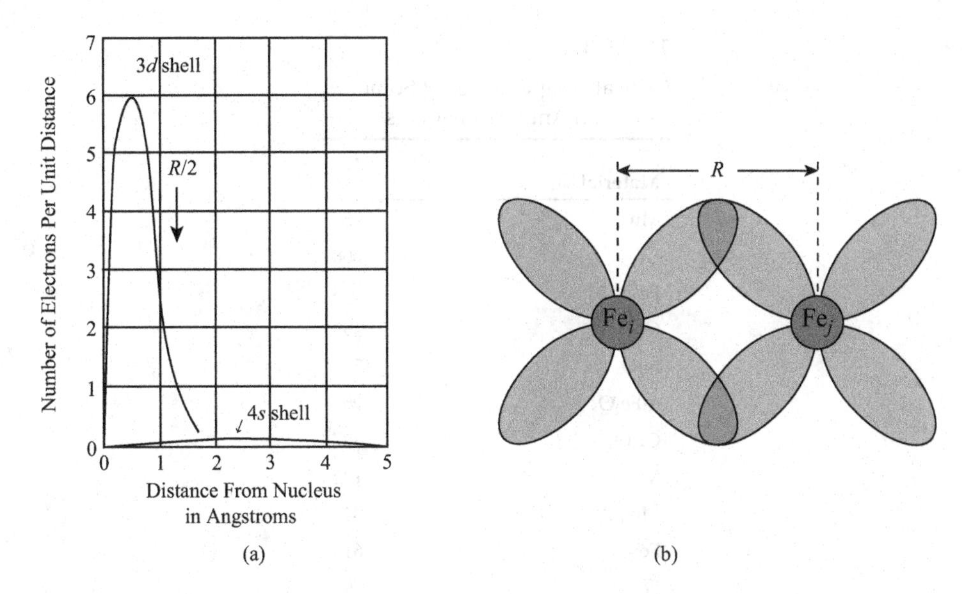

FIGURE 3.6
(a) $3d$ and $4s$ electron charge densities as a function r, the distance from the nucleus, in metallic iron. (b) Depiction of $3d$ orbit overlap in direct exchange. The arrow marks half the interatomic distance. (From R. M. Bozorth, *Ferromagnetism*, 1951, p. 438. Copyright Van Nostrand Reinhold Co. reprinted with permission).

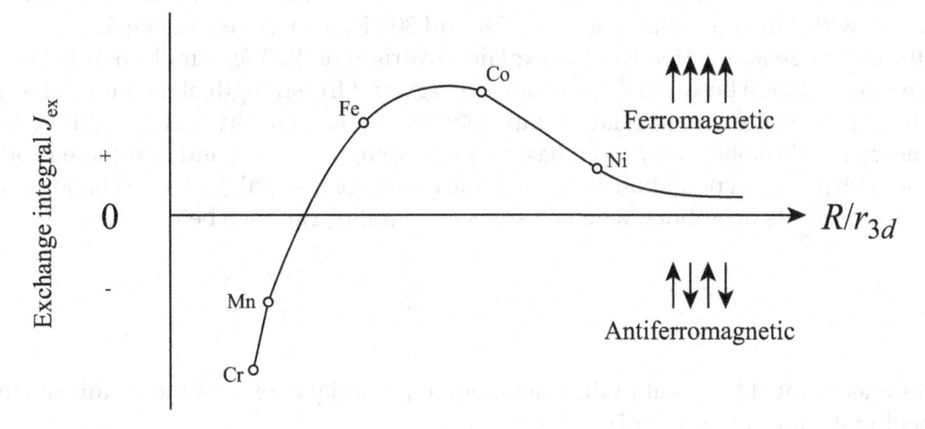

FIGURE 3.7
Depiction of the famous Bethe–Slater curve for transition metals. It depicts the variation of the exchange interaction integral as a function of interatomic distance normalized to the $3d$ shell radius, assumed the same in all metals.

altering the radius of the $3d$ electron shell. When the interatomic distance is large the exchange integral is small and positive. At closer distances, the exchange interaction becomes stronger due to greater $3d$ orbital overlap, reaching a maximum. At even closer distances, the overlap of the $3d$ shells becomes so great that the electrons are forced to occupy the same space, that is, share the same spatial wave function, and the exchange integral becomes negative favoring antiparallel spin alignment.

For positive exchange, we expect the magnitude of J_{ex} to be proportional to the Curie temperature, where thermal energies can compete with the strong exchange interactions between spins to bring about spin disorder. The positions of Fe, Co and Ni on the

TABLE 3.2

Critical Temperatures of Some Common Antiferromagnets

Material	T_N (K)
Mn	95
Cr	308
FeO	198
CoO	293
NiO	523
α-Fe_2O_3	950
Cr_2O_3	307
MnO	122
MnO2	84
FeS	613
FeF_2	79

Bethe–Slater curve correlate with the observed Curie temperatures, being highest for Co and lowest for Ni (see Table 3.1). The positions of the antiferromagnetic metals are also consistent with Mn and Cr having $T_N = 95$ K and 308 K, respectively (Table 3.2).

If the exchange interaction were to explain the origin of the Weiss molecular field, there must be a correlation between the exchange energy and the strength of the molecular field, H_{mf}. To get an approximate relationship between these two physical quantities, let us assume that in the solid every atom has the same spin, $S_i = S_j = S$, and that there is orbital overlap with n nearest neighbors with the same exchange integral, J_{ex}. The exchange energy of an atom with its neighbors, when all spins are parallel, can then be written as

$$E_{ex} = -2nJ_{ex}S^2 \tag{3.34}$$

This energy must be equal to the orientational potential energy of the atomic magnetic moment in the molecular field H_{mf},

$$U = -\vec{\mu} \cdot \vec{H}_{mf} \tag{3.35}$$

or

$$U = -g\mu_B S H_{mf} \tag{3.36}$$

at magnetic saturation. Equating Eq. (3.34) and Eq. (3.36), we obtain

$$H_{mf} = \frac{2nJ_{ex}S}{g\mu_B} \tag{3.37}$$

indicating that the Weiss molecular field does scale with J_{ex}. To relate the exchange integral to the Curie temperature, we use Eq. (3.7) with $J = S$, to obtain Eq. (3.38), where we conclude that J_{ex} also scales with T_C.

$$J_{ex} = \frac{3kT_C}{2n\mu_0 S(S+1)} \tag{3.38}$$

3.3.2 Itinerant Electron Theory of Ferromagnetism

Despite the above-mentioned successes of "direct exchange" between localized magnetic moments in predicting ferromagnetism and antiferromagnetism in transition metals, it fails to account for the values of the atomic moments observed experimentally. Table 3.1 indicates that in the ferromagnetic metals Fe, Co and Ni, the magnetic moment per atom is observed to possess non-integer values, which is incompatible with any theory that assumes localized atomic moments. In free atoms, such as in rarefied monatomic gases, the electrons occupy well-defined, discrete energy levels in accordance with the Pauli Exclusion Principle. Each energy level can accommodate up to two electrons, which must have opposite spins. When atoms are brought together to form a solid, the positions of these discrete energy levels are severely modified.

Let us consider the iron atom; when isolated, its electronic configuration is $1s^2$, $2s^2$, $2p^6$, $3s^2$, $3p^6$, $3d^6$, $4s^2$, with 2 electrons occupying the outer-most 4*s* electronic shell. Notice that due to the proximity of the 4*s* and 3*d* shells in energy, the 4*s* shell is populated before the 3*d* shell is completely filled. When two iron atoms are isolated, their 4*s* levels have exactly the same energy. As they approach each other, the 4*s* electron clouds start to overlap. The Pauli Exclusion Principle applied to the two-atom unit prevents them from maintaining a single 4*s* level occupied by four electrons. Instead, the 4*s* level splits into two levels with two electrons each, in analogy with Figure 3.5(a) where the 1*s* level of the hydrogen atom split into two levels (bonding and antibonding orbitals). Extending this reasoning to the case of *N* iron atoms brought together to form a solid, the 4*s* level of the free atom must now split into *N* levels, each accommodating two electrons of opposite spin, as the Pauli Exclusion Principle now applies to the whole *N*-atom unit. The electrons occupying these levels are no longer confined to move around one particular atom but travel in a region of space extended over all *N* atoms, a process that lowers their kinetic energy. As $N \rightarrow \infty$, a continuous band of closely spaced electronic states is formed. As the distance between the atoms is further reduced, the 3*d* levels also start to overlap, similarly splitting into a continuous band. This is shown schematically in Figure 3.8. For the lattice spacing, a_0, of iron metal, the inner electronic shell 1*s*, 2*s* and 2*p* energy levels remain discrete as they are too close to the nucleus to overlap. Experimental evidence supporting these statements comes from X-ray emission spectra of solids. When electronic transitions occur between two inner levels, radiation of a single frequency is emitted, known as the *K*, *L*, *etc*., X-ray lines. In contrast, when electronic transitions occur between an outer and an inner electronic level the emitted radiation contains a broad range of frequencies.

Thus, one talks about the number of energy states per unit energy, known as the density of states $N(E)$. Then, $N(E)dE$ gives the number of energy levels contained within the energy range between E and $E+dE$. The inverse $1/N(E)$ gives the average energy separation between states in the band. The exact functional dependence of $N(E)$ on E is complex and challenging to calculate. In Figure 3.9, we give only a schematic representation of the

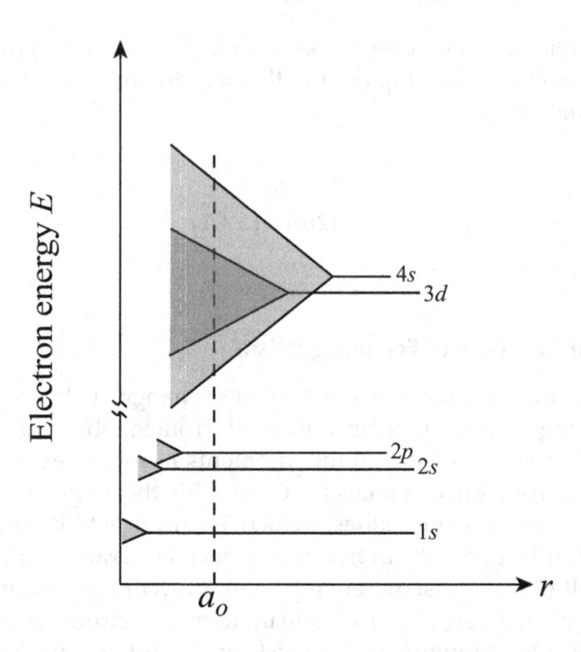

FIGURE 3.8
Schematic representation of the evolution of electronic energy bands from discrete atomic orbitals as a number of N atoms are brought closer together to form a solid. The electron energy is depicted as a function of r, the interatomic distance, with a_0 corresponding to lattice spacing or equilibrium distance between atoms in the crystal.

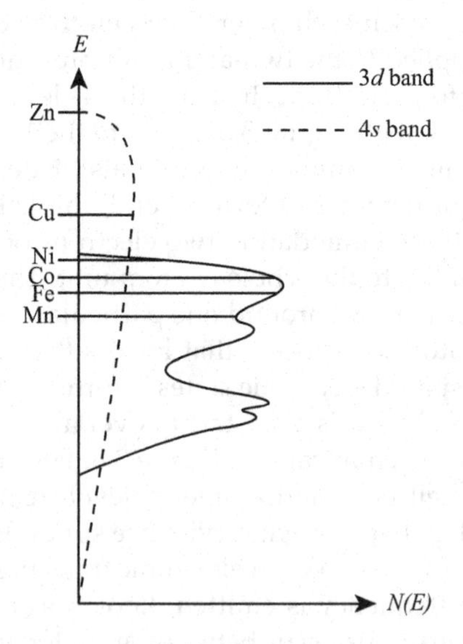

FIGURE 3.9
Schematic representation of the 3d and 4s band density of states in transition metals. The relative positions of the corresponding Fermi levels are indicated by the horizontal lines next to each metal. (Adapted from B. D. Cullity and C. D. Graham, *Introduction to Magnetic Materials*, 2009, p. 136. Copyright Wiley IEEE Press, reprinted with permission).

relationship between *N*(*E*) and *E* for the 3*d* and 4*s* electronic levels. Energy *E* is plotted on the vertical y-axis, while the number of states at energy *E* is given on the horizontal x-axis. The density of 3*d* states is given by the solid line, while the broken line represents the density of the 4*s* states. The density of states is greater for the 3*d* levels because there are five 3*d* electronic states capable of accommodating 10 electrons per atom, compared to the 2-electron capacity of the 4*s* level. The area delimited by the *E vs*. *N*(*E*) curve and the vertical axis is equal to the total number of electronic energy levels in the band.

Due to the fact that the 3*d* and 4*s* band states overlap in energy, the 4*s* levels can start to populate before the 3*d* levels are completely filled. The highest energy level filled is called the Fermi level. The horizontal lines in the schematic of Figure 3.9 indicate the relative position of the Fermi levels for the ferromagnetic transition metals Fe, Co and Ni, and the non-magnetic metals Cu and Zn. The magnetic properties of the ferromagnetic metals are due to the partially filled 3*d* band. Their Fermi level depends on the total number of (3*d* + 4*s*) electrons. Nickel has ten (3*d* + 4*s*) electrons, cobalt nine and iron eight. Thus, the Fermi level for Ni is higher than that of Co, which is higher than that of Fe. Note that when the 3*d* and 4*s* bands are partially filled, they must be filled to the same level. For Cu and Zn, the Fermi level lies entirely within the 4*s* band, that is the 3*d* band if completely filled. These metals exhibit no ferromagnetism.

Each energy level in a band accommodates two electrons of opposite spin. Thus, a band can be considered to consist of two sub-bands, a spin-up and a spin-down sub-band, shown schematically in Figure 3.10(a). Filled energy levels do not contribute to the magnetic moment since the two electrons have opposite spin and thus cancel each other. If, however, a population imbalance between sub-bands existed, a net magnetic moment per atom would form, as there would be, say, more spin-up electrons compared to spin-down electrons in the band (Figure 3.10(b)).

In the ferromagnetic metals Fe, Co and Ni, the ferromagnetism is due to a spin imbalance in the 3*d* band, as shown in Figure 3.10(b). The mechanism that produces this imbalance is the exchange interaction between neighboring atoms in the metal. Due to the high density of states of the 3*d* band, the energy required to reverse a spin and raise the electron to the next higher level in the band is small. In contrast, the density of states of the 4*s* band is low, implying that there is a greater separation between energy levels in the 4*s* band, requiring a stronger exchange to reverse a spin. Thus, the 4*s* band electrons do not contribute to the moment. Suppose that 10 atoms are brought together to form a solid. If one electron in the 3*d* band reverses spin, a population imbalance of two electrons between spin-up and spin-down sub-bands is created corresponding to a moment of $2/10 = 0.2\ \mu_B$ per atom. Thus, the formation of energy bands can explain the occurrence of non-integer values of magnetic moments per atom in a ferromagnetic metal (Table 3.1). In the rare earths, which exhibit ferromagnetism below room temperature, it is a spin imbalance in the 4*f* band that is responsible for spontaneous magnetization.

Thus, Fe, Co and Ni meet three criteria necessary for ferromagnetism: 1) the 3*d* electrons responsible for spin imbalance reside in an incompletely filled band, 2) the density of states in the 3*d* band is high and 3) the interatomic distance is of the right length to produce a strong, positive direct exchange interaction between atoms to bring about spin imbalance in the 3*d* band. This readily explains why the metals Cu and Zn show no ferromagnetism. Their Fermi levels lie high into the 4*s* band. Thus, they fail to meet the first criterion; their 3*d* bands are full.

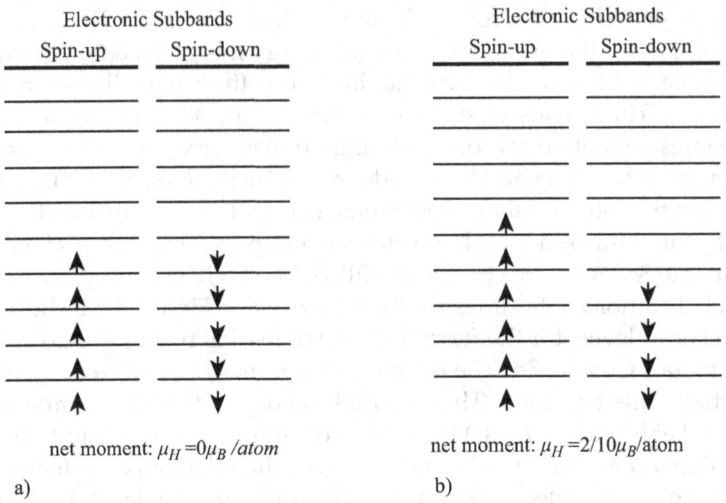

Electronic Subbands

Spin-up Spin-down

net moment: μ_H=0μ_B /atom

a)

Electronic Subbands

Spin-up Spin-down

net moment: μ_H=2/10μ_B/atom

b)

FIGURE 3.10
Schematic representation of 3*d* sub-band spin imbalance produced by flipping spins close to the Fermi level for a 10-atom aggregate. (a) Spin balanced with zero net moment, $\mu_H = 0\ \mu_B/$atom. (b) Spin imbalanced with fractional net moment at saturation, $\mu_H = 0.2\ \mu_B/$atom.

Example 3.1: Let us use Band Theoretical Concepts to Correlate the Variation of the Magnetic Moment per Atom for the Room-Ttemperature Ferromagnets Fe, Co and Ni

Fe, Co and Ni have 8, 9 and 10 (3*d* + 4*s*) electrons, respectively. Assuming that the 3*d* band is the same in all transition metals, known as the "rigid-band approximation", and that the exchange interaction in each of these metals results in five spin-up 3*d* electrons with the remaining electrons being spin-down, we can write an expression of the maximum possible magnetic moment per atom, that is, the moment at saturation, μ_H

$$\mu_H = \left[5 - (n - x - 5)\right]\mu_B \tag{3.39}$$

where n is the total number of (3*d* + 4*s*) electrons and x is the number of 4*s* electrons. The experimental value of μ_H for metallic nickel is 0.6 μ_B per atom or formula unit (Table 3.1), which corresponds to $x = 0.6$ electrons in the 4*s* band. Assuming transition metals close to Ni all have the same number of 4s band electrons, Eq. 3.39 gives

$$\mu_H = (10.6 - n)\mu_B \tag{3.40}$$

The magnetic moments of Ni, Co and Fe are then given by

$$\mu_H(Ni) = 0.6\mu_B \tag{3.41}$$

$$\mu_H(Co) = 1.6\mu_B \tag{3.42}$$

$$\mu_H(Fe) = 2.6\mu_B \tag{3.43}$$

The values predicted for Co and Fe are close to the experimentally obtained values given in Table 3.1.

Because band electrons do not populate orbitals associated with one particular atom, the "band theory" of ferromagnetism is also known as the "itinerant-electron theory" or "collective-electron theory". Itinerant electrons move about over the whole solid. On physical grounds, the itinerant electron theory is the more realistic one for describing ferromagnetism in metals. It has been also tested in the case of single-phase solid solution alloys of transition metals, in which case the total number n of ($3d + 4s$) electrons per atom can take on non-integer values. Its degree of success in describing the ferromagnetism of transition metals and their alloys is depicted in the well-known Slater–Pauling curve shown in Figure 3.11. This curve (J. C. Slater, J. Appl. Phys. 8 (1937) 385, L. Pauling Phys. Rev. 54 (1938) 899) gives the magnetic moment per atom as a function of ($3d + 4s$) electrons in the transition metal ferromagnets and various intra-3d alloys.

The theory is most successful for $n > 8.3$ when the alloys are derived by solid solutions of transition metal elements occupying adjacent positions in the Periodic Table. The theory correctly predicts that the magnetization goes to zero at $n = 10.6$, Eq. (3.40), which occurs in a Ni-Cu alloy with 60% Cu.

An important drawback of the itinerant electron theory is that it is not amenable to easy theoretical modeling of cooperative magnetic interactions to compete with the simplicity of the localized moment theories modeled by the Weiss molecular field and the Heisenberg Exchange Hamiltonian. In addition, transition metal ferromagnets exhibit features associated with both, the localized moment theories and the band theory of ferromagnetism. This is reconcilable by assuming that even though the outer electrons are delocalized, a particular atom in the metal keeps attracting electrons of the same spin, so that there is always a magnetic moment associated with that atom. This intermediate approach between

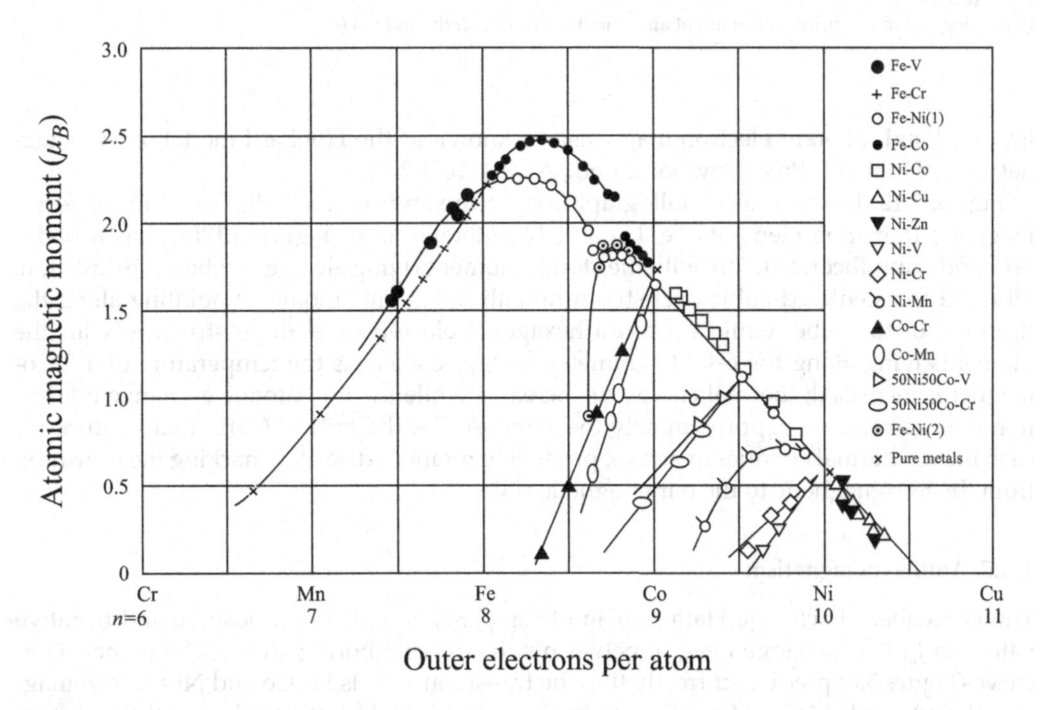

FIGURE 3.11
The Slater–Pauling curve depicts the atomic magnetic moment, in Bohr magnetons, of ferromagnetic transition metals and their alloys, as a function of ($3d + 4s$) electrons per atom. (Adapted from S. Chikazumi, *Physics of Magnetism, 1964, p. 7.* Copyright John Wiley & Sons, with permission).

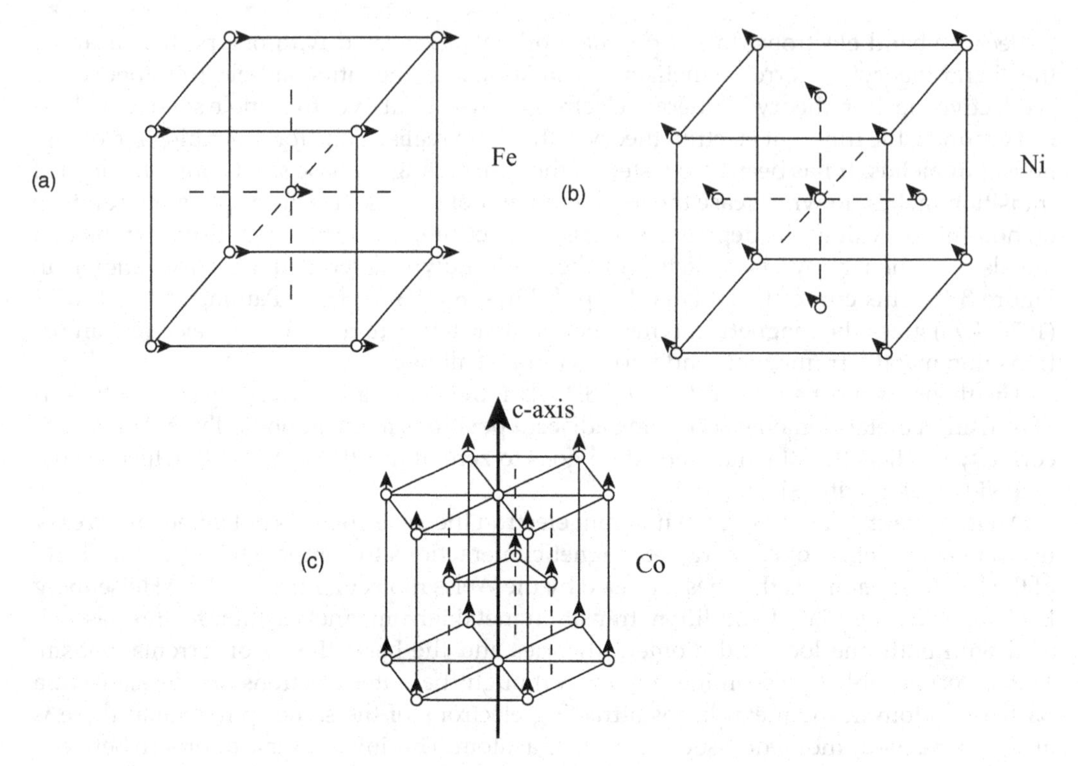

FIGURE 3.12
Crystallographic structure and moment alignment for Fe (a), Ni (b) and Co (c).

localized and itinerant electron magnetism is known as the Hubbard model of ferromagnetism (J. Hubbard, Proc. Roy. Soc. Lond., A276 (1963) 238).

Figure 3.12 depicts the crystallographic structure and moment alignment in the ferromagnetic transition elements Fe, Co and Ni. Metallic iron, Figure 3.12(a), has a body-centered-cubic (bcc) structure with the atomic moments lying along the cube edge direction. Ni has a face-centered-cubic (fcc) structure with the atomic moments pointing along the diagonal of the cube, while Co has a hexagonal close-packed (hcp) structure with the moments lying along the 6-fold symmetry axis, or c-axis. As the temperature of a ferromagnet is increased, thermal energy kT increases while the interatomic exchange interaction strength remains approximately the same. At T_C, the critical Curie temperature for each metal, thermal energies overtake, producing moment disorder, marking the transition from the ferromagnetic to the paramagnetic state.

3.3.3 Antiferromagnetism

The Heisenberg Exchange Hamiltonian of Eq. (3.32) accepts both, positive and negative values of J_{ij}, the exchange integral between pairs of neighboring atoms. The Bethe–Slater curve (Figure 3.6) predicts correctly that the transition metals Fe, Co and Ni are ferromagnets ($J_{ex} > 0$), while Mn and Cr are antiferromagnets ($J_{ex} < 0$). Negative values of the exchange integral lead to antiparallel spin alignment between neighboring atoms. We first introduced the concept of antiferromagnetism in our discussion of the Curie–Weiss Law of

Eq. (3.1) by noting that constant θ is positive for ferromagnets and negative for antiferromagnets. Thus, studying the magnetic susceptibility of solids above their critical temperature for ordering provides a ready means for the classification of the magnetic properties of materials. Figure 3.13(a) shows schematically the temperature dependence of the susceptibility for all three cases $\theta > 0$ (ferromagnet, $H_m > 0$), $\theta = 0$ (paramagnet, $H_m = 0$) and $\theta < 0$ (antiferromagnet, $H_m < 0$), where H_m is the molecular field.

The magnetic susceptibility χ for antiferromagnets exhibits a sharp cusp at their critical temperature for ordering, T_N, while for ferromagnets the susceptibility tends to infinity as the temperature is lowered to T_C (Figure 3.13(a)). The connection between the physical quantity of the ordering temperature T_N and parameter θ for an antiferromagnet is obtained from plotting $1/\chi$ against T (Figure 3.13(b)). The value of θ corresponds to the temperature at which, upon extrapolation, the inverse of the susceptibility becomes zero.

The theory of "antiferromagnetism" was developed by Louis Néel in a series of publications starting in 1932 (L. Néel, Ann. de Physique 18 (1932) 5). A strong negative exchange interaction leads to the spontaneous antiparallel alignment of neighboring atomic moments below T_N, overcoming the randomizing effects of thermal energies. Néel, who was a student of Weiss, introduced the Weiss model of antiferromagnetism by considering two interpenetrating sublattices of opposite magnetization. Thus, he visualized the crystallographic structure of the material as the superposition of two interpenetrating identical sublattices, A and B, of magnetic atoms or ions with opposite magnetic moments, as shown in Figure 3.14. At $T = 0$ K, each sublattice is spontaneously magnetized to saturation, in complete analogy to the spontaneous magnetic saturation of a ferromagnet due to the Weiss molecular field. The ordering temperature T_N separates the two physical regimes of antiferromagnetic and paramagnetic behavior of the material, in complete analogy to T_C, which separates the ferromagnetic and paramagnetic regimes.

To induce antiferromagnetic ordering, Néel assumed negative values of the Weiss constant λ that connects the molecular field with the local magnetization. If we use the

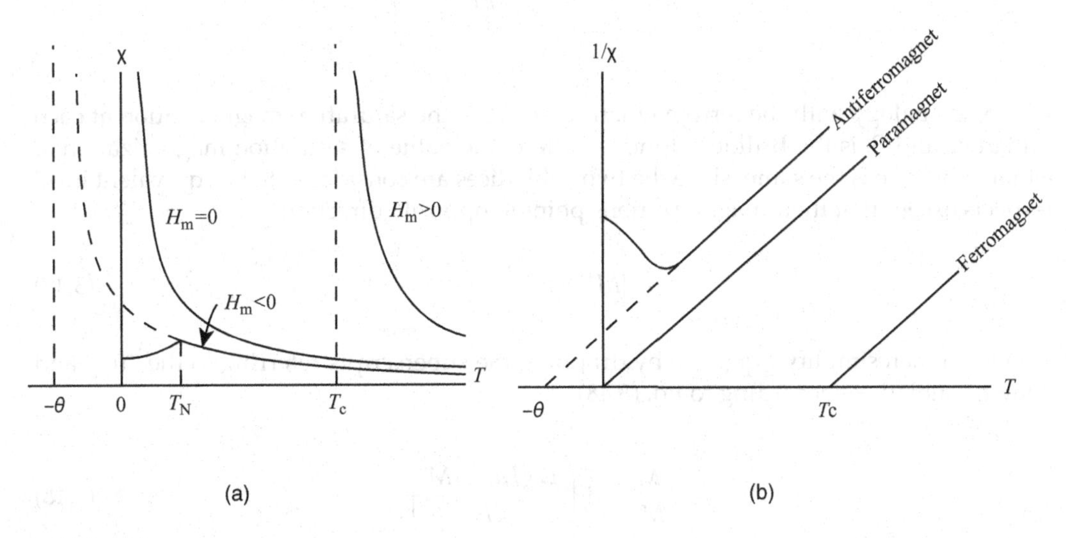

FIGURE 3.13
Schematic representation of (a) χ *vs.* T and (b) $1/\chi$ *vs.* T, for a ferromagnet ($H_m > 0$), paramagnet ($H_m = 0$) and antiferromagnet ($H_m < 0$).

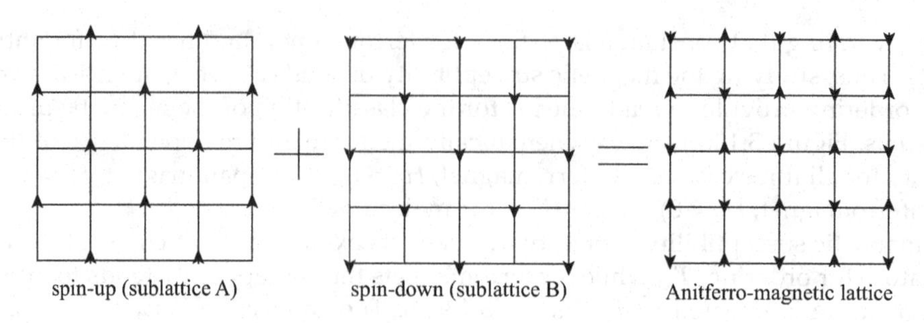

FIGURE 3.14
Schematic representation of an antiferromagnetic lattice arising from two interpenetrating ferromagnetic sublattices, A and B, of equal but opposite magnetizations.

symbols plus (+) and minus (–) to denote the "up" and "down" sublattices the molecular field on each sublattice can be expressed according to Eqs. (3.44) and (3.45).

$$H_{mf}^{+} = -|\lambda| M^{-} \tag{3.44}$$

$$H_{mf}^{-} = -|\lambda| M^{+} \tag{3.45}$$

Using Eqs. (3.3) and (3.5) for this case, the temperature dependence of the reduced magnetization at each sublattice would follow Eq. (3.46)

$$\frac{M^{\pm}}{M_s} = B_J\left(-\frac{\mu_0 g J \mu_B |\lambda| M^{\mp}}{kT}\right) \tag{3.46}$$

where, in analogy with the ferromagnetic case, M_s is the saturation magnetization at each sublattice and B_J is the Brillouin function. Here, the value of saturation magnetization at either sublattice is the same, since the two sublattices are considered to be equivalent in all respects except that their magnetizations point in opposite directions.

$$|M^{+}| = |M^{-}| = M \tag{3.47}$$

We can then simplify Eq. (3.46) by dropping the superscripts referring to the "up" and "down" sublattices according to Eq. (3.48).

$$\frac{M}{M_s} = B_J\left(\frac{\mu_0 g J \mu_B |\lambda| M}{kT}\right) \tag{3.48}$$

In analogy with Eq. (3.7), we conclude that the critical temperature for antiferromagnetic ordering is given by Eq. (3.49)

$$T_N = \frac{N\mu_0 {\mu_{eff}}^2 |\lambda|}{3k} \tag{3.49}$$

where N is the number of magnetic atoms or ions per unit volume and μ_{eff} is the effective magnetic moment on each ion. Therefore, T_N is a constant characteristic of the material. In Table 3.2, we have listed the Néel temperatures of some common antiferromagnets.

From Eq. (3.48), we conclude that the temperature dependence of the reduced sublattice magnetization, M/M_s for different J values, follows the same Brillouin curve as a function of reduced temperature T/T_N presented in Figure 3.3 for the ferromagnetic case, with T_C replaced by T_N. Figure 3.15 gives, schematically, the dependence of the spontaneous magnetization of each sublattice against the reduced temperature T/T_N for $J = 1/2$. At each temperature, $\vec{M}_A = -\vec{M}_B$, resulting in zero net magnetization for the material. Application of an external magnetic field can induce magnetization imbalance between the two sublattices, and, thus, induce a weak magnetization. An interesting class of antiferromagnets, called "canted" antiferromagnets, exhibit weak ferromagnetism in the absence of an external magnetic field, due to a slight misalignment from strict antiparallelism between the directions of the sublattice magnetizations.

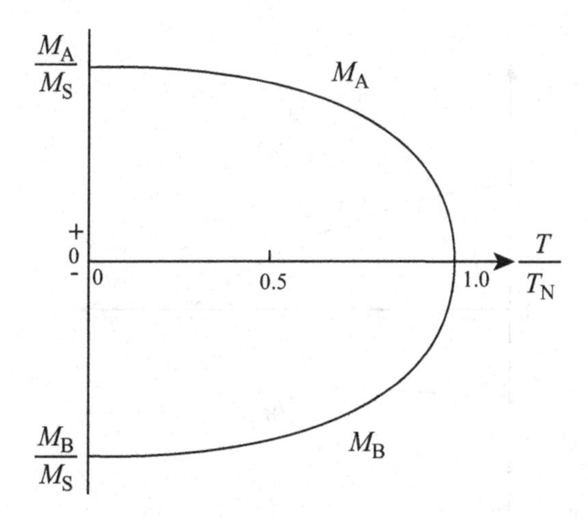

FIGURE 3.15
Schematic representation of reduced spontaneous magnetizations of the A and B sub-lattices *vs.* reduced temperature of an antiferromagnet at temperatures below T_N.

3.3.4 Ferrimagnetism

Ferrimagnetism derives from antiferromagnetism by lifting the requirement that the two sublattices A and B be equivalent. The moments on the two sublattices, while still pointing in the opposite directions, have different magnitudes. Thus, on the macroscopic scale, the material exhibits a net spontaneous magnetization, $|\vec{M}_{net}|$, where

$$|\vec{M}_{net}| = |\vec{M}_A| - |\vec{M}_B|$$

Some ferrimagnets exhibit substantial spontaneous magnetization at room temperature, just like ferromagnets, and thus have important technological applications. Sublattice inequivalence can arise when the magnetic ions associated with the A and B sublattices are not the same, or even if identical, they reside on inequivalent crystallographic sites. Then, the molecular fields for the two sublattices would differ and could have quite different temperature dependences. The net magnetization of the material can, therefore, exhibit quite complicated temperature dependence. In some cases, one sublattice may dominate the magnetization at low temperatures and another at high temperatures. Thus, the magnetization could even be reduced to zero and change signs at some temperature, called the "compensation temperature". Figure 3.16 shows schematically the temperature dependence of the spontaneous magnetization of each sublattice and the resulting net magnetization. The magnetic properties of ferrimagnets are, therefore, more complex compared to antiferromagnets. Ferrimagnets do not obey the Curie–Weiss Law. Developing a molecular field theory for ferrimagnetic order is inherently more difficult than that for antiferromagnetic order. In a classical paper published in 1948, Néel developed the molecular theory of ferrimagnetism and the name "ferrimagnetism" is owed to him (L. Néel, Ann. Phys. 3 (1948) 137). He was awarded the 1970 Nobel Prize in Physics for this achievement.

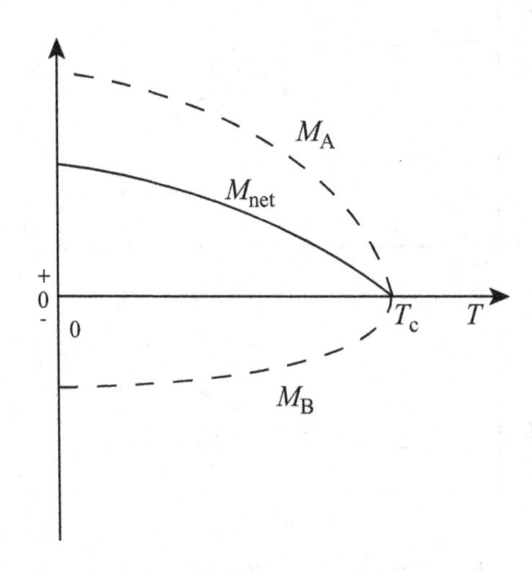

FIGURE 3.16
Schematic representation of the A and B sub-lattice magnetizations and resultant net magnetization of a ferrimagnet as a function of temperature.

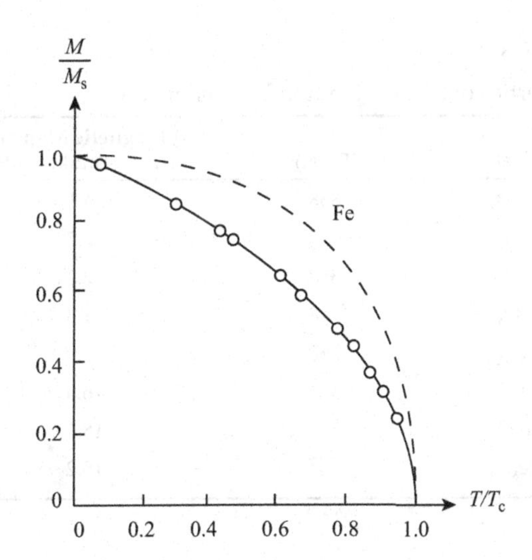

FIGURE 3.17
Schematic of reduced magnetizations *vs.* reduced temperature for a typical ferrimagnet compared to that of ferromagnetic iron, Fe. The experimental data points (o) are for $NiFe_2O_4$.

"Ferrites" constitute an important family of ferrimagnets, grouped into cubic or hexagonal, according to their crystallographic structure. Cubic magnetic "ferrites" have the general formula $MO{\cdot}Fe_2O_3$, where M is a divalent metal ion such as Mn, Ni, Fe, Co or Mg. The metal ions occupy two different crystallographic sites, whereby they are either octahedrally or tetrahedrally coordinated to six or four oxygen atoms, respectively. The tetrahedral sites are known as (A) sites and the octahedral as [B] sites. There is no moment associated with the oxygen ions; the magnetic moments are localized at the metal ions, which exhibit spin-only moments due to the effect of the local crystalline fields, which quench the orbital angular momentum, as discussed in Section 2.9. This crystallographic structure is known as the "spinel" structure. There are twice as many [B] sites than (A) sites. In "normal spinels", the M^{2+} cations sit in the (A) sites and the Fe^{3+} cations in the [B] sites. In "inverse spinels", the M^{2+} occupy half of the [B] sites while the Fe^{3+} occupy the rest of the [B] sites and the entire (A) sites. The Fe^{3+} is in the $^6S_{5/2}$ state and therefore carries a moment of $5\mu_B$. In inverse spinels, the iron ions in the (A) and [B] sites are antiparallel to each other, and thus their moments cancel. The magnetization, therefore, is due to the M^{2+} ions. Figure 3.17 gives schematically the dependence of the reduced spontaneous magnetization M/M_s on reduced temperature, T/T_c, for a typical ferrimagnet, $NiO{\cdot}Fe_2O_3$. The dashed curve is that of the Brillouin function for $J = ½$, which the ferromagnetic metals Fe, Ni and Co follow (see Figure 3.3). One observes a faster decrease in magnetization with temperature than that of the ferromagnets. The dependence of $1/\chi$ *vs.* T for ferrimagnets above their critical temperature deviates considerably from a straight line, indicating that the Curie–Weiss Law is not obeyed; in contrast to ferromagnets and antiferromagnets (Figure 3.13(b)).

Among the hexagonal ferrites, barium ferrite ($BaO{\cdot}6Fe_2O_3$) and strontium ferrite ($SrO{\cdot}6Fe_2O_3$) are well known for their industrial applications. Another class of ferrimagnets is the "garnets" with the general formula $R_3Fe_5O_{12}$, where R is a trivalent rare-earth ion. Table 3.3 gives properties of some common ferrimagnets.

TABLE 3.3

Properties of Some Common Ferrimagnets

Material	T_C (K)	Magnetic Moment (μ_B/Formula Unit)
Fe_3O_4	858	4.1
$CoFe_2O_4$	793	3.7
$NiFe_2O_4$	858	2.3
$CuFe_2O_4$	728	1.3
$Y_3Fe_5O_{12}$	560	5.0
$Gd_3Fe_5O_{12}$	564	16.0
$Dy_3Fe_5O_{12}$	563	18.2
$Ho_3Fe_5O_{12}$	567	15.2

Example 3.2: Iron oxides

The most familiar ferrimagnet is "magnetite", Fe_3O_4, the oldest magnetic substance known to mankind as "lodestone". It derives from the general formula for cubic ferrites $MO{\cdot}Fe_2O_3$, for M = Fe. Another ferrimagnet of considerable interest is "maghemite", or γ-Fe_2O_3, which derives from magnetite oxidation according to

$$2\,Fe_3O_4 + \tfrac{1}{2}\,O_2 \rightarrow 3\,Fe_2O_3$$

In maghemite, all the iron ions are high-spin Fe^{3+}. The crystallographic structure is similar to magnetite "inverse spinel", but the sites occupied by Fe^{2+} ions in magnetite are vacant in maghemite. Thus, oxidation takes place by removing iron rather than adding oxygen. Maghemite converts to the canted antiferromagnet hematite (α-Fe_2O_3) when heated above 400°C. Canted antiferromagnets have a spin structure that deviates slightly from perfect antiparallelism, leaving a small resultant moment per formula unit. We will return to the magnetic properties of iron oxides in the latter chapters, as they have played an important role in the introduction and development of magnetic recording media (Chapter 10) and novel nanomaterials for applications to biotechnology and medicine (Chapter 12).

3.4 Indirect Exchange

The Heisenberg Hamiltonian that models the "exchange interaction" in solids requires direct overlap between electron orbits localized at adjacent magnetic atoms or ions. Such overlap is possible only if the interatomic distance between magnetic atoms in the crystallographic structure of the solid is short enough so that there is considerable overlap of the outer electrons to produce a strong exchange force. The strongest orbital overlap occurs in the formation of molecular bonds, a process that we utilized to define the "exchange interaction" in the formation of bonding and antibonding molecular orbitals in the hydrogen molecule. In studying the magnetic properties of different materials, one finds that many order magnetically below a certain temperature even though the magnetic ions are situated too far apart for any direct orbital overlap to take place. Certainly, this is the case for the antiferromagnetic compounds listed in Table 3.2. One observes that some metal oxides and halides undergo a paramagnetic to antiferromagnetic phase transition, even though the metal ions, the seat of the magnetic moments, are kept apart by the intervening non-magnetic ligands. This is also true for the ferrimagnets listed in Table 3.3. Could the intervening ligands mediate exchange?

3.4.1 Superexchange

Let us consider a specific example. Manganese(II) Oxide, MnO, is paramagnetic at room temperature but undergoes a paramagnetic to antiferromagnetic phase transition at 122 K. Figure 3.18(a) depicts its crystallographic structure, which is fcc with each Mn ion octahedrally coordinated to six oxygen ligands. The manganese and oxygen ions form Mn-O-Mn linear chains. Figure 3.18(b) depicts schematically a process, known as "superexchange", whereby the oxygen ligands can mediate antiferromagnetic coupling between the two high spin Mn^{2+} ions. The $3d_{x^2-y^2}$ orbitals of the Mn ions are shown, reminding us that the 5-fold degeneracy of the $3d$ orbitals of the free ion is raised by the octahedral environment of the oxygen ligands, as discussed in Section 2.9. For small crystal field splitting, a high-spin ground electronic state of $S = 5/2$ is favored; each of the t_{2g} and e_g orbitals of Mn^{2+} holds one electron of same spin. The doubly occupied px-orbital of the oxygen atom contains two electrons, which must have opposite spins in order to satisfy the Pauli Exclusion Principle. The lobes of the oxygen px-orbital point directly toward the $3d_{x^2-y^2}$ lobes of the Mn ions.

The energy is minimized if the electrons in these orbitals are allowed to overlap, forming a partially covalent bond. Let us assume that the Mn ion on the left has a half-filled $3d$ orbital with 5 "spin-up" electrons. Figure 3.18(c) indicates that overlap with the oxygen p electron could take place only if the p orbital contained a "spin-down electron", as all $3d$ spin-up states of the Mn^{2+} ion are already occupied. This would mean that the right lobe of the oxygen p orbital must contain a "spin-up electron". Thus, the orbital overlap on the right could only take place if the Mn^{2+} ion on the right side contained 5 "spin-down" electrons. The net effect is the antiferromagnetic coupling of the two Mn^{2+} ions. Thus, superexchange effectively enables spin–spin interaction processes, as modeled by the Heisenberg Hamiltonian, to take place over extended distances. This is an example of indirect exchange.

3.4.2 The *RKKY* Interaction

Superexchange is the most common mechanism of indirect exchange responsible for ordering in magnetic insulators. In metals, the conduction electrons, which travel over long distances, can mediate the indirect exchange. Electrons in the conduction band move under the action of the electrostatic attraction of all nuclei and electrostatic repulsion of all other electrons in the solid, which cancel each other out, and the electrons move around within the volume of the solid freely, constituting a "free electron gas". In this case, the density of states takes on a simple functional dependence on E given by the proportionality relation of Eq. (3.50), depicted in Figure 3.19. In this case, we plot separately the two sub-bands, with the spin-down sub-band depicted on the left and the spin-up sub-band on the right.

$$N(E) \propto E^{1/2} \tag{3.50}$$

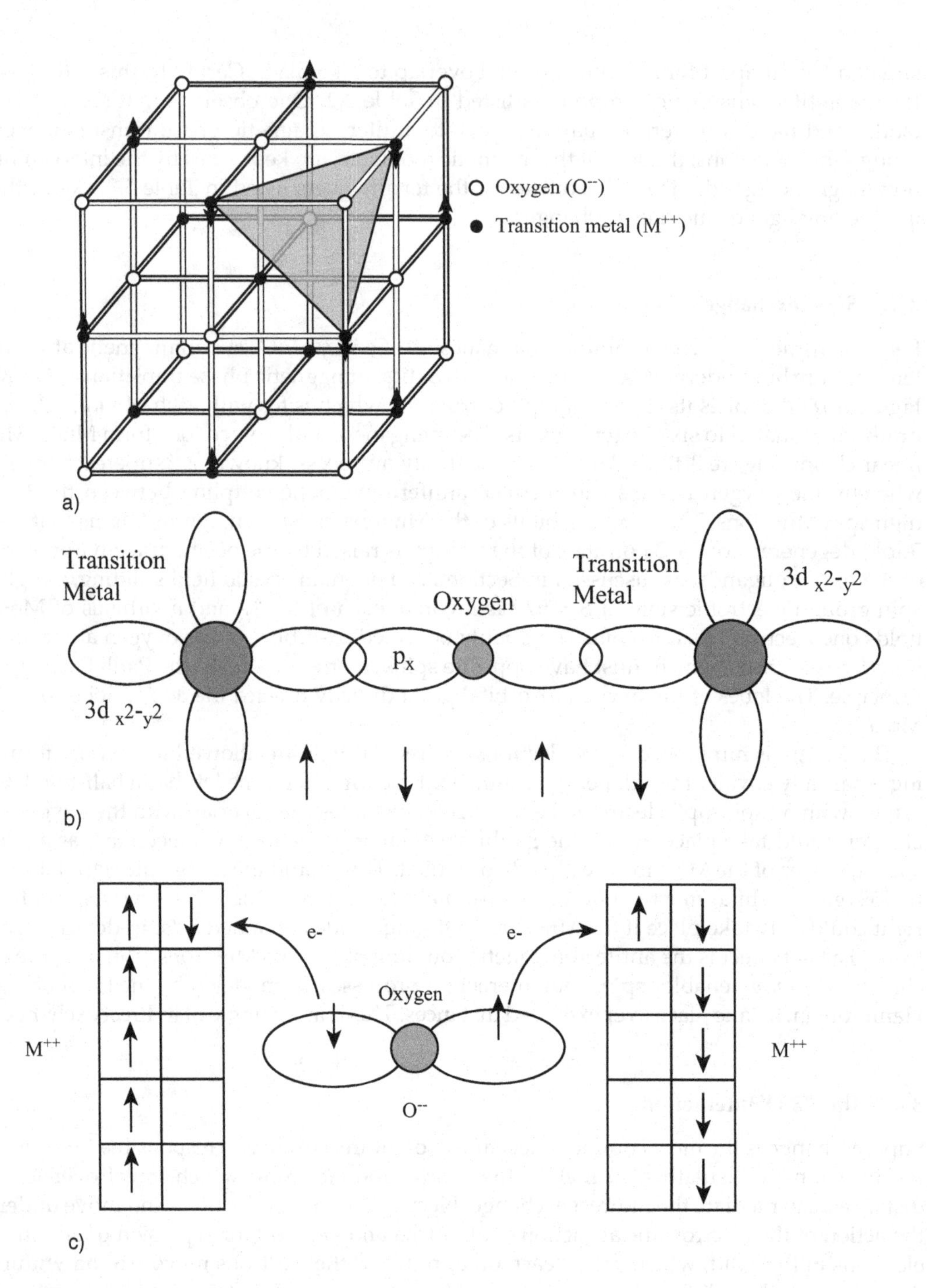

FIGURE 3.18
(a) Crystallographic and spin structure of MnO, (b) and (c) scheme of the superexchange interaction mechanism between two Mn^{2+} ions mediated by an oxygen ligand. (Adapted from *R. C. O'Handley, Modern Magnetic Materials, 2000*, Principles and Applications, pp. 123, 124. Copyright Wiley Inter-science, reprinted with permission).

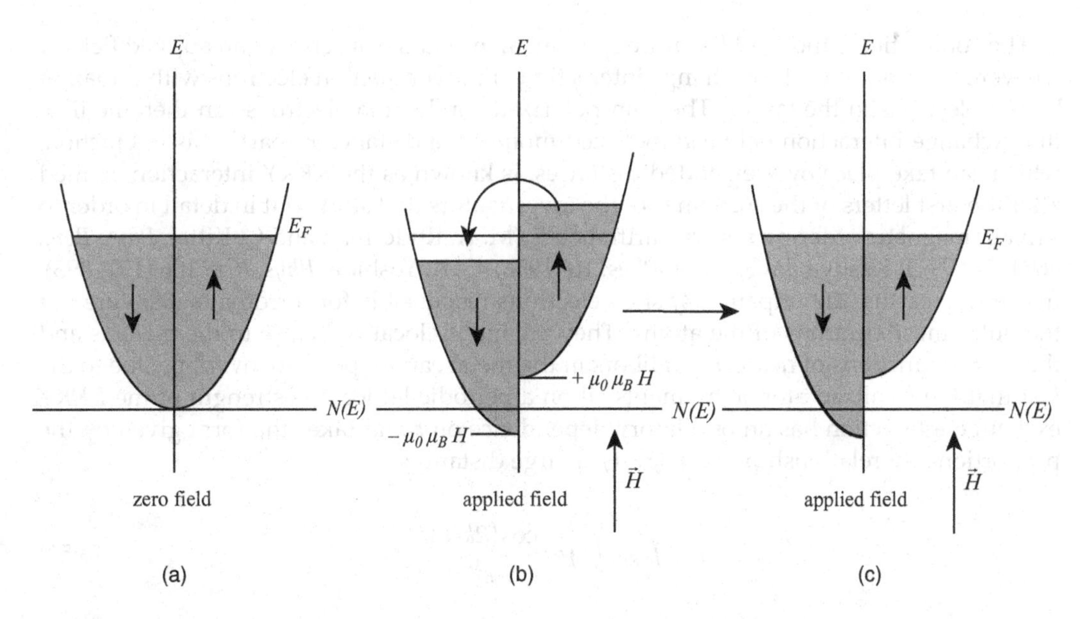

FIGURE 3.19
Pauli paramagnetism. (a) In the absence of an applied field, the spin-up and spin-down conduction electron bands have the same energy and are equally populated. (b) In the presence of an applied field, the spin-up sub-band is raised in energy. (c) A spin imbalance is created. Arrows indicate the direction of the spins.

At $T = 0$ K, electrons populate the spin-up and spin-down sub-bands according to Fermi-Dirac statistics, up to the Fermi level, Figure 3.19(a). If an external magnetic field, $\vec{B} = \mu_0\vec{H}$, were applied the energy of each electron will be changed by an amount $-\mu_0\vec{\mu}\cdot\vec{H}$, where $\vec{\mu}$ is the magnetic moment of the electron. So the electrons with their magnetic moment parallel (spin antiparallel) to the field will be lowered in energy by an amount $\mu_0\mu_B H$, while electrons whose magnetic moments are antiparallel (spin parallel) to the external field will be higher in energy by the same amount as shown in Figure 3.19(b). To minimize energy, up-spin electrons will fall into the down-spin subband, establishing a new Fermi level, close to the original, leaving an overall spin imbalance in the conduction band. This produces a weak spin-paramagnetism known as "Pauli paramagnetism" (W. Pauli, *Z. Phys.* 41 (1926) 81).

The mechanism outlined above responsible for "Pauli paramagnetism" is entirely different than that of "Langevin paramagnetism" we discussed in Chapter 2. Consider a diamagnetic metal such as Cu. Prior to the application of the magnetic field, the up- and down-spin sub-bands are equally populated, and thus each atom has zero magnetic moment. The applied magnetic field induces a small magnetic moment per atom by reversing the spin of only a few electrons close to the Fermi level. In the Langevin theory, even in the absence of a magnetic field, the paramagnetic atoms possess magnetic moments, albeit randomly oriented. The external magnetic field induces magnetization by merely orienting the atomic moments in the direction of the applied field. Furthermore, Langevin paramagnetism shows a strong temperature dependence due to the disordering effect of thermal energies, kT. In contrast, Pauli paramagnetism is temperature-independent; increasing the temperature above 0 K excites electrons close to the Fermi level into higher energy empty levels, but since electrons in the spin-up and spin-down sub-bands are equally affected, the magnitude of the field-induced Pauli paramagnetism is not altered.

The conduction band could acquire spin imbalance in the absence of an applied field, if we were to account for the exchange interaction of the conduction electrons with unpaired bound electrons in the lattice. The spin-polarized conduction electrons can then mediate the exchange interaction between localized moments a distance r apart. This interaction, which can take place over extended distances, is known as the *RKKY* interaction, named after the first letters of the surnames of the investigators that studied it in detail in order to explain magnetic ordering in rare-earth metals (M. A. Ruderman and C. Kittel, *Phys. Rev.*, 96 (1954) 99; T. Kasuya, *Prog. Theor. Phys.* 16 (1956) 45; K. Yoshida, *Phys. Rev.* 106 (157) 893). In the rare earths, the unpaired 4*f* shell electrons responsible for ferromagnetism are not the outer-most electrons of the atoms. They are highly localized close to the nucleus and thus wave functions of nearest neighbors in the metal cannot possibly overlap. Due to the fact that the localized atomic moments sit on a periodic lattice, the strength of the *RKKY* exchange interaction has an oscillatory dependence on r and takes the form given by the proportionality relationship of Eq. (3.51) at large distances r.

$$J_{RKKY}(r) \propto \frac{\cos(2k_F r)}{r^3} \tag{3.51}$$

Here, k_F is the wave vector of the conduction electrons at the Fermi level. A detailed treatment of the *RKKY* interaction is beyond the scope of this book. Suffice it to say that this is a long-range interaction that can lead to ferromagnetic, antiferromagnetic, helical and other types of spin ordering under different conditions. Figure 3.20 summarizes the possible magnetic ordering schemes introduced in this chapter.

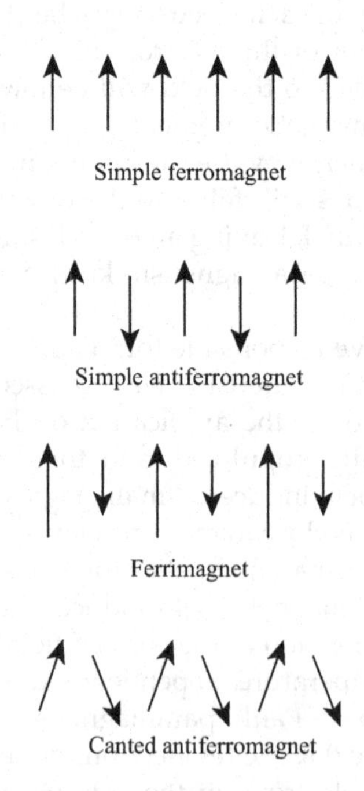

FIGURE 3.20
Possible ordered arrangements of electron spins discussed in the text.

3.5 Magnetic Microstructure

The theories presented in the previous sections have elucidated the fundamental, underlying physical processes responsible for magnetism in matter at the atomic or *micro*scopic level. In this section, we examine the magnetic properties of materials at the *macro*scopic level. Experimentally, the study of magnetism in materials proceeds by examining the induced magnetization as a function of the strength of an externally applied magnetic field. In principle, such an experiment can be carried out by wrapping a tightly wound conducting wire in the form of a solenoid around the sample to be studied, as depicted in Figure 3.21. Passing a current in the clockwise direction creates an applied field $\vec{H} = nI\,\hat{z}$ (Eq. (2.40)) at the center of the solenoid, which induces a magnetization in the $\hat{z}$-direction (to the left). The strength of the applied field can readily be varied by varying the magnitude of the current in the solenoid wire. In addition, the direction of the applied field can be reversed (to the right) by changing the sense of the current from clockwise to counterclockwise.

3.5.1 The Hysteresis Loop

Figure 3.22 sketches the magnetic response of various types of materials to the application of an external magnetic field. Two types of behavior are observed depending on whether the materials are (a) diamagnets, paramagnets and antiferromagnets or (b) ferromagnets and ferrimagnets. Important qualitative and quantitative differences are observed. Note that large applied magnetic fields are required to induce a small magnetization in (a), while relatively small magnetic fields are sufficient to attain magnetic saturation, M_s, in (b) (note difference in axes scales). Furthermore, while in dia-, para- and antiferromagnets, the *M vs. H* curve is linear, that is the slope dM/dH is a constant independent of the strength of the applied field, for ferro- and ferrimagnets, dM/dH is a function of the field. Most importantly, the induced magnetization is reversible in the former but irreversible in the latter. In Figure 3.22(a), the linear *M vs. H* curve is retraced upon reducing the strength of the magnetic field and the magnetization returns to zero when the field is removed, while in Figure 3.22(b), the initial *M vs. H* curve is not retraced, the magnetization does not return to zero upon removal of the magnetic field, rather a remnant magnetization, M_r, remains. The material is permanently magnetized.

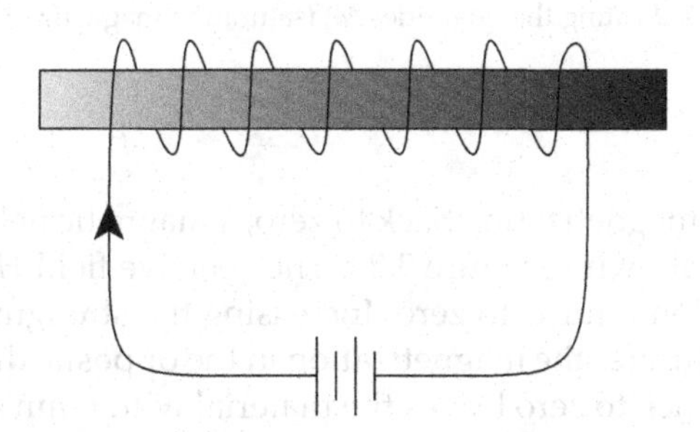

FIGURE 3.21
The principle of experimental arrangement for determining the induced magnetization in a material as a function of applied field strength.

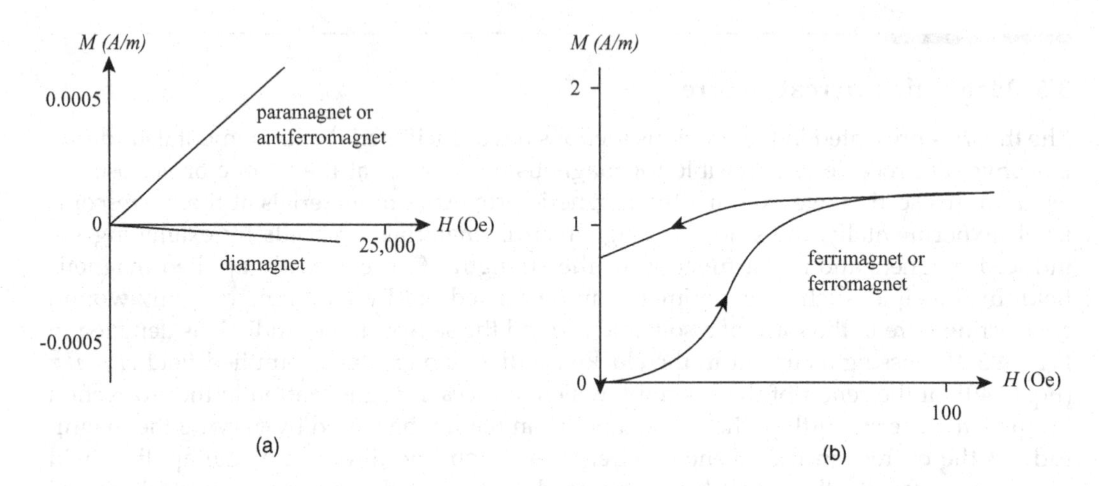

FIGURE 3.22
Schematic representation of magnetization curves for (a) diamagnets, paramagnets and antiferromagnets, and (b) ferromagnets and ferrimagnets.

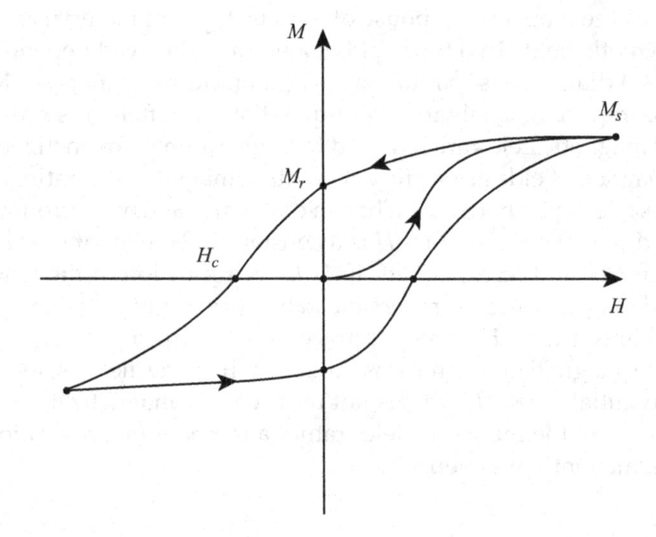

FIGURE 3.23
Schematic of a hysteresis loop indicating the quantities M_s (saturation magnetization), M_r (remnant magnetization) and H_c (coercive field).

In order to reduce the magnetization back to zero, a magnetic field in the opposite direction must be applied, as shown in Figure 3.23. The coercive field H_c is defined as the field at which the magnetization returns to zero. Increasing the strength of the applied field in the negative direction saturates the magnetization in the opposite direction, $-M_s$. Reducing the strength of the field back to zero leaves the material with remnant magnetization, $-M_r$. Applying an increasingly larger magnetic field in the positive direction brings the material back to its original saturation magnetization, M_s. The initial magnetization curve is never retraced. The closed-loop curve obtained is known as the "hysteresis loop", which is characteristic of the material. The word hysteresis derives from Greek and means to "lag behind". The material thus remains permanently magnetized, like a bar magnet.

All three room-temperature ferromagnetic metals Fe, Co and Ni exhibit this behavior at temperatures below their respective Curie temperature for spontaneous magnetization (Table 3.1). In order to demagnetize the material, one must either heat it above its Curie temperature and then let it cool in the absence of an applied magnetic field, or cycle it through successive hysteresis loops by using an alternating magnetic field, with incrementally decreasing the amplitude of the applied field, from above the magnet's coercive field to zero. Hammering or dropping the magnet several times will also induce demagnetization.

The "hysteresis loop" defines the technological properties of a magnetic material. Soft magnetic materials have small coercive fields with $H_c \leq 10^3$ A/m (or ~12 Oe), which can easily be generated with a modest current through a solenoid of a few turns of wire. Some very soft NiFe alloys have a coercivity of only 1 A/m, which is 30 times smaller than the Earth's magnetic field (~0.5 Oe). Hard magnetic materials have large values of coercivity up to 2×10^6 A/m. These find wide application in the production of permanent magnets. Common permanent magnet alloys are Co_5Sm and $Nd_2Fe_{14}B$. Materials with intermediate values of coercivity $10^4 < H_c < 10^5$ A/m have important applications in magnetic recording media, where the stability of magnetization under ambient conditions in an electronic component is needed, but the ease of writing and erasing information with the application of reasonably small fields is desired. For example, cobalt-covered γ-Fe_2O_3 nanoparticles have been used in magnetic recording tape media. We will see in the following chapters that nanomagnetism affords novel and powerful means of control over the coercivity of nanostructured materials.

3.5.2 The Demagnetizing Field $\vec{H}_d$

The experimental observations introduced in the previous section raise two fundamental questions. (*i*) Why is it that a ferromagnetic material is not permanently magnetized when cooled below its Curie temperature for spontaneous magnetization in the absence of an external field? It becomes permanently magnetized only after it is exposed to an external magnetic field. (*ii*) Why is a ferromagnet magnetized so easily compared to a paramagnet? To answer these questions, we must consider the total magnetic energy stored in a *macroscopic* piece of permanently magnetized material and the ways and means to minimize it.

According to Eq. (2.37), the magnetic field inside a magnetized material is given by $\vec{B} = \mu_0\left(\vec{H} + \vec{M}\right)$. The third of Maxwell's equations (Table 1.1) asserts that the divergence of the magnetic field $\vec{B}$ is identically equal to zero ($\vec{\nabla} \cdot \vec{B} \equiv 0$) everywhere. This guarantees the non-existence of magnetic monopoles; the $\vec{B}$-field lines close upon themselves. Figure 3.24(a) depicts the magnetic field $\vec{B}$-lines of a bar magnet. Note that field lines are continuous at the surface of the bar magnet.

Let us inquire into whether this holds true for the $\vec{H}$-field lines as well. Inside a magnetized material, the relationship of Eq. (3.52) leads to that of Eq. (3.53)

$$\vec{\nabla} \cdot \vec{B} = \mu_0\left(\vec{\nabla} \cdot \vec{H} + \vec{\nabla} \cdot \vec{M}\right) \equiv 0 \tag{3.52}$$

$$\vec{\nabla} \cdot \vec{H} = -\vec{\nabla} \cdot \vec{M} \tag{3.53}$$

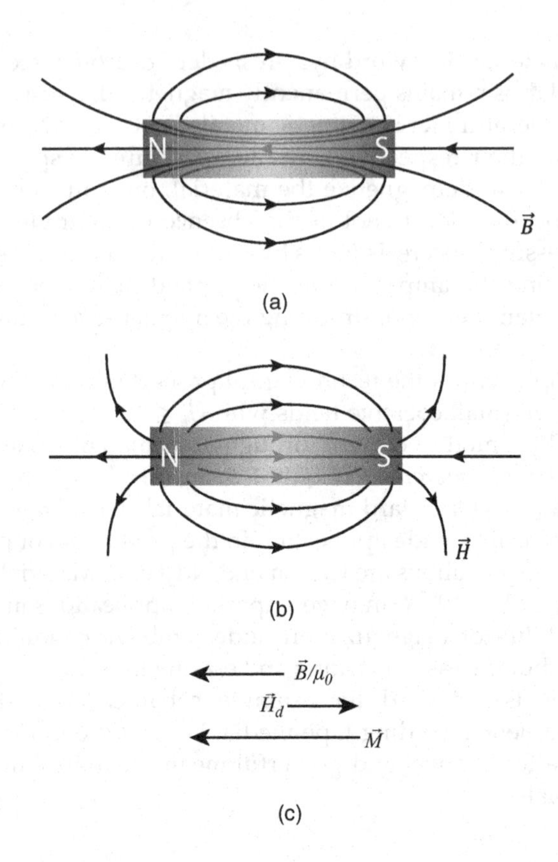

FIGURE 3.24
Depiction of magnetic field lines of a bar magnet: (a) B-field lines are continuous at the poles, (b) H-field lines are discontinuous at the poles. (c) Relative values and directions of the B-field, the demagnetizing field H_d, and the magnetization vector M inside the magnet.

If $\vec{M}$ is uniform, $\vec{\nabla} \cdot \vec{M} = 0$, making the $\vec{H}$-field also divergence-less, just like the $\vec{B}$-field. However, when the uniform magnetization $\vec{M}$ inside the magnetized material meets the surface, the magnetization must suddenly drop to zero, creating a large magnetization divergence at the surface, leading to $\vec{\nabla} \cdot \vec{H} = -\vec{\nabla} \cdot \vec{M}$. In this case, the $\vec{H}$-field has an equal and opposite divergence to $\vec{M}$. It appears as if the magnetic poles at the surface act as sources and sinks of $\vec{H}$-field lines, where $\vec{H}$ points in the opposite direction than the magnetization that produced it in the first place, as depicted in Figure 3.24(b). This field is therefore known as the "demagnetizing field", $\vec{H}_d$. Figure 3.24(c) depicts schematically the relationship between $\vec{B}$, $\vec{H}_d$ and $\vec{M}$ inside the bar magnet, while Figure 3.24(a) depicts the resulting $\vec{B}$-field lines.

The effect of the demagnetizing field inside the magnetized material is clearly seen close to the poles. The $\vec{B}$-field lines inside the bar magnet bend away from its central axis and leak out from the sides of the bar, resulting in a lower density of $\vec{B}$-field lines close to the poles. Since the poles act as magnetic charges, the strength of the demagnetizing field diminishes inversely proportionally with the square of the distance away from the poles, $\left|\vec{H}_d\right| \propto \frac{1}{r^2}$, making its effect on $\vec{B}$ most pronounced close to the poles.

When one considers the magnetic field $\vec{B}$ at the center of an infinitely long rod magnetized along its axis, one needs not worry about $\vec{H}_d$, as the magnetic poles at the two ends of the rod are located very far away, assumed at infinity. In such a case, when the applied magnetic field is reduced back to zero after magnetically saturating the material, the magnetic field inside the material is given by $\vec{B} = \mu_0\vec{M}$, where $\vec{M}$ is the remnant magnetization. In the case of a finite rod, however, the poles must be accounted for. Inside the material, $\vec{B} = \mu_0\left(-\vec{H}_d + \vec{M}\right)$; that is, the *B*-field strength is reduced due to the proximity of the poles. The demagnetizing field inside the material points from the north pole to the south pole, opposite to $\vec{M}$ and to $\vec{B}$. Since poles lie at the surface of a permanently magnetized body, $\vec{H}_d$ must depend on the size and shape of the magnetized body. In the presence of an applied field, $\vec{H}_{app}$, the magnetic field inside a bar magnet is given by $\vec{B} = \mu_0\left(\vec{H}_{app} - \vec{H}_d + \vec{M}\right)$, where now $\vec{H} = \vec{H}_{app} - \vec{H}_d$. When the applied field is removed, the specimen is under its own magnetic field $\vec{H}_d$

Example 3.3: Demagnetizing Factors

Let us consider a ferromagnetic thin film depicted as a permanently magnetized, infinite flat plate in Figure 3.25, shown in cross section. If the magnetization lies in the plane of the film, Figure 3.25(a), the demagnetizing field is negligible since the magnetic poles lie infinitely far away, and thus, they do not contribute to the *H*-field. If on the other hand, the magnetization is normal to the plane of the plate, Figure 3.25(b), a large demagnetizing field $\vec{H}_d$ is present pointing opposite to $\vec{M}$, Figure 3.25(c).

The demagnetizing field can be a very complicated function of position inside a magnetized body of arbitrary shape. It is generally given by Eq. (3.54)

$$\vec{H}_d = -N\vec{M} \tag{3.54}$$

where N stands for a 3 × 3 tensor, the "demagnetizing tensor". It allows determination of the components of $\vec{H}_d$ in terms of the components of $\vec{M}$ using matrix multiplication,

$$\left(H_d\right)_i = -\sum_{j=1}^{3} N_{ij}M_j \tag{3.55}$$

Calculation of $\vec{H}_d$ becomes easily tractable in the case of a magnetized object of ellipsoidal shape. If an ellipsoidal unmagnetized piece of iron metal is placed in a uniform

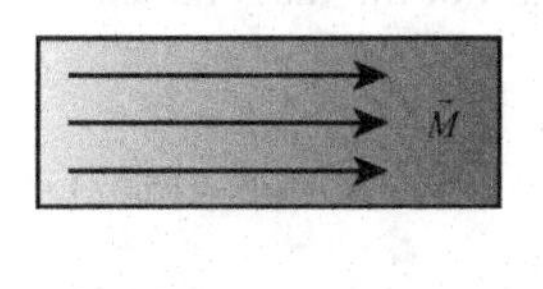

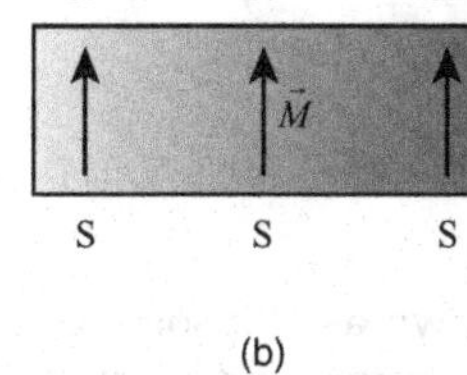

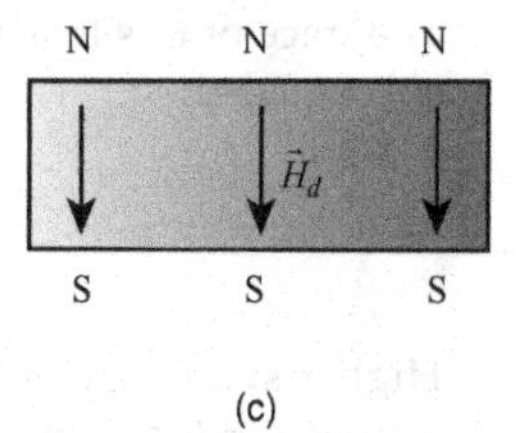

FIGURE 3.25
Demagnetizing fields in an infinitely large thin film. (a) If the magnetization $\vec{M}$ lies in the plane of the film, the magnetization is uniform everywhere within the film except at the edges, which lie far away. No magnetic poles are created on the surface of the film. (b) If the $\vec{M}$ lies normal to the film's plane, $\vec{M}$ diverges at the surface which produces magnetic poles. (c) A demagnetizing field $\vec{H}_d$ exists within the film pointing in the opposite direction than $\vec{M}$.

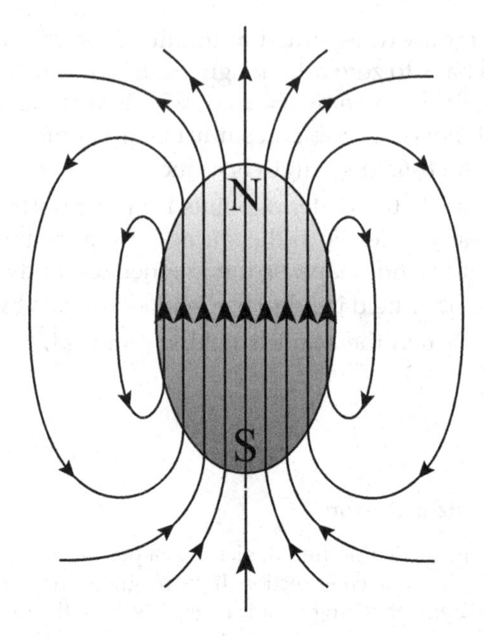

FIGURE 3.26
Depiction of the B-field lines of a magnetically saturated ellipsoidal ferromagnetic specimen in zero field.

magnetic field pointing along its major symmetry axis, it becomes uniformly magnetized throughout, as shown in Figure 3.26. The $\vec{B}$ lines are uniform inside the material due to the uniformity in $\vec{M}$ and $\vec{H}_d$. Compared to the bar-shaped magnet of Figure 3.24(a), the ellipsoidal shape carves out at the two ends of the magnet the regions in which the $\vec{B}$-field lines leak out from the side of the magnet, leaving only a uniform $\vec{B}$ inside the magnet.

Figure 3.27(a) shows a general ellipsoid with its three principal axes of different lengths ($a \neq b \neq c$). If the magnetization $\vec{M}$ points along one of the three principal axes of the ellipsoid, the demagnetization tensor is diagonal, as in Eq. (3.56), where N_a, N_b and N_c are the demagnetizing factors along the a-, b- and c-directions, respectively.

$$N = \begin{pmatrix} N_a & 0 & 0 \\ 0 & N_b & 0 \\ 0 & 0 & N_c \end{pmatrix} \tag{3.56}$$

The trace of this tensor satisfies equation Eq. 3.57; that is, the sum of the diagonal elements equals 1.

$$TrN = N_a + N_b + N_c = 1 \tag{3.57}$$

Higher symmetry ellipsoids, known as ellipsoids of revolution, such as the prolate spheroid ($a = b < c$) and the oblate spheroid ($a < b = c$), are also depicted in Figure 3.27(b) and (c).

Equation 3.57 allows deduction of the demagnetizing factors along different directions for magnetic particles of various highly symmetric shapes, which correspond to limiting cases of ellipsoid spheroids of revolution. In Exercise 3.10, you are asked to calculate the demagnetizing factors for a uniformly magnetized sphere, a thin film and a long cylindrical rod.

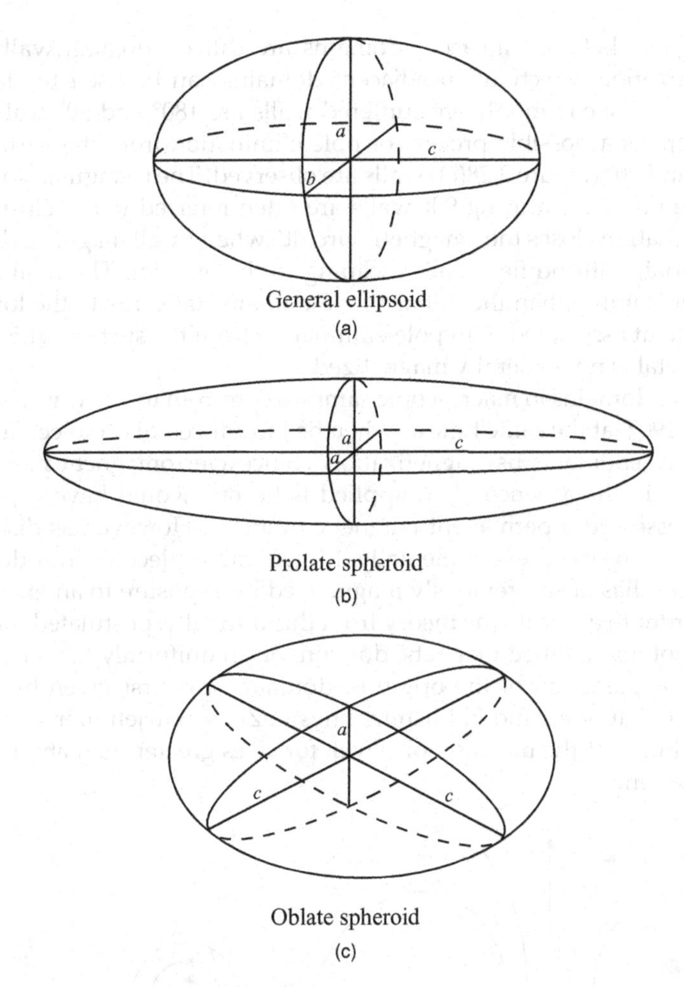

FIGURE 3.27
Depiction of various ellipsoids discussed in the text.

3.5.3 Magnetic Domains and Magnetic-Domain-Wall Formation

Magnetic poles at the surface of a ferromagnet produce strong fields in the surrounding space. Large magnetostatic energy content U_B is associated with uniformly magnetized bodies given by Eq. (3.58),

$$U_B = \frac{1}{2\mu_0} \int_{allspace} B^2 d\tau = \frac{\mu_0}{2} \int_{allspace} H_d^2 d\tau \tag{3.58}$$

since a region of space where B exists contains magnetostatic energy per unit volume given by Eq. (1.33)

$$u_B = \frac{1}{2\mu_0} B^2$$

This large magnetostatic energy can be minimized by eliminating all magnetic poles from the surface. The material subdivides spontaneously into patches of uniform magnetization, called "magnetic domains", oriented in such a way that no poles are left at the surface. The regions between adjacent domains are called "domain walls". The angle between magnetization directions in adjacent domains can be used to classify various types of domain walls; commonly encountered walls are 180° and 90° walls. Figure 3.28 schematically depicts a possible process of pole elimination from the surface. Both 180° (Figure 3.28(b)) and 90° (Figure 3.28(c)) walls are observed. The triangular domains formed at the end of the process featuring 90° walls are often referred to as "closure domains", because their formation closes the "magnetic circuit" whereby all magnetic flux is confined within the material, with no field lines escaping to the exterior. The final multi-domain state is of greater stability than the initial single-domain state, due to the lower magnetostatic energy content associated with pole elimination from the surface. This explains why a piece of iron metal is not generally magnetized.

The existence of domains in macroscopic samples of ferromagnets was first proposed by Weiss himself in 1907, at the same time as when he introduced his "molecular field" model and the concept of "spontaneous magnetization". Since a ferromagnetic metal is spontaneously magnetized in the absence of an applied field, one would have expected that the whole sample possessed a permanent magnetic moment. However, as discussed earlier, this is not what is observed experimentally. In general, a piece of iron does not attract another, except if it has been previously magnetized by exposure to an external magnetic field. Weiss, in order to extricate the theory from this difficulty, postulated that a ferromagnet is composed of non-aligned magnetic domains, each uniformly magnetized to saturation. Theoretical explanation of the origin of domains was first given by Laundau and Lifshitz in 1935 (L. Laundau and E. Lifshitz, Physik. Zeits. Sowjetunion 8 (1935) 153). The theory explained how at the macroscopic level, for sizes greater than about 1 μm, there is no net magnetization.

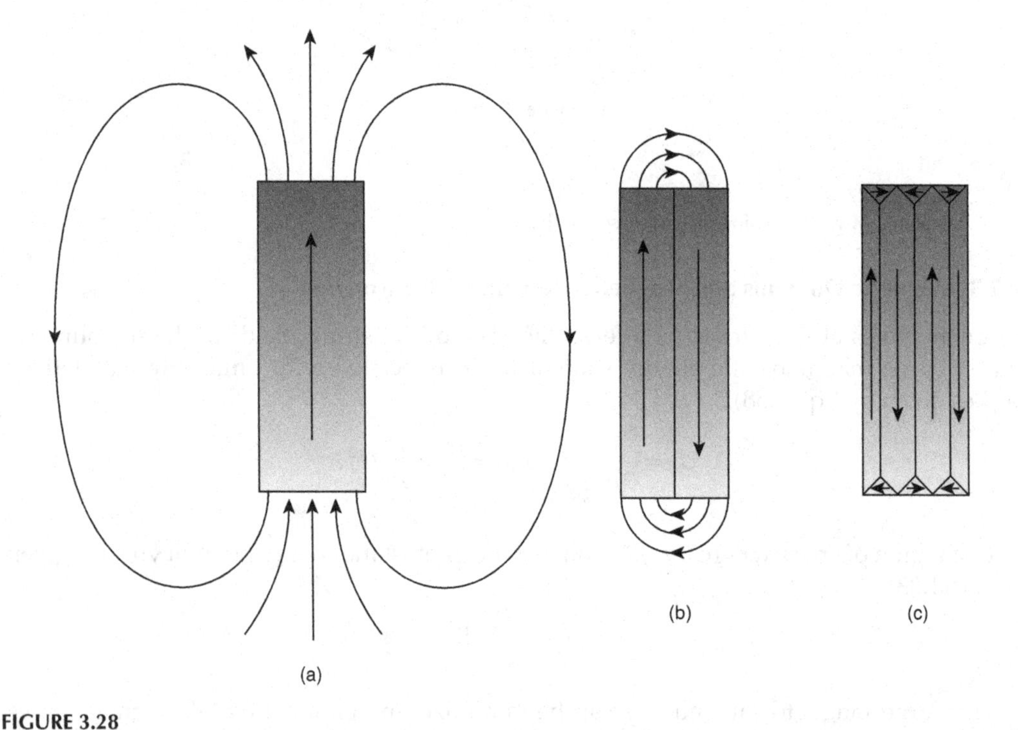

FIGURE 3.28
Schematic representation of the subdivision of a uniformly magnetized body into magnetic domains. Stray fields are confined around the magnet, reducing magnetostatic energy (a) and (b). The triangular domains in (c) are called closure domains; they eliminate all stray fields outside the magnet.

The presence of domains also explains the ease with which a ferromagnet is magnetized compared to a paramagnet (Figure 3.22). Magnetic saturation is reached by the progressive growth of domains favorably oriented relative to the applied field, brought about by the motion of domain walls. Figure 3.29 depicts schematically the process for a sample containing only two domains (Figure 3.29(a)). In reality, a ferromagnet in its demagnetized state would contain a large number of domains, but the simple two-domain hypothesis is sufficient to demonstrate the principle. As the field H is increased, a small number of atomic moments in domain 2 at the vicinity of the wall are turned over into the direction of the magnetization in domain 1. This allows domain 1 to continuously grow with increasing applied fields at the expense of domain 2 (Figure 3.29(b)). Finally, domain 2 is completely eliminated when the magnetic wall is pushed entirely to the surface (Figure 3.29(c)). Complete saturation is achieved at one final step that involves rotation of the magnetization vector in the direction of the applied field (Figure 3.29(d)). The physical domain as a whole does not rotate in the direction of the field, only its magnetization vector does. Figure 3.29(e) gives the M *vs.* H curve for the magnetization process.

It follows that a demagnetized ferromagnet, composed of a large number of domains each of which contains saturated, ferromagnetically aligned moments due to spontaneous magnetization, is easily saturated magnetically *via* the process of magnetic wall movement, a process not as energetically demanding as alignment of atomic moments in the direction of the applied field, as is the case in the magnetization process of a paramagnet.

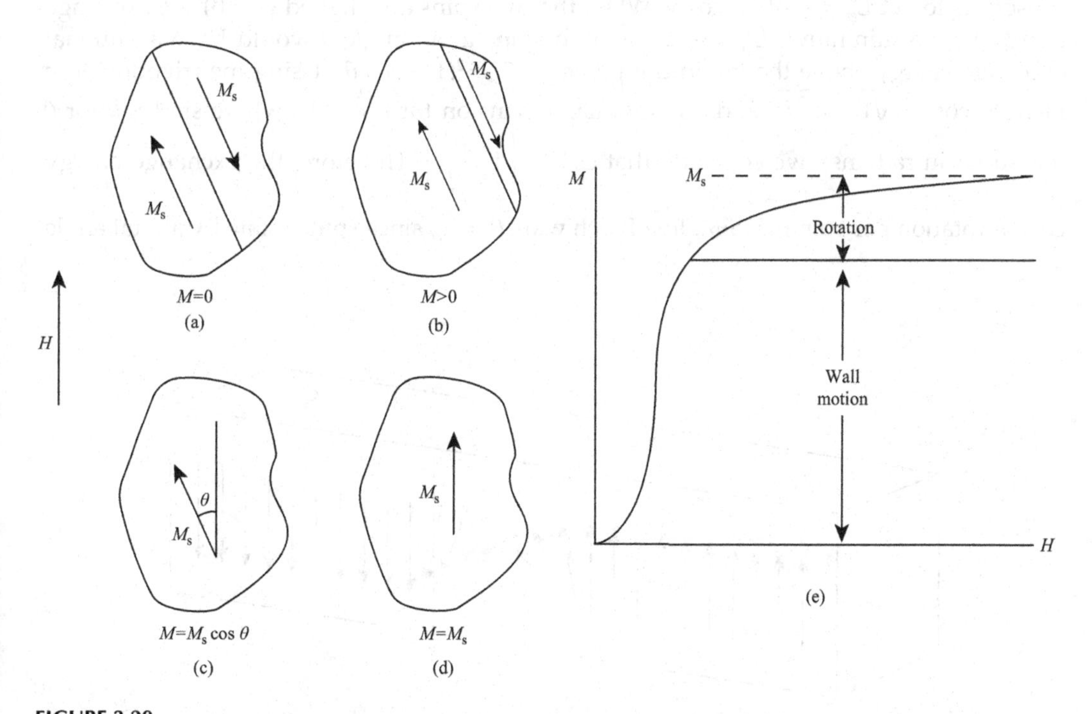

FIGURE 3.29
The magnetization process. (a) Demagnetized specimen in zero field. (b) Displacement of magnetic-domain walls; growth of domain size with magnetization direction at an angle θ, closest to the applied field direction. (c) Single-magnetic domain at an angle θ relative to the direction of the applied field. (d) Rotation of the magnetization in the direction of the applied field. (e) The resulting magnetization curve.

3.5.4 Magnetocrystalline Anisotropy and Easy Axes of Magnetization

So far, we have depicted domain walls as thin straight lines separating magnetic domains. If you were to look at a domain wall through a magnifying glass, however, you would see that a wall has a finite width within which a gradual rotation of the magnetization vector takes place, as shown in Figure 3.30. The figure depicts a continuous rotation of spins by 180° across the wall, known as the "Bloch wall". In the Bloch wall, which is the most commonly encountered 180° wall, the magnetization rotates within the plane of the wall. Magnetization rotation by 180° in the plane perpendicular to the wall defines the "Néel wall".

Questions then arise as to what is the typical size of a magnetic domain, the typical thickness of a domain wall and what are the forces that determine them? The formation of domains minimizes "magnetostatic energy", but the concomitant formation of magnetic-domain walls increases "exchange energy". In a ferromagnetically ordered material, it takes work to rotate adjacent spins away from alignment. Across a Bloch wall, named after F. Bloch who was the first to study the nature of the transition layer between domains (F. Bloch, Zeits. F. Physik 74 (1932) 295), the direction of magnetization rotates by 180° or π radians. The rotation of magnetization direction does not occur in one discontinuous jump across one single lattice spacing. Rather, the rotation is gradual and occurs over a number of lattice planes. Let us calculate the exchange energy stored in a Bloch wall of thickness Na, where N is the number of lattice planes over which the 180° rotation takes place, and a is the lattice constant, or distance between adjacent lattice sites.

The Heisenberg exchange energy between two adjacent spins $\vec{S}_1$ and $\vec{S}_2$ is given by $E_{ex} = -2J_{ex}\vec{S}_1 \cdot \vec{S}_2 = -2J_{ex}S_1S_2\cos\theta$, where θ is the angle between the spins. In an elemental ferromagnet, such as metallic iron, all atoms have the same spin. We can then drop the subscripts to get $E_{ex} = -2J_{ex}S^2 \cos\theta$. When the two spins are aligned ($\theta = 0$) the exchange energy is at a minimum, $E_{ex} = -2J_{ex}S^2$. Two spins at an angle θ would have additional exchange energy above the minimum given by $2J_{ex}S^2(1 - \cos\theta)$. Using the trigonometric identity $\cos\theta = \sqrt{1-\sin^2\theta}$, and its binomial expansion for $\theta << 1$, (where $\sin\theta \approx \theta$ for θ measured in radians), we conclude that $\cos\theta \approx \left(1-\frac{\theta^2}{2}\right)$. Therefore, the exchange energy cost of rotation per spin is $J_{ex}S^2\theta^2$. In a Bloch wall, $\theta = \frac{\pi}{N}$, since spins rotate by a total angle

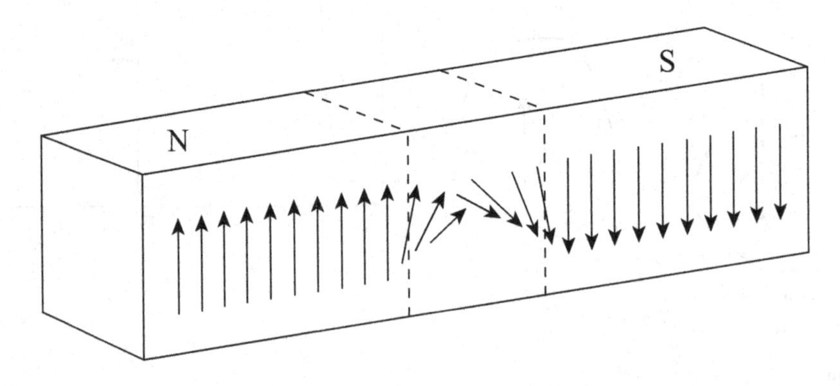

FIGURE 3.30
Rotation of the magnetization within a Bloch wall. The width of this region is typically a few hundred atomic spacings.

π over N lattice planes. The energy associated with a line of N spins across the wall is given by $N\left(J_{ex}S^2\left(\frac{\pi}{N}\right)^2\right) = J_{ex}S^2\frac{\pi^2}{N}$. In a unit area of wall, there are $\frac{1}{a^2}$ lines of spins and, therefore, the exchange energy per unit area of Bloch wall is given by

$$\sigma_{BW}\left(exchange\right) = J_{ex}S^2\frac{\pi^2}{Na^2} \tag{3.59}$$

The equation derived above indicates that the energy cost per unit area of wall formation is inversely proportional to the number of lattice planes N over which the complete 180° rotation occurs. Thus, the driving force for wall energy minimization would tend to unwind the spin rotation within the wall, as the energy per unit area is minimized for $N \to \infty$, making the wall infinite. This is in contradiction with experimental observations.

There are various methods to experimentally image magnetic-domain walls, the simplest of which is the powder pattern method developed by F. Bitter (F. Bitter, Phys. Rev. 38 (1931) 1903). It consists of dropping a small amount of a colloidal suspension of fine particles of magnetite on a highly polished surface of the ferromagnetic crystal to be studied. The magnetite particles in the suspension concentrate along domain boundaries where strong local magnetic fields exist and attract the magnetite particles. This and other imaging methods indicate that walls exist and occupy narrow regions of space separating relatively large magnetic domains. In metallic iron, for instance, a magnetic wall spreads over about 300 lattice sites, while the magnetic domains extend over a volume containing 10^{12} to 10^{18} lattice sites. There must, therefore, be some additional force or energy, hitherto unaccounted for, that contributes to domain-wall stabilization, stopping spin unwinding. This is the "magnetocrystalline" or "magnetic anisotropy" energy.

3.5.5 Magnetic Anisotropy and Easy Axes of Magnetization

Magnetocrystalline anisotropy energy derives from the electrostatic interaction of the magnetic electrons with the local crystalline environment; a concept already discussed in Section 2.9. The response of the electronic orbital motion to the local crystalline electric field (CEF) couples the electronic orbital angular momentum to the lattice, that is, local crystallographic axes. This coupling is mediated to the electronic spin through the spin–orbit interaction, often referred to as $\vec{l}\cdot\vec{s}$ coupling. Magnetocrystalline anisotropy is associated only with crystalline materials. In liquids and amorphous solids, the random distribution of local surroundings leads to macroscopically isotropic materials.

Figure 3.12, shown earlier, depicts the crystallographic structure of the three room-temperature ferromagnetic metals, Fe, Ni and Co and the directions of the atomic magnetic moments relative to the crystallographic axes of the lattice. Within their spontaneously magnetized state, the spins are not only parallel to each other, but are also parallel to certain crystallographic axes. These are called "easy axes" of magnetization. Their number and direction depend on the symmetry of the crystal. For example, in a Co single crystal with a hexagonal structure, as shown in Figure 3.12(c), the magnetization tends to be parallel to the *c*-axis, which is the "easy axis", with all directions within the basal plane, the plane perpendicular to the *c*-axis, being equally "hard axes" of magnetization. A sample of Co is therefore spontaneously sub-divided into two types of domains corresponding to the two possible orientations of the magnetic moments along this axis, as depicted in Figure 3.31(a). In contrast, an iron single crystal has a cubic structure. The magnetic

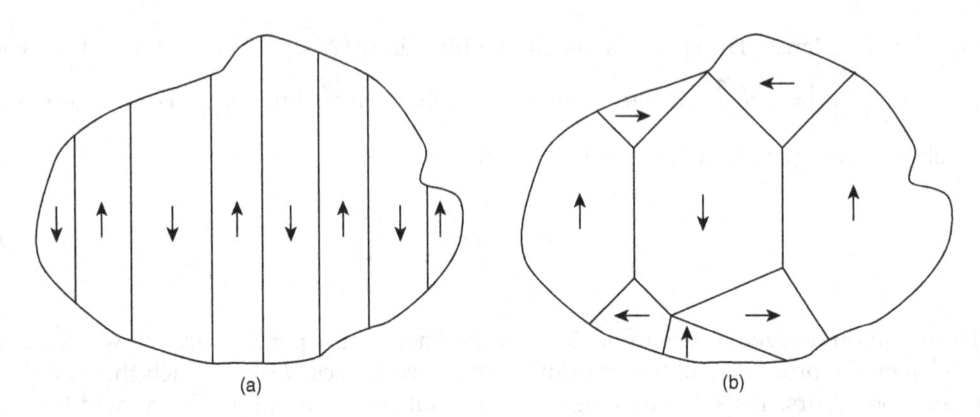

FIGURE 3.31
Schematic representation of magnetic-domain distributions in (a) a hexagonal crystal such as Co, and (b) a cubic crystal such as Fe.

moments spontaneously orient along one of the cube sides, the [100], [010] or [001] directions, which are all equivalent easy axes of magnetization. Then a macroscopic piece of metallic iron is divided into six types of magnetic domains corresponding to the six possible orientations of the magnetic moments, as depicted in Figure 3.31(b). Thus, magnetic anisotropy must play a role in domain-wall formation since it influences the type and number of domains formed. We need to better understand magnetic anisotropy in order to appreciate its contribution to magnetic wall formation and stabilization.

Example 3.4: Magnetic Anisotropy of the Ferromagnetic Metals Fe, Ni and Co

Magnetocrystalline anisotropy influences the magnetic properties of materials. The field dependence of magnetization when measured on single crystals strongly depends on the orientation of the crystal relative to the direction of the applied field. Figure 3.32 shows the dependence of induced magnetization for oriented single crystals of Fe, Ni and Co with the applied field direction along different crystallographic axes. For iron, the directions along the cube sides are the easy directions of magnetization as manifested by the ease of magnetic saturation of the sample in a small magnetic field applied along the cube sides. The body diagonal direction [111] is the hard axis of magnetization with the face diagonal, direction [110], being an intermediate axis. Single crystal magnetization measurements on Ni show that in this case, it is the body diagonal of the cube that is the easy axis of magnetization. Measurements on Co single crystals clearly indicate that magnetic saturation is readily achieved when the field is applied along the *c*-axis, as compared along any direction in the basal plane. These experimental findings indicate that it takes work to rotate spins away from the easy axis of magnetization. The amount of work needed to rotate a spin from the easy axis onto the hard axis of magnetization is a measure of the magnetocrystalline anisotropy energy of the material.

Returning to our discussion of energy cost associated with domain-wall formation, we now realize that not only does the process require the misalignment of adjacent spins, which costs "exchange energy", but also the rotation of spins away from the easy axis of magnetization, which costs "anisotropy energy". It must, therefore, be within the interplay between exchange and anisotropy energy that we must seek the fundamentals of magnetic-domain-wall stabilization; to do so, we must quantify magnetocrystalline anisotropy energy.

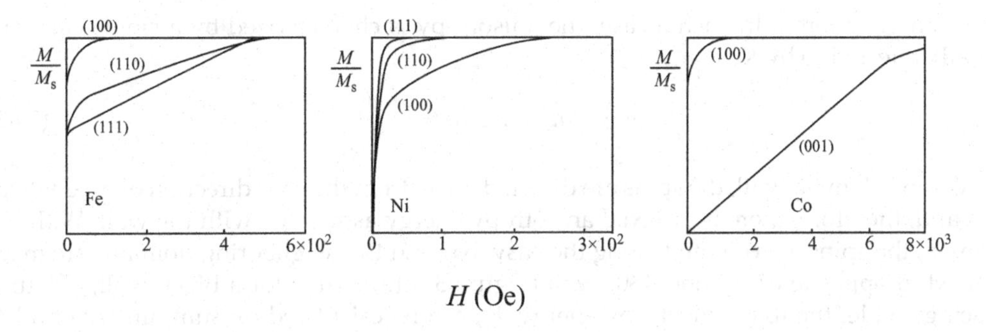

FIGURE 3.32
Magnetization of Fe, Ni and Co single crystals for an applied field along different directions, showing anisotropy.

Since magnetic anisotropy energy is associated with the orientation of the saturation magnetization within the crystal, its measure must involve the angles between $\vec{M}_s$ and the easy axes. Iron is a cubic crystal, and the cube edges are the directions of easy magnetization. If the iron is magnetized along an arbitrary direction, other than the cube edges, then there is anisotropy energy stored within the iron crystal. This anisotropy energy can be expressed in terms of the cosines of the angles the magnetization vector forms with the crystal axes. These cosines are referred to as "direction cosines" and are usually denoted by α_1, α_2 and α_3. Because opposite ends of a crystal axis are equivalent magnetically, the expression for the anisotropy energy must involve even powers of each α_i and remain invariant when the α_i's are interchanged. The lowest order combination that satisfies the symmetry requirements is the sum $\alpha_1^2+\alpha_2^2+\alpha_3^2$, but this sum is identically equal to unity and cannot describe anisotropy effects. The next acceptable combination is of the form $\alpha_1^2\alpha_2^2+\alpha_1^2\alpha_3^2+\alpha_2^2\alpha_3^2$, followed by the combination $\alpha_1^2\alpha_2^2\alpha_3^2$. This indicates that the anisotropy energy in cubic symmetry can be represented by an equation of the form

$$u_K = K_1\left(\alpha_1^2\alpha_2^2+\alpha_1^2\alpha_3^2+\alpha_2^2\alpha_3^2\right)+K_2\alpha_1^2\alpha_2^2\alpha_3^2 \tag{3.60}$$

where u_K gives the anisotropy energy density and K_1 and K_2 are constants measured in J/m³. The constants K_1 and K_2 are characteristic of the material, they are temperature-dependent, and tend to zero as $T \rightarrow T_C$. For iron, at room temperature, $K_1 = 4.2 \times 10^4 \text{J/m}^3$ and $K_2 = 1.5 \times 10^4 \text{J/m}^3$. For Ni, the room temperature value of $K_1 = -5 \times 10^4 \text{J/m}^3$.

In the case of Co, the anisotropy is uniaxial, as there is only one easy axis of magnetization. In this case, the anisotropy energy is often expressed in terms of a series expansion in $\sin^2\theta$, where θ is the angle the magnetization vector makes with the hexagonal *c*-axis. Thus, for Co, the anisotropy energy density, keeping the first two terms in the expansion, is given by

$$u_K = K_1 \sin^2\theta + K_2 \sin^4\theta \tag{3.61}$$

Again, due to symmetry considerations, only even powers of $\sin\theta$ enter the expansion signifying the presence of a uniaxial, as opposed to a unidirectional symmetry axis. At room temperature, the values of the anisotropy constants for Co are given by $K_1 = 4.1 \times 10^5 \text{J/m}^3$ and $K_2 = 1.0 \times 10^5 \text{J/m}^3$. In all cases, $K_2 < K_1$, and to a first approximation, the K_2

term can be ignored. In such a case, the anisotropy is characterized by a single constant usually referred to by K.

$$u_K = K\sin^2\theta \tag{3.62}$$

Within a domain wall, the spins are directed away from the easy direction of magnetization and, therefore, there is uniaxial anisotropy energy associated with the wall. Within a domain, the spins are oriented along the easy axis and the neighboring domains are magnetized in opposite directions, 180° wall (Figure 3.31(a)). Then for a Bloch wall, N lattice spacings wide, the total anisotropy energy, E_a, can be calculated by summing over all N spins in the wall.

$$E_a = \sum_{i=1}^{N} K\sin^2\theta_i \approx \frac{1}{d\theta}\int_0^{\pi} K\sin^2\theta d\theta = \frac{1}{\pi/N}\int_0^{\pi} K\sin^2\theta d\theta = \frac{1}{\pi/N}K\frac{\pi}{2} = \frac{NK}{2}$$

In going from the discrete sum to the continuous integral, we have multiplied and divided by $d\theta \sim \pi/N$, the increment in angle between adjacent spins in the wall. Since K gives the anisotropy energy per unit volume, to calculate the anisotropy energy per unit area of the wall, we multiply by a^3/a^2, where a is the lattice constant, to obtain the anisotropy energy contribution per unit area of the wall

$$\sigma_{BW}(anisotropy) = \frac{NKa}{2} \tag{3.63}$$

The total energy cost per unit area to form a Bloch wall would then be given by

$$\sigma_{BW} = \sigma_{BW}(exchange) + \sigma_{BW}(anisotropy)$$

Using Eqs. (3.59) and (3.63), we obtain

$$\sigma_{BW} = J_{ex}S^2\frac{\pi^2}{Na^2} + \frac{NKa}{2} \tag{3.64}$$

The expression of Eq. (3.64) is minimized with respect to N when

$$\frac{\partial\sigma_{BW}}{\partial N} = -\left(\frac{\pi^2 J_{ex}S^2}{N^2a^2}\right) + \frac{1}{2}Ka = 0$$

yielding a value for N

$$N = \pi S\sqrt{\frac{2J_{ex}}{Ka^3}} \tag{3.65}$$

Substituting this value of N in Eq. (3.64), we get the total energy per unit area of the wall

$$\sigma_{BW} = \pi S\sqrt{\frac{2KJ_{ex}}{a}} \tag{3.66}$$

The width of the wall is then given by

$$\delta_w = Na = \pi S\sqrt{\frac{2J_{ex}}{Ka}} \tag{3.67}$$

Thus, anisotropy energy would tend to limit the thickness of the wall and plays a crucial role in magnetic wall stabilization.

It is customary to express the energy per unit surface area and the wall thickness in terms of the "exchange stiffness" A of the lattice, defined as

$$A = \frac{2J_{ex}S^2z}{a} \tag{3.68}$$

where z is the number of sites in the unit cell ($z = 1$ for simple cubic, $z = 2$ for bcc and hcp and $z = 4$ for fcc). For a simple cubic crystal ($z = 1$), we obtain

$$\sigma_{BW} = \pi\sqrt{AK} \tag{3.69}$$

and

$$\delta_w = \pi\sqrt{\frac{A}{K}} \tag{3.70}$$

3.5.6 Magnetostatic Energy and Shape Anisotropy

The equilibrium magnetic microstructure of a *macro*scopic piece of a magnetic material is established by minimizing the sum of magnetostatic and domain-wall energy. To get an expression for the magnetostatic or "self-energy" of a magnetized piece of a ferromagnetic material with magnetization $\vec{M}$, let us consider its orientational potential energy in the presence of an externally applied field, $\vec{H}_{app}$. Each infinitesimal volume element $d\tau$ carrying magnetization $d\vec{M} = \vec{M}d\tau$ will have an orientational potential energy $du = -\mu_0\vec{H}_{app}\cdot d\vec{M}$.

In the absence of an externally applied field, the magnetized material is subjected to its own demagnetizing field, $\vec{H}_d = -N_d\vec{M}$, where N_d is the demagnetization factor in the direction of $\vec{M}$ (Eq. (3.54)). Substituting for $\vec{H}_{app}$, we obtain

$$du(magnetostatic) = \mu_0N_d\vec{M}\cdot d\vec{M} \tag{3.71}$$

Graphical solutions of the total magnetostatic energy per unit volume of a magnetized specimen in zero applied field give

$$u(magnetostatic) = \frac{1}{2}\mu_0N_dM^2 \tag{3.72}$$

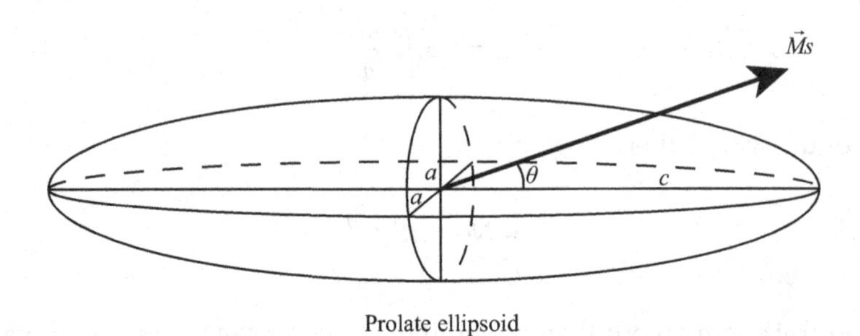

FIGURE 3.33
A ferromagnetic specimen in the shape of a prolate spheroid with its magnetization $\vec{M}$ at an angle θ, relative to the direction of the major axis c.

Thus, the magnetostatic self-energy density is proportional to the square of the magnetization. The dependence of the magnetostatic energy on shape can easily be recognized by considering a specimen in the shape of a prolate spheroid, with major axis c and minor axis a of equal length (Figure 3.27(b)). If the magnetization $\vec{M}$ of the specimen lies at an angle θ relative to the c-axis, as shown in Figure 3.33, it can be decomposed into components parallel and perpendicular to the c-axis.

This casts the magnetostatic energy density of the specimen in terms of the demagnetization factors along the major axes of the prolate spheroid N_c and N_a (Eqs. (3.55) and (3.56))

$$u(magnetostatic) = \frac{1}{2}\mu_0\left[N_c(M\cos\theta)^2 + N_a(M\sin\theta)^2\right] \tag{3.73}$$

Substituting $\cos^2\theta = 1 - \sin^2\theta$, we get

$$u(magnetostatic) = \frac{1}{2}\mu_0 N_c M^2 + \frac{1}{2}\mu_0(N_a - N_c)M^2\sin^2\theta \tag{3.74}$$

The first term on the right side of the above equation is a constant independent of angle θ. The second term depends on θ and thus imparts uniaxial anisotropy to the magnetostatic energy density, $u_{sh} = K_{sh}\sin^2\theta$, where in analogy to the uniaxial magnetocrystalline anisotropy energy of Eq. (3.62), we have defined a corresponding uniaxial anisotropy energy density constant due to shape

$$K_{Sh} = \frac{1}{2}\mu_0(N_a - N_c)M^2 \tag{3.75}$$

Thus, the magnetization is easy to be induced along the c-axis and equally hard along any of the axes perpendicular to the c-axis. The more elongated the ellipsoid, the larger the magnetic anisotropy energy due to shape. In contrast, if c decreases so that $c = a$, the specimen would be of a spherical shape and $K_{Sh} = 0$.

3.5.7 Magnetostriction and Magnetoelastic Energy

Before we leave the subject of energy, we should mention another type of magnetic lattice energy content due to the phenomenon of "magnetostriction". It is well known that the length of a rod of iron is increased when the rod is magnetized lengthwise by the application of a small magnetic field, an observation first made by Joule in 1842 (J. P. Joules, *Ann.*

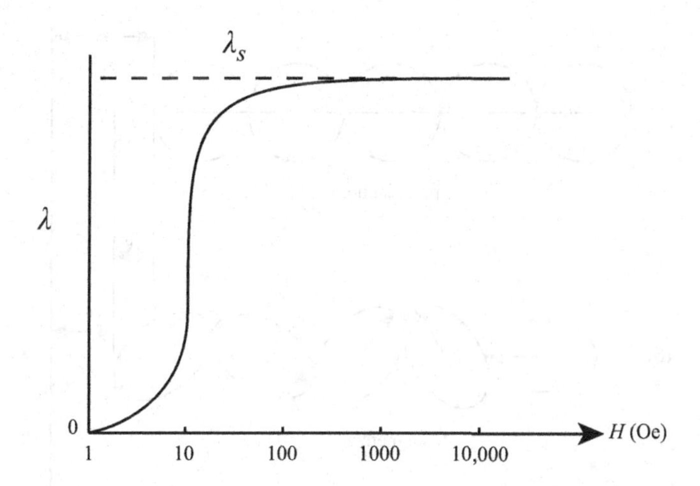

FIGURE 3.34
Schematic representation of the dependence of magnetostriction on the strength of the applied magnetic field. Note that the horizontal axis is logarithmic.

Electr. Magn. Chem. 8 (1842) 219). The fractional change in length, λ, represents a specific type of strain called magnetostriction, defined by Eq. (3.76), where l is the length of the rod in the absence of the field.

$$\lambda = \frac{dl}{l} \tag{3.76}$$

Magnetostriction occurs in all pure substances. Its value is generally small, of the order of 10^{-5}, and depends on the strength of the magnetic field applied, as shown in Figure 3.34.

When the specimen is magnetically saturated, in the sense that it is in the form of a single-magnetic domain with $\vec{M}_s$ in the direction of the applied field, the magnetostriction reaches its saturated value, λ_S. Between the demagnetized state and the magnetically saturated state, the volume of the specimen remains very nearly constant, implying that along the transverse direction, perpendicular to the applied field, the magnetostriction must have a value

$$\lambda_t = -\frac{1}{2}\lambda \tag{3.77}$$

When a ferromagnetic material becomes spontaneously magnetized as it cools from above to below its Curie temperature, spontaneous magnetostriction is created within domains, in the absence of an applied field. Both, spontaneous and field-induced magnetostriction, are defined by Eq. (3.76).

To develop the concept of magnetostriction more precisely, let us consider an isotropic material with no preferred crystallographic axes. Within a magnetic domain, the magnetostrictive strain has its maximum value along the direction of the magnetization of the domain. Let us denote the strain associated with the fractional change in length of a single-magnetic domain by e, and assume uniaxial strain anisotropy. Then, the strain would vary with angle θ according to

$$e(\theta) = e\cos^2(\theta) \tag{3.78}$$

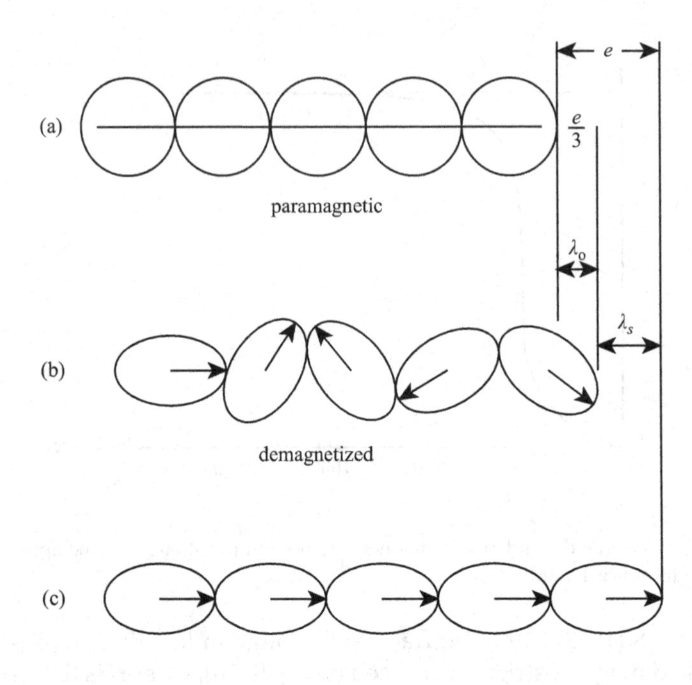

FIGURE 3.35
Schematic representation of the physical origin of λ_0 and λ_s. Scheme (a) represents the unstrained paramagnetic state, (b) represents the spontaneously strained state below the Curie temperature of the material in its ferromagnetic demagnetized state and (c) the maximally strained ferromagnetic state magnetized to saturation.

Assuming random orientation of domains, the average fractional dimensional change throughout the solid would be

$$\lambda_0 = \int_{-\pi/2}^{+\pi/2} e\cos^2\theta \sin\theta d\theta = \frac{e}{3} \tag{3.79}$$

This is the spontaneous magnetostriction due to the onset of ferromagnetic order as the material passes from the paramagnetic to the ferromagnetic demagnetized state. On the other hand, the saturation magnetostriction, λ_s, is defined as the fractional change in length of the specimen as the material passes from its demagnetized ferromagnetic state to its saturated ferromagnetic state due to the application of an external magnetic field. Thus,

$$\lambda_s = e - \lambda_0 = \frac{2}{3}e \tag{3.80}$$

A schematic representation of the physical origin of λ_0 and λ_s is given in Figure 3.35.

The dependence of the saturation magnetostriction on the angle θ, for the isotropic case (polycrystalline samples) we are considering, can be derived by combining the last three equations, which yields

$$\lambda_s(\theta) = \frac{3}{2}\lambda_s\left(\cos^2\theta - \frac{1}{3}\right) \tag{3.81}$$

TABLE 3.4

Experimental Magnetostriction Constants (× 10^6) for Various Polycrystalline Materials at Room Temperature

Material	λ_s	λ_{100}	λ_{111}
Fe	−7	+21	−21
Ni	−34	−46	−24
Co	−62	−140	50
Fe_3O_4	+40	−15	56
$MnFe_2O_4$	−5	−54	10
$CoFe_2O_4$	−110	−670	120

In the Eq. (3.81), λ_s is the saturation magnetostriction in the direction of the magnetization ($\theta = 0$), often referred to as longitudinal magnetostriction, $\lambda_{s\parallel}$. This readily predicts the relationship given in Eq. (3.77), for the transverse magnetostriction ($\theta = \pi/2$),

$$\lambda_{s\perp} = -\lambda_{s\parallel}/2 \tag{3.82}$$

If one were to measure the saturation magnetization of a specimen along mutually perpendicular directions, the internal strain on a magnetic domain e can be determined

$$\lambda_{s\parallel} - \lambda_{s\perp} = \lambda_s + \lambda_s/2 = e \tag{3.83}$$

The study of magnetostriction in single crystals, where preferred crystallographic axes exist, becomes more complex. The magnetostriction coefficients vary along different crystallographic axes. For cubic crystals, such as iron and nickel, the strain e associated with a magnetic domain must be defined as a 3 × 3 tensor with different tensor elements along different crystallographic axes. Table 3.4 gives the magnetostriction constants for some cubic substances and their polycrystalline values. It is observed that the magnetostriction of the ferrites is of the same order of magnitude as that of the metals, with the exception of cobalt ferrite. In this system, the crystal unit cell is severely distorted, from cubic to tetragonal, a distortion that has been observed by X-ray diffraction.

Magnetostrictive materials change shape when subjected to a magnetic field, and thus, in essence, they convert magnetic energy to kinetic energy. They can be used as sensors and actuators. As seen in Table 3.4, pure metals exhibit very small magnetostriction to make them useful in sensor devices. Alloys containing iron and rare-earth metals have been found to exhibit much higher magnetostriction. Of these alloys, $Tb_xDy_{1-x}Fe_2$ exhibits $\lambda_s = 2{,}000$ in a field of 2 kOe (160 kA/m) at room temperature and is the most commonly used engineering magnetostrictive material.

The physical implication of magnetostriction is the coupling of the magnetic properties of the material to lattice deformations. If one were to deform the lattice by applying an external mechanical stress, σ, a change in magnetization would also be induced due to magnetostriction. The applied stress could be positive, $+\sigma$ (compressive stress), or negative, $-\sigma$ (tensile stress). If the value of σ does not exceed the plastic deformation limit of the solid, the energy associated with the coupling of the magnetization to lattice deformation is referred to as "magnetoelastic coupling". For isotropic magnetostriction, the magnetoelastic energy is given by

$$u(magnetoelastic) = \frac{3}{2}\lambda_s\sigma\sin^2\theta \tag{3.84}$$

Eq. (3.84) is used to determine the effect of stress on magnetic properties. We note that the equation contains the product of $\lambda_s\sigma$. Thus, a material with positive magnetostriction would behave under tension as a material with negative magnetostriction does under compression. The above equation associates magnetoelastic anisotropy energy density with stress, with a uniaxial anisotropy energy density constant

$$K_\sigma = \frac{3}{2}\lambda_s\sigma \tag{3.85}$$

where

$$u(magntoelastic) = K_\sigma \sin^2\theta \tag{3.86}$$

Large magnetoelastic energies are usually associated with closure domains. Figure 3.36 shows the closure domains in a cubic crystal, such as iron. Since λ_{100} is positive for iron, the [100] closure domain would tend to elongate along its magnetization direction, as depicted by the broken lines. However, it is restrained from doing so by the main domains on either side. This makes the closure domains strained; the magnetoelastic energy stored in them is proportional to their volume.

The microscopic domain configuration finally established within a magnetic specimen would therefore be the result of total energy minimization. The total energy would be comprised of the sum of magnetostatic, exchange, magnetocrystalline and magnetoelastic energies. The process of subdivision to smaller and smaller domains will continue until further subdivision no longer results in total energy reduction. The above considerations would indicate that the average size of a single-magnetic domain is not a fundamental property of the material, but is also a rather sensitive function of the size and shape of the specimen.

If one were to consider increasingly smaller size specimens, there comes a certain size below which the energy associated with domain-wall formation would exceed any energy savings received in magnetostatic energy. At that size, the specimen will exist in a single-magnetic-domain state and it will be magnetically saturated in the absence of an external magnetic field. Thus, nanometric size magnetic particles must necessarily exist as single-magnetic domains. It is the study of the magnetic properties of such particles and their ensembles that constitutes a major part of the field of nanomagnetism.

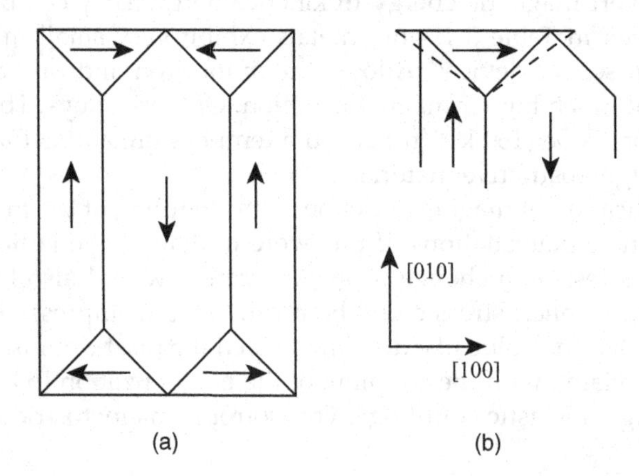

FIGURE 3.36
Closure domains in a cubic crystal (a) with <100> easy axes (along the cube edges) (b).

Exercises

1. Use Eqs. (3.7) and (3.49) to estimate the strength of the Weiss molecular field constant λ for the ferromagnetic metals Fe, Co, Ni, and Gd. (b) Do the same for the antiferromagnetic metals Mn and Cr. (c) What is the value of the molecular field in each metal? (d) Give an estimate of the Curie–Weiss constant C for each metal.
2. Use Eq. (3.38) and data given in Table 3.1 to estimate J_{ex} for Fe, Co and Ni in SI units.
3. Estimate the ratio of the exchange interaction energy to the dipolar interaction energy of two adjacent Fe atoms in metallic iron.
4. A magnetic dipole μ_1 is located at the origin of a Cartesian coordinate axis system. A second dipole μ_2 is brought in and placed at point (x = 0, y = 5, z = 0). What is the relative orientation of the two dipoles at equilibrium if (a) both dipoles are free to rotate and (b) only μ_2 is free to rotate while μ_1 is fixed pointing in the z-direction?
5. Use Figure 3.4(a) to get a rough estimate of the energy required to ionize a hydrogen molecule from its ground state compared to that needed to ionize a hydrogen atom from its ground state. Give your answer in eVs and in Joules. The difference in energy gives an estimate of the strength of electronic exchange in the hydrogen molecule.
6. *Helical Order:* Consider a linear chain of N spins as shown in Figure 3.37. Let S be the value of the spin and J_1 the exchange interaction coupling constant between nearest-neighbor and J_2 the exchange interaction coupling constant between next-nearest-neighbor spins.

   If the angle between successive spins is denoted by θ, prove that the energy of the system is given by the following expression:

$$E = -2NS^2 \left(J_1 \cos\theta + J_2 \cos 2\theta \right)$$

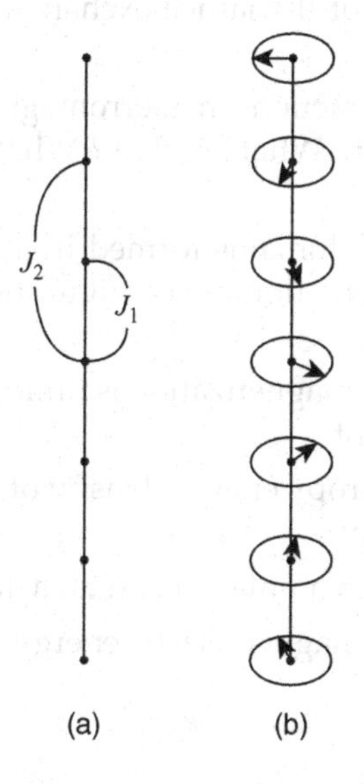

FIGURE 3.37
(a) A linear chain of spins with first- and second-order interactions. (b) The spiral spin structure in the linear chain.

Show that the energy is minimized for $\theta = 0$ or $\theta = \pi$, leading to ferromagnetic or antiferromagnetic order, and for

$$\theta = \cos^{-1}\left(-\frac{J_1}{4J_2}\right)$$

leading to helical order. Show that helical order, also known as helimagnetism, is favored for $|J_1| < 4|J_2|$. Helimagnetism is a special case of antiferromagnetism as it also leads to vanishing macroscopic magnetization for the material.

7 According to the Slater–Pauling curve of Figure 3.11, the maximum magnetization for any ferromagnetic system corresponds to an alloy rather than a pure elemental ferromagnet. (a) Which alloy does the maximum in the curve correspond to? (b) What is the corresponding maximum magnetization per unit mass of the alloy?
8 Discuss direct Heisenberg exchange in ferromagnetism and give examples where it succeeds in explaining experimental observations and examples where it fails.
9 Discuss the itinerant electron theory of ferromagnetism and give examples where it succeeds in explaining experimental observations that the direct exchange theory of ferromagnetism fails to explain.
10 Considering highly symmetric shapes as limiting cases of an ellipsoid spheroid of revolution, calculate the demagnetization factors for (a) a sphere, (b) an infinite thin film, and (c) a long cylindrical rod of length l and radius r, assuming $l >> r$.
11 (a) What is the primary driving force responsible for magnetic-domain formation in a ferromagnet? (b) What is the nature of hysteresis in multi-domain ferromagnetic materials? (c) What is the difference between soft and hard magnetic materials?
12 Which material would you use for a hard drive? And for a power generator?
13 Use Eq. (3.67) to calculate the magnetic wall thickness in metallic iron and nickel.
14 Give estimates of the values of the lattice exchange stiffness for the three ferromagnetic metals Ni, Fe and Co?
15 Two competing magnetic interactions in a ferromagnetic single crystal are responsible for magnetic wall stabilization. What are they? What determines the width of a magnetic Bloch wall?
16 What determines the type of domains formed in a single crystal of a ferromagnetic material? Contrast the magnetic-domain configuration expected in a single crystal of Co as compared to that of Fe.
17 It is usually assumed that the magnetization is uniform throughout a single-magnetic domain. Qualify this statement.
18 Calculate the magnetic anisotropy energy density of a long cylindrical rod of Co magnetically saturated along its axis.
19 What is the difference between a uniaxial and an axial or unidirectional symmetry?
20 Define magnetostriction and magnetoelastic energy.

4

Single-Magnetic-Domain Particles

The magnetic properties on a very small scale are not the same as on a large scale; there is the *domain* problem involved. A big magnet made of millions of domains can only be made on small scale with one domain.

Richard Feynman, **There is Plenty of Room at the Bottom. Lecture delivered at the annual meeting of the American Physical Society, December 29, 1959**

When a large ferromagnetic crystal is cooled from above its Curie temperature in a field-free space, the demagnetized state is the stable state. In the demagnetized state, the magnetic domains are randomly oriented so that the magnetic poles are eliminated from the surface. The magnetic flux lies almost entirely within the specimen, minimizing the magnetostatic energy. As the dimensions of a specimen are diminished, the relative contributions of volume and surface to the magnetostatic energy are altered. At very small particle sizes, volume energy contributions decrease relatively faster than surface energy contributions, as seen in Figure 4.1.

The energy associated with Bloch walls forming the boundaries of the domains is surface energy, which decreases as the square of the radius of the particle, while the magnetostatic self-energy is volume energy, which decreases as the third power of the particle radius. At very small dimensions, it becomes energetically favorable not to form magnetic-domain walls. Below a critical radius R_c, the specimen would exist as a single domain and act as a permanent magnet.

4.1 Critical Particle Size for Single-Domain Behavior

An exact calculation of the critical particle size below which the single-magnetic-domain state ought to be favored is challenging due to the large number of parameters involved. The magnetostatic energy depends on the saturation magnetization of the material and the geometry of the particle, while the wall energy at the interface between magnetic domains depends on the magnetocrystalline anisotropy constant and the exchange stiffness of the crystalline material. To a first approximation, one can attempt to arrive at a rough estimate of the critical size for single-magnetic-domain behavior by considering only magnetostatic and Bloch-wall energy contributions to a spherical, ferromagnetic single crystal of radius R. The magnetostatic energy density associated with the single-domain configuration is given by Eq. (3.72),

$$u(magnetostatic) = \frac{\mu_0}{2} N_d M_s^2 \tag{4.1}$$

DOI: 10.1201/9781315157016-5

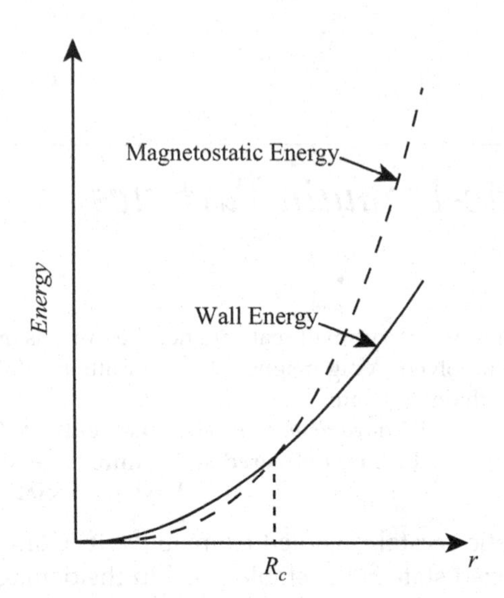

FIGURE 4.1
Magnetostatic and Bloch-wall energy dependence on particle radius.

where M_s is the saturation magnetization density and N_d is the demagnetizing factor. Using $N_d = 1/3$ for a sphere, the total magnetostatic energy of a uniformly magnetized sphere of radius R is given by

$$U(magnetostatic) = \frac{\mu_0}{6}\left(\frac{4\pi}{3}R^3\right)M_s^2 \tag{4.2}$$

For metallic Fe, $M_s = 1710$ kA/m, resulting in a magnetostatic energy density of ~6×10^5 J/m^3. For a metal iron sphere of $R = 1$ cm, for example, the corresponding energy is 2.4 joules, while for $R = 10^{-6}$ cm, the magnetostatic energy is only 2.4×10^{-18} joules. As the value of R decreases, the associated magnetostatic energy is reduced drastically.

Let us now consider the cost in surface energy associated with the formation of a single Bloch wall passing through the center of the sphere. The energy per unit surface area stored in the wall of a metal iron sphere, $\sigma_{BW}(Fe)$, using Eq. (3.68) with $z = 2$ (bcc crystal structure), is

$$\sigma_{BW}(Fe) = \pi\sqrt{AK} \tag{4.3}$$

while the area of the wall is πR^2. The simplest estimate of the critical radius for the single-magnetic-domain state, R_{SD}, can be obtained by equating the magnetostatic energy of the sphere to the surface energy of the wall. This leads to

$$\pi\sqrt{AK}\left(\pi R_{SD}^2\right) = \frac{\mu_0}{6}\left(\frac{4\pi}{3}R_{SD}^3\right)M_s^2 \tag{4.4}$$

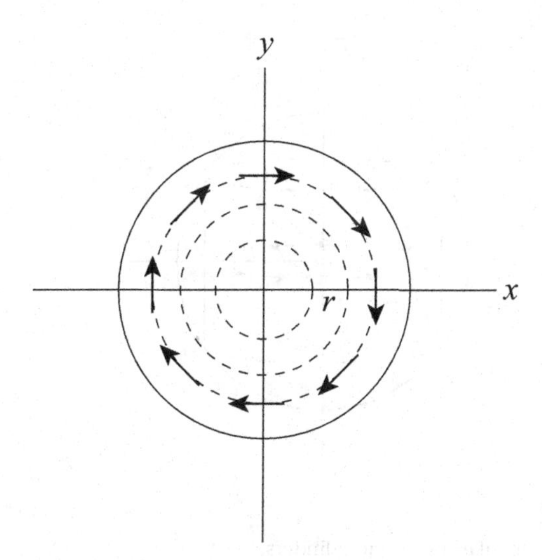

FIGURE 4.2
Schematic of flux closure configuration in a spherical particle with no or negligible anisotropy.

Solving Eq. (4.4) for R_{SD}, we obtain

$$R_{SD} \approx \frac{9\sqrt{AK}}{\mu_0 M_s^2} \text{ (strong anisotropy)} \tag{4.5}$$

For metallic iron, this equation predicts a critical particle radius of only 3 nm, which is smaller than the wall width. In the above discussion, we have made the assumption that in small particles, σ_{BW} has the same value as in a large crystal, implying that the structure of the wall is not affected by size restriction, as the size of the particle does not enter into the calculation of σ_{BW}. Such an assumption holds to a good approximation when the magnetocrystalline anisotropy is strong enough to keep $\vec{M}_s$ along the easy magnetocrystalline axis, despite the poles formed on the surface, that is, when $K \geq \mu_0 M_s^2/6$, where K is the magnetocrystalline anisotropy constant. If K is small, as in the case of iron, the magnetization will tend to follow the surface of the particle, rather than a crystallographic easy axis.

For isotropic materials, the flux-closure-domain configuration depicted in Figure 4.2 would be favored. Let us calculate the exchange energy per unit volume stored in such a domain configuration and compare it to the magnetostatic self-energy density for a magnetically saturated domain.

For a simple crystalline material with a lattice constant or interatomic separation a, there would be $2\pi r/a$ spins on a circular ring of radius r. The total change in spin angle around a ring is 2π radians, and the angle between two successive spins is $\theta = a/r$ radians. As discussed in Chapter 3, for small θ the exchange energy per pair of spins is given by $J_{ex}S^2\theta^2$, giving for the amount of exchange energy stored in a ring

$$U_{ex}(ring) = \frac{1}{2}J_{ex}S^2\left(\frac{a}{r}\right)^2\left(\frac{2\pi r}{a}\right) = \frac{\pi J_{ex}S^2 a}{r} \tag{4.6}$$

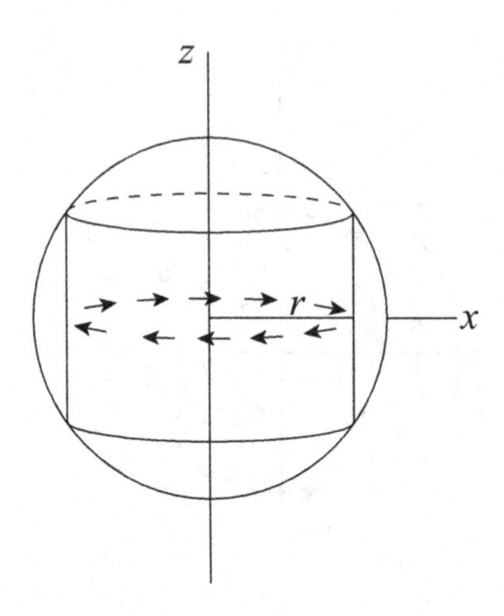

FIGURE 4.3
Decomposition of a sphere in circular concentric cylinders.

To calculate the total exchange energy we must integrate over the sphere. If we decompose the sphere into a superposition of circular concentric cylinders of radius r (Figure 4.3), each one unit cell in thickness, the number of rings per cylinder would be $2\sqrt{R^2 - r^2}/a$ and the exchange energy associated with it is

$$U_{ex}(cylinder) = \frac{2\pi J_{ex} S^2 \sqrt{R^2 - r^2}}{r} \tag{4.7}$$

The total exchange energy of the sphere is then obtained by integrating over r,

$$U_{ex}(sphere) = \frac{2\pi J_{ex} S^2}{a} \int_a^R \frac{\sqrt{R^2 - r^2}}{r} dr \cong \frac{2\pi J_{ex} S^2 R}{a} \left[\ln\left(\frac{2R}{a}\right) - 1 \right] \tag{4.8}$$

Using a, the lattice constant, as the lower bound of the integration, rather than zero, avoids the singularity introduced at the origin due to the logarithmic dependence of the result on r. Dividing by the volume of the sphere, we obtain the energy per unit volume stored as exchange energy

$$u_{ex} = \frac{3 J_{ex} S^2}{2aR^2} \left[\ln\left(\frac{2R}{a}\right) - 1 \right] \tag{4.9}$$

This energy density depends on the size of the sphere. For metallic iron, you can easily verify that for a sphere of $R = 1$ cm, the circular flux-closure domain has by far the lower energy, while for $R = 10^{-6}$ cm it is the saturated magnetic domain that has the lower energy.

Thus, for the case of single crystals with small crystalline anisotropy, the critical radius for saturated single-domain behavior would correspond to the solution of Eq. (4.10)

$$\frac{\mu_0}{6}\left(\frac{4\pi}{3} R_{SD}^3\right) M_s^2 = \frac{2\pi J_{ex} S^2 R_{SD}}{a} \left[\ln\left(\frac{2R_{SD}}{a}\right) - 1 \right] \tag{4.10}$$

where R_{SD} is the radius at which the magnetostatic energy for a magnetically saturated sphere equals the exchange energy of the sphere with a circular flux-closure-domain structure. In terms of the exchange stiffness A of a bcc crystallographic structure

$$R_{SD} = \sqrt{\frac{9A}{2\mu_0 M_s^2}\left[\ln\left(\frac{2R_{SD}}{a}\right) - 1\right]} \ (\text{weak anisotropy}) \tag{4.11}$$

Since R_{SD} appears on both sides of the equation, the above equation must be solved graphically. For a spherical Fe particle, it predicts a critical radius ~25 nm.

For prolate spheroids, the larger aspect ratio reduces the magnetostatic energy compared to the same spherical volume and introduces shape anisotropy energy. Thus, for acicular particles magnetically saturated along the major axis of the ellipsoid, the critical volume for single-domain behavior would be larger. For disk-shaped particles, with in-plane magnetic saturation, the critical radius is larger than that of a sphere made of the same material; for perpendicular magnetization, however, the magnetostatic energy is higher and the formation of walls would be favored.

The above rough calculations indicate that the critical size for single-magnetic-domain particles falls in the nanometer range for most materials, with sizes extending to ~1 μm for highly acicular particles. The critical radius for single-magnetic-domain behavior was first addressed by Kittel and by Néel in the 1940's (C. Kittel *Phys. Rev.* 70 (1946) 965, L. Néel, *Comptes Rendus* 224 (1947) 1488). More detailed calculations should additionally account for magnetoelastic energy. The latter becomes increasingly important as the size of the particle decreases due to lattice deformation at the surface and the large surface-to-volume ratio at small particle volumes.

4.2 Coercivity of Uniaxial Small Particles

In single-domain particles, the absence of domain boundaries implies that in an external magnetic field changes in magnetization cannot proceed by the energetically "easy process" of wall displacement, but must instead proceed exclusively by the energetically demanding "hard process" of the rotation of the total magnetic moment of the particle. In order to reverse the direction of the particle's magnetization vector, the magnetic orientational potential energy due to an external field must exceed the particle's effective magnetic anisotropy energy, which opposes the rotation of the magnetization vector away from the particle's easy axis of magnetization. Factors we have encountered that contribute to the effective anisotropy energy density, K_{eff}, of the particle are the magnetocrystalline anisotropy, the shape anisotropy, if the particle is elongated, and the strain anisotropy, if the particle experiences an anisotropic strain. It was first pointed out by Kittel in 1946 that increasing K_{eff}, by imposed shape or strain anisotropies, can lead to large increases in coercivity. This is of great technological value since, as we discussed in Chapter 3, it is the hysteresis loop or coercivity of the material that largely determines its technological applications. Assuming uniaxial anisotropy, we discuss below the dynamics of magnetization rotation and determine the coercive field due to magnetocrystalline, shape and strain anisotropies acting separately in the case of coherent spin rotation, whereby the magnitude of the magnetization remains constant throughout the rotational process.

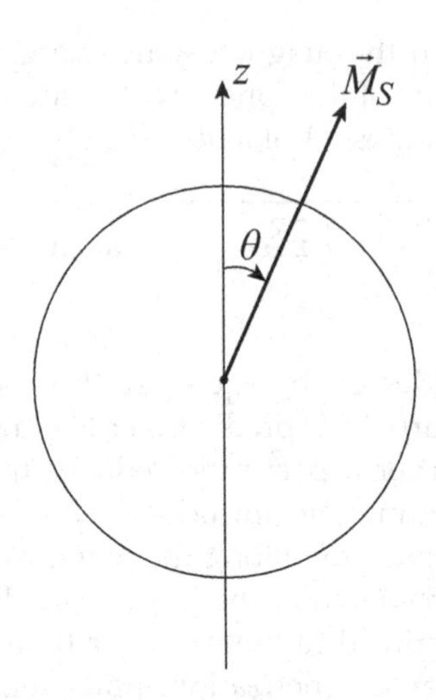

FIGURE 4.4
Schematic of the spherical sample with crystalline magnetic anisotropy axis, or easy axis of magnetization, along the z-direction and saturation magnetization deviating from the z-axis by angle θ.

4.3 Coherent Spin Rotation of Stoner and Wohlfarth Particles

4.3.1 Coercivity due to Uniaxial Magnetocrystalline Anisotropy

Using Eq. (3.61) and assuming $K_1 = K >> K_2$, the magnetic anisotropy energy density derived from uniaxial magnetocrystalline anisotropy is given by

$$u_k = K\sin^2\theta \tag{4.12}$$

where θ is the angle between the saturation magnetization vector $\vec{M}_s$ and the crystalline easy axis of magnetization assumed along the z-axis (Figure 4.4).

The magnetocrystalline anisotropy energy density is minimized when $\vec{M}_s$ points along either direction of the crystalline easy axis ($\theta = 0$ or π). The particle exhibits magnetic bistability. For any other value of θ, $\vec{M}_s$ experiences a counterclockwise, for $\theta < \pi/2$, or clockwise, for $\theta > \pi/2$, torque of magnitude

$$\tau = \left|\pm\frac{\partial u_K}{\partial\theta}\right| = 2K\sin\theta\cos\theta \tag{4.13}$$

Under the action of this torque, the magnetization vector aligns along the anisotropy axis. The magnetization dynamics would be completely equivalent if a fictitious magnetic field existed along the anisotropy axis direction exerting a similar torque on $\vec{M}_s$. This fictitious field is usually referred to as the anisotropy field and denoted by $\vec{H}_K$. The torque per unit volume due to such a field would be

$$\vec{\tau} = \mu_0\vec{M}_s \times \vec{H}_K \tag{4.14}$$

with magnitude

$$\tau = \mu_0 M_s H_K \sin\theta \tag{4.15}$$

Equating (4.13) to (4.15), we obtain

$$2K\cos\theta = \mu_0 M_s H_K \tag{4.16}$$

and for $\theta = 0$,

$$H_K = \frac{2K}{\mu_0 M_s} \quad \text{(Anisotropy field)} \tag{4.17}$$

Equation (4.17) gives the strength of the magnetic anisotropy field. It depends on the values of the uniaxial magnetocrystalline anisotropy constant and the saturation magnetization. Let us now consider an external field $\vec{H}$ applied at an angle α relative to the easy crystalline axis, as shown in Figure 4.5. The magnetization vector $\vec{M}_s$ lies in the plane defined by the direction of the applied field and the direction of the easy axis, as there is no other torque to rotate it out of this plane. The magnetic orientational potential energy density is given by

$$u_H = -\mu_0 \vec{H} \cdot \vec{M}_s = -\mu_0 H M_s \cos(\alpha - \theta) \tag{4.18}$$

While the total magnetic energy density is given by the sum of Eqs. (4.12) and (4.18)

$$u = K\sin^2\theta - \mu_0 H M_s \cos(\alpha - \theta) \tag{4.19}$$

The magnetization vector experiences two opposing torques, one due to $\vec{H}_K$ that tends to align it along the crystal axis and the other due to $\vec{H}$ that tends to rotate it away from it.

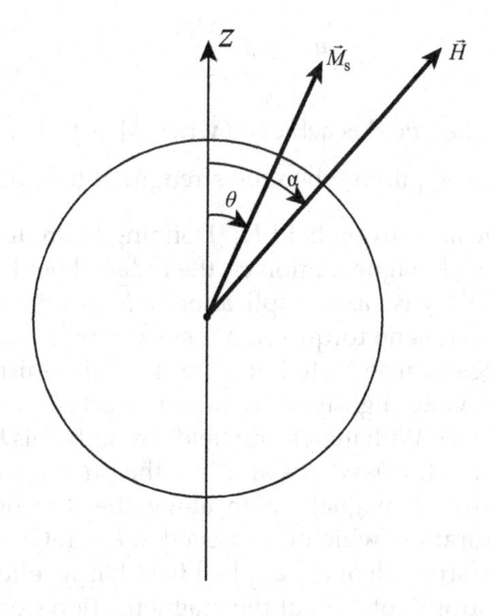

FIGURE 4.5
Schematic of geometry considered in the calculation of the coercivity of a spherical specimen due to uniaxial magnetocrystalline anisotropy.

The equilibrium condition is obtained when the magnitudes of the torques are equal or equivalently when the energy density is minimized

$$\frac{\partial u}{\partial \theta} = 2K \sin\theta \cos\theta - \mu_0 H M_s \sin(\alpha - \theta) = 0 \tag{4.20}$$

The component of the magnetization along the direction of the applied field is given by

$$M = M_s \cos(\alpha - \theta) \tag{4.21}$$

We will consider the rotational dynamics in detail for two special cases: (a) for $\alpha = \frac{\pi}{2}$, when the applied field is normal to the easy axis and (b) for $\alpha = 0$ or π, when the field is along the easy axis.

For $\alpha = \frac{\pi}{2}$, Eq. (4.20) reads

$$2K \sin\theta \cos\theta - \mu_0 H M_s \cos\theta = 0 \tag{4.22}$$

while Eq. (4.21) becomes

$$M = M_s \sin\theta \tag{4.23}$$

Using $\sin\theta = \frac{M}{M_s}$ in Eq. (4.22)

$$2K \frac{M}{M_s} = \mu_0 H M_s \tag{4.24}$$

Defining the reduced magnetization, $m = \frac{M}{M_s}$, we obtain

$$m = \mu_0 H \frac{M_s}{2K} \tag{4.25}$$

Saturation along the applied field is achieved when $\vec{M}_s$ is fully rotated into the direction of $\vec{H}$, or when $m = 1$. This is acquired when the strength of the applied field equals $\frac{2K}{\mu_0 M_s}$, which is the strength of the anisotropy field H_K. Defining the reduced field, $h = H/H_K$, we plot in Figure 4.6, the reduced magnetization *vs.* the reduced field. If initially, the magnetization vector $\vec{M}_s$ lies along the easy axis, application of $\vec{H}$ in a direction normal to the easy crystallographic direction exerts no torque on $\vec{M}_s$, since $\sin\pi/2 = 0$. However, at any finite temperature, thermal energies cause $\vec{M}_s$ to fluctuate about the anisotropy crystal axis causing the magnetization to deviate slightly away from perfect alignment, resulting in a non-vanishing torque exerted by $\vec{H}$. With increasing field strength, this torque gradually rotates the magnetization away from the easy axis and into the direction of $\vec{H}$, or the "hard axis" of magnetization. The reduced magnetization along the direction of the applied field increases gradually till saturation, which is achieved at $h = 1$. Once m has reached saturation, further increase of the strength of the applied field has no effect on m. Figure 4.6 indicates that in uniaxial anisotropy rotation of the magnetization from the "easy axis" onto a "hard axis" is reversible, no hysteresis. The reduced magnetization curve is linear with a slope of one and zero coercivity, $H_c = 0$. Starting from complete magnetic saturation at

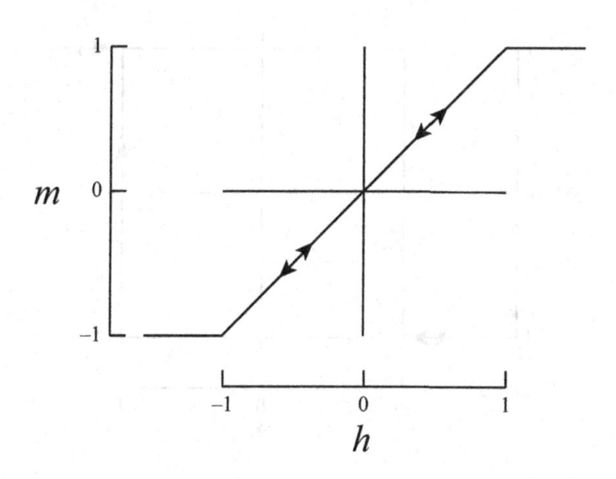

FIGURE 4.6
Reduced magnetization *vs.* reduced field; field applied in a direction perpendicular to the easy axis of magnetization ($\alpha = \pi/2$). Reversible behavior is observed.

$h = +1$, a reduced field $h = -1$ is required in order to reverse or switch $\vec{M}_s$ in the opposite direction. This is also known as the critical field for magnetization switching, H_{sw}.

For the case of $\alpha = 0$ or π, Eq. (4.19) for the energy density becomes

$$u = K\sin^2\theta \mp \mu_0 H M_s \cos\theta \tag{4.26}$$

Thus, the energy density is minimized when

$$\frac{\partial u}{\partial \theta} = 0 = 2K\sin\theta\cos\theta \pm \mu_0 H M_s \sin\theta \tag{4.27}$$

For this case, Eq. (4.21) gives

$$M = \pm M_s \cos\theta \tag{4.28}$$

Substituting $\cos\theta = \pm\frac{M}{M_s}$ into Eq. (4.27) we obtain

$$2K\frac{M}{M_s} = \pm\mu_0 H M_s \tag{4.29}$$

Indicating that saturation is reached for

$$H = \frac{2K}{\mu_0 M_s} = \pm H_K \tag{4.30}$$

where the plus sign corresponds to $\theta = 0$ and the minus sign to $\theta = \pi$. In order to appreciate the physical significance of this result, let us consider the case where initially the magnetization vector M_s points along the easy axis at $\theta = 0$ and so does the applied field $\vec{H}$. The magnetization is saturated, or $m = +1$. In this case, the applied field and anisotropy field point in the same direction as $\vec{M}_s$ and neither exerts a torque on $\vec{M}_s$. Any thermal fluctuations of the magnetization vector away from the anisotropy axis introduce torques from $\vec{H}$

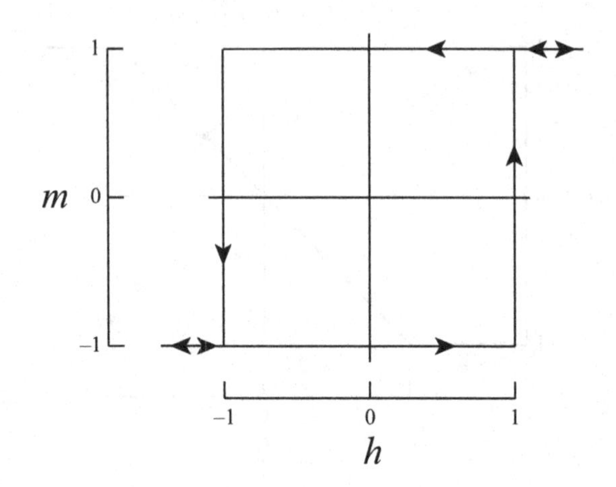

FIGURE 4.7
Reduced magnetization *vs.* reduced field; field applied in a direction parallel to the easy axis of magnetization ($\alpha = 0, \pi$). A square hysteresis loop is obtained.

and $\vec{H}_K$, both of which restore the initial equilibrium condition, maintaining magnetic saturation along the anisotropy axis. Now, consider decreasing the strength of the magnetic field to zero, while its direction is maintained parallel to the direction of the anisotropy crystal axis. The magnetization remains in stable equilibrium at $\theta = 0$. However, when the applied field increases in the negative direction, the equilibrium along $\theta = 0$ becomes unstable; any slight deviations of $\vec{M}_s$ away from the anisotropy axis would produce a torque due to $\vec{H}$ that tends to reverse the magnetization direction. If $H < H_K$, the anisotropy field maintains the initial equilibrium position. However, when the applied field reaches and exceeds the magnitude of the anisotropy field, the magnetization direction reverses abruptly and irreversibly into the negative direction. The hysteresis loop has a square shape as shown in Figure 4.7. The coercive field is then given by Eq. (4.31). In this case, the coercive and switching fields have the same value.

$$H_c = \frac{2K}{\mu_0 M_s} \tag{4.31}$$

4.3.2 Coercivity due to Particle Shape

In Chapter 3, we derived the magnetostatic energy density of a prolate spheroid with major axis *c* and minor axes *a*, as given by Eq. (3.73)

$$u(magnetostatic) = \frac{1}{2}\mu_0\left[N_c(M_s\cos\theta)^2 + N_a(M_s\sin\theta)^2\right] \tag{4.32}$$

where θ is the angle between $\vec{M}_s$ and the particle's easy axis of magnetization (the axis of revolution of the prolate spheroid) as shown in Figure 4.8.

In the presence of an applied field at an angle α with respect to the anisotropy axis, the total magnetic energy density is given by

$$u = \frac{1}{2}\mu_0 M_s^2\left(N_c\cos^2\theta + N_a\sin^2\theta\right) - \mu_0 H M_s\cos(\alpha - \theta) \tag{4.33}$$

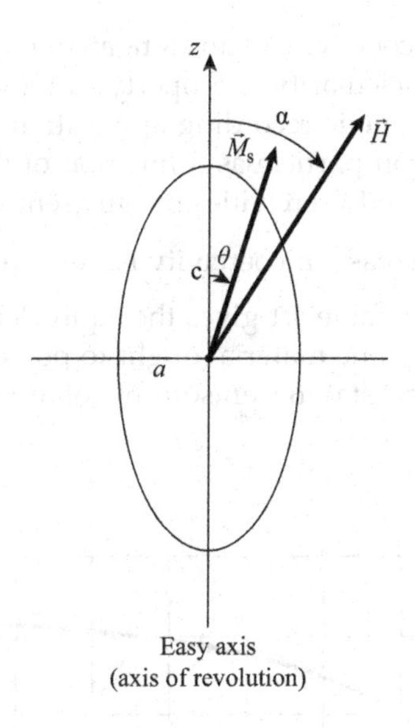

FIGURE 4.8
Schematic of geometry considered in the calculation of the coercivity due to shape anisotropy of an ellipsoidal specimen.

The first term in the equation above is the magnetostatic, self-energy density and the second term is the magnetic orientational potential energy density due to the applied magnetic field. The energy is minimized with respect to θ when

$$\frac{\partial u}{\partial \theta} = 0 = \mu_0 M_s^2 (N_a - N_c)\cos\theta \sin\theta - \mu_0 H M_s \sin(\alpha - \theta) \tag{4.34}$$

Equation (4.34) is identical to Eq. (4.20) if one identified $\mu_0 M_s^2 (N_a - N_c)$ with $2K$. The particle exhibits uniaxial anisotropy due to shape with the semi-major axis c being the easy direction of magnetization with a corresponding magnetic anisotropy energy density $K_{sh} = \frac{1}{2}\mu_0 (N_a - N_c) M_s^2$, in agreement with Eq. (3.75). This anisotropy energy density is the difference in magnetostatic energy per unit volume between a magnetically saturated specimen along its hard and easy axes. The mathematical problem is identical to that of uniaxial magnetocrystalline anisotropy with corresponding anisotropy field

$$H_a = (N_a - N_c) M_s \ \text{(Anisotropy field)} \tag{4.35}$$

The reduced magnetization, $m = M/M_s$, exhibits the same dependence on the reduced field

$$h = \frac{H}{H_a} \tag{4.36}$$

as that shown in Figures 4.6 and 4.7 for $\alpha = \frac{\pi}{2}$ and $\alpha = \pi$, respectively. The coercive field in the case of $\alpha = \pi$ is then given by

$$H_c = (N_a - N_c) M_s \tag{4.37}$$

This result indicates that the coercivity of an elongated magnetic particle can be tailored by manipulating its degree of acicularity, a property extensively utilized in the development of granular media for magnetic recording applications. Using Eq. (4.37), Figure 4.9 plots the coercive field of an iron particle as a function of the axial ratio of the particle, while Table 4.1 gives the calculated coercivities for different values of the ratio *c*/*a*, assuming $M_s = 1{,}714\frac{kA}{m}$. A 10-fold increase in coercivity is predicted as the axial ratio increases from 1 to 20. The last column of Table 4.1 gives the equivalent magnetocrystalline anisotropy a spherical particle of the same material ought to possess in order to exhibit similar coercivity. Here, the magnetocrystalline anisotropy of iron is not taken into account, assumed to be zero.

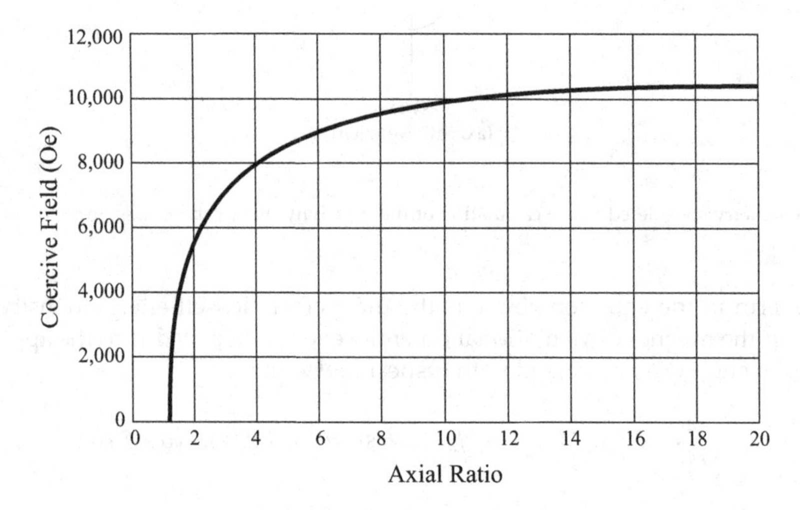

FIGURE 4.9
Coercivity of an iron particle due to shape anisotropy as a function of the axial ratio of the particle.

TABLE 4.1

Calculated Coercivities Due to Shape for Iron Single-Domain Particles as a Function of Axial Ratio and Equivalent Crystal Anisotropy of Spherical Particles

Axial Ratio *c*/*a*	$2(N_a - N_c)$	Coercive Field (Oe)	Coercive Field (kA/m)	Equivalent Crystal Anisotropy K (10^4J/m^3)
1.0	0	0	0	0
1.1	0.075	810	64.5	7
1.5	0.301	3,240	258	28
5	0.833	8,950	712	77
10	0.939	10,100	804	87
20	0.980	10,500	836	90
∞	1	10,800	859	93

Coherent spin rotation, easy axis aligned with the field.

Using the limiting values of $N_c \cong 0$ and $N_a = \frac{1}{2}$ for an infinitely long cylinder, the coercivity of a very long circular cylinder, is given by

$$H_c = \frac{M_s}{2} \text{ (long cylinder)} \tag{4.38}$$

This is the theoretical maximum value of coercivity due to shape.

The general problem of the coercivity of a prolate spheroid for an applied field at various angles α has been treated in a classic paper by Stoner and Wohlfarth (E. C. Stoner and E. P. Wohlfarth, *Phil. Trans. Roy. Soc.* (London) A-240 (1948) 599). The results are demonstrated in Figure 4.10. The Stoner and Wohlfarth model assumes that the magnitude of the magnetization vector remains constant throughout the rotational process, as we have also implicitly assumed in our discussion of coercivity. This assumption implies that all atomic moments within the particle rotate in unison. For this reason, the Stoner and Wohlfarth model is also called the "coherent spin rotation" model.

According to Figure 4.10, the reduced coercivity h_c, the value of h that reduces m to zero, decreases from 1 for $\alpha = 0°$, to 0.5 for $\alpha = 45°$, to zero for $\alpha = 90°$. The reduced switching field at which $\vec{M}_s$ flips to the opposite direction decreases from 1 for $\alpha = 0°$ to 0.5 for $\alpha = 45°$, and then increases back to 1 again for $\alpha = 90°$. For an ensemble of particles with anisotropy axes randomly oriented relative to the applied field, Stoner and Wohlfarth obtained the hysteresis loop shown in Figure 4.11. Note that in this case, the reduced magnetization is not saturated at $h = 1$, the reduced remnant magnetization, or reduced "retentively" m_r, equals 0.5 and the average reduced coercivity $\langle h_c \rangle_{Av} = 0.48$.

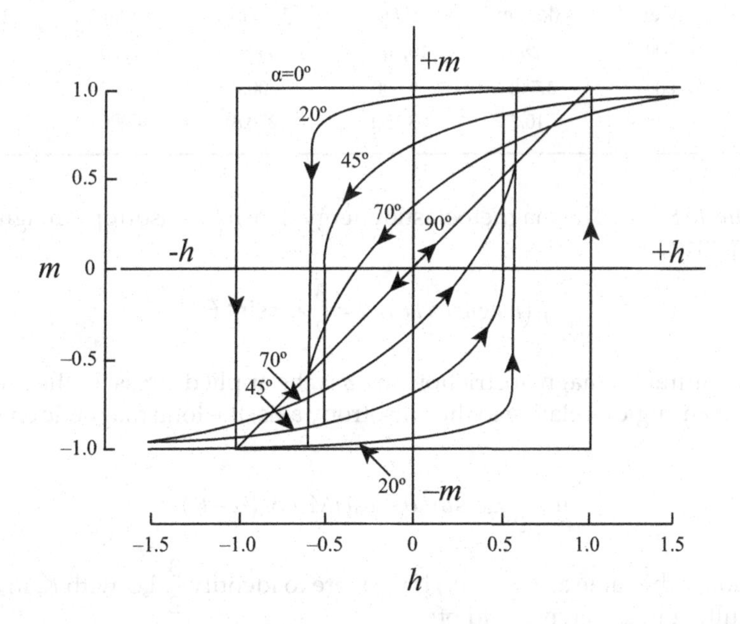

FIGURE 4.10
Hysteresis loops for coherent rotations of Stoner and Wohlfarth particles. (adapted from E. C. Stoner and E. P. Wohlfarth, A Mechanism of Hysteresis in Heterogenuous Alloys, *Phil. Trans. Roy. Soc.* (London) A-240 (1948) 599).

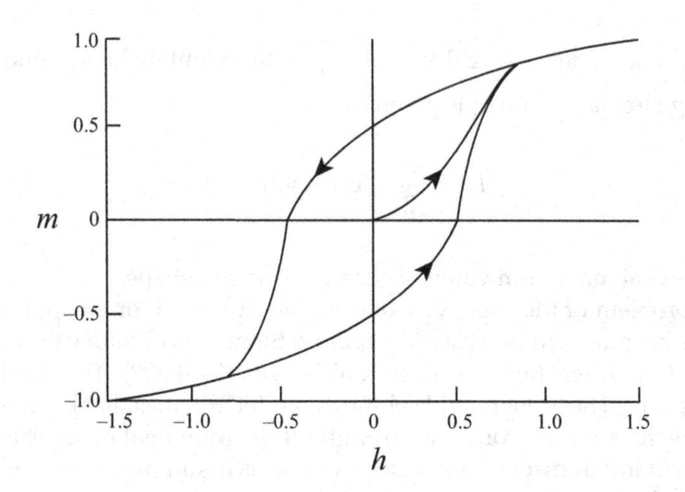

FIGURE 4.11
Hysteresis loop for an ensemble of uniaxial, non-interacting particles with random easy axes directions (according to Stoner and Wohlfarth).

TABLE 4.2
Maximum Coercivity of Small Particles Due to Magnetocrystalline, Shape and Stress Anisotropy, According to Stoner and Wohlfarth Coherent Spin Rotation Process

	Crystal Anisotropy		**Shape**		**Stress($\sigma = 2 \times 10^9$N/m²)**	
Expression	$H_c = \dfrac{2K}{\mu_0 M_s}$		$H_c = \dfrac{M_s}{2}$		$H_c = \dfrac{3\lambda_s\sigma}{\mu_0 M_s}$	
	(Oe)	(kA/m)	(Oe)	(kA/m)	(Oe)	(kA/m)
Fe	500	39.8	10,700	852	600	47.8
Co	6,000	477.8	8,800	700.7	600	47.8
Ni	135	10.7	3,150	250.8	4,000	318.5

Coercivity due to Stress. The magnetoelastic energy density for isotropic magnetostriction is given by Eq. (3.84)

$$u(magnetoelastic) = \frac{3}{2}\lambda_s\sigma\sin^2\theta$$

where λ_s is the saturation magnetostriction and σ is the applied stress. In the presence of an external field at an angle α relative to the anisotropy axis, the total magnetic energy density is given by

$$u = \frac{3}{2}\lambda_s\sigma\sin^2\theta - \mu_0 H M_s\cos(\alpha - \theta) \tag{4.39}$$

This equation is the same as Eq. (4.19) if we were to identify $\frac{3}{2}\lambda_s\sigma$ with K, in accord with Eq. (3.84), resulting in a coercive field of

$$H_c = \frac{3\lambda_s\sigma}{\mu_0 M_s} \tag{4.40}$$

for $\alpha = 0$ or π. Table 4.2 gives the maximum coercivities for Fe, Co and Ni for the three causes of coercivity we have considered separately. In real systems, all three causes of anisotropy may co-exist.

The above considerations imply that the materials engineer has at his/her disposal, the means to design single-domain-magnetic nanoparticles with tailored coercivities to meet specific technological demands, provided that these fine particles when embedded into some support or matrix for device applications do not lose their designed coercive properties.

4.4 Non-Coherent Spin Rotation Modes

The large coercivities calculated in Tables 4.1 and 4.2 all assume the Stoner and Wohlfarth model of coherent spin rotation. In real single-domain particles, other non-coherent rotation modes may occur, known as "curling" and "buckling" as depicted schematically in Figure 4.12. In the coherent rotation mode, the spins rotate in unison keeping the magnetization constant at its saturation value through the flipping process. This costs magnetostatic energy because the stray fields increase as the moment flips through the hard axis configuration, Figure 4.12(a). Reversal by curling avoids the creation of stray fields by passing through a vortex state where the magnetization lies parallel to the surface of the particle during the flipping process, Figure 4.12(b). This, however, costs exchange energy. The buckling process, Figure 4.12(c), may occur in long prolate spheroids and creates less stray fields than the coherent rotation process.

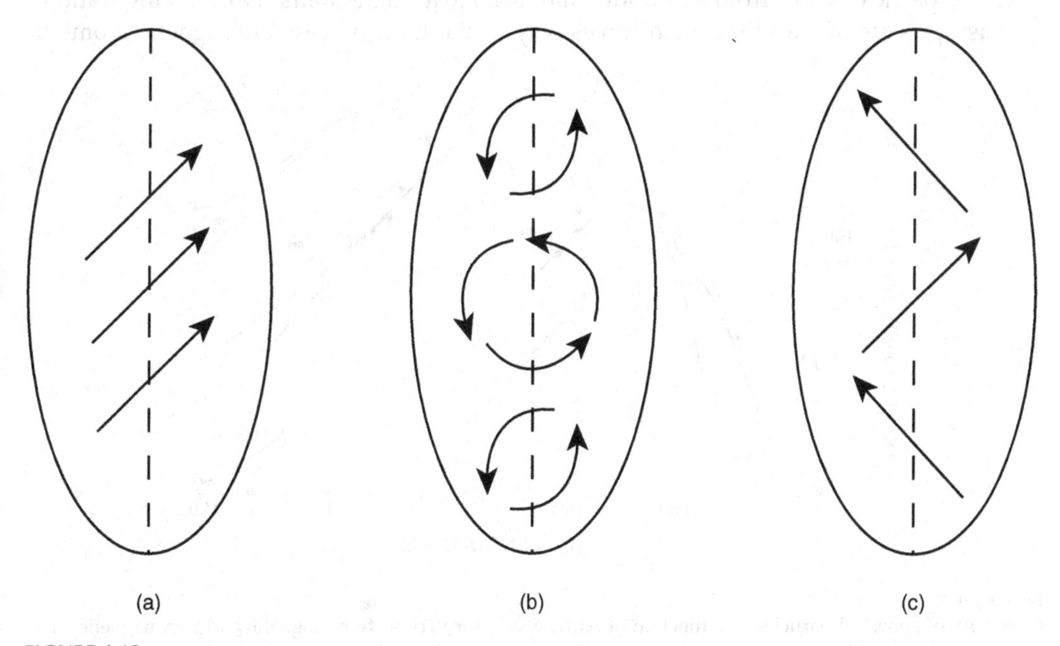

FIGURE 4.12
Schematic representation of coherent (a), curling (b) and buckling (c) spin reversal mechanisms in elongated particles.

The simplest mode of magnetization reversal is the coherent rotation mode used by Stoner and Wohlfarth in calculating the coercivities of single-domain particles. It is often the favored theoretical framework as it offers an analytical solution to the problem of coercivity. However, the actual magnetization reversal does not necessarily have to be coherent.

4.5 Size Dependence of Coercivity

Measured physical quantities whose values depend on the size of the specimen are said to exhibit "finite-size effects". Experimental observations show that the coercivities of small particles exhibit a strong dependence on particle size. The coercivity typically increases with decreasing size passing through a maximum before it decreases to zero at very small particle diameters. Figure 4.13, reproduced from the classic paper by Luborsky (F. E. Luborsky, *J. Appl. Phys.* 32 (1961) 171S), shows the correlation between diameter and coercive force for particles deriving their coercivity primarily from magnetocrystalline anisotropy energy. All data were collected at room temperature except for the data on Fe and Co which was taken at T = 76 K, as indicated.

Particle diameters range from a few angstroms to 100 μm, while measured coercivities range from ~20 Oe to about ~10,000 Oe, over three orders of magnitude. For the particle size range shown, the curves for iron, cobalt and cobalt ferrite go through a maximum at the vicinity of ~300Å, while for magnetite, $BaO{\cdot}6Fe_2O_3$ and MnBi, the coercivity increases steeply with decreasing particle size. This characteristic behavior is understood on the basis of multidomain *vs.* single-domain behavior, as depicted schematically in Figure 4.14.

As the particle passes from a multidomain (MD) to a single-domain (SD) configuration, the easy process of wall movement gives way to the hard process of magnetic moment

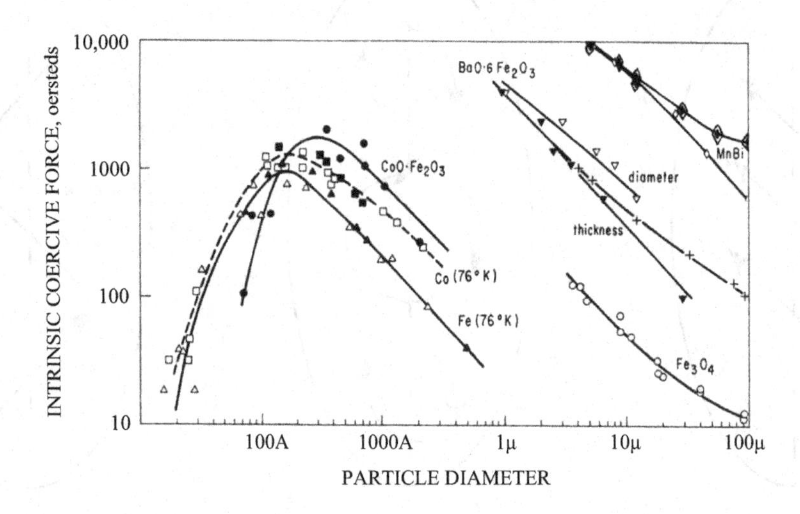

FIGURE 4.13
Coercivity of spherical particles as a function of particle size, for various ferromagnetic and ferrimagnetic substances, deriving their coercivity primarily from magnetocrystalline anisotropy energy. (A = Angstroms, μ = microns). Adapted from F. E. Luborsky, High Coercive Materials, Development of Elongated Particle Magnets. *J. Appl. Phys.* 32 (1961) S171, (Copyright American Institute of Physics, with permission).

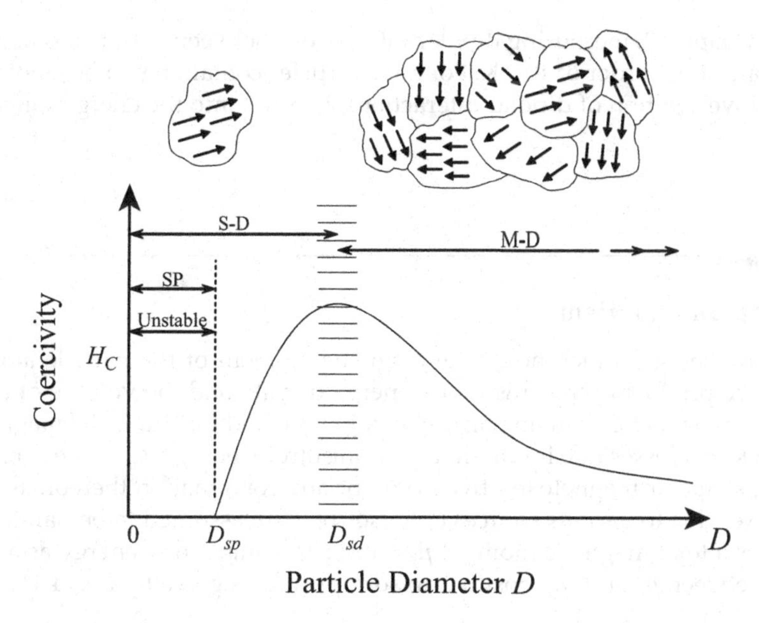

FIGURE 4.14
Schematic representation of the dependence of coercivity on particle diameter (M-D, multidomain, S-D, single-domain, SP, superparamagnetic).

rotation, with concomitant increase in coercivity. The coercivity continues to increase rapidly with decreasing size, reaching a maximum value at the single-domain-particle diameter D_{sd}. Further decrease in diameter induces a steep decrease in coercivity till a value D_{sp} where the coercivity vanishes. The exact value of D_{sp} depends on the temperature of measurement, and the experimental technique used to detect the magnetization, as we discuss in greater detail in the following section. Below D_{sp}, the particle is magnetically unstable due to thermally induced reversals of the magnetization density even in the absence of an external field. This behavior is known as "superparamagnetism".

It is important to note that the maximum coercivity measured for small particles of metallic iron and cobalt shown in Figure 4.13 is about 1,000 Oe. This is higher than that predicted by the Stoner and Wohlfarth model for Fe but much lower than that predicted for Co, due to crystal anisotropy alone (Table 4.2). This indicates that in real experimental samples, contributions from all three sources of coercivity in Table 4.2 could contribute which may reinforce or oppose each other. In addition, the rotation of the magnetization may not be coherent. The large coercivities calculated in Tables 4.1 and 4.2, all assume the Stoner and Wohlfarth model of coherent spin rotations. In real particles, the non-coherent rotation modes of Figure 4.12 can also occur, both of which lower the maximum attainable coercivity. Thus, the magnetization reversal process is generally complex, involving both coherent and non-coherent spin reversal processes. In the coherent process, the particle remains uniformly magnetized during magnetization reversal, while in the incoherent reversal process, the particle passes through an intermediate state of non-uniform magnetization.

The presence of interparticle interactions, ignored in the Stoner and Wohlfarth model, would also tend to reduce the measured coercivity. Particles in contact tend to interact magnetically through two distinct processes: (a) short-range exchange interactions through grain boundaries and (b) long-range dipole–dipole magnetostatic interactions. As we discussed in Chapter 2, measuring dipolar interactions between atomic moments requires an energy scale of the order of 10^{-3} K. For nanoparticles containing an order of magnitude of 10^5 spins, investigation of dipolar interactions is raised into the energy range of tens of Kelvin.

4.6 Superparamagnetism

The anisotropy energy, which holds the magnetic moment of the particle along an easy direction, is the product of the anisotropy energy density and the volume of the particle. That is, a single-magnetic-domain particle of volume V and total uniaxial magnetic anisotropy density K_u possesses maximum magnetic anisotropy energy K_uV. The constant K_u can be of crystal, shape or magnetostrictive origin or any combination thereof, as long as the anisotropy axes due to various sources of anisotropy are assumed to be parallel. The particle possesses a total magnetic moment $\vec{\mu}_p = V\vec{M}_s$. The anisotropy energy as a function of θ, the angle between $\vec{\mu}_p$ and the positive anisotropy axis, is given by Eq. (4.41) and is plotted in Figure 4.15.

$$E_a(\theta) = K_uV\sin^2\theta \tag{4.41}$$

The orientational potential energy landscape exhibits two energy minima, at $\theta = 0$ and $\theta = \pi$. At 0 K, $\vec{\mu}_p$ would lie along either direction of the anisotropy axis. An ensemble of spatially dispersed identical particles cooled to zero temperature in the absence of an external magnetic field would exhibit zero total magnetization, as the moment of each particle has an equal probability of pointing along the $\theta = 0$ or $\theta = \pi$ direction of the anisotropy axis. The particles exhibit magnetic bistability and they are thus trapped, in equal numbers, within one of the two potential energy wells of Figure 4.15 (solid line), unless there were mechanisms that allowed the moment to surmount the energy barrier $\Delta = K_uV$ or tunnel through it. Both processes have been observed to occur. Quantum tunneling of

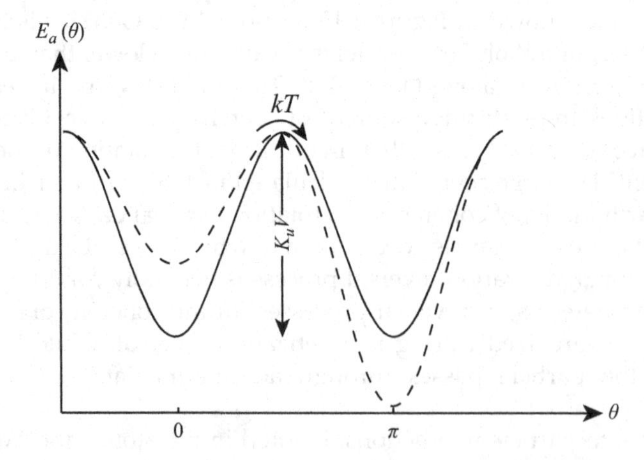

FIGURE 4.15
Orientational potential energy landscape of a particle of uniaxial anisotropy in zero (solid line) and in an external magnetic field (broken line) applied along the easy axis of magnetization.

magnetization in small magnetic particles, a process known as "macroscopic quantum tunneling" or "MQT", has been observed and is of great theoretical interest (E. M. Chudnovsky and L. Gunther, Phys. Rev. Letts. 60 (1988) 661) but of low probability of occurrence and can only be observed at extremely low temperatures. Of practical interest are moment reversals over the energy barrier activated by thermal energies $kT > K_uV$. In zero fields, the energy barrier is symmetrical for transitions in either direction between the two energy minima. Application of an external field changes the energy landscape making the potential well in the direction of the applied field deeper (Figure 4.15 broken line) and the energy barrier for moment reversals asymmetric $\Delta^{\pm} = K_uV \pm \mu_0\mu_pH \cos\theta$. A particle ensemble cooled in the presence of a magnetic field will exhibit magnetization as the deeper well has a higher probability of been populated. Zero-field-cooled and field-cooled processes are widely utilized in order to probe the characteristics of the magnetic potential energy landscape in small particle magnetism.

As the volume is reduced, the energy barrier K_uV becomes comparable to ambient thermal energy kT. Tiny magnetic particles become unstable to thermally driven spontaneous reversal of the magnetization. If the particle diameter is small enough, less that D_{sp} in Figure 4.14, rapid moment fluctuations between $\theta = 0$ and $\theta = \pi$ can occur at room temperature. An attempted measurement of the magnetization of the sample yields its average over the time needed to take the measurement. A typical DC magnetometer records the magnetization of a sample in a characteristic measuring time $\tau_m \approx 100$s. This is a relatively long time. Other spectroscopic techniques used in magnetic investigations of nanoparticles, as, for example, AC susceptibility, Mössbauer, neutron scattering and ferromagnetic resonance have much shorter characteristic measuring times that range from 10^{-2} to 10^{-12} s. We review these experimental techniques and their application to studies of nanomagnetism in Part III of this book. If the frequency f of moment fluctuations between the two potential wells is rapid, the relaxation time $\tau = 1/f$ can be very short. For $\tau < \tau_m$, the particle moment fluctuates during the time of the measurement; the particle exhibits "superparamagnetism" with no hysteresis.

Let us consider an ensemble of identical, superparamagnetic particles dispersed in a non-magnetic medium or matrix so that interparticle dipole–dipole interactions may be neglected. At finite temperatures, an applied magnetic field will tend to align the particle moments along its direction, while thermal energy will tend to disalign them. The ensemble will exhibit paramagnetic behavior as discussed in Chapter 2, with one notable difference; the atomic moments μ have now been replaced by the particle moments μ_p in the ensemble. Atomic moments are of the order of a few Bohr magnetons, while particle moments are typically of the order of several thousands of Bohr magnetons. A spherical particle of metallic iron ~5 nm in diameter contains ~ 5,500 iron atoms, resulting in a particle moment of $\mu_p \approx 12,100\mu_B$. Compare this with the moment $\mu = 2.22\mu_B$ per iron atom in metallic iron. In all practical purposes, the ferromagnetic particle acts as a giant paramagnetic "macro-spin" or "super-spin", from which the term "superparamagnetism" was coined by Bean (C. P. Bean and J. M. Livingston, *J. Appl. Phys.*, 30 (1959) 120S) in order to stress the similarities and differences in the two cases.

If the particles are isotropic, that is if $K_u = 0$, the particle moments are free to point into any direction. Upon application of a magnetic field, the magnetization of the assembly would obey the classical theory of paramagnetism developed by Langevin (Eq. (2.50)) since the situation corresponds to the classical limit of infinite spins

$$\frac{M}{M_{sat}} = L(x) = \coth x - \frac{1}{x} \tag{4.42}$$

In Eq. (4.42), M is the magnetization of the particle assembly, M_{sat} is its saturation magnetization given by $M_{sat} = N\mu_p$, where N is the number of particles per unit volume, and the dimensionless variable $x = \mu_0\mu_p H/kT$. Due to the giant value of μ_p relative to that of an atomic moment, the variable x can attain large values at ordinary fields and temperatures. As a consequence, the full Langevin magnetization curve up to saturation can easily be observed for an ensemble of superparamagnetic particles, unlike ordinary paramagnets where extraordinarily high fields and low temperatures are required to observe saturation. The low field magnetic susceptibility follows the Curie Law of Eq. (2.63) with μ_p replacing the atomic effective magnetic moment μ_{eff}

$$\chi = \frac{M}{H} = \frac{N\mu_0\mu_p^2}{3kT} \tag{4.43}$$

If the particles have finite anisotropy and the easy axes in an ensemble are all aligned in the direction of an applied field, the "macro-spin" moments could only point along or opposite to the magnetic field, reminiscent of a $J = 1/2$ quantum system. The magnetization follows the Brillouin function, Eq. (2.65) with $J = 1/2$ given by

$$\frac{M}{M_{sat}} = \tanh x \tag{4.44}$$

In the general case of non-aligned anisotropic particles, the problem is not amenable to a simple theoretical functional dependence. In addition, most experimental systems contain a distribution of particle sizes further complicating the development of an exact theoretical framework.

Moment fluctuations between the two potential energy wells of Figure 4.15 is a spin relaxation phenomenon first studied in detail by Néel who developed in 1949, a theory of superparamagnetic relaxation of ferromagnetic fine particles (L. Néel, *Ann. Géophys.* 5 (1949) 99), and extended it in 1961 to antiferromagnetic fine particles (L. Néel, *CR Acad. Sci.* 252 (1961) 4075). Néel proposed that the spin reversal frequency is determined by the product of an attempt frequency f_0 and the Boltzmann probability factor $\exp(-K_uV/kT)$ that the particle has the thermal energy necessary to surmount the barrier. Then, the relaxation time τ which is the inverse of the frequency is given by

$$\tau = \tau_0 \exp\left(\frac{K_uV}{kT}\right) \tag{4.45}$$

where $\tau_0 = \dfrac{1}{f_0}$ is the characteristic attempt time for moment reversal, of the order of 10^{-9} to 10^{-12} s, depending on the material. The fluctuations, therefore, slow down and the relaxation time τ increases rapidly with decreasing temperature, as shown in Figure 4.16.

When the relaxation time becomes much longer than the characteristic measuring time ($\tau > \tau_m$), the magnetization of the particle ensemble appears static, in the sense that thermally driven moment reversals are "blocked" from occurring during the time of the measurement; the magnetization curve becomes hysteretic. There exists, therefore, a temperature T_B at which the moment relaxation time becomes equal to the measuring time, $\tau = \tau_m$, as determined by Eq. (4.46)

$$\tau_m = \tau_0 \exp\left(\frac{K_uV}{kT_B}\right) \tag{4.46}$$

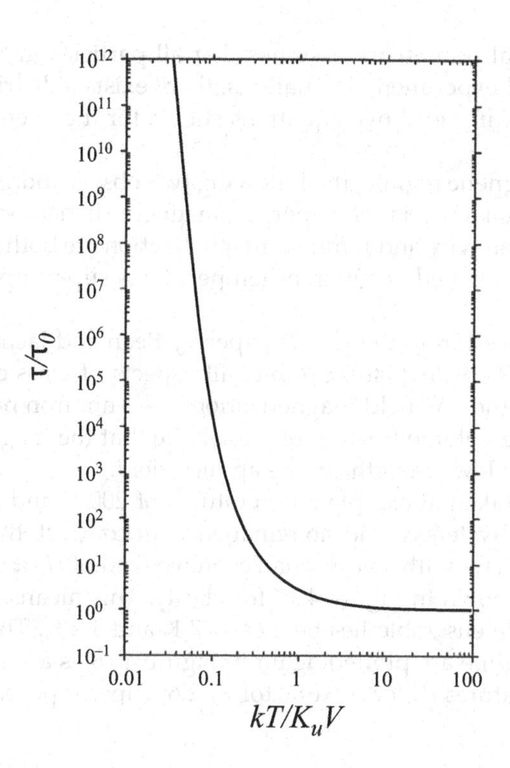

FIGURE 4.16
Scaled temperature dependence of the relaxation time as a function of temperature according to Eq. (4.45).

Solving for T_B, we obtain

$$T_B = \frac{K_u V}{k \ln(\tau_m / \tau_0)} \tag{4.47}$$

Equation (4.47) defines the "blocking temperature" of the particle ensemble for a technique with characteristic measuring time τ_m. Below T_B, thermally induced moment reversals are blocked and the magnetization becomes hysteretic. Thus, there exist three temperature regions with distinctly different magnetic behavior for the particle ensemble. At very low temperatures and up to the blocking temperature ($0 < T < T_B$), the magnetization is hysteretic; at temperatures above the blocking temperature but below the Curie temperature of the material ($T_B < T < T_C$), the magnetization is superparamagnetic, and above the Curie temperature ($T > T_C$), the magnetization is paramagnetic.

Equation (4.45) can alternatively be used to determine the volume V_{sp} at which a particle ensemble becomes superparamagnetic at a specific temperature T.

$$\tau_m = \tau_0 \exp\left(\frac{K_u V_{sp}}{kT}\right) \tag{4.48}$$

$$V_{sp} = \frac{kT}{K_u} \ln\left(\frac{\tau_m}{\tau_0}\right) \tag{4.49}$$

The above theoretical framework assumes that all particles in the ensemble have the same volume V. In real experimental situations, there exists a distribution of particle volumes, and, therefore, V in the above equations stands for the average particle volume in the distribution.

In the superparamagnetic regime, the following two observations always hold and constitute the experimental criteria of superparamagnetic behavior: (a) no hysteresis is observed, that is the coercivity and remnant magnetization are both zero and (b) measured magnetization curves obtained at different temperatures superimpose when plotted as a function of H/T.

Figure 4.17, reproduced from the classic paper by Bean and Jacobs (C. P. Bean and I. S. Jacobs *J. Appl. Phys.* 27 (1956) 1448), exhibits all aspects of classical superparamagnetic behavior. It reports on the low field magnetization of ~4-nm iron nanoparticles dispersed in mercury at a particle volume fraction of ~2%. Note that the magnetization curves start to saturate at extremely low strengths of the applied field.

The measurements taken at sample temperatures of 200 K and 77 K show superparamagnetic behavior, no hysteresis and no remanence. In contrast, the measurements at 4.2 K show hysteretic behavior with a very small coercive field of $H_C \approx 0.3 \times 10^{-3}$ Oe. Only half the hysteresis loop is shown in Figure 4.17 for clarity. This means that the blocking temperature for the particle ensemble lies between 77 K and 4.2 K. The data obtained in the superparamagnetic regime are plotted again in Figure 4.18 as a function of H/T. The two curves, whose temperatures differ by a factor of 2.6, superimpose. This observation was

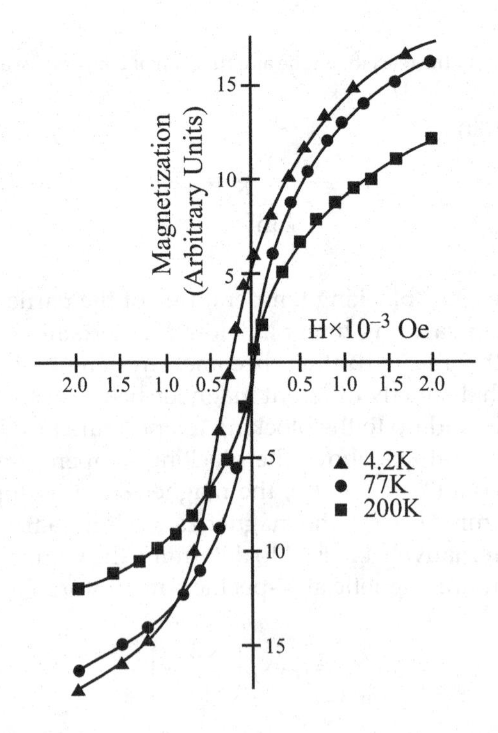

FIGURE 4.17
Low field portion of magnetization curves of an iron amalgam composed of 2% Fe by volume in the form of iron particles with diameters estimated to be in the 3–4.5 nm range. For clarity, only one branch of the hysteresis loop obtained at 4.2 K is shown. (From Bean and Jacobs, Magnetic Granulometry and Super-Paramagnetism, *J. Appl. Phys.* 27 (1956) 1448, Fig. 1, Copyright American Institute of Physics, reproduced with permission).

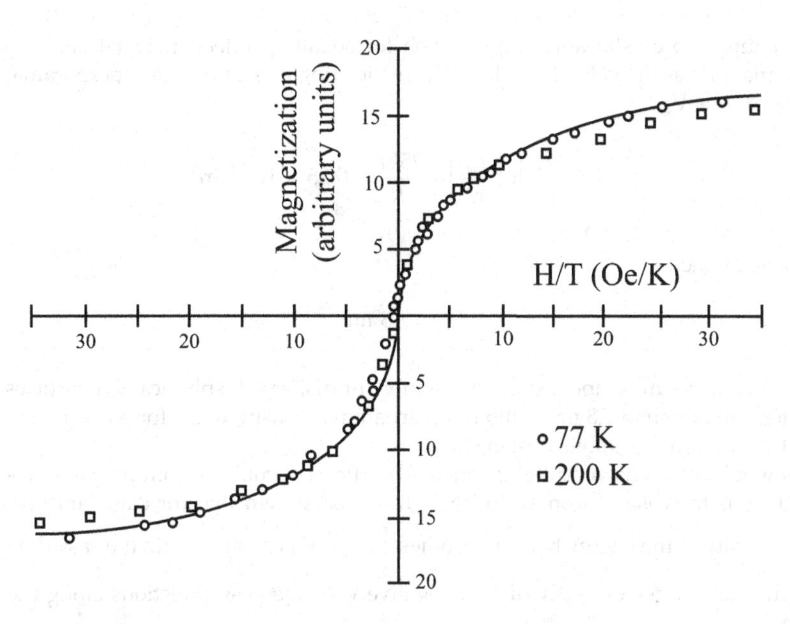

FIGURE 4.18
Proof of superparamagnetic behavior. Magnetization data for $T = 77$ and 200 K plotted as a function of H/T superimpose. (From C. P. Bean and I. S. Jacobs, Magnetic Granulometry and Super-Paramagnetism, *J. Appl. Phys.* 27 (1956) 1448, Fig. 4, Copyright American Institute of Physics, reproduced with permission).

the first graphic demonstration of superparamagnetism and the first experimental verification of Néel's theory of superparamagnetic relaxation of the Arrhenius law of Eq. (4.45).

We will discuss superparamagnetic relaxation further and how blocking temperatures are experimentally determined in Chapter 9, where different magnetic characterization techniques are reviewed. What you should appreciate at the moment is that T_B does not correspond to a change in any physical property of the ferromagnetic particles. It is not a fundamental physical quantity such as the Curie temperature of a ferromagnet (or the Néel temperature of an antiferromagnet) at which a magnetic phase transition takes place leading to spontaneous spin ordering. It is instead a property of the particle ensemble as a whole, relative to the magnetic measurement technique used.

Example 4.1: Calculate the Limiting Superparamagnetic Particle Diameter D_{sp} for an Assembly of Spherical Cobalt Particles, Highly Dispersed in a Non-magnetic Matrix

The high dispersion of the particles ensures that interparticle interactions are negligible, making Eq. (4.49) applicable in calculating the limiting superparamagnetic volume V_{sp}, from which calculation of D_{sp} is straightforward. Since no temperature or magnetic measurement technique is specified, we assume room temperature and standard DC magnetometer characteristic measuring time $\tau_m = 100$s. The attempt time is usually taken to be $\tau_0 = 10^{-9}$s, making the logarithmic factor of Eq. (4.49)

$$\ln\left(\frac{\tau_m}{\tau_0}\right) \cong 25 \tag{4.50}$$

The particles are spherical, so there is no shape anisotropy. If we were additionally to assume that the as-deposited particles within the non-magnetic matrix are free of stress, K_u

is due entirely to crystal anisotropy, which for cobalt is indeed uniaxial and to a first approximation has the value $K = 4.1 \times 10^5$ J/m^3, leading to the limiting superparamagnetic particle volume V_{sp}

$$V_{sp} = \frac{kT}{K_u} \ln\left(\frac{\tau_m}{\tau_0}\right) = \frac{25kT}{K} = 0.25 \times 10^{-24} \text{ m}^3 \tag{4.51}$$

or particle diameter

$$D_{sp} = 7.8 \text{ nm}$$

Therefore, at room temperature, an assembly of dispersed, spherical Co particles with diameters smaller than 7.8 nm exhibits superparamagnetism, while for particle diameters larger than 7.8 nm, it exhibits hysteresis.

How would your answer differ if spherical particles of cubic anisotropy were to be considered, as is the case of iron and nickel? It can be shown that for cubic anisotropy of energy density K, the energy barrier becomes $\frac{1}{4}KV$ if K is positive with the easy directions along the cube edges or $\frac{1}{12}KV$ if K is negative with the easy directions along the cube diagonal.

Example 4.2: Due to their Magnetic Bistability, the Ground State of an Ensemble of Uniaxial Magnetic Nanoparticles has Zero Magnetization. If a Magnetization M_0 is Induced by the Application of a Magnetic Field, how Long Would it take for the Ensemble to Return Back to Zero Magnetization After the Applied Field is Removed?

The rate of magnetization decay is given by

$$-\frac{dM}{dt} = Mf_0 \exp\left(-\frac{KV}{kT}\right) = \frac{M}{\tau} \tag{4.52}$$

where M is the magnetization at a later time t, f_0 is the attempt frequency for reversals, $\exp(-KV/kT)$ is the Boltzmann probability factor that the particle has enough thermal energy to surmount the anisotropy barrier and τ is the relaxation time. Integrating, we can obtain the time dependence of the magnetization of the particle ensemble (Eq. 4.53)

$$\int_{M_0}^{M} \frac{dM}{M} = -\int_0^t \frac{dt}{\tau}$$

$$\ln\left(\frac{M}{M_0}\right) = -\frac{t}{\tau}\bigg|_0^t$$

$$M(t) = M_0 \exp\left(-\frac{t}{\tau}\right) \tag{4.53}$$

The magnetization exhibits exponential decay requiring a theoretically infinite amount of time to return to its ground state of zero magnetization. Equation (4.53) gives the physical significance of the relaxation time τ. It is the time required for the magnetization to decrease to $1/e$ or 37% of its original value.

The exponential dependence of the relaxation time on the volume of the particle and the temperature of the ensemble in Eq. (4.45) indicates that the value of τ has a much stronger dependence on V and T compared to τ_0. This allows reliable predictions to be made of the stability of magnetic nanoparticles even if τ_0 is not accurately known. A small change in particle diameter can induce a huge change in τ. For example, cobalt particles of about 6.8 nm diameter have a relaxation time at room temperature of about $\tau \cong 0.1$ s, while those of 9 nm diameter have $\tau \cong 100$ years! The degree of stability against ambient thermal demagnetization is an important factor to be considered in designing magnetic nanoparticles for applications in magnetic storage media as we discuss in Chapter 10.

4.7 Collective Magnetic Excitations

Nanoparticles with volume $V \geq V_{sp}$ cannot be thermally excited above the energy barrier needed for spin reversal. However, thermal energy can still induce fast fluctuations of the magnetization direction about the easy axis, or, equivalently, precession of the magnetization vector about the anisotropy axis at an angle θ where

$$K_u V \sin^2\theta = kT \tag{4.54}$$

The observed magnetization given by the component of the magnetization along the anisotropy axis is

$$M = M_s \cos\theta$$

For small values of θ, $\sin\theta \approx \theta$, reducing the above equation to

$$M \approx M_s\left(1 - \frac{\theta^2}{2}\right)$$

Using Eq. (4.54), we can estimate $\theta \approx \sqrt{\dfrac{kT}{K_u V}}$ leading to

$$M \approx M_s\left(1 - \frac{kT}{2K_u V}\right) \tag{4.55}$$

Equation 4.55 holds for particle volumes above the superparamagnetic limit, or temperatures below T_B. These low energy thermal excitations are known as "collective magnetic excitations" as the individual atomic moments within the particle precess in unison, or collectively, about the anisotropy axis. This process was first proposed by Mørup and Topsøe in 1976 (Mørup and Topsøe, *Appl. Phys.* 11 (1976) 63).

The coercive forces for magnetization reversals calculated in Section 4.3 correspond to maximum values at essentially absolute zero temperature. At finite temperatures, the applied field is assisted by thermal energy in reversing the magnetization vector. Thus, the reversal will occur at lower fields with increasing temperature, according to

$$H_c = \frac{2K_u}{\mu_0 M_s}\left[1-\left(\frac{25kT}{K_u V}\right)^{1/2}\right] \tag{4.56}$$

In the limiting case of $V \to \infty$ or $T \to 0$, the coercivity becomes equal to $2K_u/\mu_0 M_s$, as expected when the field is unaided by thermal energy.

4.8 Interparticle Interactions

In the above discussion, "isolated" magnetic particles have been assumed and their response to temperature and external fields has been studied. The term "isolated" refers to magnetic isolation in the sense that magnetic interactions between particles within the ensemble have been minimized by dispersing the nanoparticles within a non-magnetic medium or matrix. In practice, nanoparticulate materials used in device applications, or in the construction of nanocrystalline permanent magnets, contain nanoparticles compacted together, with or without a non-magnetic matrix. This brings the particles into close proximity, making the isolated particle picture invalid. Detailed theoretical considerations of interparticle interactions, however, present a formidable challenge. Figure 4.19 depicts schematically the situation that must be considered.

As the particles are brought closer together, in addition to the applied field, each particle also experiences the magnetic field due to neighboring particles. The closer the particles are, the stronger the dipolar interparticle interactions. A measure of the average distance between particles is given by the volume fraction of the nanoparticles in the composite material, also known as the "packing fraction" p. When the anisotropy of the particles is due primarily to shape, Néel (L. Néel, *Compt. Rend.* 224 (1947) 1550) derived the relationship for the coercivity as a function of p

$$H_c(p) = H_c(0)(1-p) \tag{4.57}$$

The above equation indicates that the coercivity decreases due to particle interactions, as often observed experimentally. The value $p = 0$ corresponds to the isolated particle. As the packing fraction increases, the particles may come into direct contact increasing the probability that exchange coupling across particle boundaries leads eventually to bulk behavior and to domain-wall motion from particle to particle. Experimentally, Eq. 4.57 has been observed to be obeyed by some nanoparticulate systems, but not all.

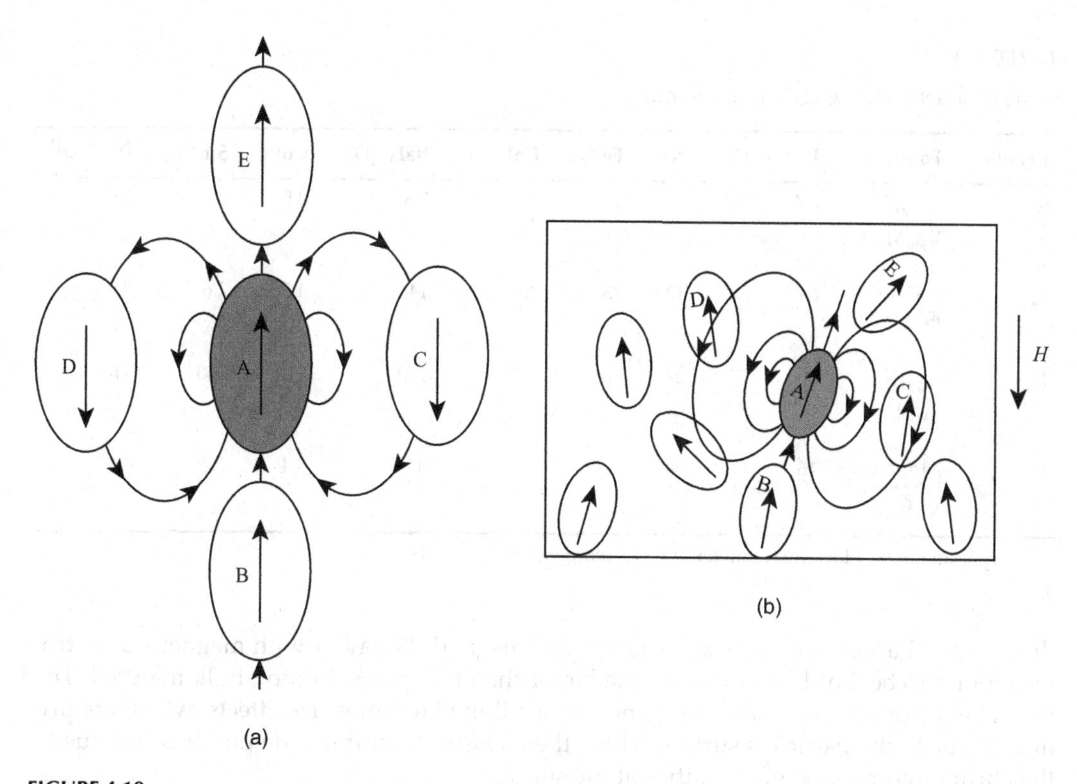

FIGURE 4.19
Schematic of an assembly of magnetic nanoparticles dispersed in a non-magnetic medium. For high particle volume fraction, in addition to the external magnetic field $\vec{H}$, each particle experiences a non-homogeneous magnetic field due to neighboring particles. (a) Schematic of arrangement of neighboring particle moments under the influence of the magnetic field due to particle A, assuming that particles are free to rotate. (b) Field due to A aids the external field H in rotating the moments of particles D and C. The opposite is true for particles B and E.

4.9 Finite-Size Effects and Characteristic Length Scales

We have established that in small particles, magnetic behavior is critically determined by the interplay between the exchange, magnetostatic and anisotropy energy densities of the material. Combinations of the various energy parameters introduce characteristic length scales, such as the exchange length, $l_{ex} = \sqrt{\frac{A}{\mu_0 M_s^2}}$, the Bloch-wall thickness, $\delta_w = \pi\sqrt{\frac{A}{K}}$ or the single-domain radius, $R_{SD} \approx \frac{9\sqrt{AK}}{\mu_0 M_s^2}$. The parameters A, K and $\mu_0 M_s^2$ represent the exchange, anisotropy and magnetostatic energy densities, respectively. Table 4.3 gives characteristic length scales for some common ferromagnetic and ferrimagnetic materials. When the particle size approaches a characteristic length scale, critical behavior in magnetic properties is expected.

Table 4.3 indicates that for most magnetic materials, the characteristic magnetic exchange length is less than 10 nm. Critical magnetic behavior is thus expected for diameters less

TABLE 4.3

Some Characteristic Length Scales in nm

Length	Formula	Fe	Co	Ni	Fe_3O_4	$CoFe_2O_4$	$BaFe_{12}O_{17}$	CoPt	$SmCo_5$	$Nd_2Fe_{14}B$
l_{ex}	$\sqrt{\frac{A}{\mu_0 M_s^2}}$	2.4	3.4	5.1	4.9	5.2	5.8	3.5	3.6	1.9
δ_w	$\pi\sqrt{\frac{A}{K}}$	64	24	125	73	20	14	4.5	2.6	3.9
R_{SD}	$\frac{9\sqrt{AK}}{\mu_0 M_s^2}$	10	56	24	38	160	280	310	560	110
D_{sp}*	$\left(\frac{12kT}{K}\right)$	16	8	34	26	10	8	3.4	2.2	3.4

* Superparamagnetic blocking diameter at room temperature.

than 10 nm. Particles of this size exhibit novel magnetic behavior with magnetic anisotropies found to be 2 orders of magnitude larger than the corresponding bulk material. The novel behavior at the nanoscale is generally attributed to finite-size effects, which are primarily due to the particle's surface where there exists structural and spin disorder due to the abrupt interruption of the lattice at the surface.

4.10 Surface Anisotropy

The importance of the surface can be demonstrated in experiments with small metallic particles ($D < 10$ nm), where it was found that their superparamagnetic properties were sensitive to different molecules chemisorbed on the surface of the particles (See for example B. S. Clausen, S. Mørup, and H. Topsøe, Influence of hydrogen chemisorption on the superparamagnetic relaxation time of Ni microcrystals *Surf. Sci.*, 82 (1979) 713). Surface anisotropy arises primarily from single-ion mechanisms, due to the coupling of surface atoms to the crystalline electric field produced by their anisotropic coordination environment. Unlike the case of bulk materials, the surface contributes to the magnetic anisotropy density of small particles. Engineering surface characteristics could therefore further govern magnetic behavior.

For spherical particles, the surface-to-volume ratio increases as $1/r$ with decreasing radius r. The crystallographic and spin–lattice structure at the surface can differ greatly from that of the bulk material due to lattice deformation and lower atom coordination symmetry at the surface. For very small particles, surface characteristics can contribute significantly to the effective magnetic anisotropy energy density of the particle. In the simplest case, surface anisotropy can be modeled by considering two contributions to the total energy barrier for magnetization reversals according to Eq. (4.58).

$$\Delta E_a = K_{eff} V = K_V V + K_S S \tag{4.58}$$

Here, K_V and K_S are the volume and surface anisotropy energy densities, respectively, and S is the surface area of the nanoparticle. For spherical particles of diameter D, we obtain

$$K_{eff} = K_V + \frac{6K_S}{D} \tag{4.59}$$

The above equation predicts effective total anisotropies that scale as $1/D$. Figure 4.20 gives experimental data of K_{eff} as a function of inverse particle diameter for metal iron particles on a carbon black support. The diameters of the particles range from 2 to 7 nm. The data follow the behavior predicted by Eq. (4.59). From the straight line fit, K_V and K_S can be evaluated. The authors of the study report $K_V = 3 \times 10^4 \text{J/m}^3$ and $K_S = 9 \times 10^{-3} \text{J/m}^2$. Surface anisotropy was first discussed by Néel, (L. Néel, *J. Phys. Radium*, 15 (1954) 255) who estimated the value of K_S to be of the order of 10^{-3} J/m^2 based on surface magnetoelastic energy, in agreement with the present results. The value of K_V is close to the magnetocrystalline anisotropy of bulk iron ($K_1 = 4.2 \times 10^4 \text{J/m}^3$).

In Fe particles with diameters as small as 2 nm, containing only about 350 iron atoms, a large fraction of atoms are found to be in a crystallographic environment similar to the bulk. Therefore, deviations of K_{eff} from those of bulk iron are presumably due to the presence of surface, shape and strain anisotropy contributions. As the surface-to-volume ratio increases with decreasing particle size, surface anisotropies can dominate the magnetic behavior of the nanoparticles. It is left for the student to prove that for nanoparticles of diameter less than 10 nm, the anisotropy energy of the particle due to the surface can exceed the value of bulk anisotropy energy by over one order of magnitude. Surface anisotropy thus becomes the most important factor after the exchange interaction in determining the magnetic properties of the particle.

Summarizing all uniaxial contributions to the effective magnetic anisotropy density of a particle encountered so far, we conclude that the total effective anisotropy constant is the sum of the magnetocrystalline anisotropy due to crystal structure, the surface anisotropy

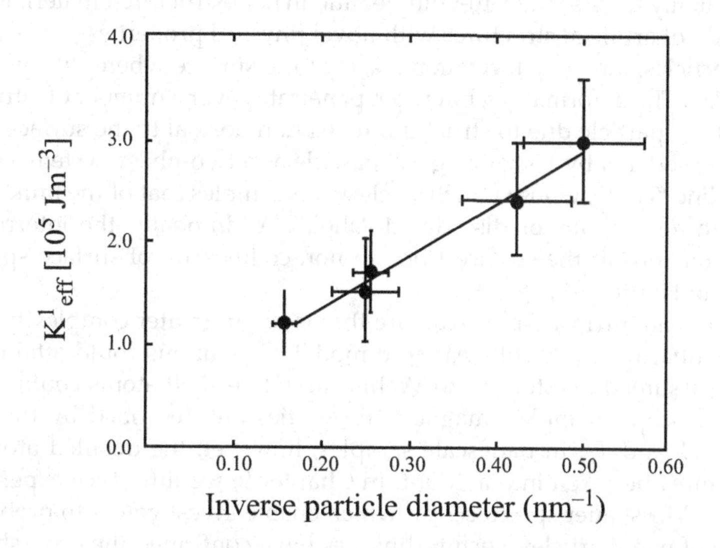

FIGURE 4.20
The magnetic anisotropy energy density of Fe nanoparticles as a function of their reciprocal diameter. (From F. Bødker, S. Mørup and S. Linderoth, Surface effects in metallic iron nanoparticles *Phys. Rev. Lett.* 72 (1994) 282, Copyright American Physical Society, reproduced with permission).

due to the lattice discontinuity at the surface, the shape anisotropy due to different demagnetization factors in different directions of a spheroid particle and the stress anisotropy due to lattice strain due to stress imposed by the support or ligands coordinated onto the surface of the particle.

$$K_{eff} = K_{mc} + K_s + K_{sh} + K_\sigma$$

The various contributions to the magnetic anisotropy density can reinforce or oppose each other. Thus, there exist multiple degrees of freedom whereby the anisotropy of nanoparticles can be controlled.

4.11 Core/Shell Nanostructures

In the case of bulk, defect-free materials, intrinsic magnetic properties, such as saturation magnetization M_s, coercivity H_c and Curie temperature T_C, depend only on the chemical composition and the crystallographic structure of the material. In bulk samples, size and surface characteristics are not of critical importance. In nanoparticulate magnetic materials, however, the magnetic properties are strongly influenced by the surface characteristics of their constituent nanoparticles.

It is generally recognized that novel properties at the nanoscale arise from the large number of atoms that lie on the surface, along grain boundaries or particle/support interfaces. Disorder at grain boundaries is due to competing interactions, broken bonds and topological disorder, surface magnetostriction and enhanced gradients of different quantities at the surface. The resulting lattice distortions at the surface trap atoms in thermodynamically non-equilibrium states not encountered in the bulk, as depicted in Figure 4.21. Thus, the unusually large surface-to-volume ratio in nanostructured materials provides for the stabilization of artificial structures with novel physical properties.

In small particles, lattice deformation starts at the surface where the symmetry of the crystal is broken. The deformation, however, penetrates over a number of lattice layers into the interior of the particle due to structural relaxation normal to the surface. These finite-size effects are modeled by visualizing the particle as a two-phase system, consisting of a highly crystalline "core", whose structure closely resembles that of the bulk material, surrounded by an amorphous or disordered "shell". Additionally, the interruption of the exchange interactions at the surface leads to non-collinearity of surface spins as shown schematically in Figure 4.22.

This produces an intrinsic spin structure that is of far greater complexity compared to the Stoner–Wohlfarth simple collinear spin model. Core atoms could attain the collinear spin structure assumed by Stoner and Wohlfarth, while shell atoms could be spin disordered. In macroscopic samples, magnetic properties are described by the macroscopic quantities M_s, H_c and T_C. In nanoscale samples, however, the detailed atomic and spin arrangement must be taken into account. In Chapter 9, we introduce experimental techniques, such as Mössbauer spectroscopy, which enable investigators to probe the internal spin structure of nanoparticles; such techniques have confirmed the core/shell picture in many experimental systems. For dimensions comparable to the characteristic length scales of Table 4.3, the fundamental magnetic properties of the material are altered due to the novel atomic-scale structures encountered only in the nanoregime.

FIGURE 4.21
Depiction of grain boundaries in a nanostructured material. (Adapted from H. Gleiter, Nanostructured materials: basic concepts and microstructure, *Acta Mater*. 48 (2000) 1, Copyright Elsevier, reproduced with permission).

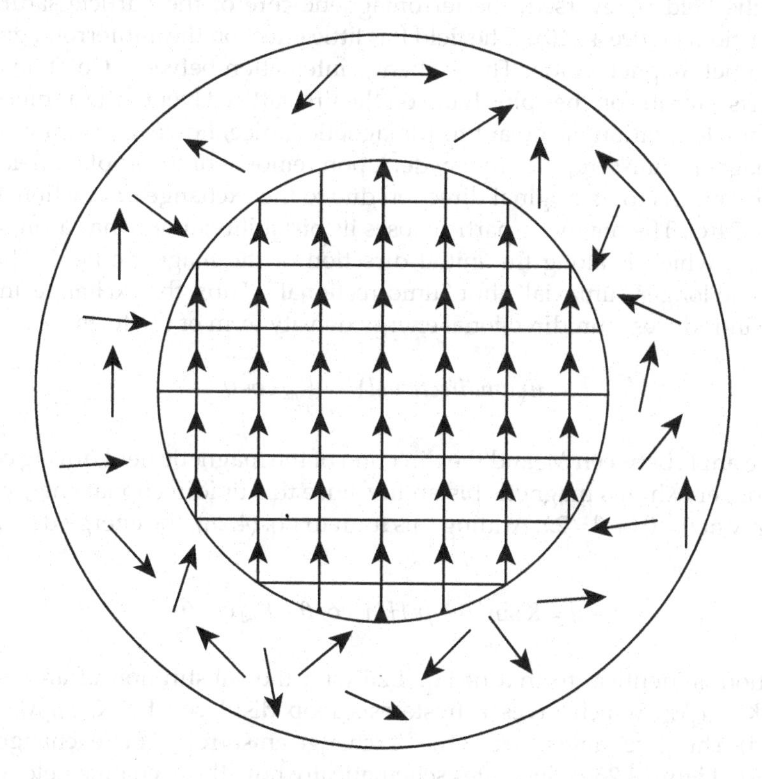

FIGURE 4.22
Schematic representation of a core/shell nanoparticle composed of a crystalline, ferromagnetically ordered interior core surrounded by an amorphous, spin-disordered shell.

4.12 Exchange Anisotropy

When metallic fine particles are exposed to air, their surface becomes oxidized resulting in core/shell nanostructures composed of a ferromagnetic metallic core surrounded by an antiferromagnetic iron oxide shell. In 1956, a new type of anisotropy was reported in a compact of core/shell nanoparticles consisting of a core of metallic cobalt ~20 nm in diameter surrounded by a thin layer of cobalt oxide (Co/CoO) (W. H. Meiklejohn and C. P. Bean, *Phys. Rev.* 105 (1957) 904). The new type of anisotropy manifested itself as a displacement of the hysteresis loop by more than 1 kOe when the material was cooled to 77 K in the presence of an applied field of 10 kOe, as shown in Figure 4.23.

The shift in the hysteresis loop was interpreted to arise from exchange coupling across the ferromagnetic/antiferromagnetic interface. It is expected to occur in structures where the Curie temperature of the ferromagnetic phase is higher than that of the Néel temperature of the antiferromagnetic phase. In this case, $T_C = 1{,}394$ K for metallic Co and $T_N = 293$ K for CoO, as indicated in Tables 3.1 and 3.2, respectively. When the temperature of the sample is cooled to low temperature from above the Néel temperature of CoO the direction of the antiferromagnetic order acquired by CoO is biased, due to the exchange interaction across the interface; the spins in the first layer of the antiferromagnetic lattice adjacent to the ferromagnetic core are aligned along the ferromagnetic axis of the core due to the positive exchange interaction between Co atoms, as shown in Figure 4.24.

Figure 4.24(a) depicts the ferromagnetic/antiferromagnetic order at the core/shell interface. When the field is reversed, the ferromagnetic core of the particle saturates in the opposite direction, Figure 4.24(b). The field has little effect on the antiferromagnetic lattice, as it has zero net magnetization. The exchange interaction between Co atoms across the interface exerts a torque on the spins lying on the first lattice layer of the antiferromagnetic oxide. A complete rotation of the antiferromagnetic lattice, however, is prevented due to the strong magnetic anisotropy of the oxide. Upon removal of the applied field, the ferromagnetic core returns to its original direction due to the exchange interaction at the interface, Figure 4.24(c). The magnetic particle loses its bistability; it now has a single direction of equilibrium, which is along the initial direction of the magnetic field. The magnetic anisotropy is no longer "uniaxial", but "unidirectional". Thus, the exchange anisotropy at the interface introduces a unidirectional energy density term of the form

$$u(unidirectional) = -K_{ud}\cos\theta \tag{4.60}$$

where θ is the angle between $\vec{M}_s$ and the direction of the magnetic field during cooling. The particle no longer exhibits magnetic bistability since this unidirectional energy term has a minimum only at $\theta = 0$ or $\theta = 2\pi$. Adding this term to Eq. (4.26), the energy density is modified to read

$$u = K\sin^2\theta + \mu_0 HM_s\cos\theta - K_{ud}\cos\theta \tag{4.61}$$

The solution is identical to that of Eq. 4.26 with the substitution of an effective field $H_{eff} = H - K_{ud}/\mu_0 M_s$, which gives a hysteresis loop displaced by $K_{ud}/\mu_0 M_s$ to the left, Figure 4.23(d). This gives a measure of the "exchange anisotropy" or "exchange bias" field $H_{ex} = K_{ud}/\mu_0 M_s$. Figure 4.23(d) indicates schematically how the exchange field and coercivity are determined experimentally from the shifted hysteresis loop. This "exchange bias" at the interface occurs irrespective of the presence or absence of the applied field upon

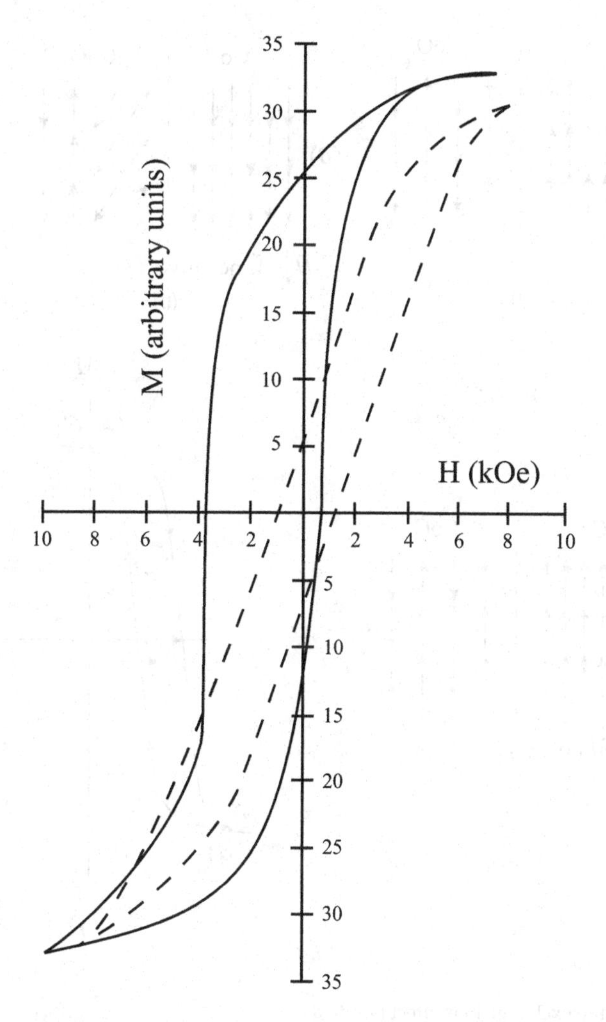

FIGURE 4.23
Hysteresis loops obtained for Co nanoparticles coated with CoO. Loop (1) (solid line) is for the sample cooled in the presence of a 10 kOe applied field. Loop (2) (dashed line) is for the sample cooled in zero field. (From W. H. Meiklejohn and C. P. Bean, New Magnetic Anisotropy, *Phys. Rev.* 105 (1957) 904, Copyright American Physical Society, reproduced with permission).

sample cooling. Due to the random orientation of the ferromagnetic axis of the particles, however, in the absence of the field, the effect is averaged out and is not observable.

Before the discovery of this effect, the known methods of manipulating the coercive force of a material were through the shape, strain and surface anisotropies, all of which are uniaxial. Exchange anisotropy affords an additional mechanism of controlling the coercivity when, for example, samples are grown in the presence of an external magnetic field. Exchange bias across interfaces adds to the materials designer toolbox for tailoring coercivities at the nanoscale.

We have introduced nanoparticle magnetism by considering increasingly smaller sample sizes of a ferromagnetic specimen. The concepts we have developed can be readily extended to ferrimagnetic nanoparticles such as particles of magnetite, maghemite and the various ferrites discussed in Chapter 3. Antiferromagnetic nanoparticles are more

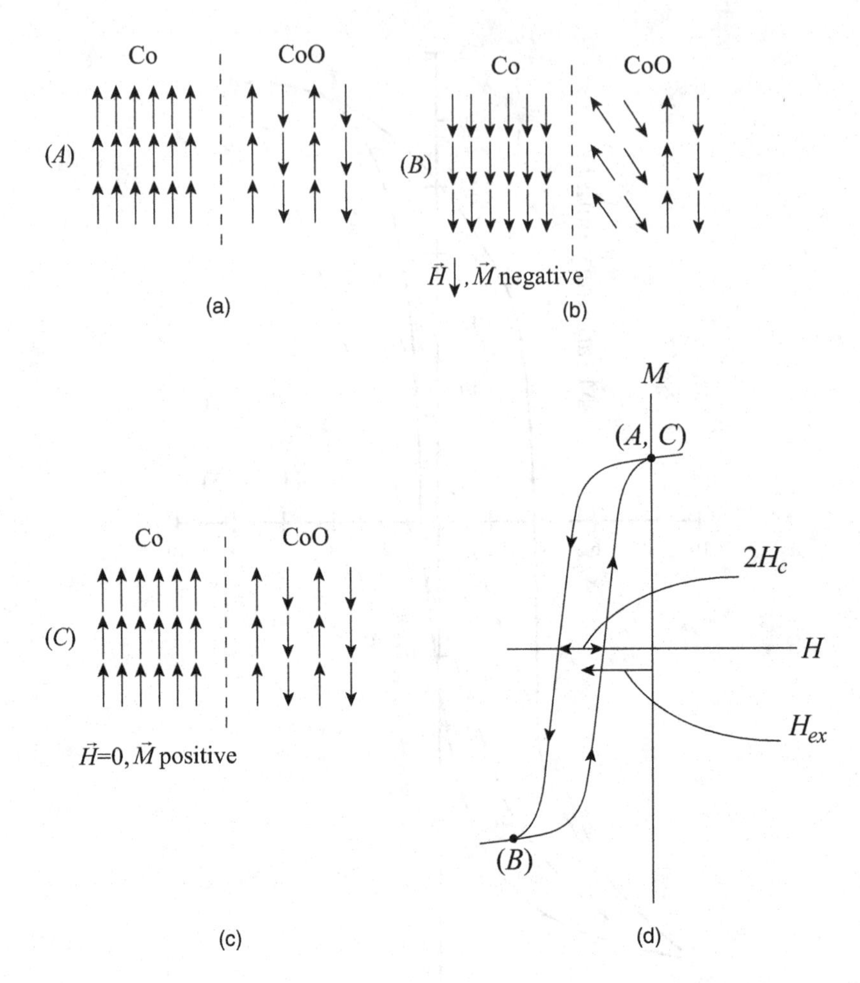

FIGURE 4.24
Schematic depiction of the exchange bias effect (see text).

complex, however, as their magnetization arises from uncompensated spins at the surface. They can be modeled by an antiferromagnetic core surrounded by a ferromagnetic shell, as originally proposed by Néel, who was the first to advance theoretical models of antiferromagnetic nanoparticles. As most metals are susceptible to oxidation, many applications in nanomagnetism make use of ferrimagnetic or antiferromagnetic oxides. In the following parts of this book, we will discuss many such systems and their applications.

4.13 Magnetic Dimensionality

Nanoparticles have all three of their dimensions reduced into the nanometer range. For this reason, nanoparticles are also known as zero-dimensional or 0-D systems, where D stands for dimensionality, not diameter. Properly coated magnetic nanoparticles can be rendered biocompatible for use in biotechnology and nanomedicine. 0-D magnetic nanocomposites can be formed by embedding the nanoparticles within a matrix. The matrix can be organic or inorganic, conducting or insulating, magnetic or non-magnetic. The resulting properties of the nanocomposite depend on the properties of the nanoparticles, the properties of the matrix chosen as the support and the details of interactions occurring along the particle/matrix interface. The properties achieved cannot be found in homogeneous bulk materials. The magnetic nanoparticles can be patterned in lines or planes, as encountered in granular magnetic recording media. Many novel properties can be designed by proper choice of materials and engineering of the interfaces between the nanoparticles and the supporting matrix, as in the case of granular materials exhibiting giant magnetoresistance. We will explore the properties and applications of such nanocomposite materials in Part IV of the book.

The magnetic nanoworld, however, is inhabited by a variety of magnetic structures not all of which are 0-D systems. If only one of the dimensions is in the range of 1–100 nm, the material is considered a two-dimensional nanomaterial or 2-D system. Magnetic "thin-films" used in modern magnetic devices are 2-D magnetic systems. Composite materials can be produced by stacking up thin magnetic films separated by non-magnetic layers known as "thin-film heterostructures" used in novel device applications such as spin valves and tunnel junctions. If two of the dimensions of a magnetic material is in the nanometer size range, it is called a 1-D system and is often referred to as "nanowire" or "nanorod". They can exist as separate entities or embedded in a matrix. Figure 4.25 depicts schematically, various low-dimensional magnetic materials. We will explore their properties in the latter chapters.

This chapter concludes our introduction to the fundamental concepts in magnetism and magnetic materials necessary for us to appreciate the diverse applications of nanomagnetism. In Part II, we turn our attention to the physical and chemical methods used in producing and stabilizing magnetic nanoparticles and nanocomposites for various applications in nanotechnology and nanomedicine.

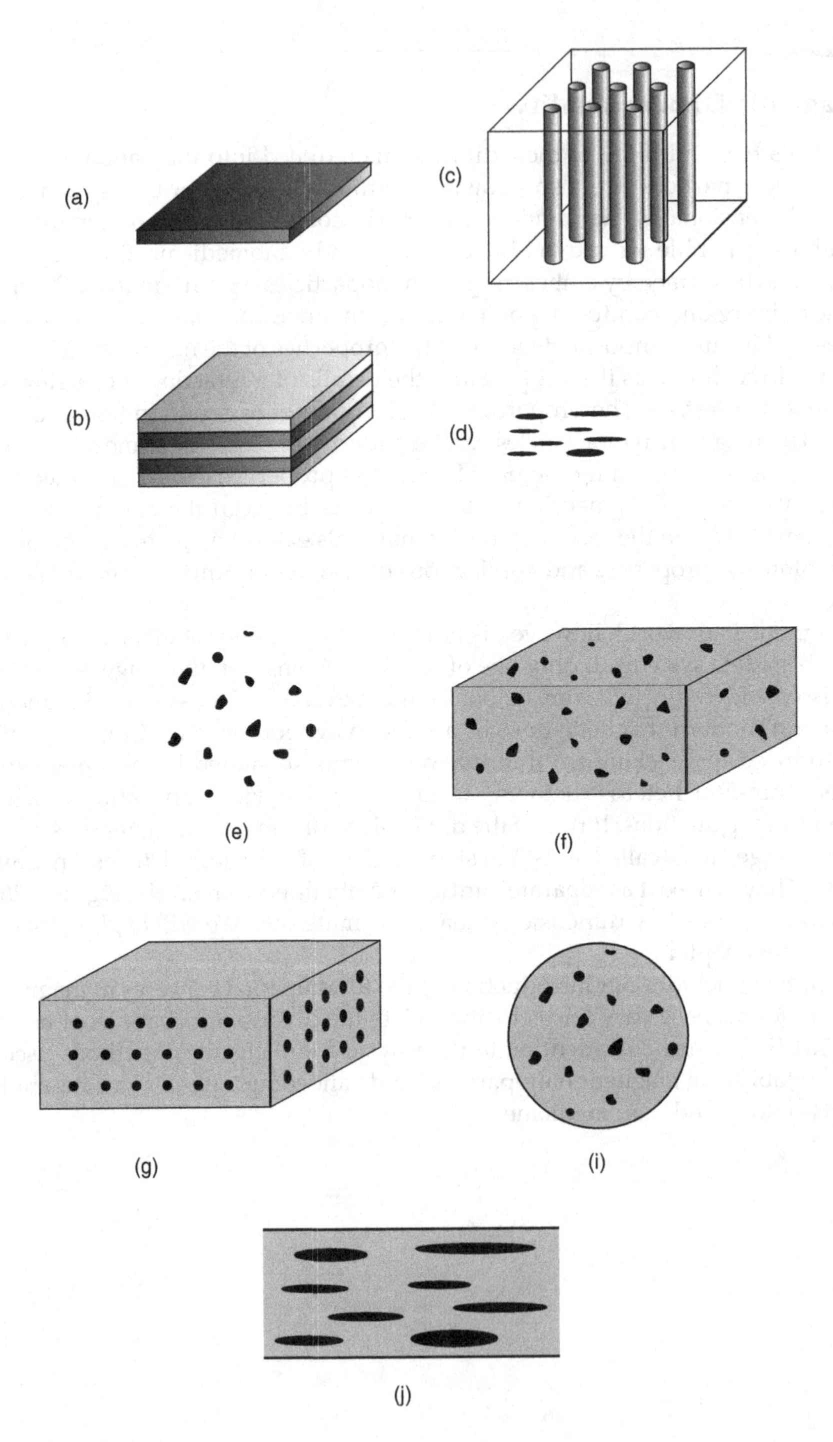

FIGURE 4.25
(a) Thin magnetic films, 2-D magnetic nanostructures. (b) Magnetic multilayers separated by non-magnetic layers. (c) Tubular magnetic structures within a non-magnetic matrix, 1-D magnetic nanocomposites. (d) Elongated magnetic particles, 0-D magnetic nanostructures. (e) Free-standing magnetic nanoparticles. (f) Magnetic nanoparticles dispersed in a non-magnetic medium. (g) Arrayed nanoparticles in a non-magnetic support. (i) Magnetic microspheres. (j) Nanoparticulate magnetic recording media.

Exercises

1 Calculate the expected single-magnetic-domain critical radius for Co and Ni in the strong anisotropy approximation.

2 What acicularity (c/a ratio) should a spheroid metallic iron particle possess in order to exhibit the same maximum coercivity as that of a spherical cobalt particle? Assume that the magnetocrystalline anisotropy axis and the axis of revolution of the spheroid are parallel.

3 What is the amount of energy needed for a magnetic field to rotate the magnetization of a spheroid iron particle from complete magnetic saturation along its major c-axis to magnetic saturation along its minor a-axis? Assume that $c = 15$ nm and $a = 10$ nm.

4 Using Figure 4.10, determine the values of the reduced coercive and switching fields predicted for coherent spin rotation of Stoner and Wohlfarth particles for α=10°, 20°, 45° and 90°.

5 Discuss the different modes of spin reversal mechanisms advanced in order to interpret coercivity values observed for single-magnetic-domain particles.

6 Single-magnetic-domain particles made of ferromagnetic or ferrimagnetic materials invariably exhibit enhanced coercivities, up to 2 orders of magnitude larger in certain cases, compared to the bulk. Why?

7 Prove that the coercivity of a magnetic nanoparticle increases according to

$$H_c = \frac{2K}{M_s}\left[1-\left(\frac{D_{sp}}{D}\right)^{1/2}\right]$$

for particle sizes above the superparamagnetic limit.

8 Prove that the dependence of the coercivity of a magnetic nanoparticle on temperature is given by

$$H_c = \frac{2K}{\mu_0 M_s}\left[1-\left(\frac{T}{T_B}\right)^{1/2}\right]$$

for temperatures below its blocking temperature.

9 Determine the temperature at which the magnetization of a spherical Co particle 10 nm in diameter is reduced to half its saturation value due to collective magnetic excitations.

10 Explain the underline mechanism responsible for exchange anisotropy. How can this phenomenon be exploited in controlling the coercivities of ferromagnetic nanoparticles coated with an antiferromagnetic material?

11 What is the significance of the characteristic length scales of Table 4.3?

12 Define magnetic dimensionality and give examples of lower-dimensional nanocomposite systems and their applications.

Exercises

1. [illegible] expected [illegible] contain [illegible] and [illegible] anisotropy [illegible]
2. [illegible] in order to [illegible] assume that the [illegible] of the [illegible] parallel.
3. What is the [illegible] energy [illegible] in a magnetic field [illegible] of a single-domain particle from [illegible] saturation magnetization [illegible] and a 20 nm [illegible]
4. Using Figure [illegible] and [illegible]
5. Discuss the different [illegible] coercivity values observed for single-domain particles.
6. Single-domain [illegible] particles [illegible] ferromagnetic or ferrimagnetic [illegible] enhanced coercivities [illegible] compared to the bulk [illegible]
7. [illegible] that the coercivity of a magnetic nanoparticle [illegible]

   for particle sizes above [illegible]
8. [illegible] dependence of the coercivity [illegible]

   at temperatures below the blocking temperature.
9. [illegible] the temperature at which [illegible]
10. Explain the underlying mechanism responsible for exchange [illegible] controlling the coercivity of [illegible] antiferromagnetic material.
11. [illegible] the characteristic length scales [illegible]
12. Define [illegible] phase system and their applications.

Part II

Production of Magnetic Nanoparticles

5

Top-Down Synthesis by Physical Methods

> The study of simple metal clusters has burgeoned in the last decade, motivated by the growing interest in the evolution of physical properties from the atom to the bulk solid, a progression passing through the domain of atomic clusters
>
> **W. A. de Heer, The Physics of Simple Metal Clusters: Experimental aspects and Simple Models, *Reviews of Modern Physics*, 65 (1993) 611**

We have established that single magnetic domain particles of nanometric dimensions possess unique magnetic properties, such as superparamagnetism and enhanced coercivities, not encountered in the bulk. Such nanoparticles constitute the fundamental building blocks of nanostructured magnetic materials and devices. Magnetic nanocrystals packed in ordered assemblies form new solids, opening up the possibility of fabricating new materials. Controlling the physical size, crystallinity, shape and interparticle distance of the constituent nanoparticles can tune magnetic materials properties. Therefore, the synthesis, stabilization and assembly of magnetic nanoparticles on macroscopic supports constitute important components of nanomagnetism, both in science and engineering. In Part II, we explore the many different approaches to the production of magnetic nanoparticles by physical, chemical, biological and biomimetic nanotemplating processes. The assembly of such nanoparticles on macroscopic supports is addressed in Part IV where specific applications to nanotechnology and nanomedicine are considered.

Broadly speaking, two diametrically opposite approaches can be considered in the production of magnetic nanoparticles, as shown schematically in Figure 5.1. In the first approach, moving from left to right, coined "top-down" synthesis, a macroscopic piece of a magnetic solid is used as the *precursor* to produce nanoparticles by physical methods, such as grinding or laser ablation. In the second approach, moving from right to left, coined "bottom-up" synthesis, increasingly larger magnetic metal clusters or magnetic ion cluster complexes are synthesized in solution by controlled reactions of *chemical precursors*, such as various salts of magnetic metal ions. A new state of matter is defined, which lies between the *micro*scopic regime of molecules and the *macro*scopic regime of bulk matter, the "mesoscopic" regime. Mesoscopic structures, 1–100 nm size range, exhibit novel physical properties due to "finite-size" effects, whereby bulk properties are modified by the onset of quantum mechanical phenomena associated with atoms and molecules. In this chapter, we present the fundamental principles involved in the production of magnetic nanoparticles by physical methods, while in Chapter 6, we address the synthesis of nanoparticles by chemical methods. The motivation is 2-fold: (a) to witness the emergence of collective magnetic phenomena in atomic clusters responsible for long-range order in solids; and (b) to develop an appreciation of the many challenges involved in stabilizing matter at the mesoscopic regime.

There are various physical processes used in the formation of magnetic nanoparticles. These include thermal evaporation, pulsed-laser ablation or ion sputtering of bulk ferromagnets followed by gas-phase condensation and deposition techniques. Another

DOI: 10.1201/9781315157016-7

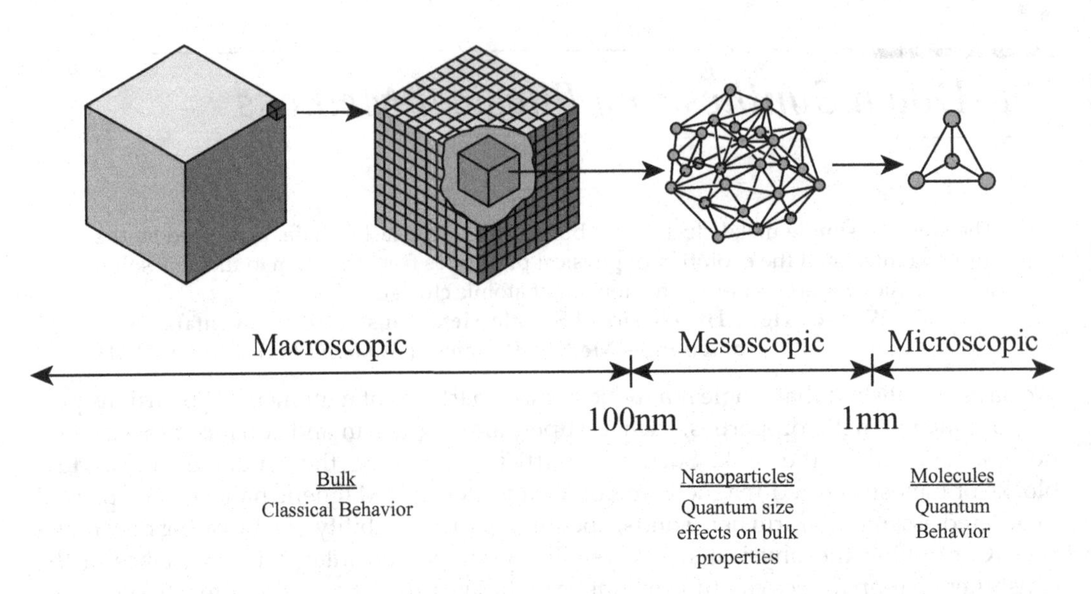

FIGURE 5.1
Top-down *vs.* bottom-up approaches to the production of mesoscopic structures. (From M. A Duncan and D. H. Rouvray, Microclusters, Scientific American, December 1989, p. 110, Credit: Hank Iken).

approach is the mechanical grinding-down of coarse metal or metal-oxide magnetic powders to nanometer dimensions. We demonstrate below some of these processes by presenting selected examples and discussing briefly the magnetic properties of the resulting nanoparticles, along with their significance in advancing our understanding of nanomagnetism.

5.1 Particle Nucleation and Growth

5.1.1 Homogeneous Nucleation

It was the study of colloids that first addressed the physical processes involved in particle nucleation and growth, a concept that permeates the subject of nanoparticle formation, whether induced from physical vapors, considered in this chapter, or from solution chemistry, described in the next chapter. Colloids are gases or solutions containing small (colloidal) particles in suspension. Pollen, water droplets or dust particles suspended in air are the most commonly encountered colloids. Colloidal particles are small enough to undergo Brownian motion due to collisions with the fluid molecules, thus, partaking of the thermal motion of the suspending fluid and resisting rapid sedimentation due to gravitational forces. Their random Brownian motion depends on the fluid's macroscopic properties, such as its viscosity and temperature. For Brownian motion to occur, the size of the particle must be at least ten times larger than the linear dimension of the fluid molecules. This requirement puts colloidal particles in the range of 1–1,000 nm diameter. The study of the nucleation and growth of colloidal particles in homogeneous solutions was first addressed in a classic publication by LaMer and Dinegar (V. K. LaMer and R. H. Dinegar, Theory, Production and Mechanism of Formation of Monodispersed Hydrosols, *J. Am. Chem. Soc.* 72 (1950) 4847) where they proposed a highly cited mechanistic model describing homogeneous particle nucleation and growth in a supersaturated solution.

According to LaMer's classical theory, in a homogeneous medium, the formation of colloids and nanoclusters is thermodynamically hindered due to the energetically demanding process of interfacial boundary formation, the surface of the nascent nucleated particle. This energy barrier to nucleation can be overcome kinetically in "supersaturated" vapors or solutions, where encounters or collisions between atoms or molecules are frequent. In its simplest form, the model considers the change in Gibbs free energy, ΔG, that results from nucleation. Bonding is energetically favored, as it liberates energy; however, the creation of the interfacial boundary consumes energy.

$$\Delta G_{(\text{nucleus})} = \Delta G_{(\text{bond formation})} + \Delta G_{(\text{surface tension creation})}$$

In the formation of a spherical cluster of radius r, the overall change in Gibbs free energy is given by Eq. (5.1), where ΔG_v is the free energy liberated per unit volume and ΔG_s is the energy cost per unit surface area created

$$\Delta G = -\frac{4\pi}{3} r^3 \Delta G_v + 4\pi r^2 \Delta G_s \tag{5.1}$$

The first term gives the energy gained in creating the volume of the particle; this is the term that *drives* nucleation. The second term gives the energy cost due to surface tension at the interface; this is the term that *hinders* nucleation. Figure 5.2 gives a sketch of the dependence of the various energy terms on r, the radius of the nucleated particle, and defines the critical radii for nucleation. For very small particles, ΔG is positive, due to the high surface-to-volume ratio, making nucleation thermodynamically disfavored. The graph of ΔG as a function of r reaches a maximum when $d(\Delta G)/dr = 0$. The maximum occurs at

$$r(kin) = \frac{2\Delta G_s}{\Delta G_v} \tag{5.2}$$

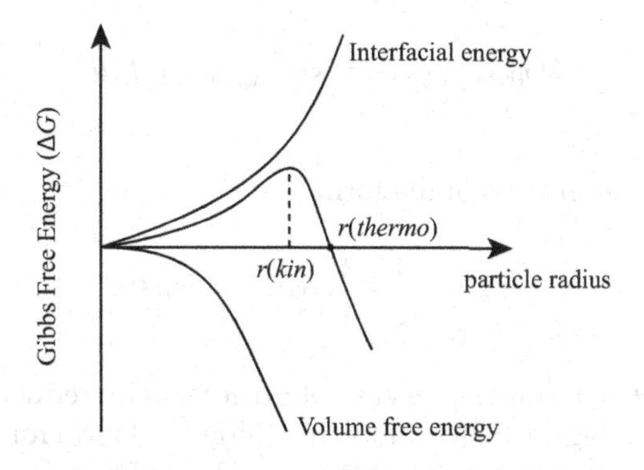

FIGURE 5.2
Graphical depiction of interfacial, volume and total Gibbs free energy, indicating the energy barrier and critical radii for particle nucleation and growth in homogenous nucleation.

Equation 5.2 defines the "kinetic critical radius" for nucleation. For particle radii larger than the "kinetic critical radius" $r > r(kin)$, further growth of the nucleated particle is kinetically favored as ΔG decreases rapidly with increasing particle size. The radius, $r(thermo)$, above which ΔG becomes negative, defines the "thermodynamic critical radius". The particle, once nucleated and grown to a radius $r \geq r(thermo)$, is thermodynamically driven to continuous growth by diffusion of additional precursor molecules onto the surface of the nucleated particle tending to bulk formation, provided additional precursor molecules are available in the nucleating medium. Thus, according to LaMer, the nucleation and growth of iron metal particles from, for example, vaporized iron metal will follow a two-stage process: (a) nucleation from supersaturated vapor ($n\text{Fe} \rightarrow \text{Fe}_n$) and (b) diffusive growth ($\text{Fe}_n + \text{Fe} \rightarrow \text{Fe}_{n+1}$). Addition of new atoms to the nucleated particle's surface releases, rather than costs, free energy. In order to stabilize and isolate nucleated particles at various stages of particle growth within the mesoscopic regime, one must use capping ligands that bond to surface atoms, or introduce surfactant molecules into the nucleating medium. The former passivate the surface energy associated with unsaturated surface-atom bonds by terminal ligation, while the latter, adsorb onto the surface of the nucleated particle lowering interfacial tension. A comprehensive discussion of such surface passivation processes necessary for nanoparticle stabilization is deferred to Chapter 6.

5.1.2 Heterogeneous Nucleation

Heterogeneous nucleation forms more readily than homogeneous nucleation and, thus, it is more common in nature. It occurs at preferential sites, such as phase boundaries or impurities present in the medium, and requires less energy than homogeneous nucleation. In the presence of such preferential sites such as, for example, the presence of dust particles or bubbles, the positive energy associated with surface tension creation at phase boundaries is greatly reduced since the nucleation on a dust particle partly eliminates the dust-particle/solution or air interface in the process of particle nucleation. Thus, the energetics of heterogeneous or "seed-mediated nucleation" are less demanding and nucleation can be more readily initiated. Surfaces promote nucleation because of "wetting". Consider Figure 5.3(a) that depicts the nucleation of a particle on an interface. Contact angles, θ, greater than zero between the nucleated particle's surface and the interface encourage nucleation. The energy barrier needed to be overcome is reduced compared to homogeneous nucleation according to

$$\Delta G_{(\text{heterogeneous})} = \Delta G_{(\text{homogeneous})} \cdot f(\theta)$$

where $f(\theta)$ is a polynomial in $\cos\theta$ of the form

$$f(\theta) = \frac{1}{2} + \frac{3}{4}\cos\theta - \frac{1}{4}\cos^3\theta$$

The wetting angle, θ, determines the ease of nucleation by reducing the energy needed for nucleation, as seen in Figure 5.3(b). Note that the critical radii for nucleation remain the same, but the energy barrier for heterogeneous nucleation is significantly reduced. However, the volume of the nucleated particle can be significantly smaller for heterogeneous nucleation due to the wetting angle affecting the shape of the cluster.

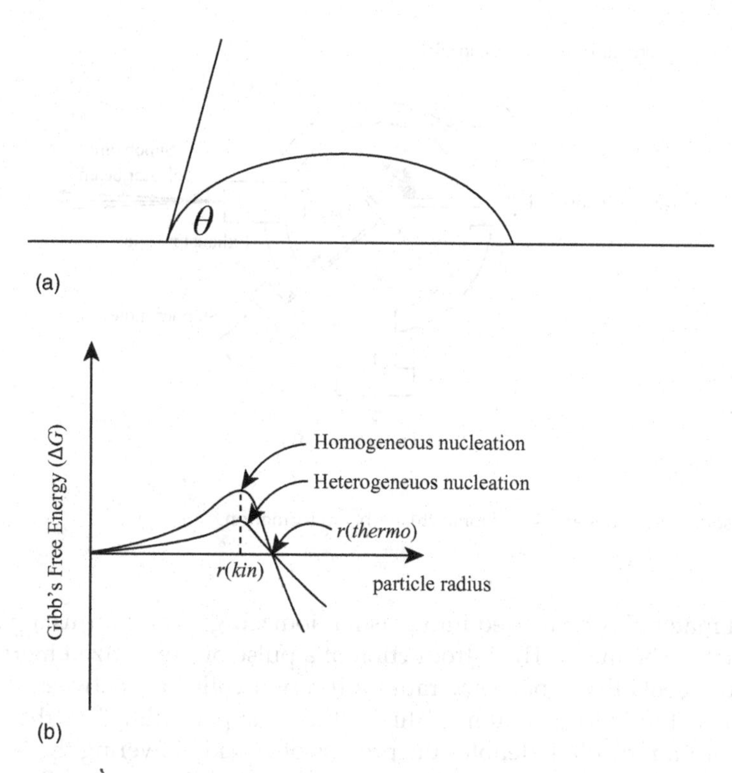

FIGURE 5.3
(a) Definition of the wetting angle θ for heterogeneous nucleation on a surface. (b) Comparison of Gibb's free energy curves as a function of particle radius for homogeneous *vs*. heterogeneous nucleation.

5.2 Gas-Phase Synthesis of Magnetic Nanoparticles

5.2.1 Synthesis of bare 3*d* Transition Metal Clusters by Pulsed Laser Ablation

As witnessed in Chapter 3, the magnetism of 3*d* transition metals is complex. Our understanding of ferromagnetism can be advanced from experimental studies of the evolution of magnetism from atoms to bulk. Such studies require the production and magnetic characterization of small metallic clusters of Fe, Ni and Co. Free clusters should preferably be investigated to avoid complications arising from particle/support interactions present in magnetic nanoparticles embedded in a matrix or coated by a protective shell. Such particles must by necessity be produced under transient, dynamic conditions with their magnetic properties investigated *in situ* as soon as they are created. One approach is the production of magnetic nanoparticles or clusters by laser ablation, a process used for removing material from a metal surface by irradiating it with a laser beam.

In this technique, a high-power pulsed laser is used to irradiate a disk (or rod) of the metal to be vaporized. The disk is situated in a vacuum chamber, as shown in Figure 5.4, and is slowly rotated and translated in order for the material to be evaporated uniformly over its surface. The amount of energy per pulse deposited on the metal surface depends on the wavelength of the laser beam, the optical properties of the metal and the duration of the pulse. The pulse duration can usually be varied from 10^{-3} to 10^{-15} seconds, allowing precise control over laser energy flux. At low laser flux, the metal heats up from the absorbed laser energy, causing the material to evaporate or sublimate. At high laser flux,

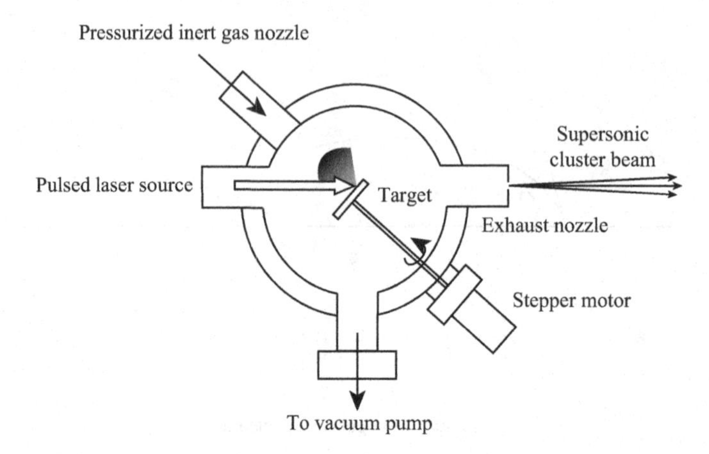

FIGURE 5.4
Schematic of the source chamber and supersonic cluster beam formation.

the vaporized material is converted into plasma (ionized gas) and forms a plasma plume over the surface of the metal. The introduction of a pulse of pressurized inert gas, such as helium or neon, cools the vapor very rapidly through collisions between the vaporized metal atoms and the inert gas atoms. This induces "supercooling" of the metal vapor. Supercooling or undercooling denotes the process of quickly lowering the temperature of a liquid or a gas below its freezing point without its becoming a solid. Supercooling the vaporized metal induces supersaturation, which leads to nucleation and growth of metal clusters. The condensed clusters quickly reach thermal equilibrium with the inert gas. Subsequently, the cluster/gas mixture is allowed to exit the vaporization chamber through a small nozzle or orifice into an adjacent vacuum chamber, producing a supersonic cluster beam, as depicted in Figure 5.4, containing very cold metal iron clusters. Thus, the inert gas serves as condensation and carrier medium and it is referred to as "carrier gas".

The magnetic properties of the metal clusters in the beam can be studied *in situ* by passing this beam of cold, neutral clusters through a non-uniform magnetic field and detect their degree of deflection away from their original path. As discussed in Section 1.6.3, a classical magnetic dipole introduced in a non-uniform magnetic field feels a pull toward the direction of the stronger magnetic field, according to Eq. (1.23). A non-uniform field can be produced by a magnet with specially shaped poles. A cross section of such a magnet is shown in Figure 5.5. This type of magnet is usually referred to as a "Stern–Gerlach" type magnet. It owes its name to the historic experiment carried out in 1922 by the scientists Otto Stern and Walther Gerlach, who first used such a magnet to detect the quantization of the magnetic moment of neutral Ag atoms whose ground state has angular momentum $S = 1/2$. Due to the fact that the atoms were not charged, there was no Lorentz force acting on them; they would pass undeflected through a homogeneous magnetic field. For the Stern–Gerlach magnet of Figure 5.5, however, one can see that the magnetic field lines get denser the closer one gets to the top pole, indicating that the magnetic field gets stronger as you move along the z-direction from the bottom pole to the top pole of the magnet. Upon entering the magnet, the atoms acquire orientational potential energy U_B due to the interaction of the magnetic moment of the silver atom, $\vec{\mu}$, with the external magnetic field, $\vec{B}$

$$U_B = -\vec{\mu} \cdot \vec{B} = \pm \mu_z B_z = \pm \mu_B B_z$$

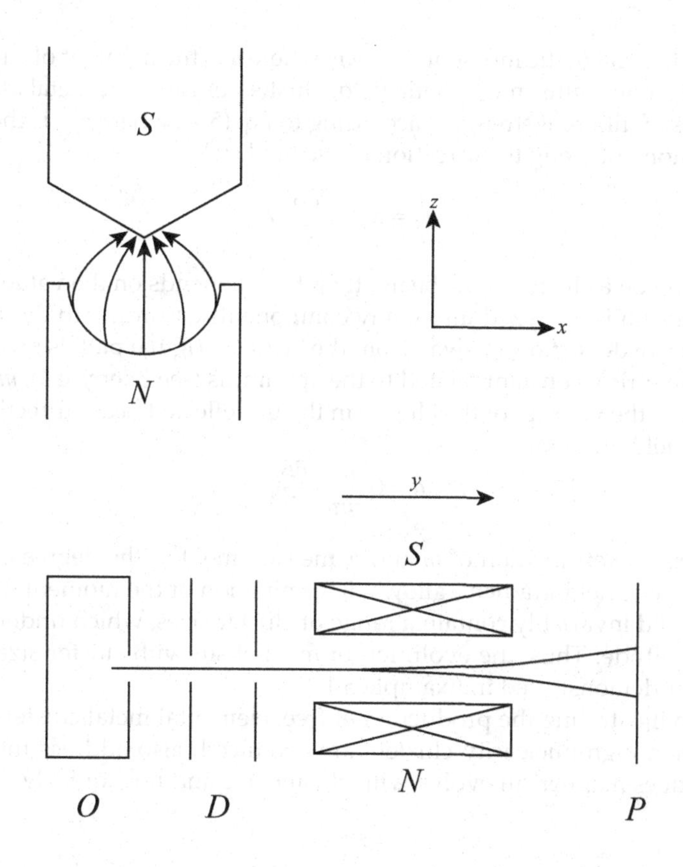

FIGURE 5.5
(Top) Schematic of the cross section of a "Stern–Gerlach" type magnet. The magnetic field gradient points along the +z-direction. The cluster beam travels along the y-direction (into the page). (Bottom) Schematic of the Stern–Gerlach experiment. An atomic beam is formed by evaporating *Ag* atoms in the oven, *O*. The beam is collimated by the diaphragms *D* and enters a magnet *N-S* of non-uniform magnetic field. The deflected beam is projected on screen *P* where the silver atoms are detected.

where, according to quantum mechanics, the atomic moments have equal probability to orient along or opposite to the direction of the magnetic field. The silver atoms also experience a force along the z-direction due to the non-uniformity of the magnetic field

$$\vec{F}_z = -\frac{\partial U_B}{\partial z}\hat{z} = \mp\mu_z\frac{\partial B_z}{\partial z}\hat{z} = \mp\mu_B\frac{\partial B_z}{\partial z}\hat{z} \tag{5.3}$$

The only force exerted on the atoms in the beam is given by Eq. (5.3), where $\pm\mu_z$ is the projection of the atom's magnetic moment along or opposite to the direction of the magnetic field and the strength of the gradient.

Figure 5.5 also depicts a horizontal beam of silver atoms as it passes through the Stern–Gerlach magnet and the observed splitting and deflection away from the initial path of the beam. Stern and Gerlach were the first to observe that the magnetic moment of the *Ag* atom had an equal probability of orienting along or opposite to the magnetic field direction, confirming spin quantization. Unlike a classical bar magnet, an atom has an equal probability of been deflected along the +z- or −z-direction.

Let us now consider a similar experiment using the supersonic cluster beam exiting the source chamber of Figure 5.4. At saturation a ferromagnetic cluster containing *N* metal atoms gives rise to a magnetic moment $\mu_N = N\mu$, where μ is the moment of each metal atom in the cluster. In a non-uniform magnetic field, clusters of different metal atom nuclearity experience forces of different strengths according to Eq. (5.4), where μ_{N_Z} is the projection of their magnetic moment along the direction of the field.

$$\vec{F}_z = \pm\mu_{Nz}\frac{\partial B_z}{\partial z}\hat{z} \tag{5.4}$$

The force acting on a cluster in the beam, therefore, depends on the value of μ_{N_Z} for that cluster, and the beam is analyzed into many components according to the various values of μ_{N_Z}. The degree of deflection, d, away from the beam's original path is given by Eq. (5.5), where C is a geometrical constant related to the apparatus (see Exercise 4), m is the mass of the cluster and v is the velocity of the cluster in the undeflected beam direction, transverse to the magnetic field gradient.

$$d = C\frac{\mu_{Nz}}{mv^2}\frac{\partial B_z}{\partial z} \tag{5.5}$$

By independent determination of m and v, measurement of the degree of deflection, d, in a known magnetic field gradient, allows determination of the moment of the cluster. A cluster beam would invariably contain a range of cluster sizes, which undergo deflections of different magnitude. Thus, the evolution of magnetism with cluster size can be determined, as further demonstrated in Example 5.1.

In addition to illustrating the production of free, elemental metal clusters, experiments of this type, when high-nuclearity clusters are produced, also address the fundamental question: How does magnetism evolve with cluster size and how quickly does bulk magnetism emerge?

Example 5.1: Magnetism of 3*d* Metal Clusters in Cluster Beam Experiments

The first experiment on free iron-metal clusters was reported in 1985 by Cox *et al.* (D. M. Cox, D.J. Trevor, R.L. Whetten, E. A. Rohlfing and A. Kaldor, *Phys. Rev. B* 32 (1985) 7290). The production of neutral iron-metal clusters containing 2–17 atoms was demonstrated in this experiment. Cluster mass determination was performed in a time-of-flight (TOF) mass spectrometer, after the clusters in the beam were selectively photo-ionized using an Excimer laser beam that could be focused down to 0.5 mm beam diameter. A schematic of the experimental setup used is shown in Figure 5.6(a) and the geometrical arrangement of the deflecting magnet and cluster beam is given in Figure 5.6(b).

In Figure 5.6(a), the cluster source chamber is depicted on the left with the deflecting magnet arrangement and TOF mass spectrometer on the right. On the left side of the cluster source chamber, the pulsed nozzle allows a pulse of inert gas to be introduced into the chamber, while the pulsed Nd:YAG (neodymium-doped yttrium aluminum garnet $Nd:Y_3Al_5O_{12}$) laser vaporizes iron atoms off a rotating metal iron rod. Through a small orifice, inert gas carrying along metal clusters enters the evacuated chamber on the right that houses the Stern–Gerlach magnet. After collimation this supersonic beam of clusters and carrier gas passes through the deflecting magnet, as shown. The incoming beam is depicted to spread out as it passes through the magnet. Just as in the case of silver atoms, clusters are deflected with equal probability above and below the undeflected beam path obtained in the absence of a magnetic field gradient. By focusing the Excimer laser (ArF) beam at a particular spot along the spread of the deflected cluster beam, different nuclearity clusters can selectively be photo-ionized. The Excimer ArF laser emits ultraviolet laser light of wavelength $\lambda = 193$ nm, of photon energy $\hbar\nu$ in the range of the ionization potentials of the metal clusters. The Double Dye used in conjunction with the Excimer laser allows for laser wavelength tunability, as different nuclearity clusters would have slightly

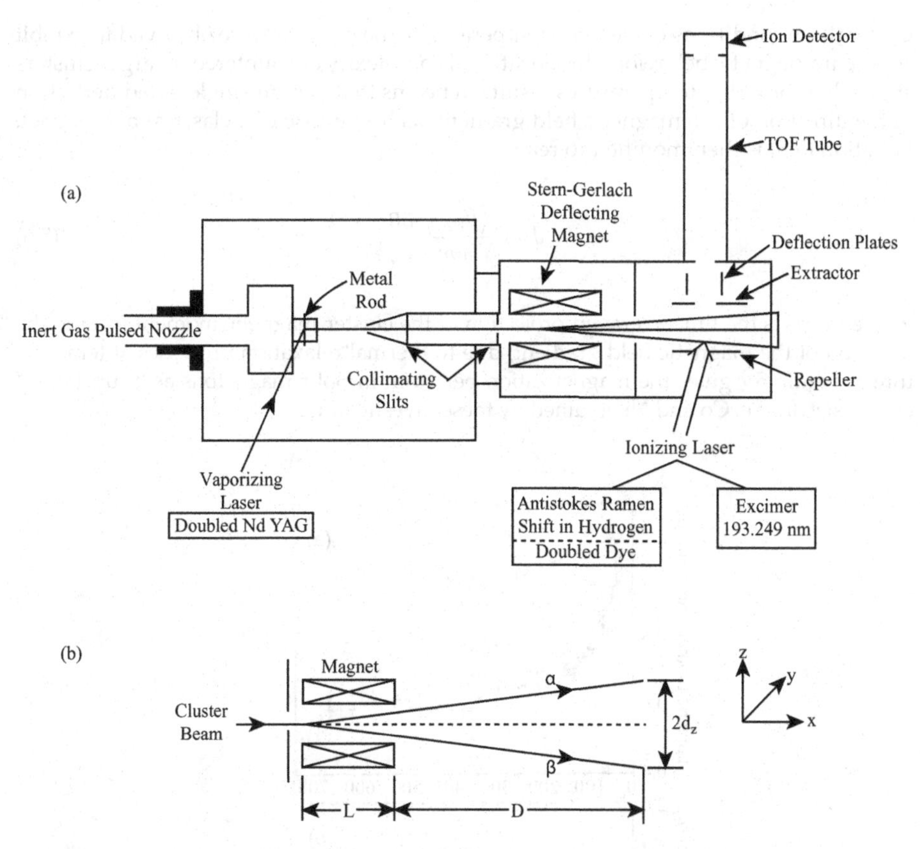

FIGURE 5.6
(a) Schematic of the pulsed cluster beam apparatus for the production of neutral iron metal clusters. (b) The geometrical arrangement for magnetic deflection of the metal-atom clusters. (From D. M. Cox, D.J. Trevor, R.L. Whetten, E. A. Rohlfing and A. Kaldor, Magnetic behavior of free-iron and iron oxide clusters, *Phys. Rev. B* 32 (1985) 7290, Copyright American Physical Society, reproduced with permission).

different ionization potentials. The selectively ionized clusters are then accelerated by an electric field produced between two parallel plates (Repeller and Extractor) held at a known potential deference ΔV, before they enter the TOF tube. For singly ionized clusters, the time it takes for the cluster to reach the end of the tube where it is detected allows determination of the cluster's mass. Heavier, high-nuclearity clusters have longer flight times (see Exercise 5). The results of the experiment indicated that small magnetic clusters of iron have magnetic moments per atom that exceed that of bulk iron, $2.2\mu_B$/atom, by as much as ~30%.

This observation was further confirmed by de Heer and co-workers (Billas, Châtelain and de Heer, *Science* 265 (1994) 1682) who produced even larger metal clusters ($20 \leq N \leq 700$) of Fe but also of Co and Ni using a similar cluster beam apparatus. As the number of atoms in the cluster increases, the internal degrees of freedom of the cluster, rotational and vibrational, play a crucial role in spin relaxation processes internal to isolated clusters. The spin of the cluster is coupled to its rotation through the anisotropy field, which is rigidly fixed to the lattice of the rotating cluster. It is the total angular momentum that is conserved. Vibrational energies are a measure of the temperature of the clusters. In this experiment, quasi-effusive cluster beams were studied, where cluster temperatures remain relatively high compared to supersonically cooled beams.

Under these conditions of elevated temperatures, the clusters were observed to exhibit superparamagnetic behavior. This additional complexity encountered in larger clusters above their blocking temperatures results in beams that exhibit single-sided deflection in the direction of the magnetic field gradient, as in the case of a classical bar magnet. Equation (5.5) is then modified to read

$$d = C\frac{\langle \mu_{Nz} \rangle_T}{mv^2}\frac{\partial B}{\partial z} \tag{5.6}$$

where $\langle \mu_{N_Z} \rangle_T$ is the time-averaged projection of the cluster magnetic moment μ_N onto the direction of the magnetic field gradient, due to thermal relaxation processes at temperature T. Figure 5.7 gives the magnetization per atom in Bohr magnetons as a function of cluster size for Fe, Co and Ni obtained by these investigators.

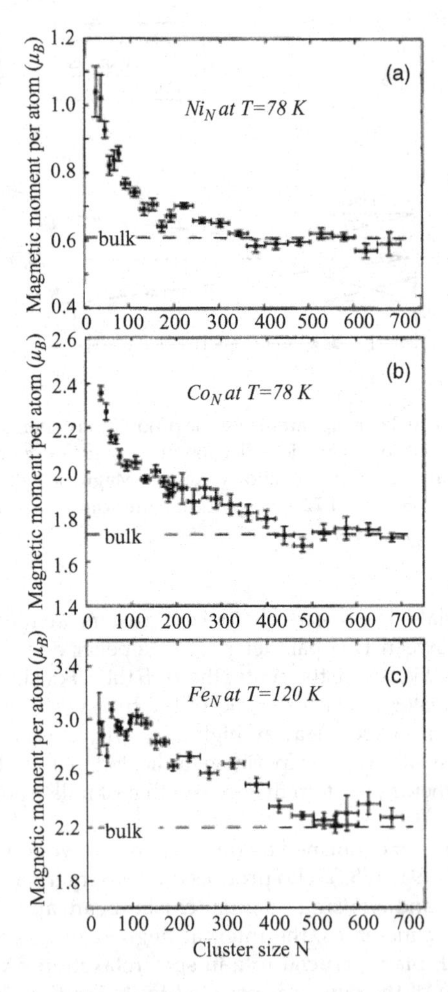

FIGURE 5.7
Average magnetic moments per atom as a function of cluster size N for (a) Ni_N clusters at $T = 78$ K, (b) Co_N clusters at $T = 78$ K and (c) Fe_N clusters at $T = 120$ K. (From Billas, Châtelain and de Heer, Magnetism from the Atom to the Bulk in Iron, Cobalt and Nickel Clusters, *Science* 265 (1994) 1682, Copyright American Association for the Advancement of Science, reproduced with permission).

Note that the smaller clusters exhibit larger magnetic moments per atom compared to the bulk, for all three elemental ferromagnets. This is the result of the greater delocalization of itinerant *d* electrons in the bulk compared to the clusters. An increase of surface atomic moments compared to interior moments is a general property of ferromagnetic solids due to the lower degree of coordination at the surface. For very small clusters, all atoms lie on the surface resulting in greater electronic localization on individual atoms. With increasing cluster size, the surface-to-volume ratio decreases. Eventually, the magnetism of the core atoms in the interior of the cluster dominates and the magnetic moment per atom approaches that of the bulk. We see that the convergence to the bulk is rather fast, within a few to several hundred atoms. Oscillations in the measured moments arise from geometrical effects in the detailed arrangement of atoms within the cluster.

The technique of pulsed laser ablation has also been applied in the production of magnetic colloids in solution by immersing the metal target in liquid, such as methanol, which acts as the carrier fluid.

5.2.2 Synthesis of Ferromagnetic Nanoparticles by Vaporization–Deposition Technique

Metal vaporization can also be achieved by simply heating a piece of metal in a crucible placed in a low-pressure chamber. In practice, the piece of metal to be vaporized is placed in a cone- or boat-shaped tungsten coil which is resistively heated by a current. The chamber is initially evacuated and then back-filled with a rarefied inert gas such as He, Ar or Ne, at controlled pressures far below atmospheric pressure. Similar to the case of the cluster beam experiments, the hot metal vapor is subsequently mixed with cold inert gas inducing supercooling of the vapor atoms, which leads to supersaturation, nucleation and growth of metal nanoparticles. The median size of the particles produced can be controlled by varying the pressure of the inert gas, with lower pressures yielding smaller median particle diameters. This method is well suited for the production of metal nanoparticles because many metals under low pressures vaporize at easily attainable temperatures.

Including reactive gases, like oxygen, in the cold inert gas nanoparticles of oxides or other compounds of the evaporated metals can be prepared. The introduction of a small amount of oxygen can yield metallic nanoparticles covered with a thin, protective shell of oxide. In addition, by vaporizing two different metals simultaneously in the same chamber, intermetallic alloy magnetic nanoparticles can be produced. The nanoparticles so produced are deposited on a cold substrate, which is usually water-cooled and located a few cm away from evaporating metal. The cold substrate can further induce particle nucleation at its surface by the process of heterogeneous nucleation. This method is also known under the name "vaporization–deposition" technique. The process is also ideally suited for the production of thin metal films appropriate for device applications. Example 5.2 further elaborates on this approach to the formation of metal magnetic nanoparticles.

Example 5.2: Ultrafine Iron Particles Produced by a Vapor-Deposition Technique

Using the vapor-deposition technique, S. Gangopadhyay *et al.* (S. Gangopadhyay, G. C. Hadjipanayis, B. Dale, C. M. Sorensen, K. J. Klabunde, V. Papaefthymiou and A. Kostikas, *Phys. Rev. B*, 45 (1992) 9778) produced iron nanoparticles of various median diameter by evaporating bulk Fe from alumina-coated tungsten crucibles in an argon atmosphere at pressures in the range 0.5–8.0 Torr (1 Torr = 1.3×10^{-3} atmospheres). During evaporation, a thermal gradient was established between the source (T ≈ 1,500°C) and the water-cooled copper substrate, placed 2 cm away from the source. The thermal gradient allowed the particles to collect on the substrate by the process of "thermophoresis". Evaporation times 5–10 minutes produced 5–10 mg of metallic iron particles. However, metal iron particles are highly reactive and burn upon exposure to air. The particles collected on the substrate were exposed for a few hours to a dilute argon–air mixture (3,000:1 argon to air volume ratio) before removing them from the apparatus in order to passivate their surface with a thin shell of iron oxide. Further exposure to air was avoided by mixing the particles with molten wax.

The collected particles were studied by Transmission Electron Microscopy (TEM) to determine their size and size distribution. TEM characterization is routinely employed in nanoparticle research. In this study, where air-sensitive particles were produced, a microscope copper grid was exposed briefly during the evaporation of the iron metal. Normally, the grid is put at the center of the substrate which, of course, would yield no information on particles deposited on other areas of the substrate. In general, when stable particles are produced, to obtain a more accurate picture of the size and size distribution of the total sample, all collected particles are removed from the substrate and thoroughly mixed. Subsequently, a small amount of the collected powder is dispersed in water with ethyl alcohol by using an ultrasonic bath. Then a drop of the solution is deposited on the microscope grid, or alternatively, the grid is briefly dipped into the nanoparticle suspension. After allowing the grid to dry, it is inserted into the electron microscope and the scattering of the microscope's electron beam off the deposited particles is observed.

Electrons are scattered more effectively from areas of high electron density, where iron metal particles reside, and less effectively from less dense regions. Thus, an iron particle appears as a dark spot on the electron micrograph. Figure 5.8 shows a typical TEM micrograph obtained in this experiment; two different magnifications are presented. The lower magnification, Figure 5.8 (left panel), indicates that nearly spherical Fe nanoparticles were produced. Under higher magnification, Figure 5.8 (right panel), an inner electronically

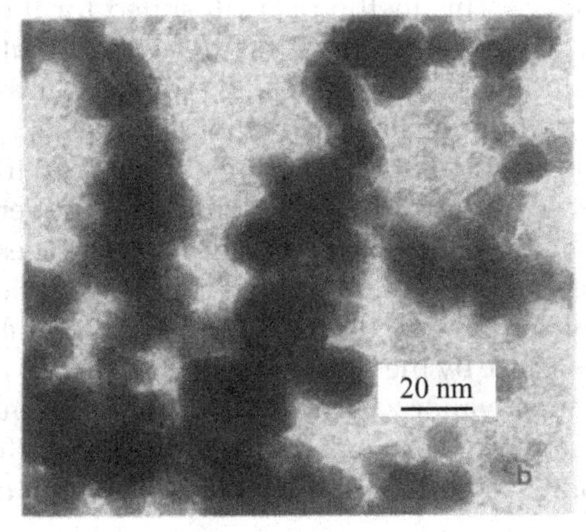

FIGURE 5.8
(Left panel) Bright-field TEM micrograph showing nearly spherical Fe particles. (Right panel) Bright-field TEM micrograph showing the core/shell structure of Fe particles (dark spots) surrounded by a thin shell of iron oxide (gray rings). (From S. Gangopadhyay, G. C. Hadjipanayis, B. Dale, C. M. Sorensen, K. J. Klabunde, V. Papaefthymiou and A. Kostikas, Magnetic properties of ultrafine iron particles, *Phys. Rev. B*, 45 (1992) 9778) (Copyright American Physical Society, reproduced with permission).

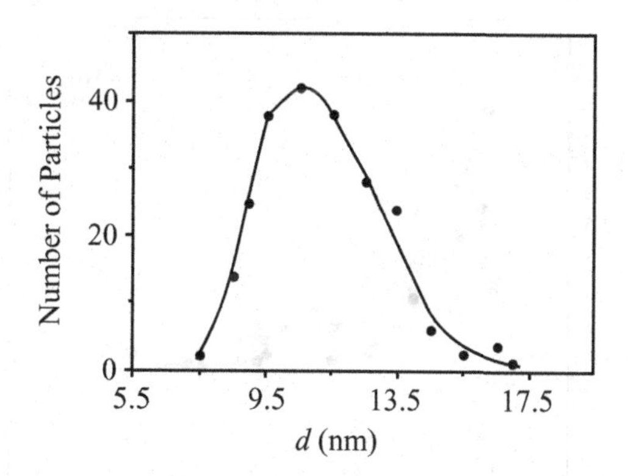

FIGURE 5.9
Particle size distribution in a typical Fe sample showing log-normal distribution. (From S. Gangopadhyay, G. C. Hadjipanayis, B. Dale, C. M. Sorensen, K. J. Klabunde, V. Papaefthymiou and A. Kostikas, Magnetic properties of ultrafine iron particles, *Phys. Rev. B*, 45 (1992) 9778, Copyright American Physical Society, reproduced with permission).

dense core of iron metal (black spots) surrounded by a less dense shell of iron oxide (gray crowns) is discernible.

Measuring the diameters of a statistically significant number of particles, typically 200–500 particles, allows a statistical analysis of the particle size distribution and polydispersity in the sample. Figure 5.9 shows that in this case, the measured diameters exhibited a log-normal distribution, modeled according to the mathematical expression

$$P(d) = \frac{1}{\sqrt{2\pi}\,d\sigma}\exp\left(-\frac{\ln^2(d/d_0)}{2\sigma^2}\right) \tag{5.7}$$

Equation (5.7) gives the probability, or frequency of occurrence, of a particular particle diameter, d, while d_0 is the median diameter and σ the polydispersity in particle diameters. The fitted data presented in Figure 5.9 indicated a median iron particle diameter $d_0 = 10.4$ nm and $\sigma = 1.22$. Such log-normal distributions are typical of nanoparticle ensembles produced by various types of synthesis routes. The presence of polydispersity in the particle sizes often presents a problem when applications are considered. Scientists are exploring a variety of methods to reduce the polydispersity, either by size selection processes or by novel synthesis techniques that result in samples containing narrow size distributions, as we explore further in Chapter 6.

At low temperatures, the coercivities of the particles showed a strong dependence on particle size, as seen in Figure 5.10. Smaller particles exhibit larger coercivities due to the relatively larger contribution of surface anisotropy to the overall effective anisotropy of the particle, consistent with Eq. (4.59). Furthermore, the coercivities of the particles showed strong temperature dependence as seen in Figure 5.11, decreasing with increasing temperature, due to collective magnetic excitations, in accord with to Eq. (4.56). The temperature at which the coercivity vanishes defines the blocking temperature T_B, as was discussed in Chapter 4. It is seen that the smaller diameter particles with core diameters 3.3 and 2.5 nm were superparamagnetic at room temperature.

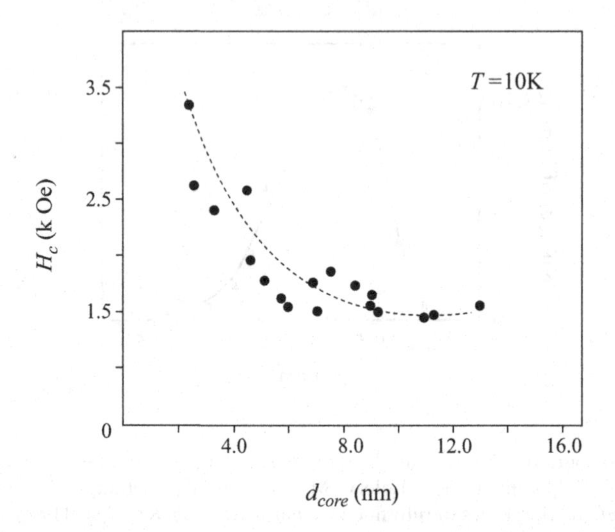

FIGURE 5.10
Coercivity as a function of median particle diameter measured at T = 10 K; the dashed line is a guide to the eye. (From S. Gangopadhyay, G. C. Hadjipanayis, B. Dale, C. M. Sorensen, K. J. Klabunde, V. Papaefthymiou and A. Kostikas, Magnetic properties of ultrafine iron particles, *Phys. Rev. B*, 45 (1992) 9778) (Copyright American Physical Society, reproduced with permission).

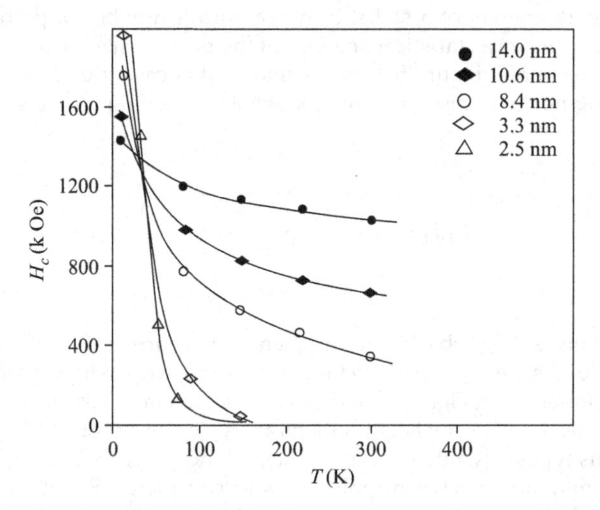

FIGURE 5.11
Variation of coercivity with temperature for various median particle diameters as indicated. (From S. Gangopadhyay, G. C. Hadjipanayis, B. Dale, C. M. Sorensen, K. J. Klabunde, V. Papaefthymiou and A. Kostikas, Magnetic properties of ultrafine iron particles, *Phys. Rev. B*, 45 (1992) 9778) (Copyright American Physical Society, reproduced with permission).

In Examples 5.1 and 5.2, we saw how pulsed laser ablation and thermal vaporization of solid metal precursors can lead to the formation of ferromagnetic nanoparticles from supersaturated physical vapors. Additional means for vaporizing a solid exist such as "spark discharge generation", in which electrodes made of the metal to be vaporized are charged, in the presence of an inert background gas, until the breakdown voltage is reached. The arc or spark formed across the electrodes vaporizes a small amount of metal from which nanoparticles are formed through subsequent gas condensation.

Bombardment of the surface of a metal by inert gas ions, a technique known as "ion sputtering", can also lead to metal vaporization. In "ion sputtering", the removal of atomized material from the solid target is affected by effective momentum transfer from the ion to a surface target atom. In a sputtering chamber, the target is kept at a high negative potential, typically −300 V. The chamber is initially evacuated. Positively charged inert gas atoms are then introduced into the chamber, which are accelerated toward the target bombarding its surface at high speeds. Figure 5.12 gives a schematic of the principle for sputtering with Ar^+ ions. The ejected atoms are emitted with a wide distribution of energies up to tens of eV, but are subsequently slowed down by collisions with inert gas atoms and move diffusively, reaching the cold substrate where they condense. The technique is widely used for the deposition of thin films (2D nanostructures) of a material on a substrate, such as a silicon wafer, in the production of magnetic data storage materials, and is known as "sputter deposition" or "physical vapor deposition" (PVD). For efficient momentum transfer and successful sputtering of the target atoms, the atomic weight of the sputtering gas should be close to the atomic weight of the target atoms. Thus, for sputtering light elements, neon is preferable while for heavy elements, krypton or xenon are used.

A variation of the ion sputtering technique described above is "magnetron sputtering". Ejection of target atoms is often accompanied by a second important process, the emission of secondary electrons from the target surface. A magnetic field can be used to *trap* these secondary electrons close to the target, as depicted in Figure 5.13. The electrons follow

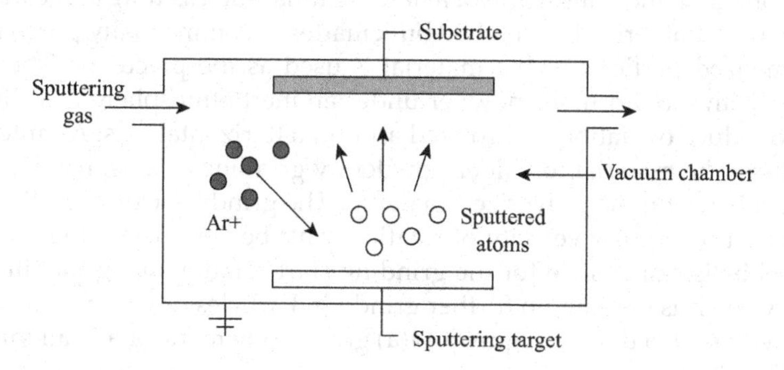

FIGURE 5.12
Schematic of an "ion sputtering" device.

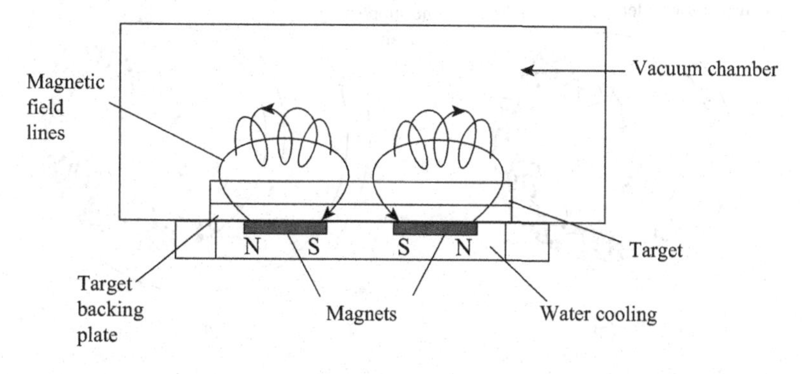

FIGURE 5.13
Schematic of a "magnetron sputtering" device.

helical paths around the magnetic field lines (see Exercise 14 of Chapter 1). This allows the electrons to undergo more ionizing collisions with neutral gaseous atoms near the target than would otherwise occur. This enhances the ionization of Ar atoms near the target leading to a higher sputter rate, enabling sputtering to be performed at lower argon pressures. The sputtered target atoms are electrically neutral and so are unaffected by the magnetic trap. Sputtering can be used to deposit all kinds of materials on a substrate. If a reactive gas is used as a sputtering gas compound materials, formed on the target surface or in flight toward the substrate, can be deposited on the substrate surface. Finally, if the target is an insulator, such as a magnetic oxide, an AC rather than a DC voltage is used in order to avoid charge build-up on the target surface leading to the technique known as "rf sputtering". Power supplies operate at ~15 MHz; for part of the cycle, Ar^+ ions bombard the target and for the rest of the cycle, electrons neutralize the built-up of positive charge.

5.3 Synthesis of Magnetic Nanoparticles by High-Energy Ball Milling

Grinding chunks of a magnetic material down to nanometer-size particles is the simplest and most direct "top-down" synthesis route of magnetic nanoparticles. In this technique, a cylindrical container and balls made of a hard material such as tungsten-carbide or hardened steel are used. Pulverized or crushed bulk matter or commercially purchased powder containing μm-sized particles of the material is used as the precursor. The container is partially filled with the precursor powder under an inert atmosphere. It is then hermetically sealed to reduce oxidation and rotated around a horizontal axis. An internal cascading effect reduces the material to a finer powder by grinding, that is, repeated impacts of the hardened spheres on the pulverized material. The grinding works on the principle of "critical speed". The angular velocity of rotation must be kept below the critical value at which the steel balls responsible for the grinding start rotating along the direction of the cylindrical device; thus causing no further grinding. Particles down to 5 nm diameter can be produced with such a device. Figure 5.14(a) gives a schematic of a "ball mill".

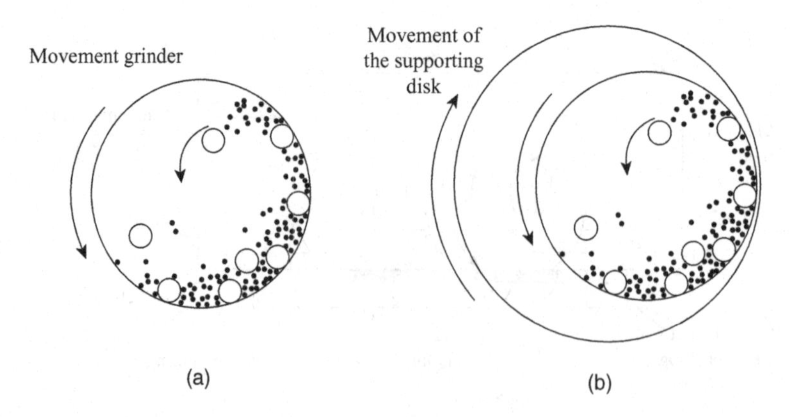

FIGURE 5.14
(a) Schematic of a "ball mill". (b) Schematic of a "planetary ball mill".

In a variation of this design, the grinder is placed eccentrically on a supporting disk which is rotated in the opposite direction than the grinder, to form what is called a "planetary ball mill". The grinding balls are then subjected to superimposed rotational motions. The difference in speeds between the ball and the grinding jar produces an interaction between frictional and impact forces releasing high dynamic energies; thus, the technique is known as "high-energy ball milling". High-energy ball mills can produce particles of diameters even smaller than 5 nm. A schematic of a planetary ball mill is given in Figure 5.14(b). A large ball to powder weight ratio is used, for example, 3:1 to 20:1 weight ratios are common. The hard balls impact the powder, transferring part of their kinetic energy to the particles, further breaking them up into smaller-sized particles. The milling is performed over various lengths of time, from several minutes to many hours.

Due to the nature of high-energy ball milling, "cold welding" processes may induce structural defects and eventually lead to an increase in particle size over prolonged milling. Thus, to improve the performance of ball milling in the production of magnetic nanoparticles, a variation of this approach has been introduced, known as "surfactant-assisted ball milling". In this approach, an organic solvent, such as methanol (CH_3OH), and a surfactant, such as oleic acid ($CH_3(CH_2)_7CH{=}CH(CH_2)_7COOH$), are introduced in the grinder along with the powder to be ground. The properties of surfactant molecules and their role in stabilizing nanoparticles are discussed in great detail in Chapter 6, where the production of magnetic nanoparticles through chemical methods is discussed. For now, suffice it to say that surfactants are organic macromolecules that act on the surface of the particles, reducing interfacial tension associated with the particle/solvent interface and aiding in sterically stabilizing and dispersing the particles in the solvent, which acts as a carrier fluid.

Ball milling has been extensively used in the production of nanoparticles of rare-earth alloys, such as $SmCo_5$, Sm_2Co_{17} and $Nd_2Fe_{14}B$, used in the production of nanostructured permanent magnets. It has also been used in the production of ferrite nanoparticles by grinding 1:1 molar mixtures of α-Fe_2O_3 and, for example, ZnO to produce zinc-ferrite ($ZnFe_2O_4$), in a process known as "mechano-chemical synthesis" of magnetic nanoparticles.

Example 5.3: Synthesis of $SmCo_5$ Magnetic Nanostructures by Ball Milling

$SmCo_5$ is a magnetic intermetallic alloy of hexagonal crystal structure with large uniaxial anisotropy of energy density $K = 1.3 \times 10^7$ J/m^3 used in the production of permanent magnets. Investigators have produced magnetic nanoparticles of this metal alloy by ball milling commercially purchased powder containing ~150 μm diameter particles. Milling was stopped every 15 min for the first 2 h and every hour thereafter to remove a small amount of material for magnetic characterization. Figure 5.15 gives the measured coercivities of the ball mill derived $SmCo_5$ magnetic nanostructures as a function of milling time. It is observed that the coercivity increases at a fast rate in the first 2 hours of milling, from 2.5 kOe for the as-purchased powder, to over 15 kOe at ~2 h of milling, before it declines back to its original value with increasing grinding time to 20 hours.

The initial increase in coercivity is associated with the decrease of particle size and the increased contribution of surface anisotropy, as was discussed in Section 4.10. The decrease in coercivity beyond 2 h milling time may be due to structural disorder, such as dislocations and vacancies that may be introduced into the structure under prolonged milling, or particle size increase due to cold welding. Interestingly, when surfactant-assisted ball milling was utilized in the production of $SmCo_5$ nanoparticles, the coercivity continued to increase after 8 h of milling, as shown in Figure 5.16. In this study, the precursor was 200-μm powder particles with heptane used as the solvent and oleic acid as the surfactant.

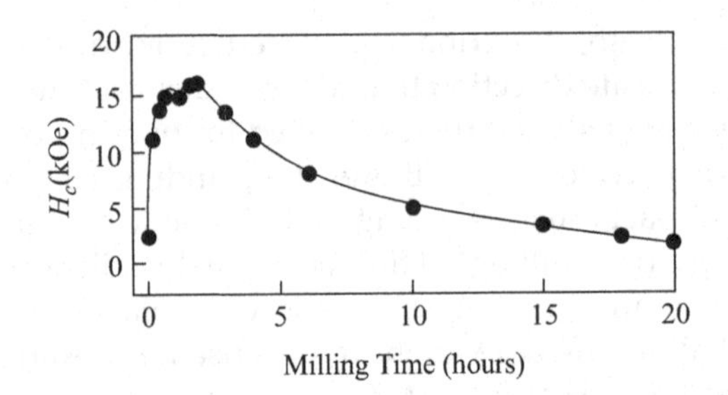

FIGURE 5.15
Room temperature coercivity of $SmCo_5$ nanoparticles as a function of milling time. (Adapted from D. L. Leslie-Pelecky and R. L. Schalek, Effect of disorder on the magnetic properties of SmCo5, *Phys. Rev. B* 59 (1999) 457, Copyright American Physical Society, with permission).

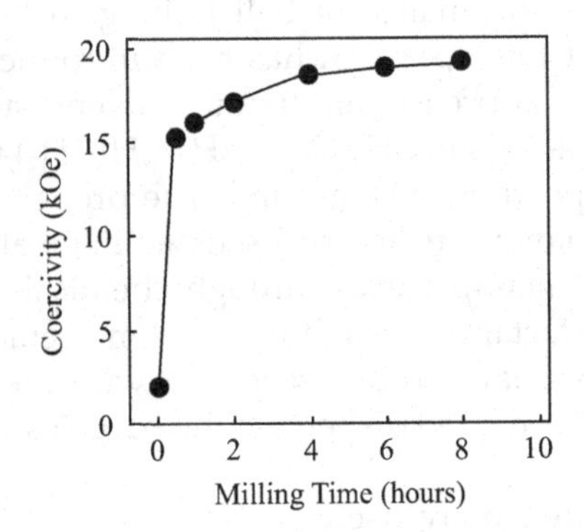

FIGURE 5.16
Coercivity of $SmCo_5$ nanoparticles produced by surfactant-assisted ball milling. (Adapted from N. G. Akdogan, G. C. Hadjipanayis and D. J. Sellmeyer, Anisotropic Sm-(Co,Fe) nanoparticles by surfactant-assisted ball milling, *J. Appl. Phys.* 105 (2009) 07A710, Copyright American Institute of Physics, with permission).

Example 5.4: Production of Fe_3O_4 Nanoparticles by Ball Milling

Nanoparticles of magnetite have been synthesized through ball milling from Fe_3O_4 powders dispersed in methanol (CH_3OH), which acts as the carrier liquid. In this study, the precursor was ~0.5 μm Fe_3O_4 particles which were dispersed in methanol at different concentrations (G. F. Goya, *Solid State Comm.* 130 (2004) 783). Between 20 and 40 mL of methanol was introduced in the grinder of a planetary mill along with the magnetite powder. Concentrations of R = 3, 10 and 50 wt% of magnetite were used. The grinder was hermetically sealed under an argon atmosphere. Partial amounts of sample were extracted at different intervals and the evolution of the median particle diameter was followed with grinding time as determined by TEM. Figure 5.17 gives an overall schematic of the experiment.

The results indicated that the particle size decreased steadily with grinding time up to 10–20 hours, depending on magnetite concentration (Figure 5.18), but the median particle size remained constant beyond this grinding time up to ~150 h. In addition, the final particle size attained of <*d*> = 7 nm was independent of concentration. The samples exhibited a log-normal distribution in particle diameter but with a correspondingly narrower polydispersity compared to other physical methods previously discussed. The polydispersity of the samples described in Figures 5.17 and 5.18 was in the range of $0.55 \leq \sigma \leq 0.77$. Dispersion in the carrier fluid appears to aid the stabilization of relatively uniformly sized particles and prevents cold working from taking place.

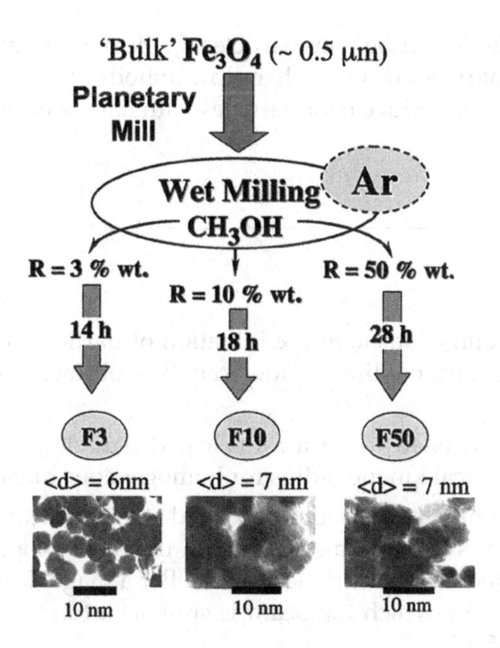

FIGURE 5.17
Schematic diagram of the ball milling experiment that resulted in a series of magnetite dispersions with different average crystallite sizes and concentrations. The TEM micrographs of the extracted samples are shown at the bottom. (From G. F. Goya, Handling the particle size and distribution of Fe_3O_4 nanoparticles through ball milling, *Solid State Comm.* 130 (2004) 783, Copyright Elsevier, reproduced with permission).

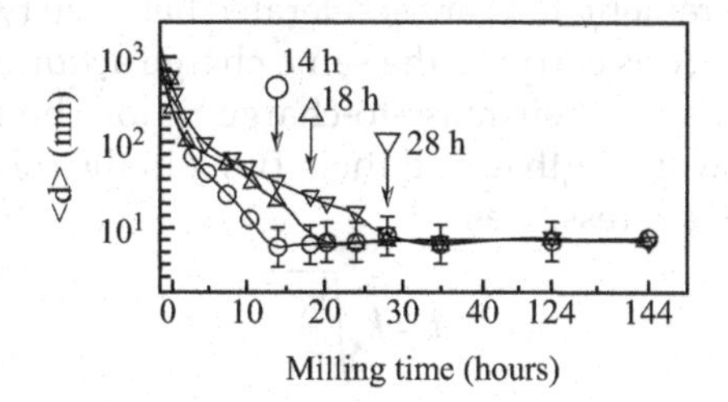

FIGURE 5.18
Semilog plot showing the evolution of the average particle size <*d*> with milling time for 3% (–o–), 10% (–Δ–) and 50 % (–∇–) of magnetic phase. (Adapted from G. F. Goya, Handling the particle size and distribution of Fe_3O_4 nanoparticles through ball milling, *Solid State Comm.* 130 (2004) 783, Copyright Elsevier, with permission).

The above-illustrated examples have allowed us to consider the basic principles involved in the production of magnetic nanoparticles by "top-down" physical methods. By no means do they represent an exhaustive review of all physical routes, variations, modifications and adaptations that investigators in the field of the production of magnetic nanoparticles have explored; this is a very interesting and rich area of nanomagnetism which continues to develop and evolve. They have, however, allowed us to appreciate how our fundamental understanding of the evolution of ferromagnetism from atom to bulk has been enhanced from the study of nascent, free ferromagnetic clusters and they have brought to our attention some limitations inherent to these approaches, as, for example, the large polydispersity of the resulting nanoparticles. In the next chapter, we explore the synthesis of magnetic nanoparticles through chemical methods and consider some experimental approaches designed to produce nanoparticles with narrower polydispersity.

Exercises

1 What presents the main obstacle to the initiation of particle nucleation in a homogeneous liquid or gaseous medium? How can this obstacle be overcome? Give an example.

2 Calculate the energy needed to form a nucleated particle of spherical shape with a radius equal to the critical kinetic radius for homogeneous nucleation.

3 A beam of Ag atoms emerges from a furnace held at a temperature $T = 1{,}000$ K. It is collimated using 0.1 mm slits and passed through the poles of a Stern–Gerlach magnet. The atoms are detected as they exit the magnet. For a magnet of length $L = 50$ cm, calculate the distance $2z$ by which the beam is split, as a function of the magnetic field gradient. See Figure 5.19.

4 Consider the cluster beam apparatus depicted in Figure 5.6(b) and a cluster beam passing through the Stern–Gerlach magnet. Derive an expression for the constant of proportionality C of Eq. (5.5) in terms of the geometry of the apparatus, that is, the length of the magnet L and the distance D the cluster beam travels after it leaves the magnet.

5 In a time-of-flight (TOF) mass spectrometer, the mass-to-charge ratio of an ion is determined *via* a time measurement. Ions are accelerated between two plates held at a known potential difference ΔV. Ions carrying the same charge acquire the same kinetic energy, but their speed depends on their mass-to-charge ratio. The ions are then allowed to travel down a TOF tube of length d, and their TOF t in the tube is measured.

(a) Prove that t can be expressed as

$$t = k\sqrt{\frac{m}{q}}$$

where the constant of proportionality k represents factors related to the instrument and experimental settings, such as d and ΔV.

(b) In a particular cluster beam experiment, a TOF tube of length $d = 1.5$ m and an accelerating voltage $\Delta V = 15$ kV were used. If singly ionized clusters were generated, what is the nuclearity, N, of an iron cluster whose TOF was detected to be $t = 4.8 \times 10^{-5}$ s?

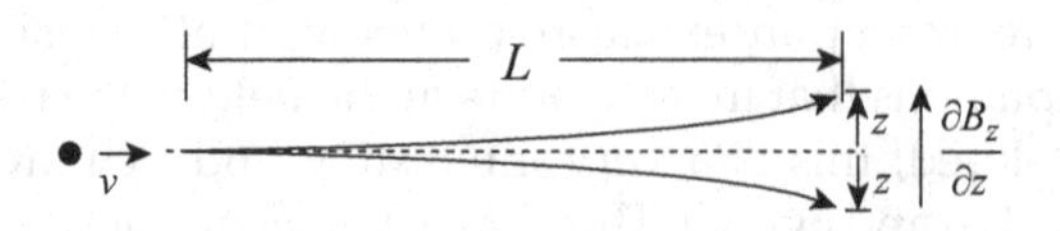

FIGURE 5.19
A beam of Ag atoms as it passes through a Stern–Gerlach magnet of length L.

6 Consider Figure 5.7. It is observed that for all three elemental ferromagnets, Fe, Co and Ni, the average magnetic moment per atom in small, low nuclearity clusters exceeds that of the corresponding bulk material. What is responsible for this effect?

7 According to the data presented in Figure 5.10, at low temperature (T = 10 K), the coercivity of the iron nanoparticles increases from a value of H_c ~ 1.56 kOe for particles of median diameter d = 14 nm, to a coercivity of H_c ~ 2.38 kOe for d = 3.3 nm. This sharp increase in coercivity is due to the contributions of surface anisotropy, K_s, to the total effective anisotropy, K_{eff}, of the particles as discussed in Section 4.10. Assuming isolated nanoparticles, with saturation magnetization independent of particle size for this particle diameter range (d between 4 and 14 nm) and uniaxial total effective anisotropy, use the relationship

$$H_c = \frac{2K_{eff}}{\mu_0 M_s}$$

to get an estimate of the factor by which the surface anisotropy constant, K_s, is stronger than the volume anisotropy constant, K_v.

8 In Exercise 8 of Chapter 4, you proved that the dependence of coercivity on T of a magnetic nanoparticle for temperatures below its blocking temperature is given by

$$H_c(T) = \frac{2K}{\mu_0 M_s}\left[1 - \left(\frac{T}{T_B}\right)^{½}\right]$$

where T is the temperature of measurement, T_B is the blocking temperature and $\frac{2K}{\mu_0 M_s} = H_c^0$ is the saturation coercivity at zero absolute temperature. Figure 5.11 gives the temperature dependence of the coercivity of iron nanoparticles derived from a vaporization–deposition technique. The coercivity of the 8.4-nm-diameter particles is seen to increase from H_c ~ 348 Oe at room temperature, T = 300 K, to H_c ~ 765 Oe at liquid nitrogen temperature, T = 80 K. Smaller nanoparticles show a sharper increase of coercivity with decreasing temperature, while larger nanoparticles show a less pronounced effect. For example, in the case of the 14-nm-diameter particles, the coercivity changes only from H_c ~ 1,040 Oe at T = 300 K to H_c ~ 1,200 Oe at T = 80 K. Use the relationship above and this experimental data extracted from Figure 5.11 to derive the blocking temperatures of the 8.4-nm-diameter and 14-nm-diameter iron particles.

6

Bottom-Up Synthesis by Chemical Methods

> The reduction of metal salts from ethereal or hydrocarbon solvents, using alkali metals as reducing agents, lead to highly reactive metal powders.
>
> **R. D. Rieke, from Accounts of Chemical Research, 10 (1977) 301**

Chemical methods are widely used in the production of magnetic nanoparticles due to their straightforward nature and cost-effective bulk quantity production. Solution chemistry offers a greater choice of precursors and greater variety and control of experimental parameters compared to physical methods. In addition, the molecular level design inherent in chemical synthesis leads to enhanced particle-phase homogeneity.

6.1 Particle Nucleation and Growth in Solution

A typical reaction vessel used in the chemical synthesis of nanoparticles is shown in Figure 6.1(a). Rapid injection of reagents through the syringes raises precursor concentration to supersaturation levels, inducing a surge of nucleation, as depicted in Figure 6.1(b). Nucleation depletes the solution of reagents to a level below supersaturation, impeding further nucleation. Subsequently, the nucleated particles grow under diffusion-limited conditions, that is, additional atoms from the solution slowly diffuse onto the particle's surface where they are adsorbed and bond to surface atoms. As long as the rate of adsorption of atoms from the solution is not exceeded by the rate of addition of reagents into the solution, no further nucleation occurs. A two-stage particle formation/growth is observed as proposed by LaMer, whereby the processes of particle nucleation and particle growth are separated in time. This temporal separation between nucleation and growth is crucial to the production of small, monodispersed particles. Many systems exhibit a second distinct particle growth phase, known as Ostwald ripening; during this phase, the high Gibb's free energy associated with the smaller nanoparticles in the solution promotes their dissolution back into solution and subsequent deposition onto the surface of the larger particles. As the particles grow with time, nanoparticles of various sizes can be harvested by periodically removing aliquots from the reaction vessel, as depicted at the bottom of Figure 6.1(b).

The harvested nanoparticles tend to agglomerate due to Van der Waals interactions and the high surface energies of individual particles. Particles in contact interact strongly magnetically, due to both dipole–dipole interactions and magnetic exchange through grain boundaries. These interactions compromise nanoscale properties, such as superparamagnetism and enhanced coercivities, which are uniquely associated with isolated magnetic nanoparticles. Various routes for the stabilization and isolation of magnetic nanoparticles in the mesoscopic regime have been advanced, all of which require surface passivation by terminal ligation or encapsulation by surfactant molecules that reduce surface tension and provide steric isolation of the magnetic cores.

DOI: 10.1201/9781315157016-8

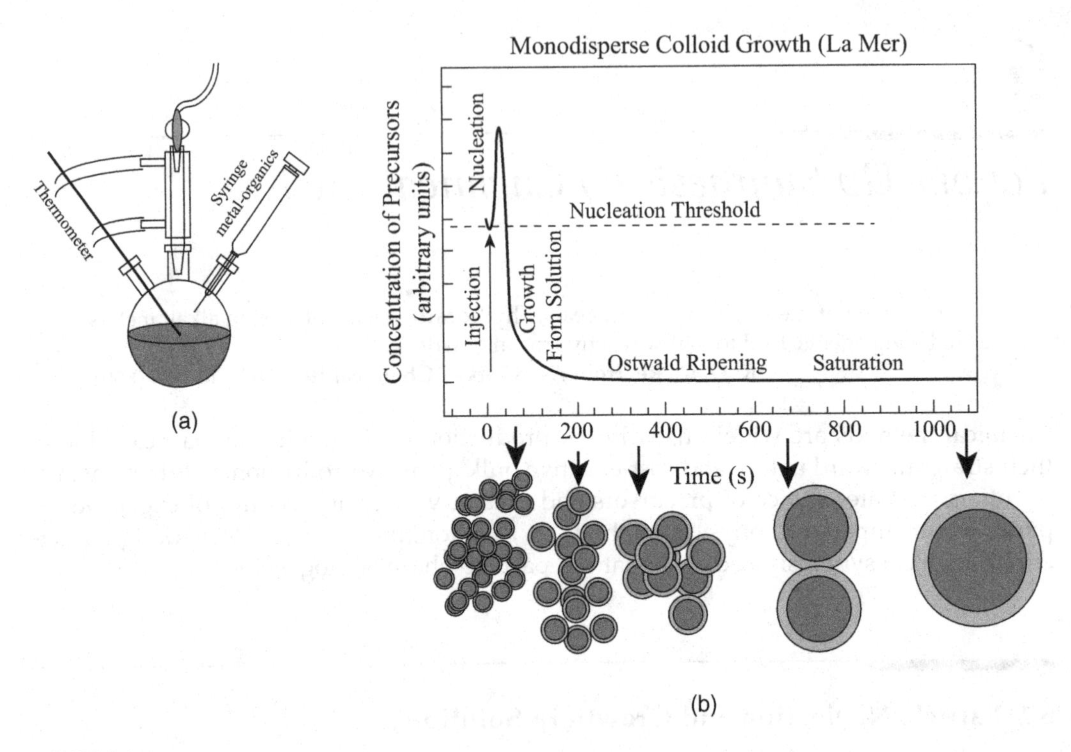

FIGURE 6.1
(a) Typical reaction vessel for the chemical synthesis of nanoparticles. (b) Schematic depiction of nucleation and growth of monodispersed nanoparticles in the framework of LaMer's theory. (From C. B. Murray, C. R. Kagan and M. G. Bawendi, Synthesis and characterization of monodispersed nanocrystals and closed-packed nanocrystal assemblies, *Annu. Rev. Mater. Sci.* 30 (2000) 545, Copyright Annual Reviews, reproduced with permission).

6.2 Cluster Stabilization by Terminal Ligation

Production of isolated, monodispersed particles or clusters in the "mesoscopic" regime requires control over (a) arresting particle growth at the desired particle size, and (b) preventing agglomeration. This is generally achieved by using competitive reaction chemistry between nucleation, growth and the capping of coordinately unsaturated surface atoms with organic ligands. Terminal surface ligation passivates tangling bonds and sterically prevents particle aggregation. In this section, we demonstrate the concept of surface passivation *via* terminal ligation, as used in traditional coordination chemistry. In particular, Example 6.1 presents the case of the hydrolytic polymerization of iron and the isolation and stabilization of nascent antiferromagnetic iron clusters at the initial stages of polymerization *via* surface atom terminal ligation with organic ligands. In Section 6.3, we describe in some detail the concept of particle surface passivation *via* surfactant molecules.

Example 6.1: Cluster Growth Onset in the Hydrolytic Polymerization of Iron

The hydrolytic polymerization of iron plays an important role in corrosion. It is well known that metallic iron in air, under humid conditions, undergoes oxidation to produce a brown flocculent powder of bulk iron oxides and hydroxides, commonly known as "rust". Rust consists of hydrated iron(III) oxides ($Fe_2O_3{\cdot}nH_2O$) and iron(III) oxo-hydroxides ($FeO(OH)$, $Fe(OH)_3$), which are produced by a sequence of iron oxidation and hydration/dehydration reactions as shown below:

$$
\begin{aligned}
&Fe \rightarrow Fe^{2+} + 2e^-. \\
&4Fe^{2+} + O_2 \rightarrow 4Fe^{3+} + 2O^{2-} \\
&Fe^{2+} + 2H_2O \rightleftharpoons Fe(OH)_2 + 2H^+ \\
&Fe^{3+} + 3H_2O \rightleftharpoons Fe(OH)_3 + 3H^+ \\
&Fe(OH)_2 \rightleftharpoons FeO + H_2O \\
&Fe(OH)_3 \rightleftharpoons FeO(OH) + H_2O \\
&2FeO(OH) \rightleftharpoons Fe_2O_3 + H_2O
\end{aligned}
\tag{6.1}
$$

Given sufficient time, oxygen and water, any metal iron mass will eventually convert entirely to rust through this process. Because of the difficulty in controlling hydrolytic chemistry, examples of discrete soluble polyiron oxo-hydroxo-complexes are rare. In order to arrest the iron polymerization process at the initial stages, and thus stabilize and isolate nascent polynuclear iron oxo-hydroxo-cluster complexes, iron polymerization must be carried out under controlled hydration reaction conditions. This requires the use of hydrocarbon solvents containing limited amounts of water and the capping of the cluster's surface by terminal ligation. The process entails the formation of iron complexes containing oxo and hydroxo bridging ligands producing superexchange-coupled iron ions in the dimer, trimer, tetramer and higher multimer clusters. In the presence of ligands containing carboxylate or imidazolyl coordinating moieties, the reaction of mononuclear $[FeCl_4]^{1-}$ and binuclear $[Fe_2OCl_6]^{2-}$ iron components in aprotic solvents can lead to the formation of $[Fe_3O]^{7+}$ and $[Fe_4O_2]^{8+}$ units containing oxo-bridged iron ions according to Scheme 6.1.

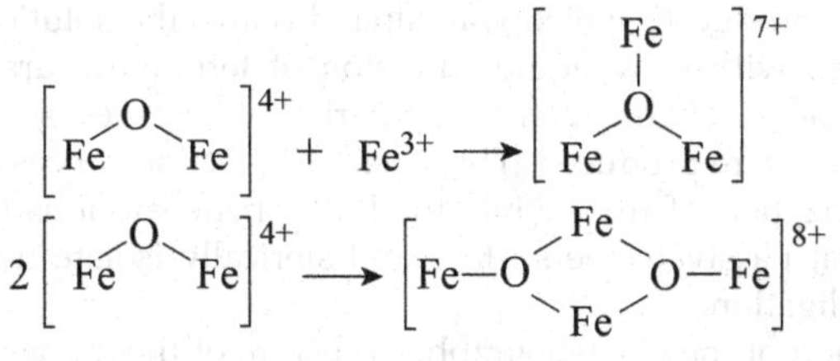

$$[Fe{-}O{-}Fe]^{4+} + Fe^{3+} \longrightarrow [Fe_2O(Fe)]^{7+}$$

$$2\,[Fe{-}O{-}Fe]^{4+} \longrightarrow [Fe{-}O(Fe)_2O{-}Fe]^{8+}$$

SCHEME 6.1
Controlled hydration reaction conditions in aprotic solvents containing limited amounts of water lead to the formation of higher iron nuclearity complexes.

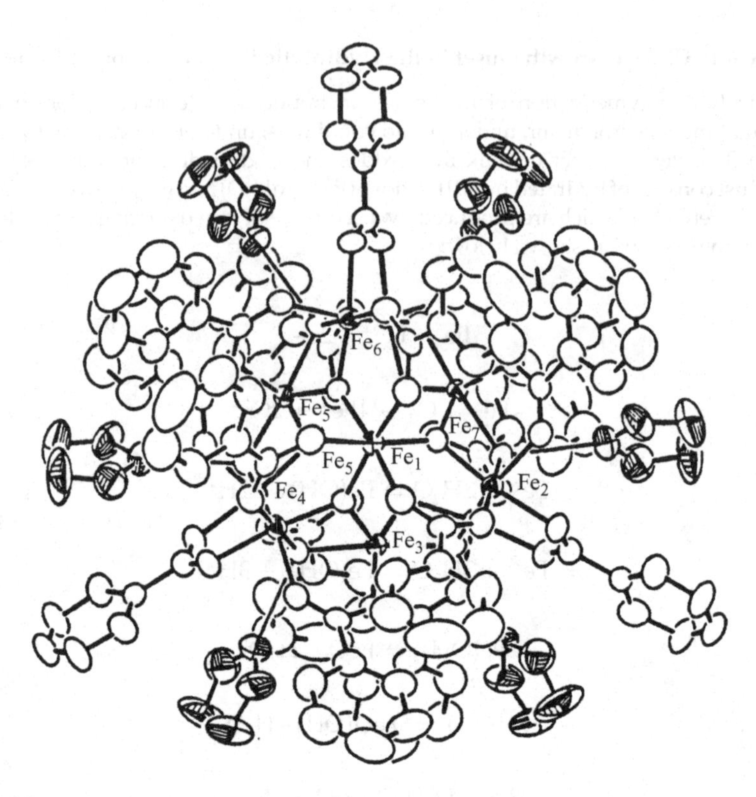

FIGURE 6.2
Crystal structure of $[Fe_{11}O_6(OH)_6(O_2CPh)_{15}]$ •6THF. (From S.M. Gorun, G.C. Papaefthymiou, R.B. Frankel and S.J. Lippard, Synthesis, structure, and properties of an undecairon(III) oxo-hydroxo aggregate: an approach to the polyiron core in ferritin, *J. Am. Chem. Soc.* 109 (1987) 3337, Copyright American Chemical Society, reproduced with permission).

The ability of the oxo-bridged diiron(III) units to aggregate and polymerize in this manner prompted investigators Gorun and Lippard (S. M. Gorun and S. J. Lippard, *Nature* 319 (1986) 666) to explore the formation of higher nuclearity iron ion clusters under controlled reaction conditions. The result was the synthesis of an unprecedented undecanuclear iron cluster of chemical formula $[Fe_{11}O_6(OH)_6(O_2CPh)_{15}]$ •6THF and of crystallographic structure as shown in Figure 6.2.

The $[Fe_{11}O_6(OH)_6(O_2CPh)_{15}]$ •6THF cluster was obtained by the hydrolysis of $[Fe_2O]^{4+}$ units following the synthesis route depicted in Figure 6.3. Briefly, the cluster was prepared by the addition of anhydrous sodium benzoate ($Na(O_2CPh)$) into an acetonitrile (CH_3CN) solution containing iron dimers in the form of $(Et_4N)_2[Fe_2OCl_6]$. Over time, a red suspension was produced, which was filtered out of the solution. The filtrate was subsequently treated with slow liquid diffusion of tetrahydrofuran (THF, $(CH_2)_4O$)) solution containing only 0.8% of water over a period of four weeks, whereby red-brown crystals of the cluster were produced. This chemical reaction process affords controlled hydrolytic polymerization of iron, while the bulky benzoate ions ($(O_2CPh)^-$) provide capping ligands that passivate the surface and sterically isolate the polynuclear iron cluster by terminal ligation.

Careful examination of the crystallographic structure of the cluster indicates the presence of oxygen atoms triply bridging Fe ions, forming a motif of FeO_6 octahedra within the $[Fe_{11}O_6(OH)_6]^{15+}$ core of the cluster. This kind of condensation is believed to occur when hydrated iron ions, $[Fe(H_2O)_6]^{3+}$, polymerize with the formation of oxo and hydroxo

$$(Et_4N)_2[Fe_2OCl_6]+2\ Na(O_2CPh) \xrightarrow{CH_3CN} \xrightarrow[\text{precipitate}]{\text{filter}} \text{red solution (57mM in iron)}$$

$$\text{Solution} \xrightarrow[\text{Liquid diffusion (1 month)}]{0.8\%\ H_2O\ \text{in THF}} [Fe_{11}O_6(OH)_6(O_2CPh)_{15}] + 6THF,\ 57\%\ \text{yield}$$

FIGURE 6.3
Synthesis of $[Fe_{11}O_6(OH)_6(O_2CPh)_{15}]\bullet 6THF$. (Adapted from S. M. Gorun and S. J. Lippard, A new synthetic approach to the ferritin core uncovers the soluble iron(III) oxo-hydroxo aggregate $[Fe_{11}O_6(OH)_6(O_2CPh)_{15}]$, *Nature* 319 (1986) 666, Copyright Springer Nature, reproduced with permission).

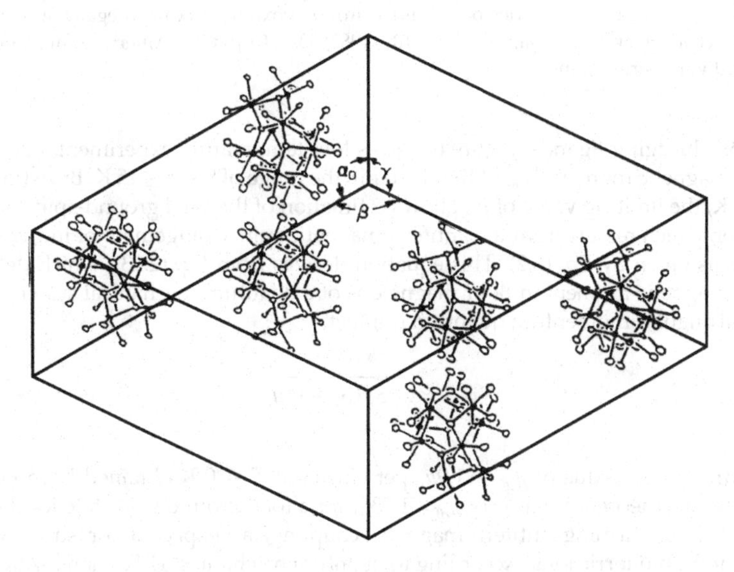

FIGURE 6.4
Crystal packing of Fe_{11} clusters. Only the inner magnetic cores $[Fe_{11}O_6(OH)_6]^{15+}$ of the clusters are shown. (From G. C. Papaefthymiou, Nanometer-Sized Structures and the Transition from the Molecular to the Solid State. *Phys. Rev.* **B 46**, (1992) 10366, Copyright American Physical Society, reproduced with permission).

bridges. The 15 benzoate groups provide terminal ligation of surface atoms and a protective sheath around the aggregate, limiting further growth and preventing aggregates from linking together via bridging ligands. On the periphery of the core, six protons associated with hydroxo bridges protrude through the benzoate sheath to form hydrogen bonds to six tetrahydrofuran solvent molecules. The bulky organic ligands form a hydrophobic shell that isolates and protects the hydrophilic iron cluster from further hydrolytic polymerization. Figure 6.4 shows the crystal packing of the Fe_{11} clusters. These are neutral molecules forming crystals by Van der Waals forces. The bulky benzoate ligands that encapsulate the central exchange-coupled ion cores, which are about 1 nm in diameter, keep the cores magnetically isolated from one another at an average distance between centroids of ~2 nm. Thus, magnetic ordering is limited to intra-molecular spin interactions rather than long-range magnetic interactions throughout the crystal.

The coordination geometry and physical properties of individual iron ions correspond to high spin $S = 5/2$, Fe^{3+} ions, antiferromagnetically coupled through superexchange oxo

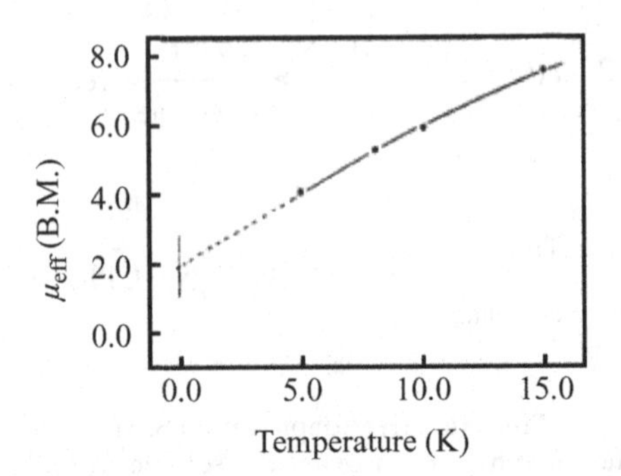

FIGURE 6.5
Temperature dependence of the effective magnetic moment per Fe11 cluster at low temperatures and extrapolation to $T \rightarrow 0$ K. (From S.M. Gorun, G.C. Papaefthymiou, R.B. Frankel and S.J. Lippard, Synthesis, structure, and properties of an undecairon(III) oxo-hydroxo aggregate: an approach to the polyiron core in ferritin, *J. Am. Chem. Soc.* 109 (1987) 3337, Copyright American Chemical Society, reproduced with permission).

or hydroxo bridging ligands. Figure 6.5 gives low-temperature experimental data on the effective magnetic moment, μ_{eff}, of the cluster in the range of $5 \leq T \leq 15$ K. By extrapolating to $T \rightarrow 0$ K, the limiting value of μ_{eff} allows estimation of the total ground-spin state, S_T, of the cluster, which results from the antiferromagnetic spin arrangement of individual iron ion spins as given by Eq. (6.2). This equation derives from Eq. (2.23), which defines the effective magnetic moment in Bohr magnetons of a quantum mechanical system in terms of its total angular momentum quantum number.

$$\mu_{eff} = g\sqrt{S_T(S_T+1)}\mu_B \tag{6.2}$$

The extrapolated value of $\mu_{eff} \rightarrow 1.9\mu_B$ per cluster at $T \rightarrow 0$ K obtained from Figure 6.5 approaches the theoretical value of $\mu_{eff} = 1.73\mu_B$ for a total ground spin state for the cluster of $S_T = 1/2$, confirming antiferromagnetic coupling, as expected for superexchange-coupled high spin ferric ions. According to quantum mechanics, 11 Fe^{3+} ions with spin $S = 5/2$ each can couple to give total spins $S_T = \frac{1}{2}, \frac{3}{2}, \frac{5}{2}$, *etc.*, up to a maximum of $S_T = 55/2$; the maximum corresponding to the case where all 11 iron spins are parallel to each other. The rapid rise of μ_{eff} with temperature in Figure 6.5 indicates the presence of low lying, excited S_T states that thermal energies, kT, cause to populate as the temperature is increased.

The polyiron complex $[Fe_{11}O_6(OH)_6(O_2CPh)_{15}]$ of Example 6.1 is a supramolecular cluster of well-characterized, singly defined molecular crystallographic structure and size. Ensembles of such clusters possess absolute monodispersity; they present themselves as ideal experimental systems to test the magnetism of isolated, monodispersed magnetic structures at the nanoscale. As indicated in Figure 6.4, due to the bulky benzoate ligands encapsulating the cluster, there are no significant dipole–dipole interactions between the clusters. Isolated multinuclear iron clusters of this type can address the fundamental question of the onset of collective magnetic interactions and particle-like behavior with increasing cluster size in antiferromagnetic iron oxides, just like the cluster beam experiments described in Chapter 5 did for the case of metallic iron clusters. We will return to this important subject in the latter chapters.

6.3 Particle Stabilization by Surfactant Molecules

Unlike the case of "supramolecular" magnetic clusters that possess a single size and unique crystallographic structure, nanoparticles may be produced in a highly crystalline, poorly crystalline or amorphous phase and they are always polydispersed to some degree. The synthesis of monodispersed (uniform size) magnetic nanoparticles is of paramount importance for technological applications and scientific inquiry. For example, the synthesis of discrete nanoparticles in the range of 2–20 nm is of significant importance in applications of high-density magnetic storage devices, biotechnology and nanomedicine. Colloidal chemistry methods have been advanced in the preparation of discrete magnetic nanoparticles stabilized by surfactants and dispersed in various solvents. In order to fully appreciate how surfactants stabilize nascent magnetic nanoparticles, we must consider the structure and properties of surfactant molecules in greater detail.

Surfactants derive their name from "surface acting" agent. They are usually long-chain organic compounds that contain a polar (hydrophilic) and a non-polar (hydrophobic) section and they can be neutral or ionic. Figure 6.6(a) gives cartoon representations of various surfactants, while Figures 6.6(b–g) give the structure of some representative surfactant molecules. Note that single-chain, double-chain and even triple-chain surfactants are encountered. The structure of oleic acid, $CH_3(CH_2)_7CH{=}CH(CH_2)_7COOH$, a commonly used neutral surfactant, is depicted in Figure 6.6(b). The carboxyl group on the right end of the molecule carries an electric dipole moment due to the uneven distribution of electronic charge. It is referred to as the "polar head" of the molecule. The long hydrocarbon tail on the left has a homogeneously distributed electronic charge and thus it is non-polar. Polar molecules tend to interact *via* dipole–dipole interactions or *via* the formation of hydrogen bonds and thus are soluble in water; they are "hydrophilic". In contrast, non-polar molecules do not dissolve in water; they are "hydrophobic". Instead, they dissolve in non-polar, hydrocarbon solvents, such as hexane for example. A surfactant molecule contains both, a water-insoluble and a water-soluble component. Surfactants, therefore, are "amphiphilic" molecules containing both a "hydrophilic head" and a "hydrophobic tail". Whether a surfactant is soluble in protic or aprotic solvents depends on which of the two sections of the molecule is more dominant. All surfactants share this property and a surfactant molecule is usually represented by the cartoons shown in Figure 6.6(a), composed of a "polar head" and a "non-polar tail".

One approach to the synthesis of uniform-sized magnetic nanoparticles is to inject the reagents into a hot surfactant solution. Upon particle nucleation, the surfactant molecules migrate onto the surface of the particle, forming a sheath around the particle as shown in Figure 6.7, depicted for the case of a hydrocarbon-solvent-based reaction medium. The polar heads of the surfactant molecules, trying to avoid the aprotic solvent, adsorb or bond onto the surface of the particle, leaving the hydrocarbon tail in close interaction with the organic solvent. This alignment and aggregation of surfactant molecules on the particle's surface act to alter the surface properties, reduce interfacial tension, and help disperse the particles in the solvent medium. The surfactant coating, thus, stabilizes the particles and prevents them from clustering by steric repulsion. If additionally, the surfactant is ionic, then electrostatic repulsion is also present.

Dynamic adsorption and desorption of surfactant molecules onto the surface of the particles allow reagents to reach the surface and thus enable particle growth after nucleation. Particle size can be controlled during synthesis by varying reaction parameters, such as reaction time, temperature, and the relative amounts of reagents and surfactant molecules

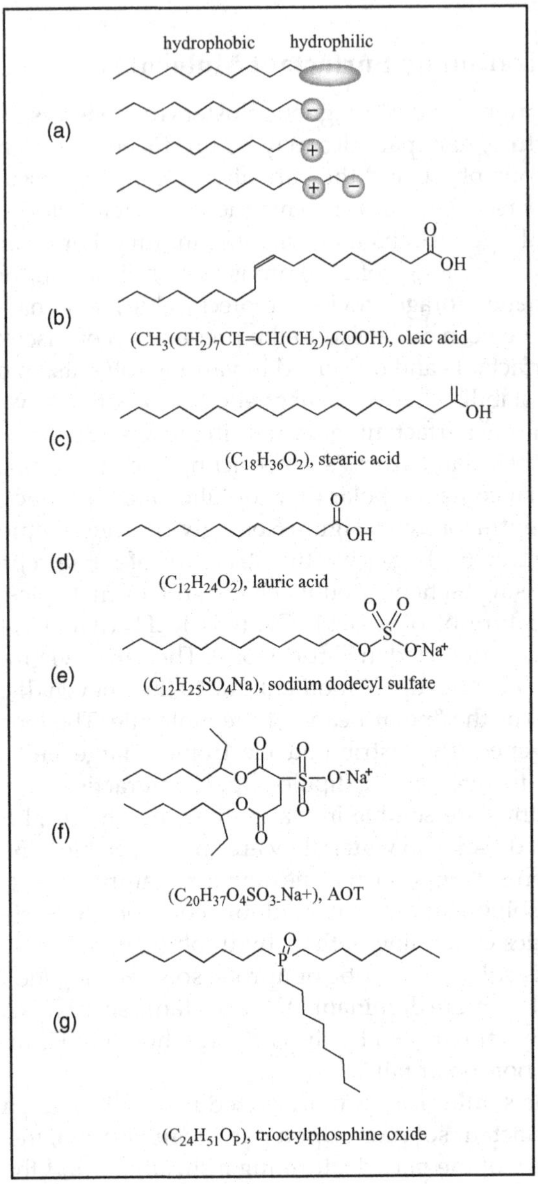

hydrophobic hydrophilic

(a)

(b) $(CH_3(CH_2)_7CH{=}CH(CH_2)_7COOH)$, oleic acid

(c) $(C_{18}H_{36}O_2)$, stearic acid

(d) $(C_{12}H_{24}O_2)$, lauric acid

(e) $(C_{12}H_{25}SO_4Na)$, sodium dodecyl sulfate

(f) $(C_{20}H_{37}O_4SO_{3-}Na+)$, AOT

(g) $(C_{24}H_{51}O_P)$, trioctylphosphine oxide

FIGURE 6.6
(a) Cartoon representation of surfactant molecules, neutral, negatively or positively charged, (b–g) chemical structures and molecular formulas of some common surfactants.

in the solution. In general, particle size increases with increasing reaction time and with increasing temperature. The derived nanoparticles can usually be dispersed in a variety of solvents and can later be retrieved as powders when the solvent is removed by drying. The resulting powders generally contain nanoparticles of polydispersity ~ 15%. Spherical nanoparticles are considered to be monodispersed if they exhibit a standard deviation in diameter less than 5%. Thus, subsequent size-selection processes are needed in order to narrow the polydispersity of the nanoparticles below 5%. Figure 6.8 depicts schematically, the general features of this synthetic process.

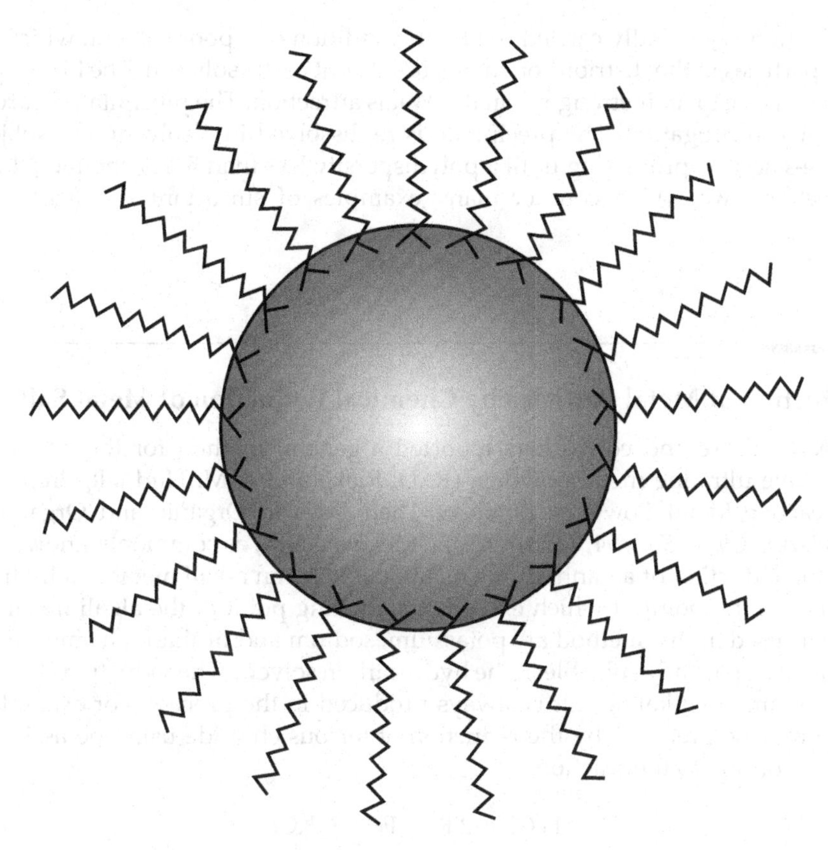

FIGURE 6.7
Depiction of a surfactant sheath around a nanoparticle in an organic solvent.

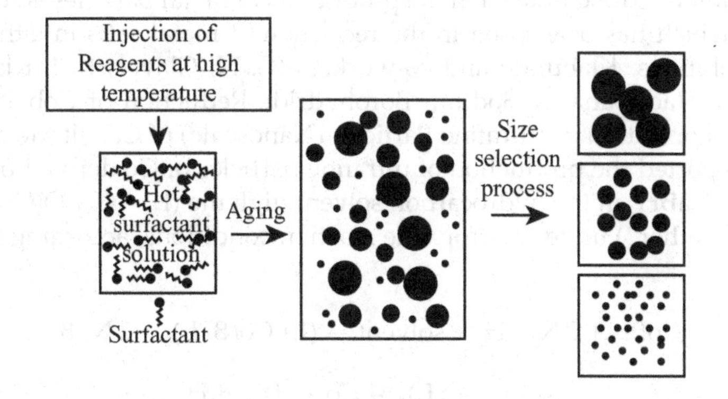

FIGURE 6.8
Generalized synthesis of monodisperse nanoparticles by the injection of reagents into hot surfactant solutions followed by aging and size-selective processes. (From T. Hyeon, Chemical Synthesis of Magnetic Nanoparticles, *Chem. Commun.* (2003) 927–935, Copyright the Royal Society of Chemistry, reproduced with permission).

Size selection is generally carried out by the addition of a poor solvent, which induces the larger particles in the distribution to precipitate out of the solution. The bigger particles flocculate first due to their stronger Van der Waals attraction. The precipitated particles are removed by centrifugation. The precipitate is re-dissolved in a solvent and subjected to further size-selection processing until a polydispersity less than 5% is obtained. In the following sections, we will encounter many examples of surfactant stabilized magnetic nanoparticles.

6.4 Production of Metal Particles by Chemical Reduction of Metal Salts

In 1972, R. D. Rieke and co-workers reported a general method for the preparation of highly reactive ultrafine metal powders (R. D. Rieke and P. M. Hudnall, Preparation of Highly Reactive Metal Powders: Some of Their Uses in Organic and Organometallic Synthesis *J. Am. Chem. Soc.*, 94, 7178 (1972)). "Rieke metals", as commonly known, are prepared by the reduction of an anhydrous metal halide by an alkali metal in a hydrocarbon solvent, the boiling point of which exceeds the melting point of the alkali metal. Typical alkali metals used in this method are potassium, sodium and lithium. The metal salt to be reduced must be partially solvable in the hydrocarbon solvent. The reduction is carried out under argon and an alkaline salt is always produced in the process. For example, Rieke iron metal may be prepared by the reduction of ferrous chloride using potassium as the reductant according to the reaction,

$$FeCl_2 + 2\,K \rightarrow Fe + 2\,KCl \tag{6.3}$$

Ultrafine metal particles of Fe, Co and Ni have been generated through this method. Reaction rates and particle sizes can be controlled by varying the temperature of the reaction solution. The particles are highly reactive and are susceptible to oxidation when exposed to air.

Another common reductant used in the production of metal particles is sodium borohydride, $NaBH_4$, which has been used in the reduction of metal salts in either aqueous or hydrocarbon solutions. Klabunde and co-workers (G. N. Glavee, K. J. Klabunde, C. M. Sorensen and G. Hadjipanayis, Sodium Borohydride Reduction of Cobalt Ions in Nonaqueous Media. Formation of Ultrafine Particles (Nanoscale) of Cobalt Metal. *Inorg. Chem.* 32 (1993) 474) reported the production of ultrafine particles of Co derived from the reduction of $CoBr_2$ by $NaBH_4$ in the hydrocarbon solvent diglyme (L = $[CH_3O(CH_2CH_2O)_2CH_3]$, "diglycol methyl ether") under appropriate reaction conditions, according to the reaction scheme:

$$CoBr_2 + 2\,NaBH_4 + \text{solvent} \rightarrow (L)_nCo(BH_4)_2 + 2\,NaBr \tag{6.4}$$

$$(L)_nCo(BH_4)_2 \rightarrow Co + H_2 + B_2H_6 \tag{6.5}$$

where $(L)_n$ indicates the number of solvent molecules bound to the cobalt borohydride produced in the first step of the reaction, Eq. (6.4). This intermediate product of $(L)_nCo(BH_4)_2$ is unstable and decomposes upon heating to metallic cobalt in the form of ultrafine particles, hydrogen and borohydride. As we can see, the solvent plays an important role in the synthesis.

TABLE 6.1
Metal Containing Products of Borohydride Reduction of Fe, Co and Ni Ions

Metal ion	Reduction Potential, E^a (V)	In Water under Argon	In Diglyme under Argon	In Water under Air
Fe^{3+}	−0.036[b]	Fe	FeB	Fe, FeO_x
Fe^{2+}	−0.41	Fe	Fe_2B	Fe, FeO_x
Co^{2+}	−0.28	Co_2B	Co	Co, $Co_3(BO_3)_2$
Ni^{2+}	−0.23	Ni_2B	Ni, Ni_2B, Ni_3B	Ni, NiO

[a] Standard condition $M^{2+} + 2e^- \rightarrow M$
[b] $Fe^{3+} + 3e^- \rightarrow Fe$.
Source: Adapted with permission from G. N. Glavee, K. J. Klabunde, C. M. Sorensen and G. Hadjipanayis, *Langmuir* 10 (1994) 4726, Copyright American Chemical Society.

The production of Ni and Fe ultrafine powders has also been demonstrated by borohydride reduction and subsequent heat treatment. Table 6.1 gives a summary of the products of the reactions. A mixture of compounds is obtained along with the metallic nanopowders. Due to high surface energy, all metal nanoparticles produced by this method are susceptible to surface oxidation and many would burn upon exposure to air.

In contrast, when the anaerobic reduction of metal salts takes place in the presence of surfactants, monodispersed magnetic nanoparticles can be produced. In a general scheme introduced by investigators, metal halides or acetates are dissolved in high boiling-temperature inert solvents, along with a combination of long-chain carboxylic acids, such as oleic acid, as stabilizing surfactants. The solution of metal salts and surfactants is vigorously stirred and heated before a solution containing a strong reducing agent is injected through the syringe into the reaction flask of Figure 6.1(a). Under these conditions, metal nanocrystals nucleate and grow till the reagent is consumed. The size of the nanocrystals can be coarsely tuned by the ratio of surfactants to metal salts. Followed by size-selection precipitation processes, nanocrystals with particle size distributions of less than ~5% standard deviation in particle diameter can be produced. The as-synthesized nanocrystals may be further processed by heating in various atmospheres, a process known as "annealing". Annealing is generally used to induce crystallization of amorphous nanoparticles or to induce crystallographic phase transitions in nanocrystals, with accompanied changes in magnetic properties.

Example 6.2: Synthesis of Cobalt Nanocrystals by Reduction of Cobalt Chloride

Sun and Murray (S. Sun and C. B. Murray, Synthesis of monodisperse cobalt nanocrystals and their assembly into magnetic superlattices, *J. Appl. Phys.* 85 (1999) 4325) reported the synthesis of monodispersed Co nanocrystals with precise control over sizes from 2 to 11 nm diameter using selective precipitation after synthesis. Each Co nanocrystal was encapsulated within a sheath of a robust organic ligand shell (surfactant) which limited oxidation and prevented aggregation of particles. The particle synthesis was achieved with the injection of 2 M dioctyl-ether super-hydride ($LiHB(CH_2CH_3)_3$ solution into a hot bath (200°C) $CoCl_2$ dioctyl-ether solution in the presence of oleic acid and trialkylphosphine (PR_3, R = n-C_4H_9 or n-C_8H_{17}), as surfactants. A reduction occurred instantly upon injection leading to the simultaneous formation of many small metal clusters. Continued heating at 200°C allowed steady growth of these clusters into nanosize single crystals of cobalt. The steric bulk of the alkylphosphine controlled the rate of particle growth. Short-chain alkylphosphines allowed faster growth as reagents could more easily reach the particle surface during dynamic adsorption and desorption of the surfactant, while heavier chain alkylphosphines reduced growth and favored the production of smaller particles. When the desired particle size was reached, further particle growth was arrested by cooling the dispersion. These organically stabilized cobalt nanocrystals were readily dispersible in aliphatic, aromatic and chlorinated solvents. Furthermore, using short-chain alcohols as poor solvents led to size-selective precipitation, allowing the isolation of monodispersed samples with extremely narrow size distribution ($\sigma < 7\%$) in diameter.

Figure 6.9 gives the X-ray-powder-diffraction (XRD) patterns of the surfactant-coated, 11-nm diameter selected cobalt nanocrystals. XRD characterization is a standard procedure in materials science and solid-state chemistry for characterizing new compounds. It is based on the fact that crystalline substances act as three-dimensional diffraction gratings for X-ray wavelength, λ, similar to the spacing, d, between planes in a crystal lattice. When a collimated, monochromatic X-ray beam is scattered off a crystal, one observes the constructive interference of the scattered radiation according to Bragg's law

$$n\lambda = 2d \sin\theta \tag{6.6}$$

where n is an integer, λ the wavelength of the electromagnetic radiation and θ the scattering angle, as seen in Scheme 6.2. This makes X-ray diffraction a common technique for the study of crystal structures and atomic spacings. When the sample is in powder form and the diffracted radiation is collected on a flat plate detector the averaging of all possible orientations of a crystallite relative to the X-ray beam leads to smooth diffraction rings around the beam axis, rather than to a diffraction pattern of regularly spaced spots observed for single crystal diffraction. The angle between the beam axis and the ring (the scattered radiation) is 2θ. Powder diffractometers can operate in either the transmission or reflection

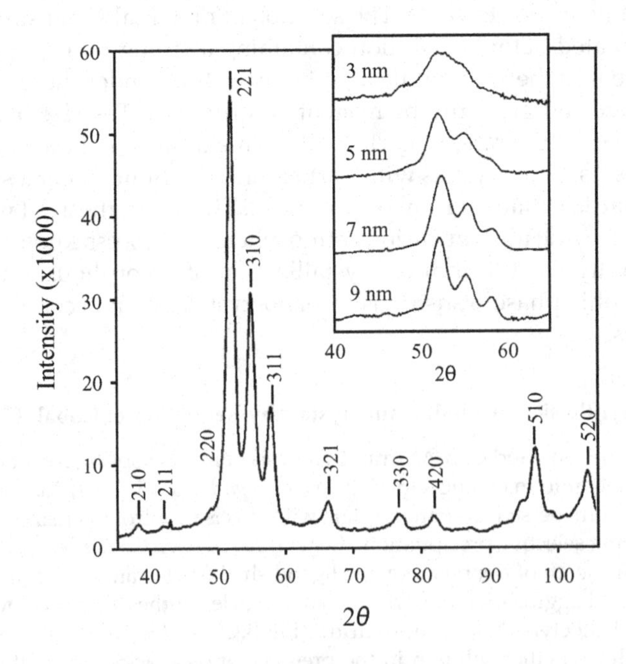

FIGURE 6.9
XRD pattern of 11-nm diameter ε-Co nanocrystals deposited on Si(100) substrate. (Inset: Size dependence of the XRD patterns from S. Sun and C. B. Murray, Synthesis of monodisperse cobalt nanocrystals and their assembly into magnetic superlattices, *J. Appl. Phys.* 85 (1999) 4325, Copyright American Institute of Physics, reproduced with permission).

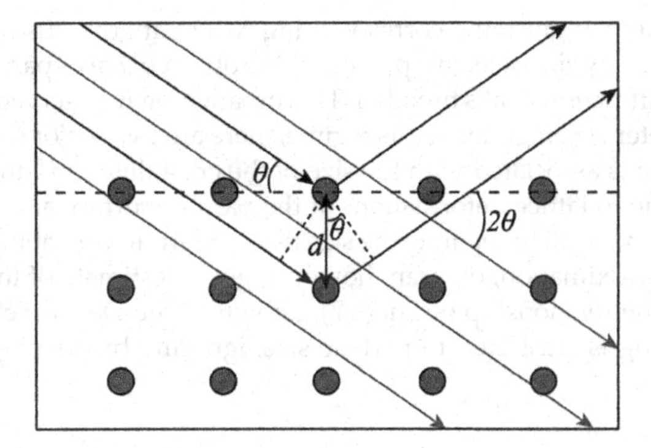

SCHEME 6.2
Constructive interference between the two indicated scattered beams takes place when the difference in their path lengths is an integral multiple of the wavelength of the radiation, according to Eq. (6.6).

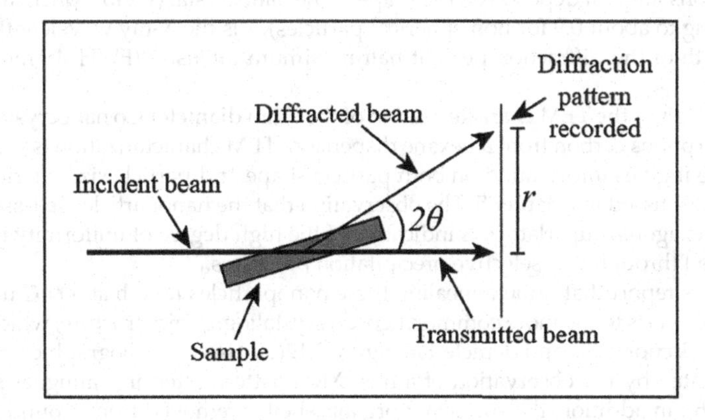

SCHEME 6.3
Schematic of an X-ray diffractometer. For a powder sample, the diffracted beam forms a ring of radius r on a flat-screen detector.

configuration, with the reflection configuration being the more common. The powder sample is filled in a small disc-like container and its surface is carefully flattened. The geometry of an X-ray diffractometer is such that the sample rotates in the path of the collimated, monochromatic X-ray beam at an angle θ while the X-ray detector is mounted on an arm to collect the reflected diffracted X-rays and rotates at an angle of 2θ (see Scheme 6.3). By scanning through a range of 2θ angles, all possible diffraction directions of the lattice should be attained due to the random orientation of the powdered material.

Each ring represents a certain lattice plane identified by three integers, known as Miller indices, a notation system for planes and directions in crystal lattices. Conversion of the diffraction peaks to d spacings allows the identification of the compound because each compound has a set of unique d spacings. Typically, this is achieved by the comparison of d spacings with standard reference patterns stored in computer data banks. The fundamental physics upon which the technique is based provides high precision and accuracy in the measurement of inter-planar spacings, sometimes to fractions of an Angstrom, resulting in authoritative identification of the crystallographic phase. The observed pattern in Figure 6.9 identifies the presence of ε-Co with a cubic crystal cell, a less commonly observed crystallographic structure of metallic Co, compared to its normal hexagonal crystallographic structure discussed in Section 3.3.2.

The inset of Figure 6.9 gives the corresponding XRD patterns of smaller diameter Co nanoparticles isolated by size-selective processes. We observe that as particle size decreases, the characteristic diffraction peaks broaden. This broadening is observed only for particles in the sub-micrometer size regime. In this regime, there are two major contributions to line broadening; the first is associated with the size of the crystallites and the second with lattice micro-strain due to lattice deformations at the particle surface and in the presence of defects. It is usually difficult to separate the size from the strain contributions to line broadening. To a first approximation, one can, however, get an estimate of the lower bound of particle size from the relationship of Eq. (6.7), known as the Debye–Scherrer formula, in which all broadening is attributed to particle size, ignoring broadening due to strain or instrumental factors.

$$\tau = \frac{k_{sh}\lambda}{\beta\cos\theta} \tag{6.7}$$

Here, τ is the linear dimension of the nanocrystal (its diameter for a spherical nanocrystal), k_{sh} is a constant that depends on the shape of the nanocrystal (1.2 for spherical particles and deviating to about 0.9 for non-spherical particles), λ is the X-ray wavelength and β is the full width of the diffraction peak at half maximum intensity (FWHM), measured in radians.

Figure 6.10 gives the TEM characterization of the 9-nm diameter Co nanocrystals deposited on amorphous carbon from a hexane dispersion. TEM characterization is very important because it gives information on both particle shape and particle size distribution, as we already discussed in Chapter 5. The observation that the nanoparticles self-assemble to form a 2D hexagonal superlattice is indicative of the high degree of uniformity in particle size, achieved through size-selective precipitation processes.

The authors report that upon annealing these nanoparticles for 3 h at 300°C under vacuum, ε-Co converts to the more common hcp-Co crystallographic structure, which we first discussed in Section 3.3.2 and depicted in Figure 3.12(c). This crystallographic transformation is indicated by the observation of a new XRD pattern after annealing, as shown in Figure 6.11(b). In addition, the surfactant organic shell is removed from around the particles during the process of annealing, making the particles vulnerable to oxidation upon exposure to air. One can avoid excessive oxidation by immersing the annealed samples in acetone prior to exposure to air. However, trace amounts of oxygen and water in the

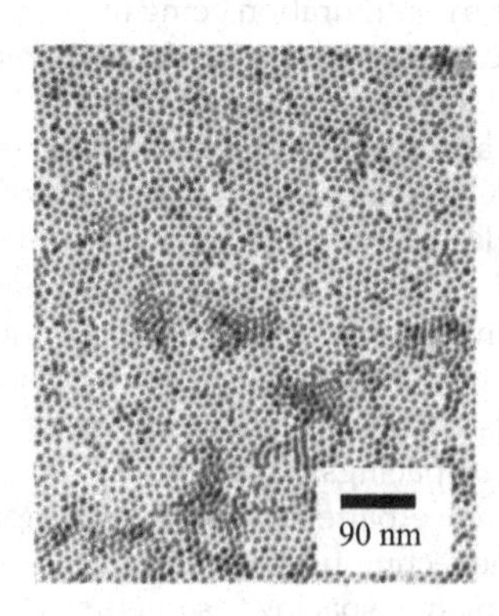

FIGURE 6.10
TEM micrograph of 9-nm diameter cobalt nanocrystals. (From S. Sun and C. B. Murray, Synthesis of monodisperse cobalt nanocrystals and their assembly into magnetic superlattices, *J. Appl. Phys.* 85 (1999) 4325, Copyright American Institute of Physics, reproduced with permission).

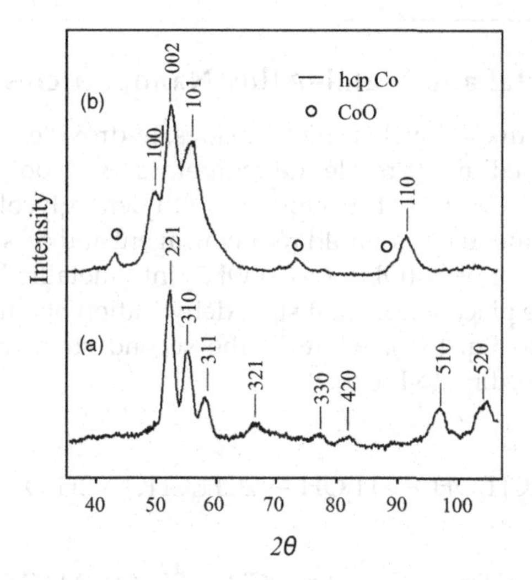

FIGURE 6.11
XRD patterns of Co-nanocrystals (a) before and (b) after annealing at 300°C. (From S. Sun and C. B. Murray, Synthesis of monodisperse cobalt nanocrystals and their assembly into magnetic superlattices, *J. Appl. Phys.* 85 (1999) 4325, Copyright American Institute of Physics, reproduced with permission).

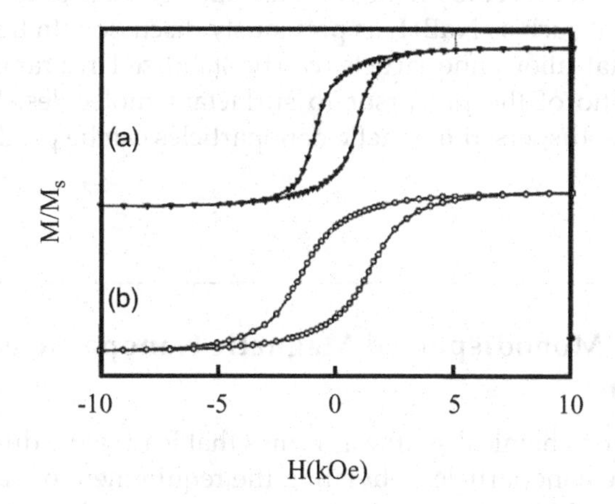

FIGURE 6.12
Hysteresis loops of Co-nanocrystals (a) before and (b) after annealing at 300°C for 3 h under vacuum. (Adapted from S. Sun and C. B. Murray, Synthesis of monodisperse cobalt nanocrystals and their assembly into magnetic superlattices, *J. Appl. Phys.* 85 (1999) 4325, Copyright American Institute of Physics, reproduced with permission).

acetone lead to some surface oxidation, as indicated by the weak CoO signature identified in the XRD pattern. The phase transition from ε-Co → hcp-Co was accompanied by a change in coercivity, as shown in Figure 6.12. Before annealing, the as-prepared 9-nm ε-Co nanoparticles showed a coercivity H_c = 500 Oe; after annealing, the coercivity increased to H_c = 1,450 Oe. TEM micrographs showed no change in particle size upon annealing, indicating that the increase in coercivity derives entirely from the crystallographic phase transition from ε-Co → hcp-Co.

6.5 Preparation of Metal and Metal–Alloy Nanoparticles in Polyol Media

Reduction of metallic precursors, such as metal oxides, hydroxides and salts, can also take place in a liquid polyol medium at moderate temperatures. A polyol is an organic compound containing multiple hydroxyl groups (*e.g.*, ethylene glycol, glycerol, diethylene glycol), which acts simultaneously as a mild reducing agent and dissolving medium. In the reduction by ethylene glycol of $Ni(OH)_2$ or $Co(OH)_2$ into metallic Ni and Co, a two-step reaction is believed to take place. In the first step, dehydration of ethylene glycol produces acetaldehyde according to Eq. (6.8), while in the second step, reduction of the metal hydroxide takes place according to Eq. (6.9)

$$2CH_2OH - CH_2OH \rightarrow 2CH_3CHO + 2H_2O \tag{6.8}$$

$$2CH_3CHO + M(OH)_2 \rightarrow CH_3COCOCH_3 + 2H_2O + M^\circ \text{ (M = Co or Ni)} \tag{6.9}$$

The process can also lead to the formation of metal–alloy Co-Ni nanoparticles. To obtain such alloys the preparation conditions must be carefully optimized to adjust the kinetics of the reaction to both constituents, producing conditions under which the direct formation of the alloy nanoparticles is thermodynamically more favorable than the separate formation of the constituent metals. Alloy nanoparticles can also be produced by the addition of a strong reducing agent such as $NaBH_4$ as previously discussed. In the presence of surfactants, the metal or metal–alloy nanoparticles can be stabilized in a range of particle sizes by varying the molar ratio of the precursor to surfactant molecules. Followed by a size-selection process, monodispersed magnetic nanoparticles can be produced.

6.6 Preparation of Monodispersed Magnetic Nanoparticles Using Microemulsions

It is desirable to develop chemical synthesis routes that lead to the direct formation of uniformly sized magnetic nanoparticles, obviating the requirement of size-selection processing after synthesis. To this end, investigators have advanced more elaborate synthesis techniques in which chemical reactions in solution take place within confined spaces derived from the elaborate architectural motifs of microemulsions.

Emulsions are mixtures of two immiscible or unblendable liquids, such as polar and non-polar solvents, usually referred to as "water" and "oil" mixtures depicted in Figure 6.13. Such mixtures form colloidal systems in which both the dispersed and continuous phases are liquid. For the emulsion to be formed, rigorous shaking of the mixture is necessary, Figure 6.13(b). Due to high interfacial tension, the colloidal phase is unstable and over time, phase separation occurs, as depicted in Figure 6.13(c). If surfactants are also added to the mixture, the surfactant molecules self-assemble at the interface between the two immiscible phases, reducing the interfacial tension and, thus, stabilizing the colloid, Figure 6.13(d).

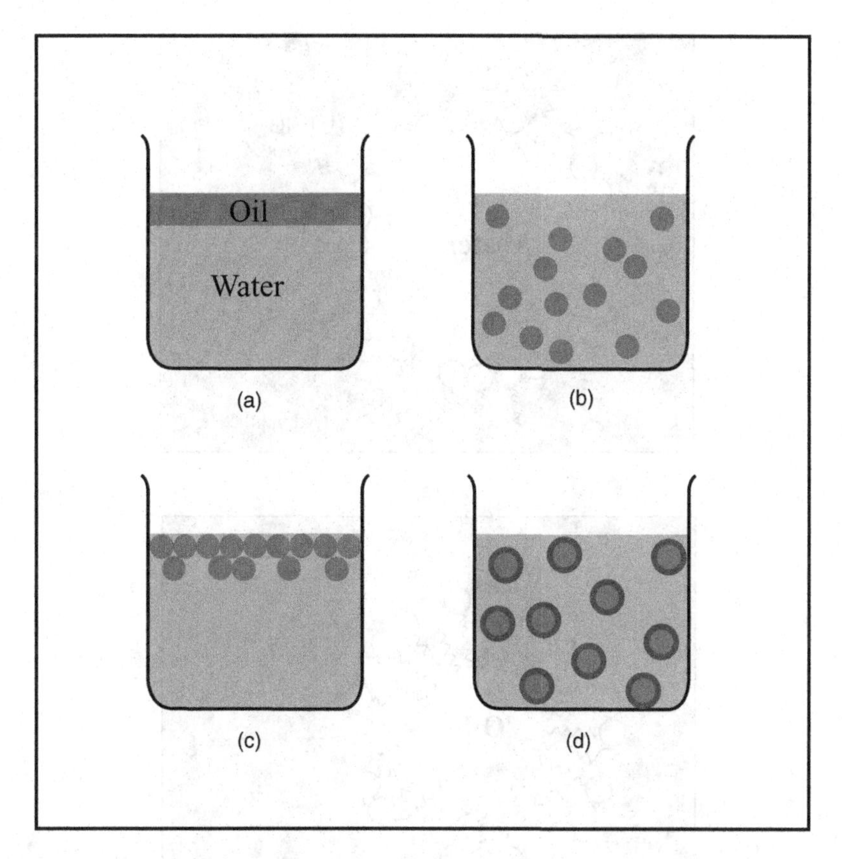

FIGURE 6.13
(a) Mixture of two immiscible liquids, (b) formation of the emulsion through rigorous shaking, (c) subsequent phase separation after cessation of shaking, (d) stable microemulsion in the presence of surfactants indicated by the dark solid line at the interface of the two phases.

In general, in a two-phase system, surfactants self-assemble at the interface of the two phases, where they introduce a degree of continuity between the two different phases, resulting in the production of a stable microemulsion.

Microemulsions are clear, stable, isotropic liquid mixtures of oil, water and surfactant. The "aqueous phase" may contain other ingredients, and the "oil" phase may actually be a complex mixture of different hydrocarbon solvents. In contrast to ordinary emulsions, due to the presence of surfactants, microemulsions form upon simple mixing of the components and do not require the high shear conditions generally needed in the formation of ordinary emulsions. The two basic types of microemulsions are direct, oil dispersed in water (o/w) and reverse, water dispersed in oil (w/o), determined by the relative amounts of the two phases present in the mixture. Figure 6.14 depicts the interface structure in a water–oil–surfactant mixture in a microemulsion. These are oil-in-water or water-in-oil droplets stabilized by a monolayer of the surfactant. Thus, even in small quantities, surfactants markedly affect the surface characteristics of the dispersed phase and this is another reason they are known as "surface-active agents".

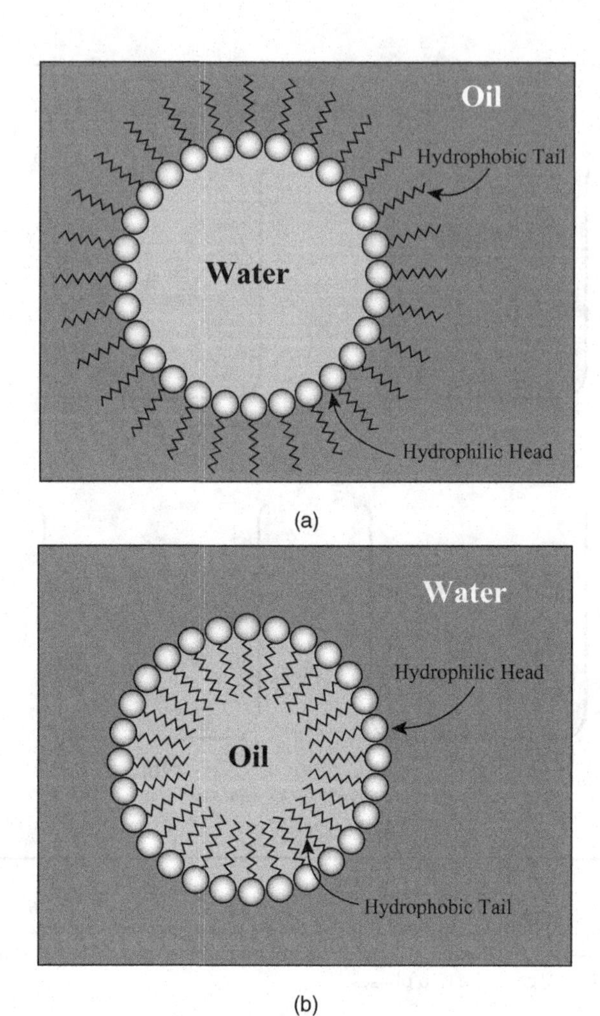

FIGURE 6.14
Depiction of interface characteristics in (a) reverse and (b) direct microemulsions.

Relative to bulk solutions, microemulsions can be used to add extra control over chemical reactions. For example, chemical reactions to produce magnetic nanoparticles can be carried out within the "pools of water" confined in reverse microemulsions. Both metallic and oxide nanoparticles can be prepared through this method. The uniform, tiny droplets, often called "nanoreactors", impose kinetic and hydrodynamic constraints on particle formation, limit particle size and lead to the production of monodispersed nanoparticles. In addition to controlling the particle size, the droplets may also influence the outcome of the reaction. Water molecules interact strongly with the polar heads of the surfactant molecules, responsible for stabilizing the microemulsion. This imposes structural order to the water molecules on the interior surface of the droplet, absent in bulk liquid-water solutions. Thus, novel magnetic nanostructures may be stabilized within reversed microemulsions compared to reactions carried out in bulk water. The synthesis of magnetic nanoparticles in reverse microemulsions is a complex and not a very-well-understood process. These are colloidal solutions; due to Brownian motion, the droplets undergo frequent collisions. During a collision, they have the ability to exchange the content of their water pools. Afterward, the droplets separate back to independent droplets of the same size. This makes possible chemical reactions to take place between compounds solubilized in different droplets. In addition, chemical reactants may be exchanged between the exterior and interior of the droplets, reminiscent of the lipid bilayer function in biological cells.

The medium of choice in forming reversed microemulsions uses anionic NaAOT (sodium bis (2-ethylhexyl) sulfosuccinate, (Figure 6.6(f)) as the surfactant, often referred to simply as AOT, and alkanes (saturated hydrocarbons that consist only of carbon and hydrogen linked together by single bonds) as oil. The ternary system Alkane/AOT/Water affords spherical water pools with a linear variation of the size of the droplet with the amount of water solubilized in the system. The water-pool diameter is related to the water content of the droplet, defined as the ratio of water-to-surfactant concentration $w = [H_2O]/[AOT]$, by

$$D(nm) = 0.3w \tag{6.10}$$

The diameter of the water droplet is thus controlled by the volume of solubilized water and can usually be varied from 0.5 to 18 nm.

By mixing two reverse microemulsions with the same water content, w, one containing the metal salt to be reduced and the other the reducing agent, metal nanoparticles can be formed by having the reduction reaction take place exclusively within the droplets as they exchange their water contents.

Example 6.3: Synthesis of Monodispersed Cobalt Nanoparticles in Reverse Microemulsions

The synthesis of cobalt magnetic nanoparticles in reverse microemulsions has been reported by Pileni and co-workers (C. Petit, A. Taleb and M. P. Pileni, Cobalt Nanosized Particles Organized in a 2D Superlattice:Synthesis, Characterization, and Magnetic Properties, *J. Phys. Chem. B* 103) (1999) 1805). They mixed two Na(AOT) microemulsion solutions of similar water content, w. The first contained $Co(AOT)_2$ (derived from $CoCl_2$) within its water pool while the second contained the reducing agent $NaBH_4$ (sodium borohydride) within its water pool. The mixing resulted in the formation of colloidal particles of metallic cobalt, which were extracted from the reverse micelles by covalent attachment to trioctylphosphine and subsequently redispersed in pyridine. Trioctylphosphine oxide ($C_{24}H_{51}OP$, Figure 6.6(g)) or TOPO is a common agent in solvent extraction of metals, due to its strong lipophilicity and high polarity. Its high polarity, which results from the dipolar phosphorus–oxygen bond, allows this compound to bind strongly to metal ions, while the hydrocarbon chains of the octyl groups (CH_3-$[CH_2]_7$-) confer high solubility in non-polar solvents. Figure 6.15 shows (a) the TEM patterns obtained and (b) the histogram of the diameter size distribution. As we noted earlier, when a drop of solution containing monodispersed nanoparticles is deposited on a TEM grid the particles spontaneously form hexagonally packed 2D networks, or superlattices, as those observed in Figure 6.15(a). This indicates that the as-extracted particles possess a narrow size distribution making subsequent size-selection treatment unnecessary. Assuming lognormal distribution, Eq. (5.7), analysis of the histogram indicates particles of mean diameter $d_0 = (5.8 \pm 0.5)$ nm with a polydispersity of 11% or $\sigma = 0.11$.

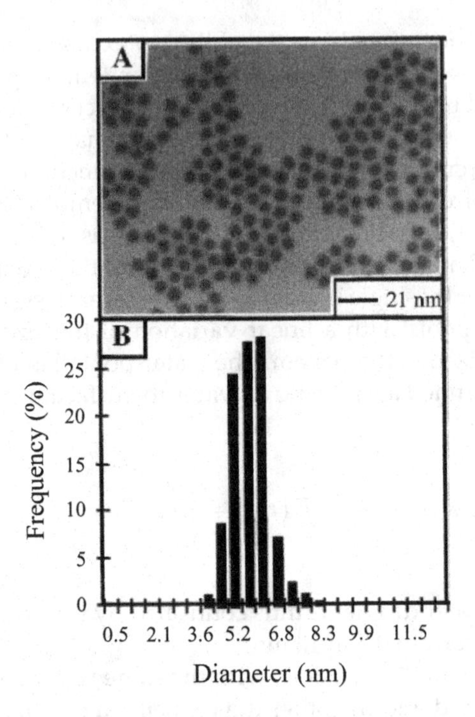

FIGURE 6.15
Monodispersed Co nanoparticles produced in a reverse microemulsion solution. (From C. Petit, A. Taleb and M. P. Pileni, Cobalt Nanosized Particles Organized in a 2D Superlattice: Synthesis, Characterization, and Magnetic Properties, *J. Phys. Chem. B* 103) (1999) 1805, Copyright American Institute of Physics, reproduced with permission).

6.7 Synthesis of Metallic Nanoparticles in Inverse Micelles

Microemulsions are three-component mixtures of water, organic solvents and surfactants. Two-component mixtures of just water and surfactants would result in the formation of surfactant aggregates with the polar heads pointing outward toward the water and the hydrophobic tails inward, avoiding contact with the water, as shown in Figure 6.16(a). These aggregates are called "micelles". Mixtures of organic solvents and surfactants also produce surfactant aggregates; only this time, it is the polar heads that point inward, as shown in Figure 6.16(b). These aggregates are called "inverse micelles". Controlled nucleation and growth of metal clusters can occur in the interior of inverse micelles in the absence of water. In this process, an ionic salt of the metal is dissolved in the hydrophilic interior of the inverse micelle. The surrounding continuous hydrophobic oil limits nucleation and growth to the micellar interior volume. Encapsulation within the inverse micelle ensures spatial homogeneity during nucleation, while the particle growth kinetics is determined by the rate of reduction of the metal salt and the diffusion properties of the micelles in the solution.

Micelle size, inter-micellar interactions and reaction chemistry are parameters that can be varied to control the final particle size. Micelle size is determined by the length of the

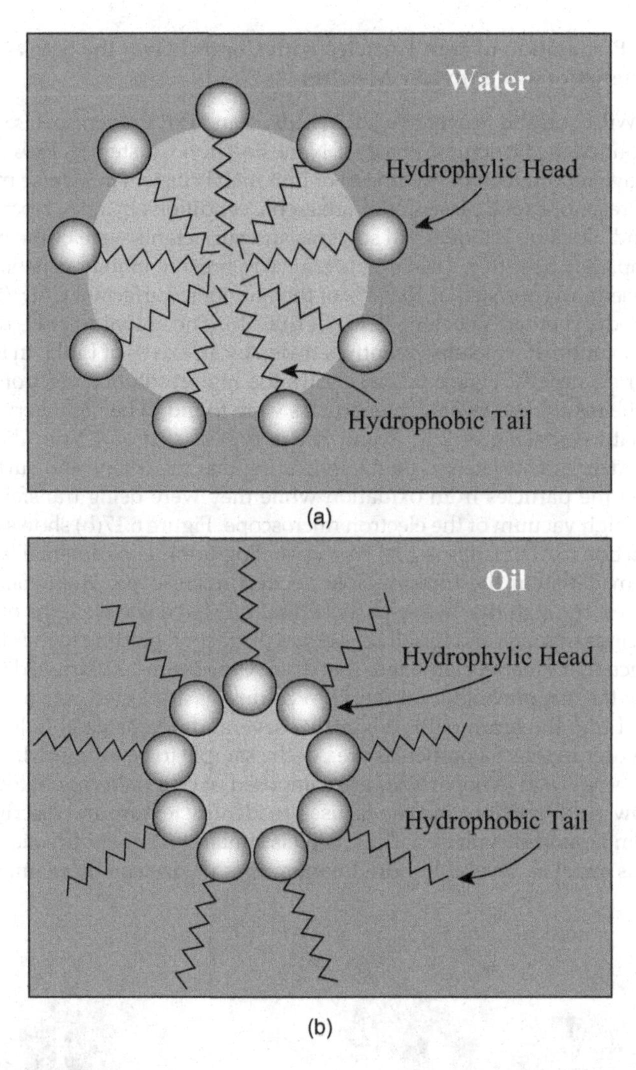

FIGURE 6.16
(a) Surfactant aggregate in water forming a micelle. (b) Surfactant aggregate in oil forming an inverse micelle.

hydrophobic chain of the surfactant, while the chemical structure of its polar head in an inverse micelle affects its binding to the growing metal cluster. Variation of surfactant type could, thus, affect particle size and the structural phase characteristics of nanosized metal clusters grown within inverse micelles. For example, the use of non-ionic surfactants in the room-temperature synthesis of metal iron clusters *via* the inverse micellar route leads to the production of γ-Fe, a phase of metal iron that is known to be stable only in high temperatures and pressures, possessing an fcc structure. In contrast, the use of cationic surfactants produces the normal bcc α-Fe iron phase. This phase variation has been observed in the case of the production of iron metal clusters *via* the inverse micellar route, while in contrast, cobalt and nickel nanoclusters obtained *via* the inverse micellar route do not exhibit such variation.

Example 6.4: Preparation of Iron Particles with Control Over the Size and Structural Phase Characteristics using Inverse Micelles

Investigators Wilcoxon and Provencio (J. P. Wilcoxon and P. P. Provencio, Use of Surfactant Micelles to Control the Structural Phase of Nanosize Iron Clusters, *J. Phys. Chem. B*, 103 (1999) 9809) have reported on the synthesis of iron metal clusters in inverse micelles. They report that all reactions took place under anaerobic conditions in a dry box with continuous oxygen and moisture removal. All solvents and surfactants were of the highest grade and were completely dust free. The latter is critical to prevent inhomogeneous nucleation. The micellar medium consisted of 10 wt % of the non-ionic surfactant $C_{12}E_4$ (tetraethylene glycol monododecyl ether) in octane ($CH_3(CH_2)_6CH_3$). The anionic salt $Fe(BF_4)_2$ was dissolved in this solution. It was subsequently reduced by an excess of $LiBH_4$ in tetrahydrofuran while stirring rapidly. Figure 6.17(a) shows the high-resolution electron micrograph (HRTEM) of the resulting clusters. The particles were found to be highly crystalline, with an observed lattice spacing of 2.12 Å and had a typical size of 4.3 nm diameter. They showed no evidence of surface oxidation, indicating that the octane and surfactant molecules protected the particles from oxidation while they were being transferred from the dry box to the high vacuum of the electron microscope. Figure 6.17(b) shows selected area electron diffraction (SAD) patterns. SAD is a crystallographic experimental technique that can be performed within the transmission electron microscope. High-energy electrons treated as waves, through the "wave-particle duality", have wavelengths of the order of the lattice spacings of solids and the atoms act as a diffraction grading to the electron beam, which is diffracted in a manner similar to that of an X-ray beam, as discussed in Section 6.4. Located below the sample holder on the TEM column is a selected area aperture, which can be inserted into the beam path. It contains several different-sized holes, and can be moved by the user to select a particular area of the sample to be examined. The SAD patterns of nanocrystals and nanoparticles are composed of ring patterns analogous to those from X-ray powder diffraction, and can be used to identify texture and discriminate nanocrystalline from amorphous phases. Thus, SAD is similar to X-ray diffraction, but unique in that areas as small as several hundred nanometers in size can be examined, whereas

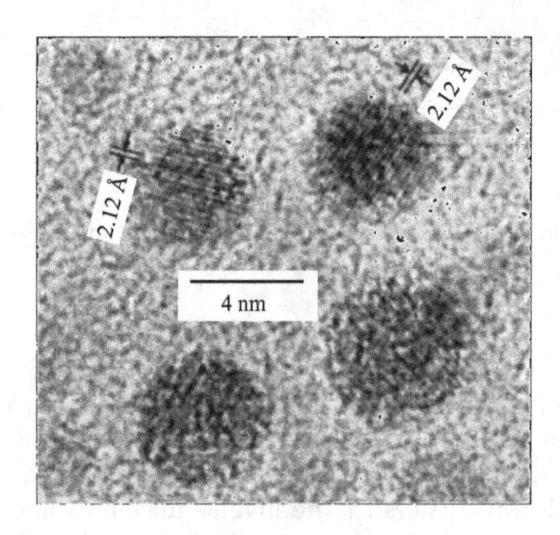

(a)

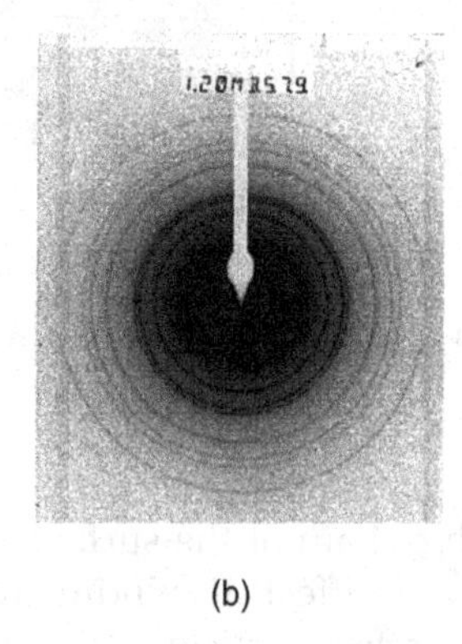

(b)

FIGURE 6.17
(a) HRTEM micrographs of 4.3-nm-diameter Fe clusters obtained in inverse micelles using non-ionic surfactants. (b) SAD patterns identify the iron phase as γ-Fe of fcc crystallographic structure. (From J. P. Wilcoxon and P. P. Provencio, Use of Surfactant Micelles to Control the Structural Phase of Nanosize Iron Clusters, *J. Phys. Chem. B*, 103 (1999) 9809, Copyright American Institute of Physics, reproduced with permission).

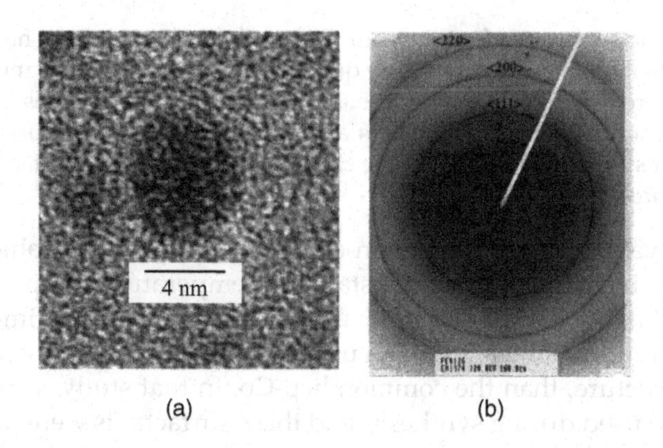

FIGURE 6.18
(a) HRTEM micrographs of 4.2-nm-diameter Fe clusters obtained in inverse micelles using cationic surfactants. (b) SAD diffraction patterns identify the iron phase as α-Fe of bcc crystallographic structure. (From J. P. Wilcoxon and P. P. Provencio, Use of Surfactant Micelles to Control the Structural Phase of Nanosize Iron Clusters, *J. Phys. Chem. B*, 103 (1999) 9809, Copyright American Institute of Physics, reproduced with permission).

X-ray diffraction typically samples areas several centimeters in size. The electron diffraction rings in Figure 6.17(b) are associated with the γ-Fe phase of fcc iron.

Figures 6.18(a) and (b) show HRTEM and corresponding SAD patterns of iron particles synthesized *via* a similar synthetic route using the cationic surfactant DTAB (dodecyltrimethylammonium bromide) 10 wt % in octane containing 10 wt % hexanol as a co-surfactant. The identical metal salt, salt concentration and reducing agent were used; only the surfactant was altered from a non-ionic to a cationic one. Figure 6.18(a) indicates highly crystalline iron particles with a typical particle diameter of 4.2 nm. However, the SAD patterns of Figure 6.18(b) indicate the presence of the bcc phase of α-Fe in this case. Table 6.2 lists the predicted and observed *d* spacings in Angstroms for the two crystallographic phases of metallic iron.

Within the micro-heterogeneous system of inverse micelle solutions, the ionic iron ions are completely confined in the interior of the micelles, since they are absolutely insoluble in the continuous non-polar oil phase of the organic solvents. In fact, the iron ions may be

TABLE 6.2

Predicted and Observed ***d*** Spacings from Figures 6.17 and 6.18

index	*d*(γ-Fe), fcc	*d*(α-Fe), bcc	*d* (Figure 6.17)	*d* (Figure 6.18)
110	2.027	none	2.00	none
111	none	2.08	none	2.08
200	1.4332	1.80	1.41	1.78
211	1.1702	none	1.165	none
220	1.0134	1.27	1.01	1.26
310	.9064	none	.896	none
311	none	1.083	none	1.089
222	.8275	1.037		1.049
400	none	.90		.92

Source: Reproduced with permission from J. P. Wilcoxon and P. P. Provencio, *J. Phys. Chem. B* 103 (1999) 9809.

coordinated to the surfactants forming complicated charge complexes. The growth of the initial nuclei into larger particles occurs by diffusion and collisions of the micelles containing the nascent iron clusters resulting in mass transfer between micelles. Because of the intimate interaction between iron clusters and surfactant molecules during the growth process, it is not surprising to find that the chemical structure of the surfactant influences the iron phase produced.

The above observation of stabilizing non-equilibrium crystallographic structures in the nanoscale, not available in the bulk under standard temperature and pressure, is an example of the wealth of new possibilities attainable in the nanometer regime. In Example 6.2, we also encountered the stabilization of an unusual phase of metallic cobalt, ε-Co with fcc crystallographic structure, than the common hcp-Co. In that study, a complex mixture of surfactants was also used during synthesis, and these surfactants were found to bind quite strongly to the Co particle surface.

6.8 Preparation of Iron Nanoparticles *via* Thermal Decomposition of Iron Pentacarbonyl

6.8.1 Thermolytic decomposition of $Fe(CO)_5$

The thermal decomposition of $Fe(CO)_5$ at 100°C in the presence of oleic acid leads to the formation of an iron–oleic acid complex. Subsequent thermal treatment at 300°C in a non-oxidizing atmosphere leads to the synthesis of monodispersed metal iron nanoparticles. Following such a synthesis procedure, Hyeon and co-workers (T. Hyeon, Chemical Synthesis of Magnetic Nanoparticles, The Royal Society of Chemistry, *Chem. Comm.* (2003) 927–934), produced monodispersed iron nanoparticles ranging in size from 4 to 20 nm without using any size-selection process, as shown by TEM imaging of the as-synthesized nanoparticles. The electron diffraction pattern of the sample, however, indicated that the nanoparticles contained iron in an amorphous phase. After annealing at 500°C, XRD phase characterization performed on the nanoparticles indicated that the annealing process induced phase transformation from amorphous iron to the α-Fe bcc crystallographic structure. Figure 6.19 shows the TEM micrograph of 11-nm α-Fe nanoparticles derived through this process.

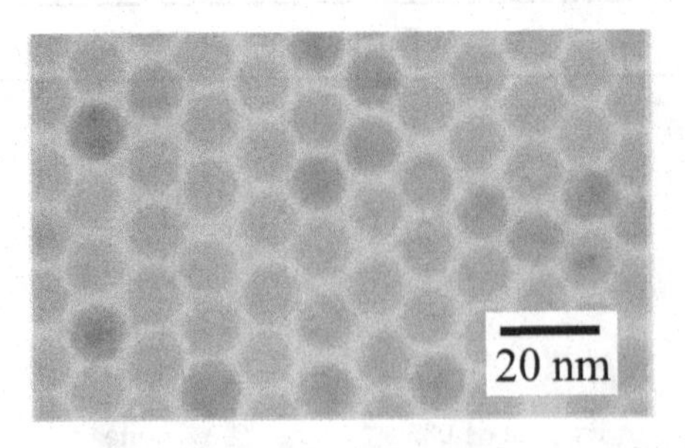

FIGURE 6.19
TEM image of 11-nm Fe nanoparticles produced from the thermal decomposition of iron pentacarbonyl followed by annealing at 500°C. (From T. Hyeon, Chemical Synthesis of Magnetic Nanoparticles, The Royal Society of Chemistry, *Chem. Comm.* (2003) 927–934, Copyright the Royal Chemical Society, reproduced with permission).

The resulting particle size could be controlled by varying the molar ratio of iron pentacarbonyl to oleic acid in the precursor solution. The 11-nm-diameter particles of Figure 6.19 were derived by using a molar ratio of $Fe(CO)_5$/oleic acid of 1:3, while a molar ratio of 1:2 led to 7-nm-diameter metal particles. The authors report that the largest particle size produced by using different molar ratios was 11 nm. The production of larger sizes necessitated the introduction of additional iron–oleic acid complexes into a solution containing previously prepared 11-nm iron particles followed by aging at 300°C. This procedure allowed for the fine-tuning of maximum particle size from 11 to 20 nm diameter.

6.8.2 Sonochemical Decomposition of $Fe(CO)_5$

Iron nanoparticles have also been produced by the thermal decomposition of $Fe(CO)_5$ induced by sonochemistry. Sonochemistry takes place by inducing cavitation, that is, the formation, growth and collapse of bubbles in a liquid when irradiated with an intense ultrasonic beam. The collapse of such bubbles generates local hot spots with transient temperatures rising to ~5,000 K and cooling rates exceeding 10^{10} K/s. Volatile organometallic compounds, such as $Fe(CO)_5$, inside the cavities are decomposed at these high temperatures yielding individual metal atoms. In alkanes and in the absence of trapping surfactant molecules, these atoms agglomerate to produce highly porous nanostructured materials including amorphous metals and alloys. However, in the presence of a stabilizing agent in the original organic solvent/iron pentacarbonyl solution, agglomeration is prevented and ultrasonic irradiation leads to the formation of iron colloids.

Suslick and co-workers (K. Suslick, M. Fang and T. Hyeon, Sonochemical Synthesis of Iron Colloids, *J. Am. Chem. Soc.* 118 (1996) 11690) have reported the formation of such colloids produced by ultrasonic irradiation of 0.2 mL of $Fe(CO)_5$ in 20 mL of octanol with 1 g of polyvinylpyrrolidone (PVP) as the stabilizing agent. The TEM image shown in Figure 6.20(a) indicates that the resulting iron particles dispersed in the polymer matrix range in size from 3 to 8 nm, that is, the process leads to polydispersed iron particles. However, when oleic acid was used as the colloid stabilizing agent, monodispersed nanoparticles were obtained. Specifically, a hexadecane solution of 2 M $Fe(CO)_5$ and 0.3 M oleic acid was sonicated at 30°C for 1h, producing a colloidal solution of iron particles with particle a diameter of ~8 nm, as shown in Figure 6.20(b).

Again, we witness the critical role that the presence of stabilizing agents can play in the formation of magnetic nanoparticles. Furthermore, we observe that the specific nature of the stabilizing agent also influences the outcome.

6.9 Synthesis of Iron Oxide and Ferrite Magnetic Nanoparticles

Transition metal oxides constitute a very important class of magnetic materials due to their stability and their wide variety of structures. While metal magnetic nanoparticles are susceptible to oxidation, nanoparticles of magnetic metal oxides are stable in the atmosphere. This makes metal-oxide magnetic nanoparticles easily amenable to technological applications. As a result, nanoparticles of magnetic oxides including many ferrites have been the subject of intense study over many years for their applications as magnetic storage media and as magnetic colloids. It has been known for a long time that adding alkali to aqueous mixtures of ferric and ferrous salts, in the mole ratio Fe^{3+}/Fe^{2+} = 2:1, induces the

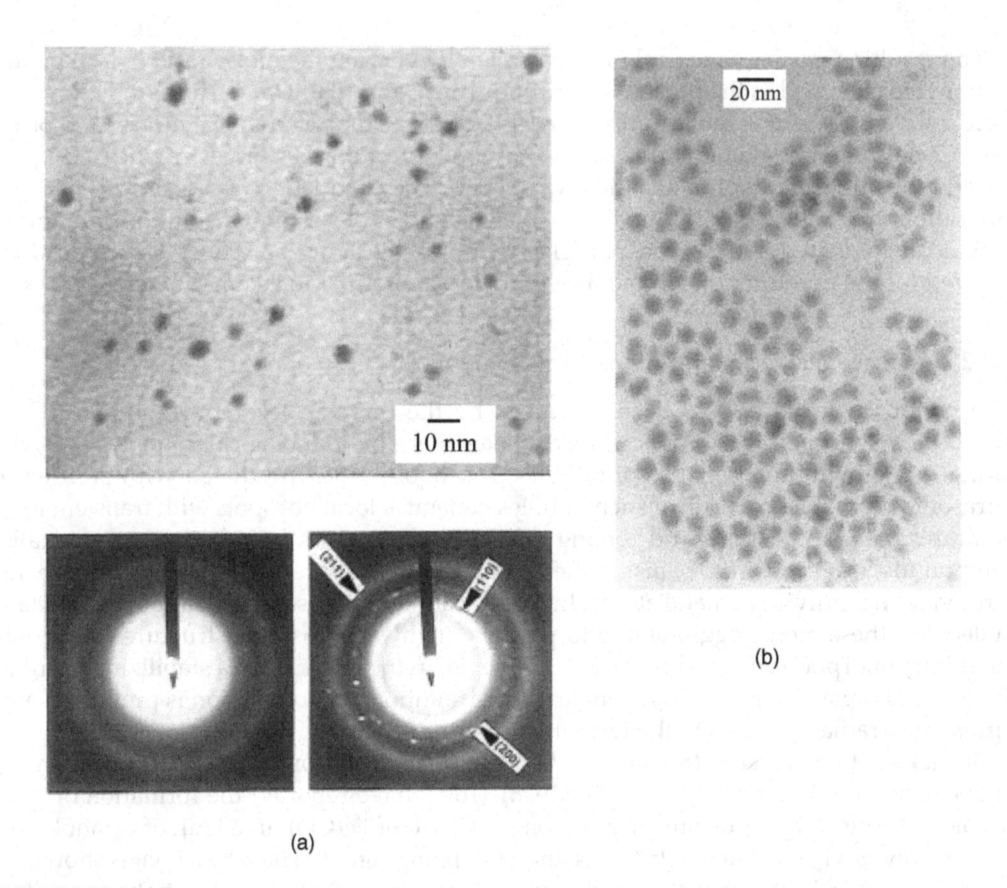

FIGURE 6.20
(a) TEM micrograph of sonochemically prepared iron colloid particles dispersed in PVP. (b) TEM micrograph of sonochemically prepared iron colloid particles stabilized with oleic acid. (From K. Suslick, M. Fang and T. Hyeon, Sonochemical Synthesis of Iron Colloids, *J. Am. Chem. Soc.* 118 (1996) 11690, Copyright American Chemical Society, reproduced with permission).

precipitation of magnetite particles. The reaction is complex and involves the intermediate formation of iron hydroxide particles which are subsequently converted to magnetite. The overall reaction can be represented by Eq. (6.11).

$$2Fe^{3+} + Fe^{2+} + 8OH^{-} \rightarrow Fe_3O_4 + 4H_2O \tag{6.11}$$

According to the thermodynamics of this reaction complete precipitation of Fe_3O_4 is expected at a pH between 8 and 14 in a non-oxidizing environment. Oxidation of ferrous ions during preparation often leads to the production of non-stoichiometric magnetite particles, with $Fe^{3+}/Fe^{2+} > 2:1$ mole ratio. The size of the particles obtained can be controlled by changing the experimental conditions, such as the temperature of the reaction medium. Magnetite particles are susceptible to oxidation over time, whereby the magnetite particles are converted into maghemite with the release of water according to Eq. (6.12).

$$Fe_3O_4 + 2H^{+} \rightarrow \gamma - Fe_2O_3 + Fe^{2+} + H_2O \tag{6.12}$$

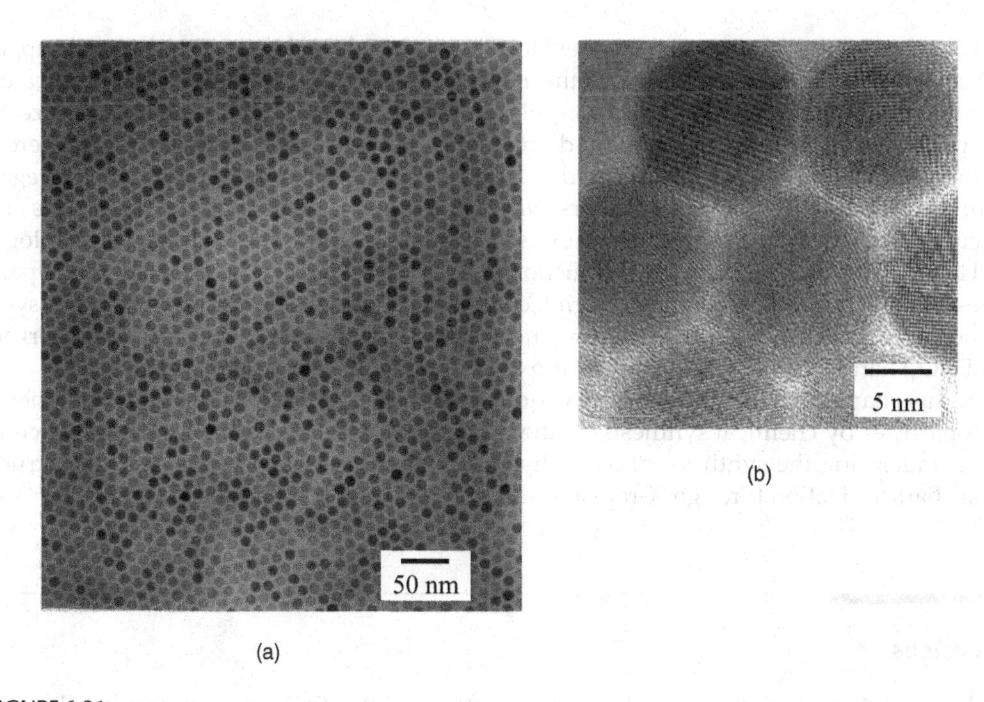

FIGURE 6.21
(a) TEM micrograph of 11-nm γ-Fe_2O_3 nanocrystals obtained by the controlled oxidation of metal iron nanoparticles of the same size. (b) HRTEM micrograph indicates highly crystalline particles. (From T. Hyeon, S. S. Lee, J. Park, Y Chung and H. B. Na, Synthesis of Highly Crystalline and Monodispersed Maghemite Nanocrystallites without Size-Selection Process, *J. Am. Chem. Soc.* 123 (2001) 12798, Copyright American Chemical Society, reproduced with permission).

The first controlled preparation of superparamagnetic iron oxide nanoparticles using alkaline precipitation of $FeCl_3$ and $FeCl_2$ was performed by Massart in 1981 (R. Massart, Preparation of aqueous magnetic liquids in alkaline and acidic media, *IEEE Trans. Magn.* 17 (1981) 1247).

The coprecipitation reactions of ferrous and ferric salts can also lead to the production of a wide range of substituted ferrite nanoparticles. The Fe^{2+} ion is simply replaced or partially replaced by another or combination of divalent metal ions such as Co^{2+}, Mn^{2+}, Ni^{2+}, Zn^{2+}, *etc.* Ions of different valences, such as Li^{+}, can also be used to prepare substituted ferrites. The interest in substituted ferrites stems from the differences in their magnetic properties. The main advantage of the coprecipitation process is that large amounts of magnetic nanoparticles can be easily produced. However, the precipitated particles suffer from a very large distribution in particle size.

In contrast, monodispersed γ-Fe_2O_3 nanoparticles can be produced by the controlled oxidation of iron nanoparticles prepared by the thermal decomposition of iron pentacarbonyl, as described in Section 6.8.1. Hyeon and co-workers (T. Hyeon, S. S. Lee, J. Park, Y Chung and H. B. Na, Synthesis of Highly Crystalline and Monodispersed Maghemite Nanocrystallites without Size-Selection Process, *J. Am. Chem. Soc.* 123 (2001) 12798) have reported the transformation of monodispersed metal-Fe/oleic-acid-covered nanoparticles to highly monodispersed γ-Fe_2O_3/oleic-acid-covered nanocrystals by controlled oxidation using trimethylamine N-oxide ($(CH_3)_3NO$) as a mild oxidant. Figure 6.21 shows the TEM and HRTEM micrographs of 11-nm γ-Fe_2O_3 nanocrystals obtained through the controlled oxidation of 11-nm metal iron nanoparticles, similar to those shown in Figure 6.19.

The particles are monodispersed as evinced by the highly ordered 2-D hexagonal superlattice formed (Figure 6.21(a)) while the HRTEM micrograph (Figure 6.21(b)) indicates highly crystalline particles. The size control from 4 to 20 nm was determined by the size of the precursor metal iron particles as discussed in Section 6.5. These oleic-acid-covered nanocrystallites can be easily dispersed in many hydrocarbons without particle aggregation. As we will see in the latter chapters, where applications of magnetic nanoparticles are discussed, maghemite nanoparticles are used in various applications in nanotechnology and bio-nanomedicine. Thus, the production of uniform highly crystalline maghemite particles discussed in this section is of high technological importance. Furthermore, this synthetic procedure has been extended to the production of monodispersed cobalt ferrite ($CoFe_2O_4$) nanocrystals by the controlled oxidation of Fe-Co alloy nanoparticles.

In this chapter, we have discussed various approaches to the production of magnetic nanoparticles by chemical synthesis. In the process, we introduced the fundamental concepts underlying the synthesis of monodispersed magnetic nanocrystals and their structural characterization through X-ray- and electron-diffraction techniques.

Exercises

1. Figure 6.4 depicts the crystallographic packing of the Fe11 supramolecular clusters within the unit cell. Internal superexchange spin coupling within an individual cluster leads to a ground total spin state for the cluster of $S_T = ½$, as discussed in the text. Use Eq. (3.14) to calculate the dipole–dipole interaction energy between neighboring clusters in the crystal. At what temperature could such an interaction lead to long-range magnetic order in the crystal?
2. What role do surfactants play in the synthesis of magnetic nanoparticles? Give examples of (a) non-ionic, (b) anionic and (c) cationic surfactant molecules and describe their mode of action.
3. Define "size-selection" processing. How is it performed? Give an example.
4. Figure 6.9 gives the XRD pattern for Co nanocrystals derived by the reduction of $CoCl_2$. TEM characterization of the sample determined monodispersed nanoparticles of 11-nm diameter. Use the Debye–Scherrer formula given in Eq. (6.7) together with information extracted from Figure 6.9 to confirm that the particle size predicted from the XRD pattern agrees with that established by TEM analysis.
5. In Examples 6.2 and 6.4, two different techniques used for phase characterization of magnetic nanoparticles were introduced, XRD and SAD. What physical principle do these techniques utilize in determining the crystallographic phase of a nanocrystal? Discuss similarities and differences.
6. Define the following entities encountered in solution chemistry for the synthesis of magnetic nanoparticles: (a) micelle, (b) inverse micelle, (c) direct microemulsion, (d) reverse microemulsion. For what purpose were such systems introduced in the synthesis of magnetic nanoparticles? Explain.
7. How is it possible to stabilize unusual or thermodynamically non-equilibrium crystallographic phases at the nanoscale? We have encountered two such examples, the production of ε-Co and γ-Fe nanocrystals, while in their corresponding bulk crystallographic structures, these metals acquire the hcp-Co and α-Fe crystallographic phases, respectively. Explain.

7

Biogenic Magnetic Nanoparticles

Bacteria with motility directed by the local geomagnetic field have been observed in marine sediments. These magnetotactic microorganisms possess flagella and contain novel structured particles, rich in iron, within intracytoplasmic membrane vesicles. Conceivably these particles impart to cells a magnetic moment. This could explain the observed migration of these organisms in fields as weak as 0.5 Gauss.

Richard Blakemore, *Science* 190 (1975) 377

As witnessed in Chapters 5 and 6, the production of magnetic nanoparticles with well-defined size, morphology and crystallinity is quite challenging. We discussed examples of elaborate physical and chemical procedures, size selection processes and high-temperature annealing methodologies developed to this end, often with limited success. In sharp contrast, living organisms can produce magnetic nanoparticles of well-defined size and crystallinity under mild, physiological conditions through the process of biomineralization, that is, the biological regulation of crystal growth, particle size, morphology and organization. Living systems produce a variety of complex composite materials, ranging from the nano- to the macroscale, by integrating inorganic (hard) matter within the organic (soft) cellular world of biology for structural support, "magnetoreception" and iron storage. Biominerals may be amorphous or crystalline forming structures of varying degrees of complexity from a single unit to numerous individual units or aggregates. The aggregated units are usually arranged in an orderly fashion and when crystalline the crystallographic axes are often aligned. The resulting highly organized bioinorganic structures exhibit excellent physical and chemical properties that often surpass those of artificial materials produced by usual synthetic methods employed in the laboratory.

In this chapter, we explore biogenic magnetic nanoparticle formation in (a) magnetotactic bacteria and (b) the iron-storage protein ferritin. The goal is to get an appreciation of the role biological molecules and/or genes play in the controlled formation of magnetic nanoparticles *in vivo* and learn from the chemistry of life how best to perfect synthesis routes in the laboratory for the production of high-quality magnetic nanocrystals.

7.1 Biomineralization of Iron

The biomineralization of iron hydroxides is widespread among organisms due to the utilization of iron atoms by proteins for oxygen and electron transport in metabolic processes. The most widely studied biomineralization product of this type occurs in the iron-storage protein ferritin. Ferritins represent a family of proteins that are ubiquitous in biological

DOI: 10.1201/9781315157016-9

systems. Although there are differences among them, their overall similarities are more important than their differences. They are all large multi-component proteins that self-assemble to form molecular cages within which a hydrated ferric oxide is mineralized. Mammalian ferritin forms a 7-nm micellar core of hydrated iron (III) oxide (ferrihydrite), with various amounts of inclusions of phosphate. It was first described by V. Laufberger in 1937 as a protein isolated from horse spleen containing about 20% iron. An iron- and phosphate-rich mineral deposit similar in composition to that of ferritin is also found in the dermal granules of *Molpadia Intermedia*, a species of marine invertebrates. These dermal granules, ranging in size from 10 to 350 μm, serve as strengthening agents in the connective tissues of the dermis; they contain inclusions of iron hydroxide deposits seen as electron-dense subunits of 9–14 nm diameter in transmission electron micrographs.

Magnetite is the most common of the known iron oxide biominerals. It was first identified by H. Lowenstam in 1962 in the denticle capping of chitons (primitive marine mollusks). Magnetite precipitation and tooth formation in chitons proceed through the biochemically controlled reduction of ferrihydrite. The mineral hardens the major lateral teeth of chitons enabling them to scrape surface and embedded algae from hard rocks in the sea for food. Unlike ferrihydrite, which is a common product of both biological and inorganic processes, inorganically magnetite is formed only at elevated temperatures and pressures in igneous and metamorphic rocks. Yet, organisms ranging from bacteria to vertebrates are capable of forming magnetite under ambient conditions. By natural selection, the chitons somehow biochemically mediate the transformation of ferrihydrite to magnetite, in order to perform a biological function, even at atmospheric temperatures and pressures. In an entirely different biological function, magnetite deposits have been identified with "magnetoreception", the ability of living organisms to sense the polarity or the inclination of the Earth's magnetic field. Some bacteria, honeybees, homing pigeons and migratory fish are known to possess such sense. Furthermore, numerous iron-reducing bacteria are known to form extracellularly precipitated single-magnetic-domain magnetite during their growth on organic substrates, a process not associated with intracellular metal accumulation.

7.2 Formation of Magnetic Nanoparticles by Bacteria

Magnetotactic bacteria synthesize membrane-enclosed intracellular crystalline magnetic particles, called magnetosomes, comprising primarily of iron oxides or, in rare cases, of iron sulfides. Magnetosomes are nanometer-sized, magnetic mineral crystal deposits enveloped by a stable lipid membrane that contains some lipids and proteins, often referred to as the membrane vesicle. They are aligned to form chains within the bacterium, thus, creating a bio-magnetic compass that enables the bacterium to orient in the Earth's magnetic field, a phenomenon known as "magnetotaxis". These are micro-aerobic bacteria endowed with flagella, which allow them to swim and migrate along oxygen gradients in aquatic environments. They were first reported in 1975 by microbiologist Richard P. Blakemore of the Woods Hole Oceanographic Institution. It was an accidental observation that led to their discovery. It was noticed that when a drop of mud was placed on a microscope slide some microorganisms contained in it moved toward one side of the slide. When magnets were placed around the slide, the movement of the microorganisms changed. Blakemore examined the bacteria with a microscope and found that some of the cells had lots of iron. When the bacteria were immobilized, the iron-rich cells oriented to the magnetic field, but the bacteria did not move toward the source of the field. That finding indicated that magnetic bacteria are not simply "pulled" toward the magnet by the magnetic field gradient; rather, iron-rich cells inside the bacteria caused them to orient in the direction of the magnetic field while another mechanism, the flagella, caused them to swim and migrate. Further studies have revealed that magnetic bacteria tightly control the synthesis of their own magnetite mediated by the magnetosome membrane which has a distinct biochemical composition and contains specific magnetosome membrane proteins (MMPs). Since 1975, a variety of strains have been found to exist in marine and freshwater habitats. Figure 7.1 shows electron micrographs of magnetosomes and bacteria representing three different strains of typical magnetotactic bacteria. Note that magnetosomes can be isolated from bacteria with intact magnetosome membranes surrounding the magnetic particles as indicated in Figure 7.1(c), where the magnetosome membranes, indicated by the arrow, are clearly visible.

Most magnetotactic bacterial strains living in marine or freshwater environments grow magnetite (Fe_2O_4) crystals within their magnetosomes. The magnetosomes are characterized by a narrow size distribution and uniform, species-specific, crystal habits. Magnetotactic bacteria living in marine, sulfidic environments grow magnetosome crystals of the iron-sulfide mineral greigite (Fe_3S_4), which is isostructural with magnetite and is also ferrimagnetically ordered at room temperature. Bacterial Fe_3O_4 appears to persist in sediments after death and lysis of cells, contributing to the fossil and paleomagnetic records.

The magnetite particles in the magnetosomes are of single-domain size and have a stable magnetic moment aligned along the {111} crystallographic axis, the easy direction of magnetization for magnetite. This maximizes the magnetic moment per particle, as the {111} direction yields approximately 3% higher saturation magnetization than do other directions. The organic supporting matrix and magnetic interactions between magnetosomes keep the magnetic moment of the individual magnetite crystals parallel to the chain axis. Along the chain, the individual magnetosome moments simply add up vectorially to maximize the total magnetic moment and, therefore, the torque exerted on the resulting "biological bar magnet" by an external magnetic field. Figure 7.2 shows an electron holograph of a section of the chain in *Magnetospirillum magnetotacticum*, along with the magnetic field lines derived from electron interference patterns superimposed on the positions of the magnetosomes. The confinement of the magnetic field lines within the magnetosomes is indicative of perfectly aligned single-magnetic domains and shows that the chain of magnetosomes acts as a single magnetic dipole. Using this extraordinary design of magnetic engineering, the bacterium builds a strong enough magnetic moment to passively orient, at ambient temperatures, in the direction of the weak terrestrial field of only ~0.5 Gauss (0.5×10^{-4} T).

Detailed studies of the crystal structure of the magnetite particles within the magnetosomes indicate that they contain highly crystalline magnetite of cubo-octahedral shape, yielding superior magnetic properties. Overall, magnetosome crystals have high chemical purity, narrow size distribution and species-specific morphologies, unattainable in inorganically precipitated magnetite. These features point to magnetosome formation under strict biological control, a process known as "biologically controlled mineralization".

Magnetosomes extracted from cells can easily disperse in aqueous solutions because of the presence of the enveloping organic membrane. Thus, they do not suffer from particle agglomeration as purely inorganically grown nanocrystals do. Figure 7.3 shows electron micrographs of magnetosomes extracted from magnetotactic bacterial cells together with bare or oleic-acid-coated inorganically prepared magnetite nanoparticles. The biogenic

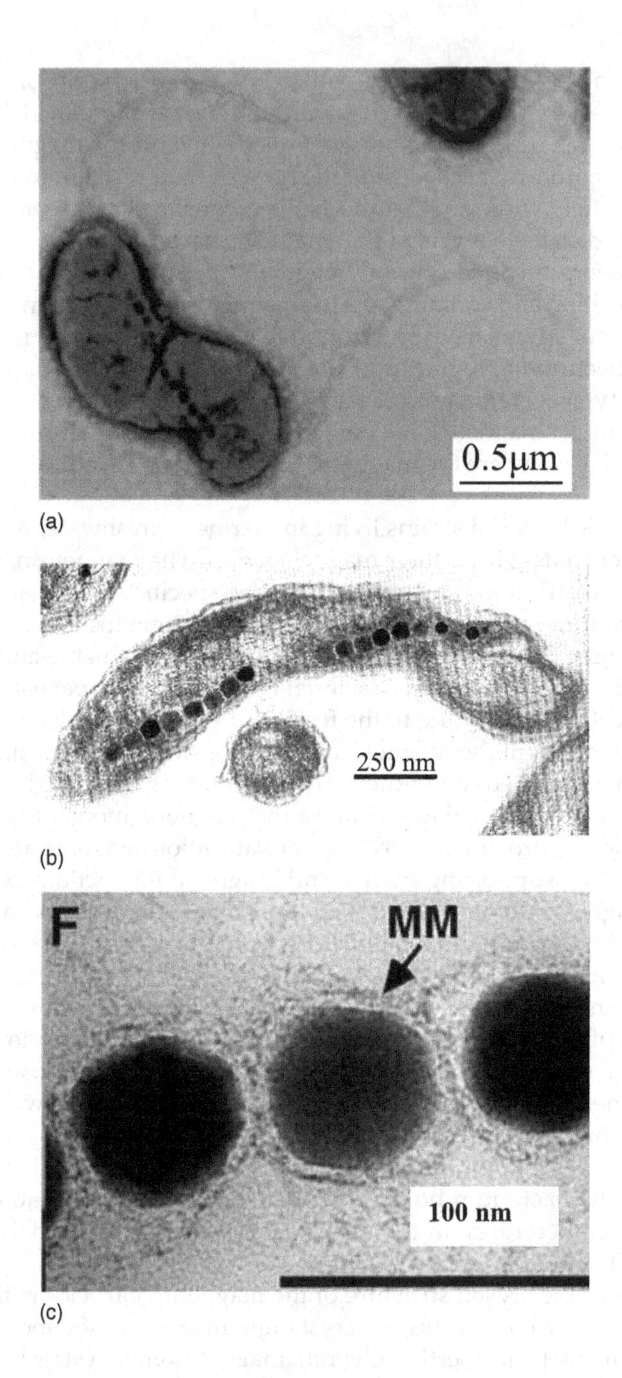

FIGURE 7.1
Transmission electron micrographs of typical magnetotactic bacteria and magnetosomes. (a) *Magnetotactic Spirillum* (MV-4) with a flagellum at each end of the cell and a chain of electron-dense, magnetite containing magnetosomes along the long axis of the cell. (b) Transmission electron micrograph of thin-sectioned magnetic cells of *Magnetotactic Spirillum* (MS-1). The chains of crystals within the cell are clearly visible. (c) Isolated magnetosomes from *Magnetospirillum gryphiswaldense*. Arrow indicates the magnetosome membrane (MM). ([a] D. A. Bazylinski and R. B. Frankel, Magnetosome formation in prokaryotes, *Nat. Rev. Microbiol.* 2 (2004) 217–230, Copyright Nature Research, reproduced with permission; [b] Richard B. Frankel, Richard P. Blakemore and Ralph S. Wolfe, Magnetite in Freshwater Magnetotactic Bacteria, *Science* 203 (1979) 1355, Copyright American Association for the Advancement of Science, USA, reproduced with permission; [c] D. Schüler, Molecular analysis of a subcellular compartment: the magnetosome membrane in *Magnetospirillum gryphiswaldense, Arch. Microbiol.* 181 (2004) 1, Copyright Springer (Germany), reproduced with permission).

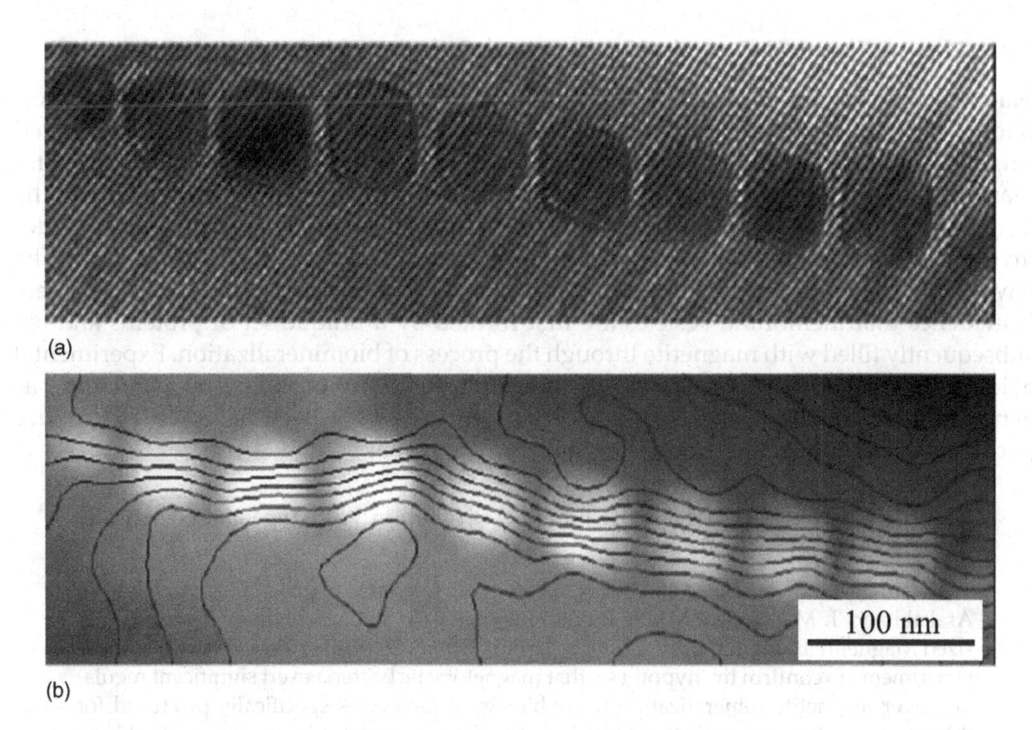

FIGURE 7.2
(a) Electron holograph of a region of the magnetosome chain in *Magnetospirillum magnetotacticum*. (b) Magnetic field lines. (From D. A. Bazylinski and R. B. Frankel, Magnetosome formation in prokaryotes, *Nature Rev. Microbiol.* 2 (2004) 217–230, Copyright Nature Research, reproduced with permission).

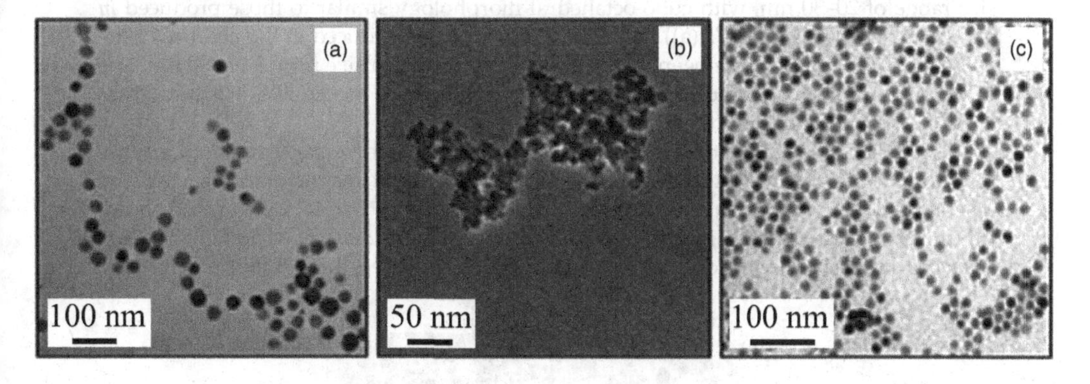

FIGURE 7.3
TEM images of (a) isolated magnetosome particles, (b) purely inorganic magnetite nanocrystals, and (c) inorganic magnetite crystals in oleic acid. (From Claus Lang, Dirk Schüler, Damien Faivre, Synthesis of Magnetite Nanoparticles for Bio- and Nanotechnology: Genetic Engineering and Biomimetics of Bacterial Magnetosomes, *Macromol. Biosci.* 2007, 7, 144–151, Copyright Wiley (Germany), reproduced with permission).

magnetites do not agglomerate because of their membrane, whereas inorganic particles need further treatment to prevent them from agglomerating. In dispersing the nanoparticles, the biological membrane is seen to play a role similar to that of surfactants in the bottom-up synthesis of magnetic nanoparticles by chemical methods; *in vivo*, however, as we are about to witness, the biological membrane of the magnetosomes has a far more active role to play in controlling and guiding magnetosome nanocrystal growth.

7.2.1 Physico-Chemical Control of Fe_3O_4 Crystal Growth within Magnetosomes

Magnetotactic bacteria provide us with an ideal model organism for investigating morphological regulation in biomineralization because they synthesize highly controlled single-crystalline magnetites. Magnetosome membrane vesicles are formed from the cytoplasmic membrane through an invagination process, with the vesicles aligned along the long axis of the bacterium. The magnetite crystal formation in the interior of the vesicles involves several processes, such as nucleation, growth and morphological regulation. The enveloping magnetosome membrane consists mainly of phospholipids and proteins. There is evidence that membrane vesicles are first formed by a unique set of proteins and are subsequently filled with magnetite through the process of biomineralization. Experimental evidence indicates that specific MMPs are implicated in morphological regulation, as genetically modified bacteria that lack certain proteins have been observed to produce poorly crystalline magnetosomes of irregular size and shape.

Example 7.1: Nanosized Magnetite Biomineralization in Wild-Type and Genetically Modified Magnetotactic Bacteria

T. Matsunaga and collaborators at the University of Tokyo (M. Tanaka, E. Mazuyama, A. Arakaki and T. Matsunaga, MMS6 Protein Regulates Crystal Morphology during Nanosized Magnetite Biomineralization *in Vivo*, J. Biol. Chem. 286 (2011) 6386) have carried out experiments to confirm the hypothesis that magnetotactic bacteria exert significant regulation over magnetite mineralization using biological molecules specifically produced for this purpose. Working with the AMB-1 strain of *M. magnetotacticum*, they were able to identify a series of proteins in the magnetosome membrane that are tightly bound to magnetite crystals. Among these, a protein known as *mms6* was shown to mediate the formation of uniform magnetite crystals of specific morphology during *in vitro* Fe_3O_4 chemical synthesis, as Figure 7.4 indicates. Fe_3O_4 crystals produced in the presence of *mms6* had a size range of 20–30 nm, with cubo-octahedral morphology similar to those produced *in vivo* by intact cells (Figure 7.4(a)). In contrast, the crystals produced in the absence of the *mms6* protein were non-homogeneous in shape and ranged in size from 1 to 100 nm. The TEM micrographs also indicated the formation of elongated, needle-like structures indicated by the arrows in Figure 7.4(b).

This led the investigators to carry out *in vivo* experiments where they compared magnetosome formation of wild-type *vs.* genetically compromised magnetotactic bacteria by deleting the gene responsible for the production of the mms6 protein. Figure 7.5 shows transmission electron micrographs of wild-type and mutant (*Δmms6*) bacteria of *M. magnetotacticum* (AMB-1) with deleted *mms6* genes. It is observed that while the wild-type

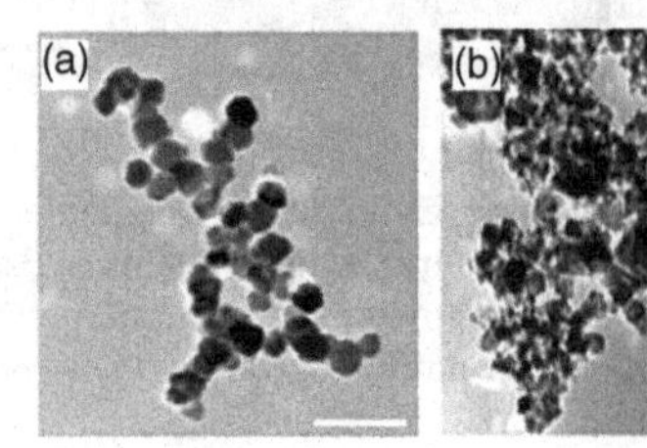

FIGURE 7.4
Fe_3O_4 nanoparticles synthesized *in vitro* in (a) the presence and (b) the absence of the *mms6* protein, including some needle-like particles indicated by the arrows. Bars = 100 nm. (From A. Arakaki, J. Webb, and T. Matsunaga, A Novel Protein Tightly Bound to Bacterial Magnetic Particles in *Magnetospirillum magneticum* Strain AMB-1 *J. Biol. Chem.* 278 (2003) 8745, Copyright American Society for Biochemistry and Molecular Biology (United States), reproduced with permission).

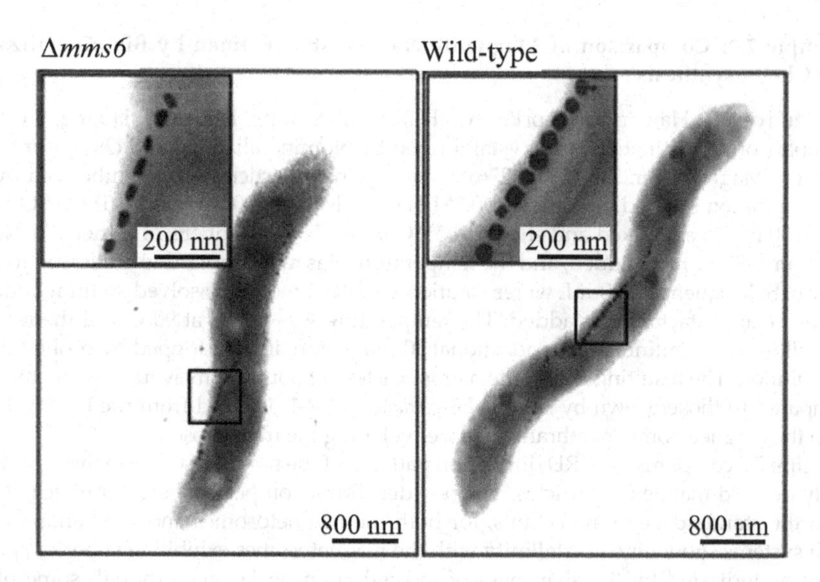

FIGURE 7.5
TEM micrographs of Mms6-deleted mutant (Δmms6) and wild-type strains of *M. magnetotacticum*. (From M. Tanaka, E. Mazuyama, A. Arakaki and T. Matsunaga, *J. Biol. Chem.* 286 (2011) 6386, Copyright American Society for Biochemistry and Molecular Biology (United States), reproduced with permission).

bacteria synthesized highly ordered cubo-octahedral crystals of magnetite, the mutant *Δmms6* bacteria formed smaller crystals with uncommon crystal faces and irregular shapes. In order to further characterize the morphological differences, the authors calculated the average size ((minor axis + major axis)/2) and shape factor ((minor axis)/(major axis)) of the particles. The average particle size for the wild-type strain was (48.3 ± 12.5) nm with a maximum size of 89.3 nm, while the mutant strain produced significantly smaller particles of an average size of (27.4 ± 8.9) nm and a maximum size of 63.7 nm. There were also differences in the shape factors with the wild–type, giving an average shape factor of 0.92 ± 0.16 compared to 0.74 ± 0.23 for the mutant cells. This confirmed that the *mms6* protein is involved in morphological regulation during crystal growth to produce highly crystalline cubo-octahedral crystals not seen in direct inorganic synthesis of magnetite. The observed crystal shape is unique to biogenic magnetite.

7.3 Magnetosomes *vs.* Synthetic Nanomagnetites: Crystallinity and Magnetism

As we discussed in Section 6.9, magnetite nanocrystals can be formed readily by coprecipitation reactions of ferrous and ferric salts in the mole ratio of Fe^{3+}/Fe^{2+} = 2:1 in aqueous solutions by the addition of alkalis at elevated temperatures. A comparison of the crystalline and magnetic properties of magnetite nanocrystals formed by biomineralization with that of chemically synthesized magnetite nanoparticles can further enhance our appreciation of the superior quality of magnetic nanocrystals derived from the process of biomineralization in magnetotactic bacteria. In a particular study presented in Example 7.2, the crystalline and magnetic properties of magnetosomes isolated from the cells of *M. gryphiswaldense* MSR-1 are compared to those of magnetite particles prepared by coprecipitation reactions.

Example 7.2: Comparison of Magnetite Nanocrystals Formed by Biomineralization and Chemosynthesis

In a study by L. Han and co-workers (L. Han, S. Li, Y. Yang, F. Zhao, J. Huang, J. Chang, Comparison of magnetite nanocrystal formed by biomineralization and chemosynthesis, J. Magn. Magn. Mater. 313 (2007) 236), magnetite nanoparticles were synthesized by the coprecipitation method. Specifically, 0.5M $FeCl_3$ solution (300 mL) and 0.5M $FeSO_4$ solution (150 mL) were mixed and stirred at 55°C under N_2 atmosphere, and then 3M NaOH solution (250 mL) was added, and the temperature was raised to 65°C and allowed to react for 1 h. Subsequently, 100-mL water solution with 0.03 mol of dissolved sodium dodecylsulfonate as surfactant was added. The temperature was raised at 90°C and the reaction was allowed to continue for an additional 30 min before it was stopped by cooling down the solution. The resulting magnetite nanoparticles of about 10 nm average diameter were compared to those grown by *M. gryphiswaldense* MSR-1, isolated from the bacterial cells with the magnetosome membrane intact, enveloping the magnetosomes.

Figure 7.6 compares the XRD diffraction patterns of the magnetosomes to those of chemically derived magnetite particles. The powder diffraction peaks observed matched well with the standard Fe_3O_4 reflections, for both the magnetosomes and the nanoparticles. Both systems show high crystallinity with the magnetosomes exhibiting superior crystallinity, as indicated by the sharpness of the reflection peaks, even though some of the broadening of the peaks associated with the synthetic Fe_3O_4 nanoparticles is due to their smaller size.

Figure 7.7 presents TEM micrographs of the samples. It is observed, as shown in Figure 7.7(a), that the magnetosomes tend to form bent chains, often forming closed loops. This configuration minimizes magnetostatic energy by concentrating the magnetic field lines within the magnetosome chain loops, whereby magnetic field lines close upon themselves minimizing any stray magnetic fields. In contrast, the synthetic magnetite particles cluster together in a random fashion. The flexibility of the magnetosomes to form bent chains is believed to arise from the elasticity of the magnetosome membrane.

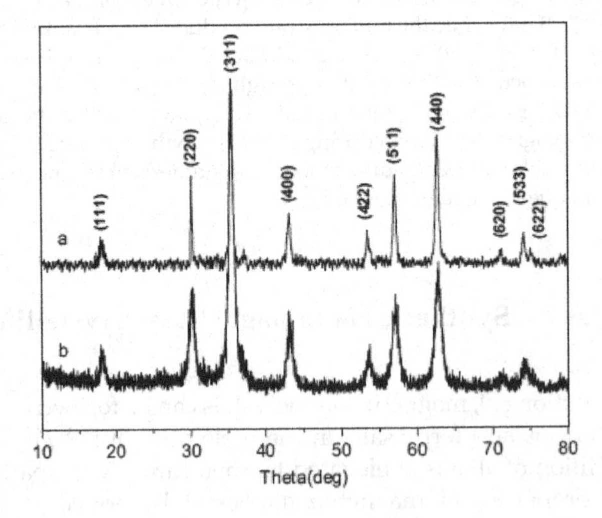

FIGURE 7.6
X-ray powder diffraction patterns of (a) magnetosomes and (b) synthetic magnetite. (From L. Han, S. Li, Y. Yang, F. Zhao, J. Huang, J. Chang, Comparison of magnetite nanocrystal formed by biomineralization and chemosynthesis, *J. Magn. Magn. Mater.* 313 (2007) 236, Copyright Elsevier B.V., reproduced with permission).

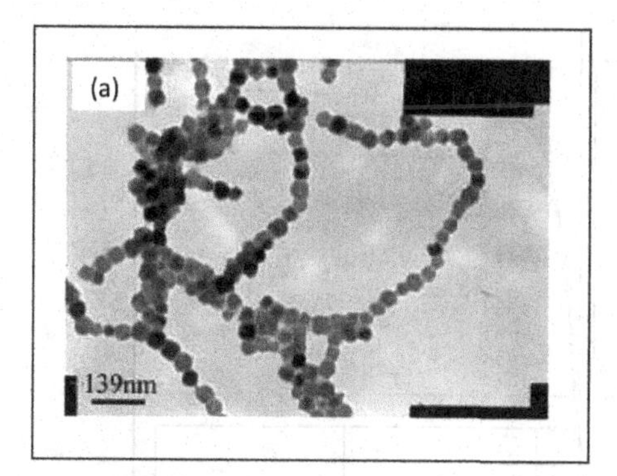

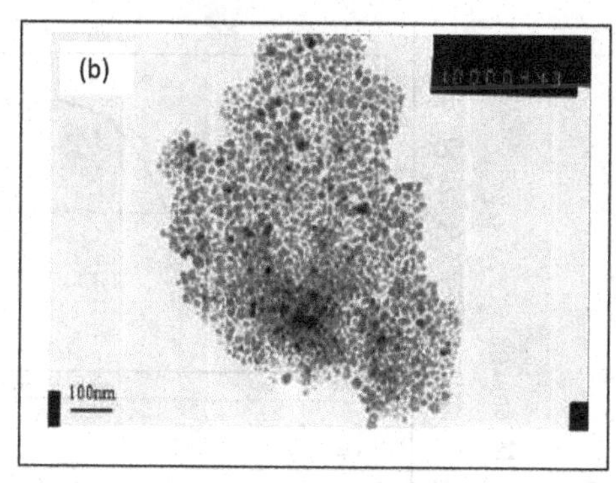

FIGURE 7.7
(a) TEM micrographs of magnetosomes and (b) synthetic magnetite particles. (From L. Han, S. Li, Y. Yang, F. Zhao, J. Huang, J. Chang, Comparison of magnetite nanocrystal formed by biomineralization and chemosynthesis, *J. Magn. Magn. Mater.* 313 (2007) 236, Copyright Elsevier B.V., reproduced with permission).

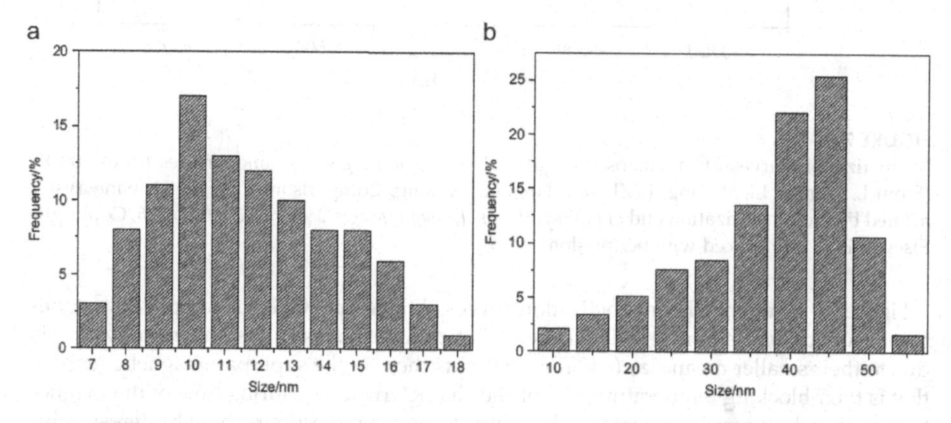

FIGURE 7.8
Particle size distributions of synthetic magnetite particles (a) and magnetosomes (b). (From L. Han, S. Li, Y. Yang, F. Zhao, J. Huang, J. Chang, Comparison of magnetite nanocrystal formed by biomineralization and chemosynthesis, *J. Magn. Magn. Mater.* 313 (2007) 236, Copyright Elsevier B.V., reproduced with permission).

The particle size distributions obtained from measuring the diameters of 100 nanoparticles of synthetic Fe_3O_4 and magnetosomes are shown in Figure 7.8(a) and (b), respectively. The particle size of the synthetic magnetite varied from 7 to 18 nm, with the maximum of the size distribution at 9–12 nm. The particle size of the magnetosomes varied from 10 nm to 60 nm, with the maximum of the size distribution at 40–50 nm. The crystal size distribution of the magnetosomes is asymmetric with a sharp cut-off toward larger particles, while that of the synthetic particles exhibits a log-normal distribution with a sharp cut-off toward smaller particles. As we have encountered previously crystal growth resulting from particle nucleation under supersaturation conditions results in log-normal particle size distributions (see, *e.g.*, Figure 5.9). In contrast, the controlled crystal growth of magnetosomes within membrane vesicles results in the characteristic particle size distribution observed in Figure 7.8(b).

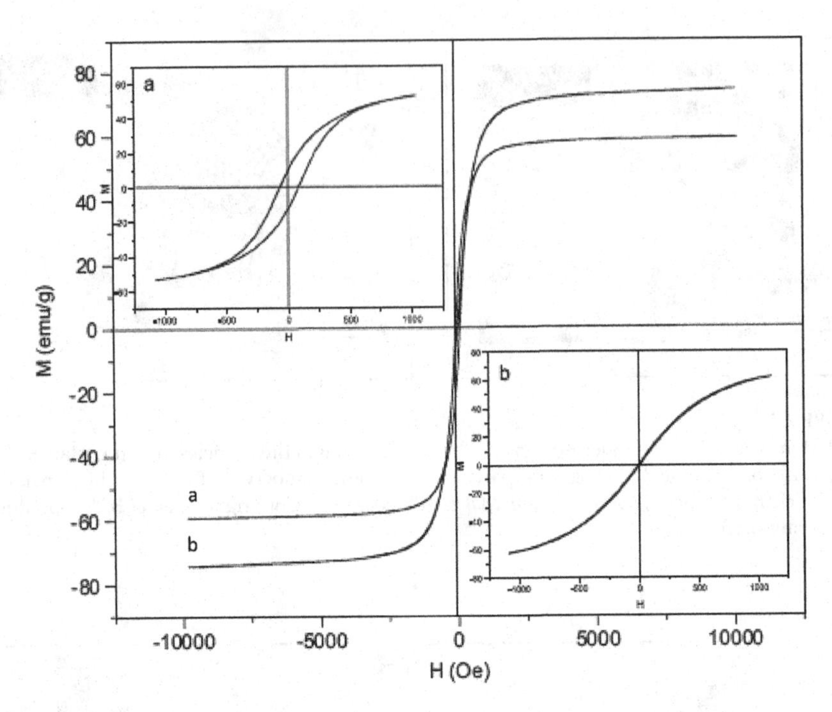

FIGURE 7.9
Magnetization curves of magnetosomes (a) and synthetic magnetite nanoparticles (b) at 300 K. (From L. Han, S. Li, Y. Yang, F. Zhao, J. Huang, J. Chang, Comparison of magnetite nanocrystal formed by biomineralization and chemosynthesis, *J. Magn. Magn. Mater.* 313 (2007) 236, Copyright Elsevier B.V., reproduced with permission).

Figure 7.9 compares the magnetization curves obtained at 300 K for samples of magnetosomes and synthetic magnetite. The synthetic magnetite particles show no hysteresis due to their smaller mean size (~10 nm) that puts them in the superparamagnetic regime; that is their blocking temperature, T_B, for the characteristic measuring time of the magnetometer is below room temperature. In contrast, the magnetosomes show hysteresis with a coercivity of 75 Oe, indicating that at room temperature, they are stable single-magnetic-domain particles due to their larger mean diameter (~45 nm) and possibly higher magnetocrystalline anisotropy.

The observation that biogenic magnetites in bacteria have dimensions that fall in the relatively narrow single-domain stability range of magnetite particles has important implications in the fields of rock magnetism and paleomagnetism because these single-domain magnetite particles are among the most stable carriers of "natural remnant magnetization" in many marine sediments and sedimentary rocks. This average grain size has been interpreted as a result of natural selection on the magnetic crystals formed by organisms that use their internally formed magnetite for geomagnetic sensitivity of some sort. Thus, these biogenic magnetite particles are often appropriately termed "magnetofossils". The search for magnetofossils in Martian meteorites and on the surface of Mars constitutes the basis of explorations for extraterrestrial life in our planetary system. Furthermore, an in-depth understanding of how MMPs exert physico-chemical control over crystal growth in magnetotactic bacteria has immediate relevance to biotechnology with respect to the tailoring of magnetic nanoparticles with desired magnetic characteristics using biomimetic approaches.

7.4 Ferritin and Its Dual Function in Biological Iron Regulation

Life has evolved to utilize iron for the performance of important metabolic functions. Iron is a central part of the heme group, a metal complex in hemoglobin that binds molecular oxygen (O_2) in the lungs and carries it to other cells in the body. In addition to hemoglobin, there are other important proteins that contain heme groups such as myoglobin that stores O_2 in the muscles and the cytochromes that catalyze important reactions by enabling one-electron oxidation–reduction processes to take place. Free iron, however, is toxic to cells as it acts as a catalyst in the formation of "free radicals" from reactive oxygen species, such as peroxides, *via* the Fenton reaction, depicted below:

$$Fe^{2+} + H_2O_2 \rightarrow Fe^{3+} + OH\cdot + OH^- \quad (7.1)$$

$$Fe^{3+} + H_2O_2 \rightarrow Fe^{2+} + OOH\cdot + H^+ \quad (7.2)$$

In the net reaction, the presence of iron is truly catalytic, and two molecules of hydrogen peroxide are converted into two hydroxyl radicals and water. The free hydroxyl radicals, OH· and OOH·, where (·) signifies the presence of unpaired electrons on an open shell configuration that renders free radicals highly chemically reactive, have a very short *in vivo* half-life of approximately 10^{-9} seconds; they can react virtually with all types of macromolecules, such as, carbohydrates, nucleic acids, lipids, and amino acids and may cause significant damage to cell structures. It is, therefore, important that the body regulate and maintain an appropriate supply of iron at all times in order to form hemoglobin and other molecules that depend on iron to function, while simultaneously control iron cytotoxicity. Poor regulation of iron levels in the body can lead to iron deficiency or to iron-overloading disorders. Ferritin is characterized by the capacity to remove Fe^{2+} ions from a solution in the presence of oxygen. It, thus, provides a necessary buffer against iron deficiency or iron overloading by storing away iron and releasing it in a controlled fashion, while simultaneously detoxifying the cell from free radicals. Thus, the role performed by ferritin as the cellular repository of excess iron is unique. In some ways, one could say that ferritin acts as a tiny organelle in its ability to secrete iron away from the delicate machinery of the cell, and then to release it again in a controlled fashion avoiding toxicity.

7.4.1 Structure of the Ferritin Molecule

Ferritin is a ubiquitous intracellular protein that is produced by almost all living organisms, including bacteria, algae, higher plants, and animals. It is a large globular protein, present in almost all cell types, whose unique molecular structure allows it to play a very important role in cellular chemistry. All proteins consist of chains of amino acids. An amino acid is a molecule containing a central carbon atom, a carboxylic acid group (COOH), an amino group (NH_2) and a variable side chain denoted by R, as indicated in Figure 7.10(a), which gives the general molecular structure of an amino acid. Twenty different amino acids are used by living organisms to synthesize proteins. The 20 amino acids vary in the identity of the side chain R, which endows each amino acid with specific properties. The side chain may consist of a single hydrogen atom as in the case of the amino acid glycine (Gly) or a bulky molecule as in tryptophan (Try); within the protein, the amino acid can be electrically neutral as in leucine (Leu) or charged as in aspartate (Asp) and glutamate (Glu), it can be polar or non-polar, thus, rendering the amino acid hydrophilic or hydrophobic. Figure 7.10(b) depicts the molecular structure of some amino acids that are present in ferritin. Table 7.1 lists the 20 amino acids and their properties.

In constructing proteins, amino acids are linked together by forming a covalent bond known as the peptide bond, as depicted in Figure 7.11. The peptide bond is formed between

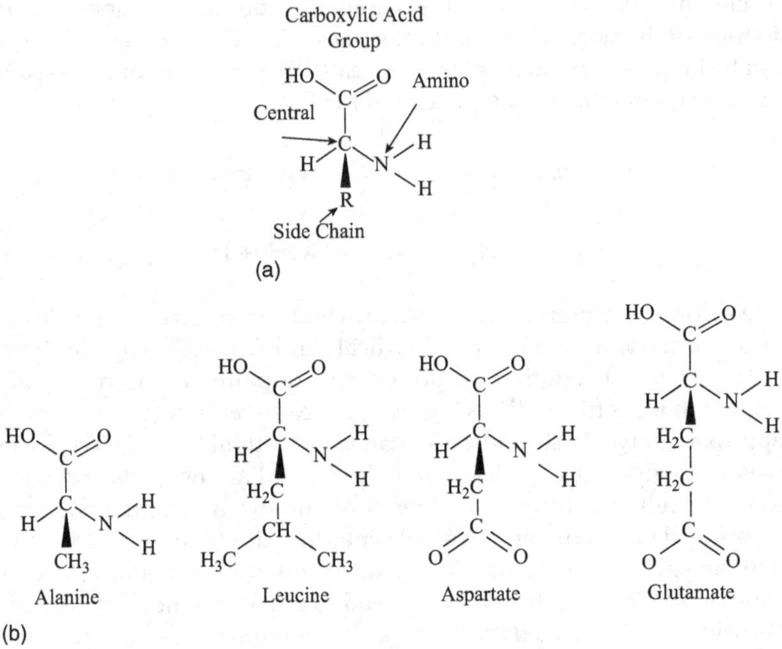

Carboxylic Acid Group
Amino
Central
Side Chain
(a)

Alanine
Leucine
Aspartate
Glutamate
(b)

FIGURE 7.10
(a) General structure of an amino acid. (b) Structure of alanine, leucine, aspartate and glutamate.

TABLE 7.1
List of Amino Acids and Their Properties

Nonpolar (Hydrophobic)	Polar (Hydrophilic)	Electrically Charged (Negative and Hydrophilic)
Glycine (Gly)	Serine (Ser)	Aspartic acid (Asp)
Alanine (Ala)	Threonine (Thr)	Glutamic acid (Glu)
Valine (Val)	Cysteine (Cys)	
Leucine (Leu)	Tyrosine (Tyr)	**Electrically Charged (Positive and Hydrophilic)**
Isoleucine (Ile)	Asparagines (Asp)	Lysine (Lys)
Methionine (Met)	Glutamine (Glu)	Arginine (Arg)
Phenylalanine (Phe)		Histidine (His)
Tryptophan (Trp)		
Proline (Pro)		

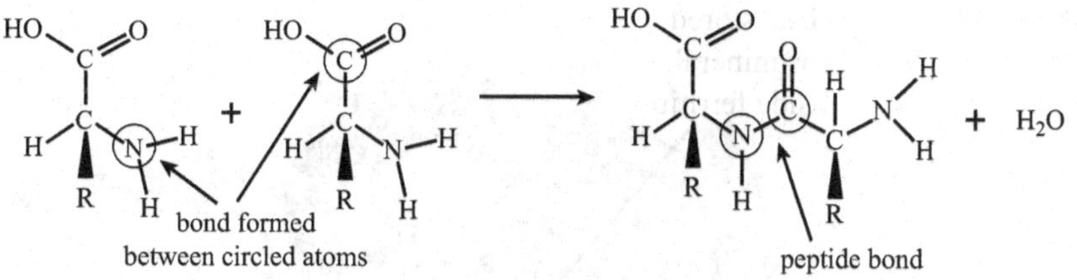

bond formed between circled atoms

peptide bond

+ H_2O

FIGURE 7.11
Definition of the peptide bond. It is formed between the carbon atom of the carboxylic acid group of one amino acid and the nitrogen atom of the amino group of the adjacent amino acid with the release of one molecule of water. Atoms participating in the bonding are circled.

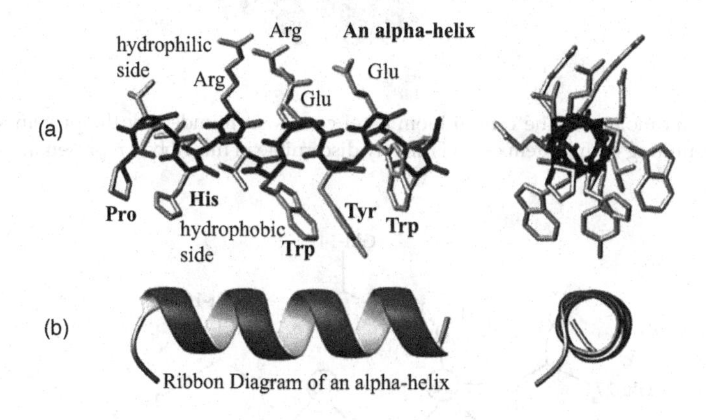

FIGURE 7.12
Two views of (a) the structure of the alpha-helix and (b) its ribbon representation. On the left, the views along the axis of the helix are shown and on the right, the views down the axis.

the carbon atom of the carboxylic acid group of one amino acid and the nitrogen atom of the amino group of the adjacent amino acid, forming the central backbone of an amino acid chain. Very long chains of amino acids can be formed, which invariably fold due to interactions between side chains to produce three-dimensional structures. One common example of folding is the formation of the alpha-helix motif, shown in Figure 7.12, usually depicted as a coiled ribbon leading to the ribbon representation of proteins. The shape and other properties of proteins are dictated by the precise sequence of amino acids within the ribbon. The amino acids are then numbered according to their position in the protein sequence. Each ferritin chain consists of a four α-helix bundle.

Vertebrate ferritins are composed of 24 amino acid or "polypeptide" chains, also called subunits, which co-assemble to form a hollow protein shell, as determined from X-ray crystallography, creating an interior cavity within which iron is sequestered, as shown in Figure 7.13. The de-mineralized protein, consisting of the protein shell void of iron, is known as apoferritin; it has an exterior diameter of 12 nm and an interior diameter of 7 nm. The 24 subunits are of two types known as light (L) and heavy (H), with a molecular weight of 19 and 21 kDA, respectively, resulting in a protein of molecular weight of about 450 kDA. The L and H subunits are structurally similar, each comprised of a bundle of four α-helices. Within the protein cavity, each ferritin molecule can store up to 4,500 iron (Fe^{3+}, $S = 5/2$) ions in the form of a solid mineral, hydrous ferric oxide or "ferrihydrite". Smaller,

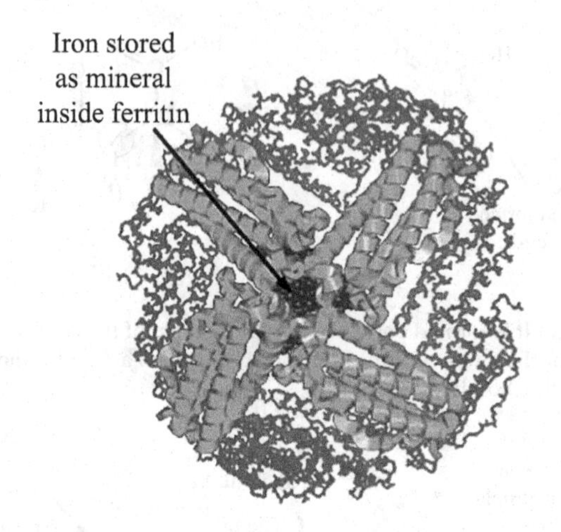

FIGURE 7.13
Structure of the ferritin molecule. The central biomineral core is surrounded by the protein sheath. The four α-helix bundles constituting each protein chain is clearly discernible in this ribbon representation of the protein.

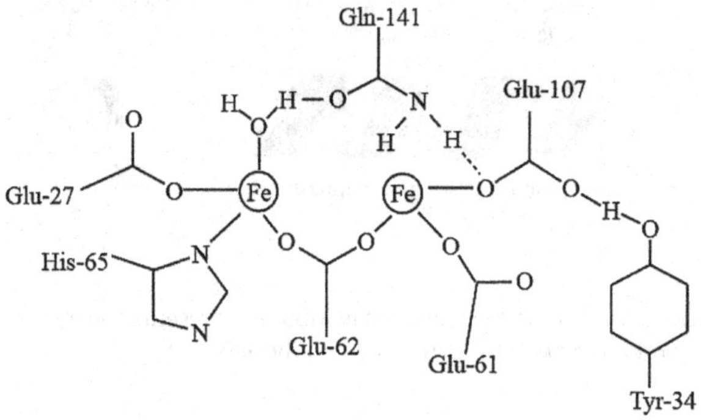

FIGURE 7.14
Structure of the ferroxidase center of the H-chains of ferritin.

ferritin-like shells have been identified in bacteria, known as Dps proteins, which are composed of only 12 subunits and likewise form mineral cores albeit of much smaller size, ~500 Fe/shell *vs.* ~4,500 Fe/shell for canonical ferritins.

In mammalian ferritin, the H and L subunits have about 55% identity in amino acid sequence. Their relative percentage in an apoferritin molecule varies depending on the cellular origin of the ferritin sample. The most important difference between the H and L chains lies in the fact that H-chains contain a catalytic site known as the "ferroxidase center" which involves amino acids Glu27, His65, Glu62, and Glu107 as metal ligands where two Fe^{2+} ions bind at a time to the protein, as depicted in Figure 7.14. The center, located in the central hydrophilic region of the H-subunit 4-helix bundle, facilitates iron incorporation into the protein cage through a unique enzymatic catalysis pathway *in vivo* and *in vitro* even under acidic conditions in which iron is not spontaneously oxidized. The center reacts with iron in its ferrous state and induces its oxidation to its ferric state, using either O_2 or H_2O_2 as the oxidant. The oxidized iron is then released from the ferroxidase center and ushered into the interior cavity of the apoferritin molecule. The center's binding sites thus become available for additional iron oxidation reactions. L subunits lack such a catalytic center but provide a higher density of negatively charged carboxyl groups from glutamic acid residues on the interior cavity surface of the protein that function as multiple nucleation sites for the controlled hydrolytic polymerization of Fe^{3+} ions, leading to the formation of the ferrihydrite biomineral core of ferritin. That is, after oxidation, the iron ions migrate from the ferroxidase center to the interior of the protein where mineral core formation is initiated at nucleation sites of the L-chain ferritin. Thus, the L- and H-chains play cooperative roles in the iron uptake and deposition mechanism. By their high capacity to promote efficient iron core nucleation and growth, the L-chains contribute to the avoidance of the non-specific hydrolysis of iron outside the protein shell and result in all iron being hydrolyzed and stored within the ferritin cage.

X-ray crystal structure determination by Pauline Harrison and co-workers in 1991 indicated that between subunits, there are small pores, or channels, of about 0.3 nm diameter, through which ions or small molecules can travel. These channels play a crucial role in ferritin's ability to uptake and release iron in a controlled fashion. They are of two types: 4-fold channels are formed at the intersection of four subunits, while 3-fold channels are formed at the intersection of three subunits. The two types of channels have different properties and thus perform different functions.

The walls of a 3-fold channel are lined with charged, polar (hydrophilic) amino acids such as aspartate (Asp) and glutamate (Glu), while the walls of a 4-fold channel are lined with non-polar (hydrophobic) amino acids, such as alanine and leucine. As long as the iron is incorporated within the solid mineral core of ferritin, the iron cannot be removed from the protein. The mineral lattice must be dissolved for an iron ion to break away from the crystalline core. This is accomplished by reducing the iron ion from the ferric to the ferrous state. Upon reduction, a hydrated ferrous ion is formed, $Fe(H_2O)_6^{2+}$, in which six water molecules are octahedrally coordinated to a central ferrous ion. The polarity of the 3-fold channels facilitates interaction with the Fe^{2+} ions enabling their exit from the protein. In contrast, the non-polar nature of the 4-fold channels cannot facilitate such ion passage from the interior to the exterior of the protein. It is believed that the function of the 4-fold channels is the transport of electrons into and out of the cavity, which allows oxidation or reduction of iron on the surface of the mineral core. The details of the mechanism responsible for electron transport are not, as yet, well understood.

7.4.2 Nature of the Ferrihydrite Core

Ferrihydrite, or hydrous ferric oxide, can be precipitated directly from oxygenated iron-rich aqueous solutions as a fine-grained, defective nanomaterial of various degrees of crystallinity. Its powder X-ray diffraction pattern can vary from that of two broad scattering bands in its most disordered state to a maximum of six strong lines in its most crystalline state, as shown in Figure 7.15. The six-line form corresponds to the nominal chemical formula $FeOOH \bullet 0.4H_2O$; however, the exact structure is fundamentally indeterminate as the water content is variable.

In native ferritin, the iron is stored as ferrihydrite with inclusions of various amounts of phosphate of general chemical formula $[FeOOH]_8[FeO(H_2PO_4)]$. One of the most readily (commercially) available sources of ferritin for experimentation is that extracted from equine spleen, or horse spleen ferritin (HoSF). Figure 7.16 gives the TEM micrograph of a sample of HoSF ferritin. The dark spots correspond to the electron-rich iron ferritin core where the electron beam is more effectively scattered away from the forward direction. The iron core is seen to have a diameter of up to ~7 nm, consistent with the internal size of the

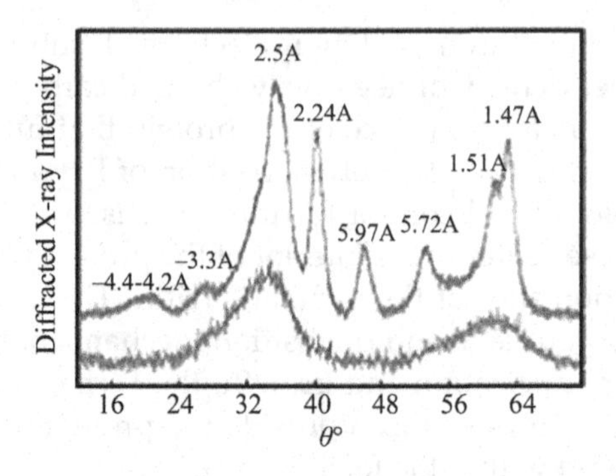

FIGURE 7.15
X-ray diffraction patterns for six-line (top) and two-line (bottom) ferrihydrite, Cu Kα radiation. (From V.A. Drits, B.A. Sakharov, A.L. Salyn, A. Manceau, Structural model for ferrihydrite, *Clay Miner.* 28 (1993) 185, Copyright Cambridge University Press, reproduced with permission).

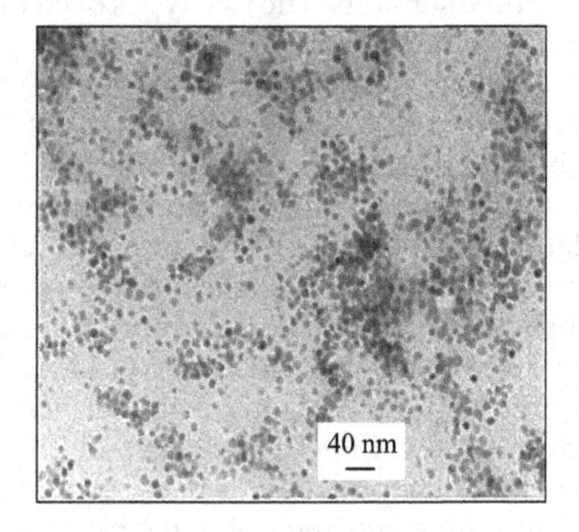

FIGURE 7.16
TEM micrograph of native, *in vivo* produced HoSF ferritin (sample purchased from Sigma-Aldrich).

protein shell. The protein shell around each core keeps the hydrous iron oxide particles isolated from each other preventing coagulation.

The in *vivo* produced biomineral core may be removed from the protein to obtain apoferritin, subject to subsequent *in vitro* reconstitution of the core under controlled laboratory conditions in the absence of phosphates. Through this process, "reconstituted ferritins" with various degrees of iron loading can be prepared as shown in Figures 7.17(a) and (b), which present the TEM micrographs of reconstituted HoSF with 500 and 3,000 Fe atoms/protein shell, respectively. Through terminal ligation, the protein sheath provides surface passivation to the nascent ferrihydrite nanoparticles and keeps the cores sterically separated, thus, preventing particle agglomeration. The XRD spectra of the derived cores indicate the formation of ferrihydrite, with the degree of crystallinity increasing with particle size, as shown in Figure 7.17(c).

When ferritin is reconstituted with iron and later incubated with phosphate, a phosphate layer is formed on the mineral core surface indicating that anions can diffuse through the protein shell and become associated with the iron mineral of ferritin by binding to the surface and stacking faults, or defects, in the iron mineral crystal.

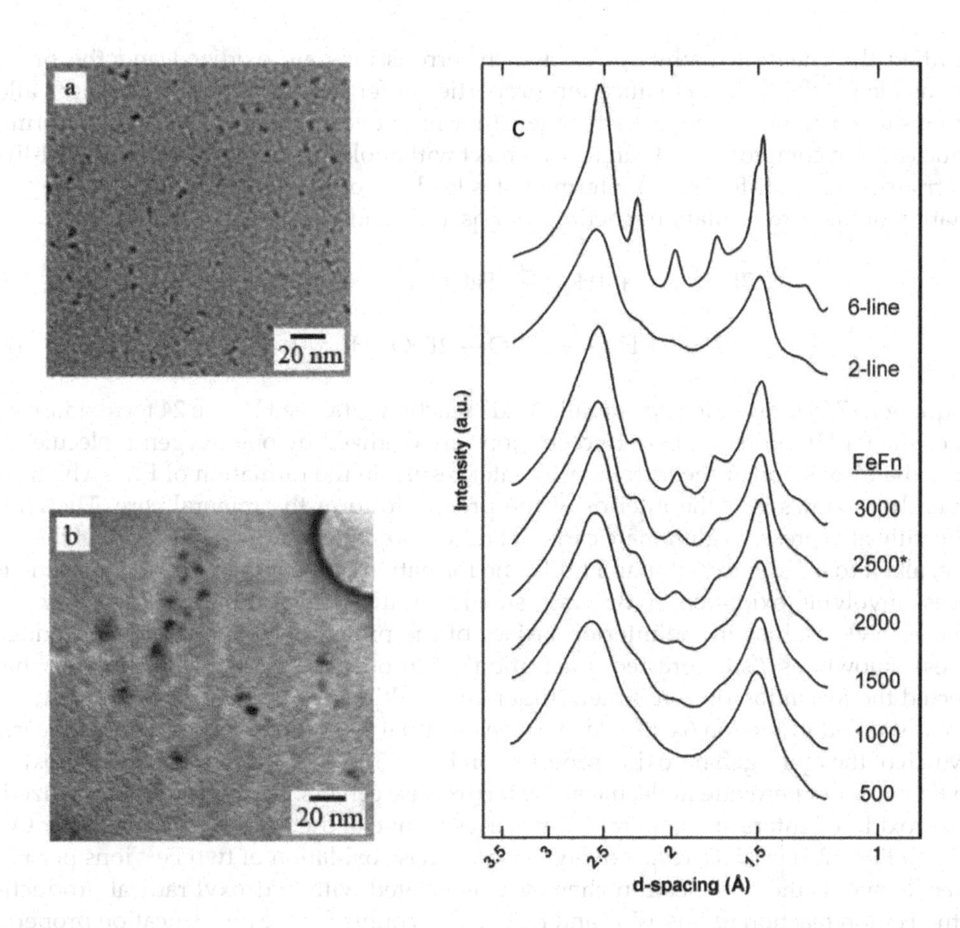

FIGURE 7.17
(a) TEM micrograph of *in vitro* reconstituted HoSF with 500Fe/protein, (b) TEM micrograph of reconstituted horse HoSF with 3,000Fe/protein and (c) variation of XRD spectra of the reconstituted HoSF mineral core with increasing iron loading (core size). (From F. Marc Michel, Hazel-Ann Hosein, Douglas B. Hausner, Sudeep Debnath, John B. Parise, Daniel R. Strongin, Reactivity of ferritin and the structure of ferritin-derived ferrihydrite, *BBA* 1800 (2010) 871, Copyright Elsevier, reproduced with permission).

In vitro studies using reconstituted ferritins have greatly contributed to the elucidation of the mechanism by which ferritin stores and releases iron. Furthermore, using recombinant DNA technology, recombinant ferritin molecules, not found in biological organisms, have been produced in the laboratory. While in living systems, ferritin invariably occurs as a heteropolymer, containing both L- and H-chains, recombinant human ferritin homopolymers, HuHF and HuLF, have also been produced and extensively studied, in order to further elucidate the different functions of L- and H-subunits in the ferritin molecule.

7.4.3 Recombinant HuHF and the Role of the Ferroxidase Center

Iron mineralization in recombinant HuHF has elucidated details of the process of biomineralization and the role played by the ferroxidase center, since only H-chains contain ferroxidase sites, specifically one ferroxidase site per subunit. The ferroxidase site, depicted in Figure 7.14, is located inside the 4-helix bundle of the H-chain ferritin subunit and endows the ferritin molecule with enzymatic activity associated with iron oxidation. Studies have identified the chemical pathways *via* which ferrous ions are oxidized and the process responsible for the H_2O_2 detoxification properties of ferritin. Hydrophilic passages allow Fe^{2+} ions to reach the ferroxidase site where they bind two at a time to the protein, forming di-nuclear iron complexes that can readily react with molecular oxygen to form short-lived di-ferric-peroxo → di-ferric-oxo intermediates leading to the oxidation of $Fe^{2+} \rightarrow Fe^{3+}$ and initiation of the core formation reactions of Eqs. (7.3) and (7.4):

$$2Fe^{2+} + O_2 + 4H_2O \rightarrow 2FeOOH_{(core)} + H_2O_2 + 4H^+ \quad (7.3)$$

$$2Fe^{2+} + H_2O_2 + 2H_2O \rightarrow 2FeOOH_{(core)} + 4H^+ \quad (7.4)$$

Equation (7.3) represents the overall, "end" reaction catalyzed by the 24 ferroxidase centers of the HuHF protein where two Fe^{2+} ions are oxidized by one oxygen molecule. The "immediate" reaction at the ferroxidase center results in the formation of Fe^{3+}, which subsequently migrates into the interior of the protein to form the mineral core. The HuHF apoferritin also presents glutamate carboxyl residues on the interior cavity wall of the molecule, albeit to a lesser extent than HuLF. The formation of the mineral core is a multi-step process involving oxidation, hydrolysis, supersaturation, nucleation and crystal growth. The carboxyl residues in the interior surface of the protein facilitate nucleation under a process known as "site-directed nucleation". Various spectroscopic techniques have detected the formation of iron dimer, trimer and multimer complexes similar in nature to those discussed in Section 6.2 that lead to the eventual formation of a solid mineral core.

Much of the hydrogen peroxide produced in Eq. (7.3) reacts through Eq. (7.4), to oxidize two Fe^{2+} ions per peroxide molecule at the ferroxidase centers. Ferrous ions are oxidized at the ferroxidase centers at comparable rates irrespective of the type of oxidant, either O_2 or H_2O_2 *via* Eqs. (7.3) or (7.4), respectively. The pair-wise oxidation of two Fe^{2+} ions per H_2O_2 molecule avoids the odd electron chemistry associated with hydroxyl radical production *via* the Fenton reaction of Eqs. (7.1) and (7.2) and accounts for the detoxification properties of ferritin and its ability to protect cells from oxidative damage. We may note that in the case of Dps proteins from bacteria, H_2O_2 is the primary oxidant for Fe^{2+} at the ferroxidase centers of these proteins and the mineralization reaction proceeds according to Eq. (7.4).

Once a sufficient core has been developed, ($\geq$ 800 Fe/shell), the surface of the nascent core exhibits autocatalytic properties for iron oxidation and additional Fe^{2+} ions are readily oxidized at the surface prior to being incorporated into the biomineral core according to Eq. (7.5), which then becomes the dominant reaction.

$$4Fe^{2+} + O_2 + 6H_2O \rightarrow 4FeOOH_{(core)} + 8H^+ \quad (7.5)$$

Equation (7.5) describes a more efficient iron oxidation process, as four Fe^{2+} ions are oxidized per oxygen molecule, as opposed to the reaction of Eq. (7.3), where only two Fe^{2+} ions are oxidized per oxygen molecule. Studies with human L-chain homopolymers (HuLH), which lack ferroxidase centers, indicate that L-chain homopolymers deposit their iron *via* Eq. (7.5). This observation may also explain the biological evolution of heteropolymer ferritin molecules; namely, that a delicate balance between the need for fast iron oxidation and iron nucleation must be satisfied. Iron-storage organs, such as the liver and spleen, contain L-chain-rich ferritins and many contain large crystalline cores, possibly emphasizing the L-subunit nucleation-driven reactions. The heart and erythrocytes contain H-rich ferritins perhaps indicating that here their major role may be in removing toxic iron rather than providing iron reserves.

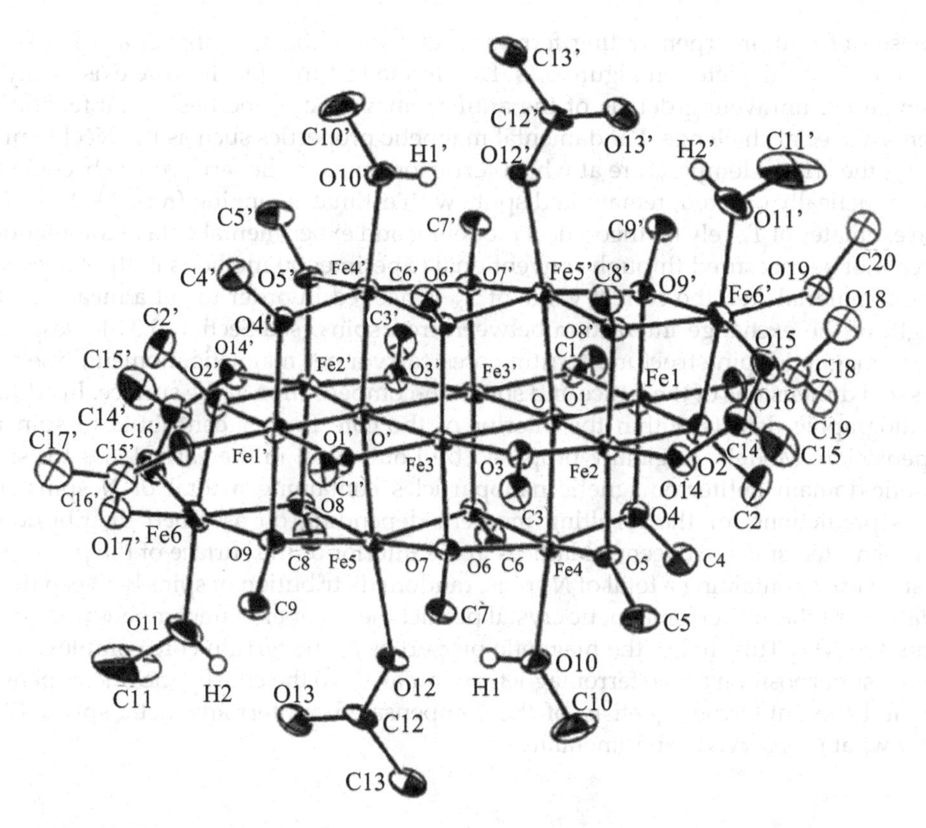

FIGURE 7.18
A model of the structure of the nascent ferritin mineral core. ORTEP depiction of $[Fe^{III}_4Fe^{II}_8(O)_2(OCH_3)_{18}(O_2CCH_3)_6(CH_3OH)_{4.67}]$ displaying 40% probability thermal ellipsoids, hydrogen atoms bound to carbon are omitted for clarity. (From K. L. Taft, G. C. Papaefthymiou and S. J. Lippard, A mixed-valent polyiron oxo complex that models the biomineralization of the ferritin, *Science*, 259 (1993) 1302, Copyright American Association for Advancement of Science, reproduced with permission).

Figure 7.18 depicts the molecular structure of a dodecanuclear iron complex proposed by scientists to constitute a model for the nascent ferritin biomineral core. This polyiron oxo complex of molecular formula $[Fe^{III}_4Fe^{II}_8(O)_2(OCH_3)_{18}(O_2CCH_3)_6(CH_3OH)_{4.67}]$ has been prepared from ferrous acetate and lithium methoxide in methanol by slow addition of O_2. Its three-dimensional close-packed layered structure closely mimics that of the inorganic core in ferritin. The small size and mixed-valent nature of this structure suggest that it is a reasonable model for intermediates formed in the biomineralization of iron during ferritin core formation.

7.5 Magnetic Properties of Ferritin

As observed in the model structure of the ferritin core depicted in Figure 7.18, iron ions are exchange-coupled to adjacent iron ions through oxo- or hydroxo-superexchange bridges to produce an antiferromagnetic lattice. In its extended form, the ferrihydrite lattice would be composed of two, interpenetrating ferromagnetic sub-lattices, A and B, as discussed in Section 3.3.3 and depicted in Figure 3.14. Due to the fact that ferrihydrite exists only as a nanomaterial, unraveling details of the antiferromagnetic properties of the ferritin core presents a special challenge. Fundamental magnetic properties such as the Néel temperature, T_N, the critical temperature at which ferric ion spins in the ferritin core become antiferromagnetically ordered, remain in dispute with estimates ranging from 200 K to 500 K. These estimates of T_N rely on theoretical modeling and experimental data extrapolation, as they cannot be measured through conventional experimental methods in the corresponding bulk material. Yet, the correct value of T_N is needed in order to get a measure of the strength of the exchange interaction between iron spins (see Section 3.3.3). Despite the antiferromagnetic spin structure, ferritin cores carry a net magnetic moment, due to the nanosized dimensions of the lattice and spin non-compensation at the surface. In addition, crystallographic defects within the interior of the core further contribute to spin non-compensation. Models, originally proposed by Louis Néel in the early 1960s for single-magnetic-domain antiferromagnetic nanoparticles containing a total of N spins make various predictions for the resulting moment depending on assumed distributions of uncompensated spins between sublattices in the interior or the surface of the particle. For a crystal lattice containing a total of N spins, random distribution of spins between the two sub-lattices of the antiferromagnetic crystal predict the number of uncompensated spins to be equal to $N^{1/2}$. This makes the magnetic properties of the ferritin core complex, arising from the superposition of the ferromagnetism imparted to the core by the uncompensated spins and the antiferromagnetism of the compensated antiferromagnetic spins. This is exactly what is observed experimentally.

Example 7.3: Magnetization Studies of Horse Spleen Ferritin

Investigators Kilcoyne and Cywinski (S. H. Kilcoyne and R. Cywinski, Ferritin: a model superparamagnet, *J. Magn. Magn. Mater.* 140–144 (1995) 1466) studied the magnetization of ferritin as a function of the applied magnetic field at various temperatures in the range of 40 K to 290 K. The results are presented in Figure 7.19. At these temperatures, the magnetic ferritin cores are expected to exhibit superparamagnetism with the magnetization following the Langevin function of Eq. (2.52). However, the isothermal magnetization curves of Figure 7.19 were satisfactorily fit to a modified Langevin function, which contained an additional term, which is linear with respect to the applied magnetic field H.

Equation (7.6) gives the theoretical model for the magnetization M, as a function of the strength of the applied field, which best fits the experimental data.

$$M(x) = M_s(\coth(x) - 1/x) + aH \tag{7.6}$$

Here, M_s is the saturation magnetization, $(\coth(x) - 1/x) = L(x)$ is the Langevin function, H is the magnitude of the applied magnetic field, $x = \mu_c H/kT$, where μ_c is the magnetic moment of the ferritin core, k is Boltzmann's constant and a is an experimentally determined constant. The authors indicate that the best fits to the experimental curves were obtained for a total particle moment of the order of $\mu_c \approx 300\mu_B$ per ferritin core. Assuming ~4,000 high-spin ferric ions ($S = 5/2$) per fully loaded HoSF protein, that is $5\mu_B$ per ferric ion, Néel's model for random spin distribution between the two interpenetrating sub-lattices predicts a value of $\mu_c \approx 316\mu_B$, in excellent agreement with the experimental data. The additional linear contribution to the magnetization is proposed to arise from the

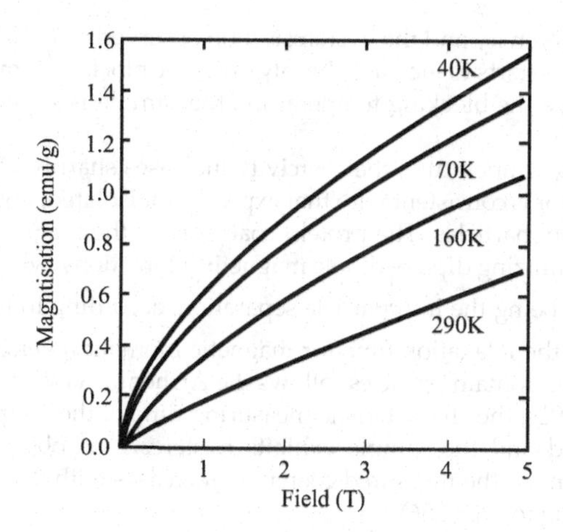

FIGURE 7.19
Isothermal DC magnetization curves of fully loaded horse spleen ferritin at various temperatures. The solid lines are least-square fits to the modified Langevin function discussed in the text. (From S. H. Kilcoyne and R. Cywinski, Ferritin: a model superparamagnet, *J. Magn. Magn. Mater.* 140–144 (1995) 1466, Copyright Elsevier B.V., reproduced with permission).

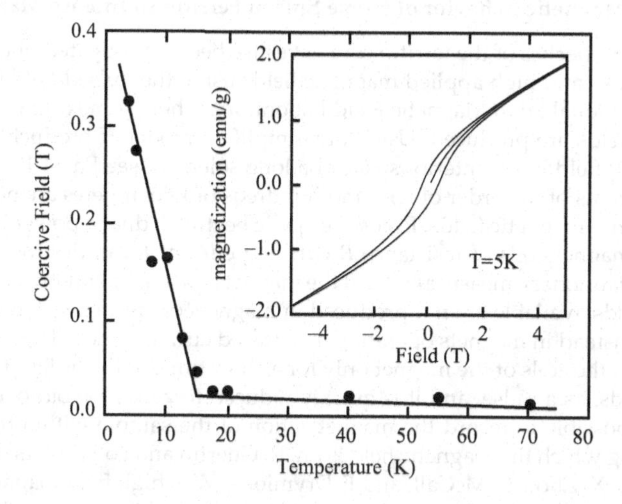

FIGURE 7.20
Temperature dependence of the coercivity of ferritin. Inset, hysteresis loop at $T = 5$ K. (From S. H. Kilcoyne and R. Cywinski, Ferritin: a model superparamagnet, *J. Magn. Magn. Mater.* 140–144 (1995) 1466, Copyright Elsevier B.V., reproduced with permission).

super-antiferromagnetic behavior of the core. "Super-antiferromagnetism" is a finite size effect, or nanomagnetic effect, first predicted by Néel in 1961. It occurs in small antiferromagnetic particles with an even number of interpenetrating ferromagnetic planes and results in an enhancement of the magnetization at large applied fields relative to the standard Weiss molecular field model. Its origin lies in the continuous rotation of the staggered magnetization within the particle when the field is applied perpendicular to the antiferromagnetic axis.

As expected, when the ferritin sample is cooled down to liquid helium temperatures, it exhibits hysteresis with a coercivity $H_c = 0.27$ T at $T = 5$ K. Figure 7.20 gives the temperature dependence of the coercivity and the hysteresis loop observed at $T = 5$ K. The temperature at which the coercivity starts to increase sharply gives the blocking temperature, $T_B \approx 15$ K, for the sample. Above the blocking temperature, the sample is superparamagnetic with zero coercivity.

Below the blocking temperature, the coercivity increases sharply with decreasing temperature. This behavior is consistent with that expected for isolated, non-interacting single-magnetic-domain nanoparticles. The protein coat isolates the magnetic cores preventing coagulation and minimizing dipole–dipole magnetic interactions, which diminish at a rate proportional to $\frac{1}{r^3}$, r being the interparticle separation, according to Eq. (3.14). As a function of temperature, the relaxation time for magnetic moment reversals for isolated, uniaxial, single-magnetic-domain particles follows the Arrhenius law of Eq. (4.45). When the relaxation time $\tau \geq 100$s, the characteristic measuring time of the magnetometer, moment reversals are blocked and the sample exhibits hysteresis, as observed in the inset of Figure 7.20. Furthermore, the measured coercivity increases with decreasing temperature as expected, according to Eq. (4.56).

Example 7.4: Magnetic Behavior of Horse Spleen Ferritin in Intense Magnetic Fields

The magnetic properties of the ferritin core have also been investigated under extremely low temperatures and high applied magnetic fields using the pulsed-field facility of the Los Alamos National High Magnetic Field Laboratory, where some of the world's highest magnetic fields are produced. Using our simplified model of producing a homogeneous magnetic field in the interior space of a long solenoid, see Eq. (1.13) or Eq. (2.40), very large currents of the order of tens and hundreds of kiloamperes are needed to produce very intense magnetic fields. Excessive "joule heating", due to power dissipation ($P = I^2R$) in the magnet coils of resistance R currying current I, can destroy the magnets, even with precautionary measures to cool the magnets during operation. For this reason, the highest fields available are not produced in magnets operating with continuous DC currents, but instead in magnets operating in a pulsed current mode. The high current is driven through the coils of the magnet only for a short time, of the order of milliseconds or microseconds, as a pulse, and it is quickly reduced to zero to avoid overheating the magnet. It is possible to record the magnetization of the sample within the short time window during which the magnetic field is on. R. Guertin and co-workers (R. P. Guertin, N. Harrison, Z. X. Zhou, S. McCall, and F. Drymiotis, Very high field magnetization and AC susceptibility of native horse spleen ferritin, *J. Magn. Magn. Mater.* 308 (2007) 97) report the isothermal magnetization of native HoSF at $T = 1.52$ K and applied fields up to 55 T shown in Figure 7.21. The temperature of 1.52 K is at the lower limit of temperatures that can be reached by pumping on the liquid helium bath reservoir. The high field measurements indicate that the characteristic curvature, observed in the isothermal magnetization *vs.* applied field data of ferritin in Figure 7.19, persists into much higher applied magnetic fields.

Additionally, these data give us a sense of the strength of the antiferromagnetic exchange within the ferritin core. The inset of Figure 7.21 compares the magnitude of the magnetization recorded at the largest magnetic fields applied to that of the estimated saturated magnetization of the core expected if individual iron moments were to be polarized along the applied magnetic field, overcoming the antiferromagnetic exchange interactions between adjacent moments. The data indicate that even at an applied field of 55 T, only 4% of the estimated full magnetization is achieved.

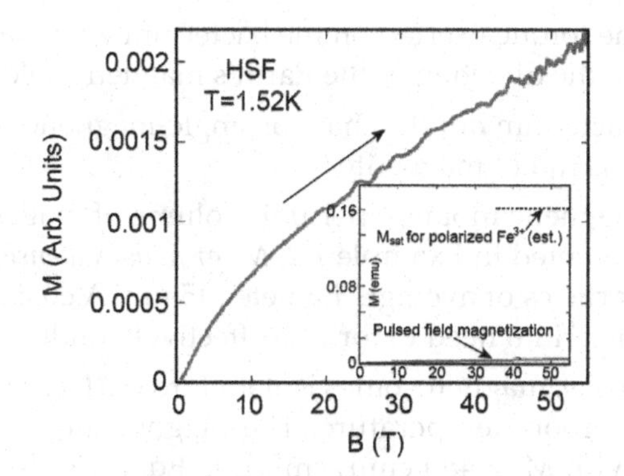

FIGURE 7.21
High-field magnetization of native horse spleen ferritin at $T = 1.52$ K up to 55 T applied field. (From R. P. Guertin, N. Harrison, Z. X. Zhou, S. McCall, and F. Drymiotis, Very high field magnetization and AC susceptibility of native horse spleen ferritin, *J. Magn. Magn. Mater.* 308 (2007) 97, Copyright Elsevier B.V., reproduced with permission).

The study of ferritin occupies a prominent place in the study of nanomagnetism. The ability to engineer the polarity and electrical properties of the wall lining of the interior protein cavity through "site-directed mutagenesis" can offer a unique handle in tailoring the nucleation and crystal growth habits of various magnetic biomineral cores.

In this chapter, we have witnessed the biologically directed growth of magnetic nanoparticles within confined spaces such as membrane vesicles and protein cages, which are genetically engineered for the production of functional magnetic bio-nanomaterials. Two prominent systems were presented in detail, magnetosomes produced by magnetotactic bacteria and wild-type mammalian ferritins, produced *in vivo* or reconstituted *in vitro*, as well as, genetically engineered recombinant ferritins. Life appears to be a natural laboratory for the study of nanomagnetism, whereby years of evolution have perfected exquisite processes in nanoscale engineering. Unlike the passive encapsulation within surfactant molecules described under the synthesis of magnetic nanoparticles by physical and chemical methods (Chapters 5 and 6), here, the biological sheaths are endowed with specific functionality and are actively involved in facilitating nucleation and in directing the crystal growth of the encapsulated magnetic nanophases. Today's nanoengineers are actively engaged in the study of such biologically directed processes in an attempt to imitate the chemistry of life in the laboratory by synthesizing bio-inspired or bio-mimetic magnetic nanostructures using nanotemplating, as we explore in the next chapter.

Exercises

1 For a bacterium to orient in a magnetic field $\vec{B}$ of a certain strength, the magnetic orientational potential energy U_B of its biomagnetic moment $\vec{m}$ must exceed thermal energy, which tends to randomize its direction.

(a) How strong a magnetic dipole moment does the bacterium need to build in order to successfully orient in the earth's magnetic field of 0.5 G at 300 K? What is the magnitude of the torque exerted on the bacterium when oriented in the direction perpendicular to the direction of the Earth's magnetic field?

(b) Why does the bacterium build a chain of single-magnetic-domain particles rather than a continuous rod of magnetite?

2. At saturation, the magnetic moment per unit volume of magnetite is M_s = 480 emu/cm³. In the study presented in Example 7.2, *M. gryphiswaldense* forms roughly spherical magnetosome particles of average diameter 45 nm. A chain of how many magnetosomes does the bacterium need to form to effectively undergo magnetotaxis?
3. According to Figure 7.9, magnetosomes isolated from *M. gryphiswaldense* show coercivity H_c = 75 Oe at room temperature. The magnetization within each magnetosome is saturated with M_s = 480 emu/cm³. Use Eq. (4.56) to get an estimate of the value of the uniaxial magnetic anisotropy constant K_u of the magnetosomes. How does your answer compare with the magnetocrystalline anisotropy constant of (bulk) magnetite?
4. For the two examples of biogenic magnetic nanoparticles discussed in this chapter, describe in general terms, similarities and differences in the functionality of the encapsulating organic sheaths in exerting biological control over crystal growth.
5. Define the terms "magnetotaxis", "magnetoreception", "peptide bond", "polypeptide chain", "alpha-helix", "ferroxidase center", "hydrated ion", "free radicals", "apoferritin", "reconstituted ferritin", "recombinant ferritin", "site-directed nucleation" and "site-directed mutagenesis".
6. Ferrihydrite possesses antiferromagnetic spin order; local ferric ion spins ($S = 5/2$) are antiferromagnetically coupled to neighboring ($S = 5/2$) iron spins. Yet, the ferritin core exhibits ferromagnetism. What is the origin of ferromagnetism?
7. Ferritin molecules as extracted from horse spleen exhibit a blocking temperature T_B = 15 K, as seen in Figure 7.20. Assuming a characteristic measuring time for the magnetometer of τ_m = 100 s and fully loaded proteins, use the Néel relaxation process of Eq. (4.46) to estimate the magnitude of the effective (uniaxial) magnetic anisotropy constant of the ferritin core (assume $\tau_0 = 10^{-9}$s for spin reversal attempt time).
8. HoSF ferritin has a coercivity of H_c = 0.27 T at T = 5 K, according to the inset of Figure 7.20. Using the value of the anisotropy constant for HoSF you derived in Exercise 7, evaluate the saturation magnetization M_s of the ferritin core.

8

Biomimetic Magnetic Nanoparticles

> The basic constructional processes of biomineralization – supramolecular preorganization, interfacial molecular recognition (templating) and cellular processing – can provide useful archetypes for molecular-scale building, or "molecular tectonics", in inorganic materials chemistry.
>
> **Stephen Mann, *Nature*, 365 (1993) 499**

The exquisite control over crystallinity, shape and form exercised in the formation of biogenic magnetic nanoparticles through the process of biomineralization has given birth to bio-inspired or biomimetic synthesis, which explores the possibility of preparing man-made materials by mimicking biological processes in nature. The biogenic magnetic nanoparticles discussed in Chapter 7, namely the bacterial magnetosomes and the biomineral core of ferritin, are examples of nature's incorporation of hard, inorganic matter within soft, organic templates, endowed with designed, end-use functionalities. It is these hard/soft interface processes, which biomimetic materials synthesis attempts to profitably utilize for the synthesis of novel hierarchical structures. Understanding the bio-molecular construction routes that give rise to the incorporation of well-controlled inorganic phases within biological entities should allow the development of biomimetic chemistry in the laboratory, where synthesis and self-assembly processes on organic templates could be coupled to produce designer's materials with controlled properties. In some respects, the precipitation reactions within surfactant micelles in microemulsions explored in Chapter 6 can be considered biomimetic, as the synthesis takes place within constrained spaces or "nanoreactors" mimicking "bacterial membrane vesicles" and "protein cages" in biology.

8.1 Electrostatic Interactions in Ferritin

As we discussed in the previous chapter, studies of iron biomineralization in "recombinant ferritin" cages indicate that both human H-chain and L-chain homopolymers (HuHF and HuLF) form mineral cores within their protein cages, utilizing the autocatalytic reaction of Eq. (7.5). The negatively charged glutamate residues lining the inner surface of the protein with COO^- groups play a crucial role in this nucleation-driven crystallization, due to their electrostatic attraction of the positively charged iron ions. This Coulombic attraction is not specific to iron alone; it would act on any other positively charged ions, enabling nucleation-driven mineralization within the ferritin cage to occur for a range of transition metal ions. Once an ionic species enters the interior of the protein cage, nucleation and hydrolytic polymerization, which leads to mineralization, can be driven by purely electrostatic effects.

DOI: 10.1201/9781315157016-10

Consider a charged surface introduced within an electrolytic solution. Oppositely charged ions, counter-ions, would migrate onto the surface due to the electrostatic interaction energy of Eq. (8.1) between a surface charge ion, q_1, and a counter-ion, q_2, in the solution.

$$U(r) = \frac{q_1 q_2}{4\pi\varepsilon_r\varepsilon_0 r} \tag{8.1}$$

Here, r is the distance of separation between the charges, ε_r is the dielectric constant of the solution and ε_0 is the permittivity of free space. This interaction results in the formation of a "double layer" of oppositely charged ions effectively neutralizing the charge on the surface. Within the double layer, the value of the electrostatic potential, $V(x)$, decreases linearly with distance from the charged surface, just like in the space between the plates of a charged, parallel plate capacitor, as depicted in Figure 8.1(a). However, under standard temperature and pressure, the counter-ions in the solution are free to move and thus the electrostatic attraction is in competition with thermal energies, kT, and Brownian motion.

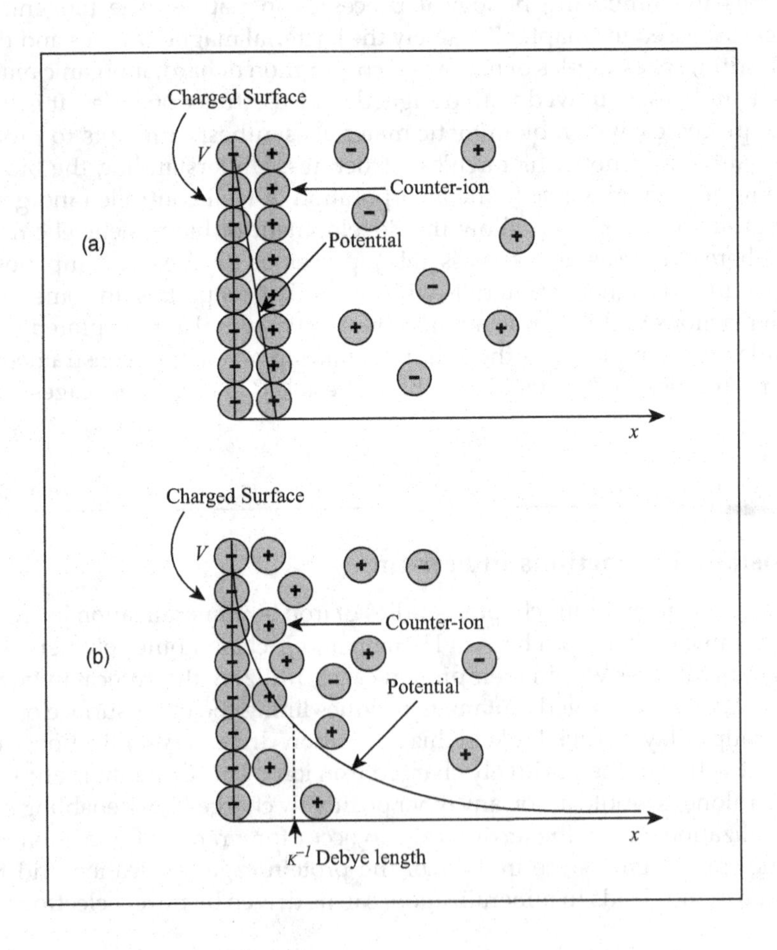

FIGURE 8.1
(a) Neutralization of surface charge by double layer formation under the action of electrostatic forces alone; (b) diffuse double-layer formation under the action of electrostatic forces and thermal agitation. The variation of the electrostatic potential with x is depicted and the Debye screening length is indicated.

The range over which two electron charges have an interaction energy $U \geq kT$ defines a characteristic length, known as the Bjerrum length, l_B. For pure water,

$$l_B = 0.7\,\text{nm} \tag{8.2}$$

The result of the competition between thermal and electrostatic energies is the formation of a cloud of excess counter-ions next to the charged surface, within which the potential drops rapidly over a wider diffuse charged layer, as depicted in Figure 8.1(b). According to the "Gouy–Chapman theory" developed in the early 1900s, within this "diffuse double layer", the potential decays exponentially with distance x away from the surface, according to Eq. (8.3).

$$V(x) = V_0 e^{-\kappa x} \tag{8.3}$$

V_0 is the absolute value of the strength of the potential at the charged surface and κ is the decay constant, characteristic of the solution and temperature; κ is measured in reciprocal meters. The distance x from the surface at which the potential decreases to $(1/e)V_0$ defines the "thickness", or "width", of the diffuse double layer, also known as the "Debye screening length" of the electrolyte. For an electrolytic solution containing various counter-ions, the screening length, or width of the double layer, can be calculated from Eq. (8.4).

$$\kappa^{-1} = \left(\frac{\varepsilon_r \varepsilon_0 kT}{e^2 \sum_{i=1}^{N} c_i z_i^2} \right)^{1/2} \tag{8.4}$$

In this equation, c_i is the mean concentration of charges of species i in the solution, z_i is the charge on the ith ionic species, e is the elementary charge; k is Boltzmann's constant and T the absolute temperature. In water, at room temperature, 1 M solution of monovalent ionic species, such as in the case of Na^+ and Cl^-, would have a Debye screening length, or double layer width of $\kappa^{-1} = 0.304$ nm, while divalent and trivalent ionic species would have smaller screening lengths. The concentration of the counter ions, $\rho(x)$, as a function of the distance from the charged surface, follows a similar exponential dependence as observed for the potential, with the same value of the decay constant κ.

$$\rho(x) = \rho_0 e^{-\kappa x} \tag{8.5}$$

We conclude that according to the Gouy–Chapman theory, the negatively charged interior surface of the apoferritin molecule lined with COO^- ions will attract and accumulate positively charged Fe^{2+} counter ions at its immediate vicinity at concentrations significantly higher than those in the bulk solution. The probability that Fe^{2+} ions bind to carboxyl groups at the inner surface of the protein is thus increased, lowering the oxidation/reduction potential and promoting the oxidation of iron from $Fe^{2+} \rightarrow Fe^{3+}$. Thus, the highly charged interior surface serves as a substrate for the stabilization of highly charged Fe^{2+}/Fe^{3+} clusters. This opens the way to the laboratory use of ferritin cages as nanotemplates for the synthesis of monodispersed ferrimagnetic nanoparticles, such as magnetite and maghemite, which exhibit much stronger magnetism than antiferromagnetic ferrihydrite.

Moreover, these electrostatically driven processes would operate equally on other cations such as Co^{2+} or Ni^{2+}, leading to the possibility of the precipitation of various magnetic phases within the ferritin cage.

Indeed, calculated charge densities on HuHF homopolymers correlate with areas of the protein thought to interact with Fe^{2+} ions. The 3-fold channels, ferroxidase sites and nucleation sites on the surface of the protein cavity, all show negative electrostatic potentials. Furthermore, along the 3-fold channels an electrostatic gradient toward the interior cavity provides a guidance mechanism for cations to enter the protein.

8.2 Magnetoferritin: Fe_3O_4 and γ-Fe_2O_3 Grown within Apoferritin

The electrostatically driven molecular control of mineral precipitation exercised by ferritin within the confined space of the protein cage is of general interest to the synthesis of magnetic nanoparticles. Under reducing conditions and in the presence of iron-chelators, ferritin can be depleted of its ferrihydrite biomineral core and reconstituted under conditions favoring the synthesis of magnetic iron-oxide phases, magnetite or maghemite. In 1992, Stephen Mann and co-workers demonstrated for the first time the synthesis of such ferrimagnetic phases within the ferritin shell *via* this process, schematically depicted in Figure 8.2.

In their experiment, the iron was removed from native horse spleen ferritin (HoSF) by dialysis, under nitrogen atmosphere, against thioglycolic acid ($HSCH_2COOH$) in sodium acetate (CH_3COONa) buffer at pH 4.5. The resulting apoferritin solution was buffered at pH 8.5 and maintained at a temperature of 55°C to 60°C under argon in a water bath. Fe^{2+} solution (prepared by the dissolution of ferrous ammonium sulfate, $(NH_4)_2Fe(SO_4)_2{\cdot}6H_2O$, in de-aerated water) was added slowly in small increments along with small amounts of air to produce slow oxidation. Control experiments in the absence of apoferritin in the solution were also performed. Remarkably different products were observed under TEM analysis, in size and morphology, as shown in Figure 8.3. Most particles derived from the ferritin-reconstituted experiments were discrete, spherical and exhibited a narrow size distribution with mean particle diameter of (6 ± 1.2) nm, as seen in Figure 8.3(a). The data confirmed that the protein cage remained intact and that the magnetic particles were formed within the protein shell, as seen in Figure 8.3(b), where the protein image forms bright, white crowns around the magnetic nanoparticles. The visualization of the protein coat is generally achieved by negative staining; a well-known technique based on the

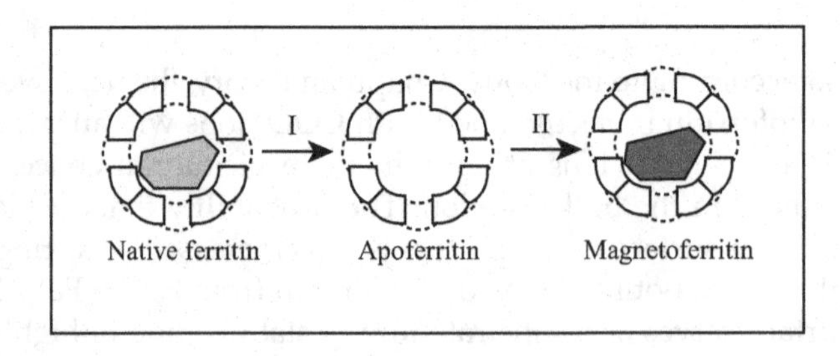

FIGURE 8.2
Depiction of the synthesis route for the formation of magnetoferritin: protein demineralization followed by remineralization under controlled conditions.

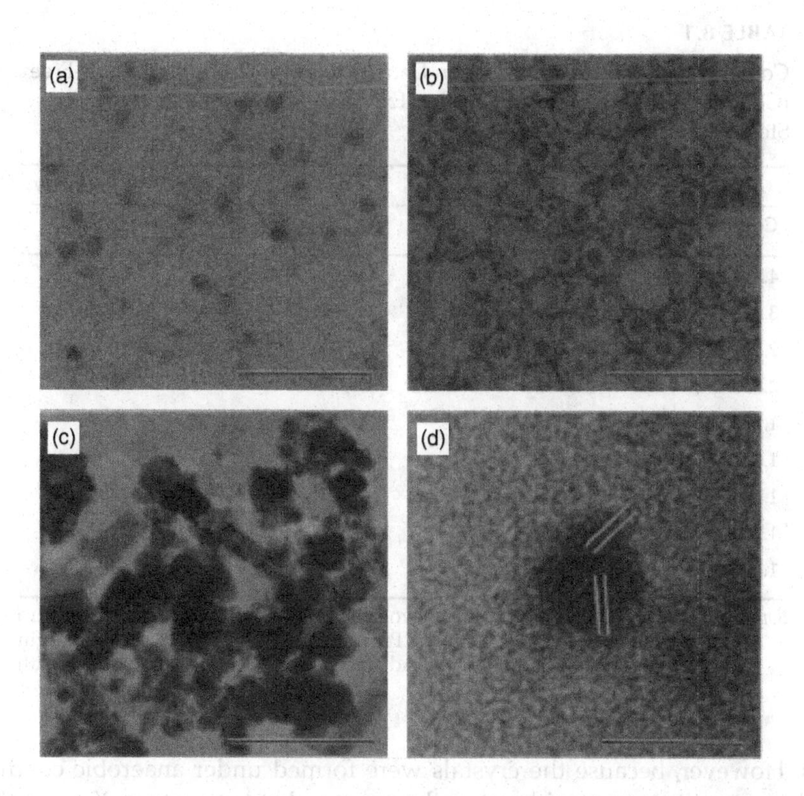

FIGURE 8.3
TEM micrographs of (a) an unstained sample of reconstituted ferritin showing distinct magnetite/maghemite particles (scale bar = 50 nm); (b) a stained sample of (a) showing an intact protein shell surrounding the magnetite cores (scale bar = 50 nm); (c) magnetite crystals formed in the control reaction (scale bar = 100 nm); the needle-shaped crystals were identified as the mineral goethite (α-FeOOH); and (d) a high-resolution lattice image of an individual, reconstituted ferritin core showing the single-crystal nature of the particle. Two sets of lattice fringes are observed corresponding to the {111} (interatomic spacing d = 0.465 nm) and {002} (d = 0.4198 nm) planes of magnetite or maghemite. The angle between these planes is 54°, consistent with cubic lattice symmetry (scale bar = 5 nm). (From F. C. Meldrum, B. R. Heywood and S. Mann, Magnetoferritin: In vitro synthesis of a novel magnetic protein, *Science* 257 (1992) 522, corrected figure on p. 729, Copyright American Association for the Advancement of Science, reproduced with permission).

addition of a drop of a solution containing a heavy metal salt, such as 2% of uranyl acetate, onto the TEM grid deposited sample. The electron-rich uranyl ions bind onto the protein shell, and due to their strong interaction with the electron beam, introduce strong contrast between the protein and the magnetic core.

Control, protein-free experiments, in contrast, yielded heterogeneous, aggregated particles, as seen in Figure 8.3(c), ranging from 4 to 70 nm in size, with morphology ranging from irregular spheroid to cubic and cubo-octahedral shape. Figure 8.3(d) is a high-resolution lattice image of an individual, reconstituted ferritin core showing the single-crystal nature of the particle. Two sets of lattice fringes are observed corresponding to the {111} (interatomic spacing d = 0.465 nm) and {002} (d = 0.4198 nm) planes of magnetite or maghemite. Table 8.1 presents a comparison of the high-resolution electron diffraction data (d spacings) of phases produced in the control (protein-free) and the ferritin-reconstituted experiments to those of stoichiometric magnetite. The data obtained could not unequivocally distinguish between magnetite and maghemite, which have similar crystallographic

TABLE 8.1

Comparison of Electron Diffraction Data (*d* Spacings) from the Protein-Free (Control) and Protein Reconstitution Experiments with that from Stoichiometric Magnetite

d Spacing (Å)			Relative Intensity
Control	Ferritin	Magnetite	I/I_0
4.81		4.85	8
3.016	2.945	2.967	30
2.528	2.558	2.532	100
2.086	2.099	2.099	20
1.722	1.715	1.715	10
1.621	1.636	1.616	30
1.484	1.499	1.485	40
1.286		1.281	10
1.094		1.093	12

Source: From F. C. Meldrum, B. R. Heywood and S. Mann, Magnetoferritin: In vitro Synthesis of a Novel Magnetic Protein, *Science* 257 (1992) 522, Copyright American Association for the Advancement of Science, Reproduced with Permission.

structures. However, because the crystals were formed under anaerobic conditions and were black, magnetite was considered as the most probable structure. X-ray diffraction of the control crystals confirmed that under analogous solution conditions, magnetite is produced, which over time oxidizes to maghemite upon exposure to air. This phase transformation is accompanied by a color change from black to red-brown.

This seminal experiment established the use of apoferritin as a confined reaction vehicle, or nanoreactor, for the production of iron oxide nanoparticles. The process takes advantage of the unusual stability of the apoferritin shell at elevated temperatures (60°C) and pH (8.5) needed for the synthesis of magnetite, as low-temperature methods failed to produce magnetite in the presence of the protein shell. It is believed that magnetite production within the ferritin cage proceeds in analogous way to that of ferrihydrite, through the controlled oxidation of Fe^{2+} ions at the ferroxidase center and subsequent migration and nucleation at the COO^- lined interior surface of the protein cage. Just like in the case of ferritin biomineralization *in vivo*, this process is presumably favored over the competing reaction in bulk solution because of the catalytic oxidation of Fe^{2+} ions at the ferroxidase center and the surface of the developing mineral core. This differentiation between "inside" and "outside" is essential to the effective *in vivo* functioning of the protein and is also central to the ferritin-nanotemplating approach for the synthesis of nanophase materials within the apoferritin shell.

8.2.1 Magnetic Moment of Magnetoferritin *vs.* Ferritin

The synthesis of magnetoferritin is now well established having been reproduced by other investigators, at various degrees of iron loading, and its magnetic properties have been widely studied. Figure 8.4 shows TEM and high-resolution TEM (HRTEM) micrographs of 7.3 nm magnetoferritin particles along with magnetization data. Superparamagnetic

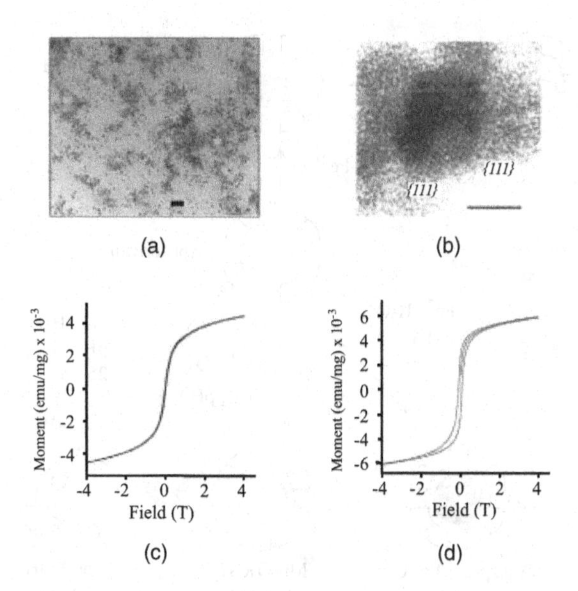

FIGURE 8.4
(a) TEM micrographs of magnetoferritin (bar = 40 nm). (b) HRTEM micrograph of a single magnetoferritin core with the {111} lattice fringes indicated (bar = 4 nm). (c) Magnetic moment *vs.* applied magnetic field for magnetoferritin at $T = 300$ K. (d) Hysteresis loop of magnetoferritin at $T = 4$ K. (From M. T. Klem, M. Young and T. Douglas, Biomimetic Magnetic Nanoparticles, Materials Today, September 2005, p. 28, Copyright Elsevier, reproduced with permission).

behavior is observed at room temperature; the magnetization *vs.* applied magnetic field data shows no hysteresis at $T = 300$ K, but it becomes hysteretic at $T = 4$ K with a coercive field of $H_c = 1{,}200$ Oe. Fitting of the superparamagnetic data to a Langevin function indicates roughly a core magnetic moment $\mu_c = 13{,}100\ \mu_B$, corresponding to a particle core containing of the order of 12,000 Fe atoms. This is a much larger moment compared to that of the native ferritin core of $\mu_c = 300\ \mu_B$, as we discussed in Example 7.3. This is due to the fact that the ferrihydrite core of ferritin is antiferromagnetic, with the particle moment arising primarily from non-compensated spins at its surface and possible structural defects within the interior of the ferrihydrite core. In contrast, the core of magnetoferritin comprises of magnetite and/or maghemite, both structures being ferrimagnetic with non-compensated spin sub-lattices throughout the particle volume.

8.2.1.1 Beyond Iron Oxides

In living systems, ferritin has evolved to exclusively sequester iron; however, in the laboratory additional synthetic pathways have been identified for the production of various inorganic nanophases within apoferritin, as depicted in Figure 8.5. For example, Mn^{2+} and Co^{2+} ions undergo oxidation and mineralization following a path analogous to Fe^{2+} by forming $Mn(O)OH \rightarrow Mn_3O_4$ and $Co(O)OH \rightarrow Co_3O_4$ phases, respectively, while bubbling of hydrogen sulfide gas into solutions of magnetoferritin results in the production of non-magnetic iron sulfide according to the known reaction of hydrogen sulfide with iron oxide

$$Fe_2O_3 + H_2O + 3H_2S \rightarrow Fe_2S_3 + 4H_2O \tag{8.6}$$

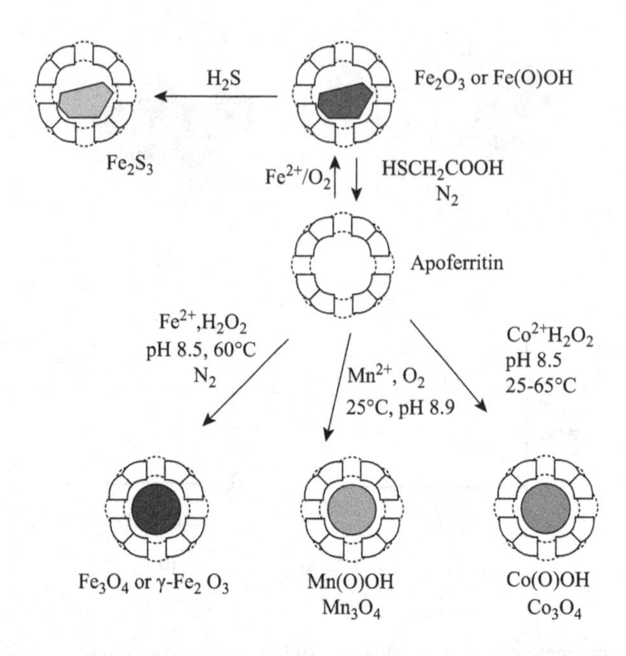

FIGURE 8.5
Schematic of various synthetic pathways possible for the production of a variety of nanophases within the apoferritin cage. (From M. T. Klem, M. Young and T. Douglas, Biomimetic Magnetic Nanoparticles, Materials Today, September 2005, p. 28, Copyright Elsevier, reproduced with permission).

8.2.2 Metal and Metal–Alloy Magnetic Nanoparticles

To date, many other phases have been synthesized within apoferritin, both magnetic and semiconducting, further supporting the notion that the reactions are not specific to iron and that the electrostatic properties of the protein play a significant role in the process of mineralization. In addition to metal oxide and hydroxide phases, metallic magnetic nanoparticles of Ni and Co have been produced within horse spleen apoferritin, as indicated schematically in Figure 8.6. Once the Ni^{2+} and Co^{2+} ions have entered the protein cavity, they can be reduced by reaction with sodium borohydride ($NaBH_4$), which is small enough to pass through the 3-fold channels and enter the apoferritin interior. Thus, the divalent metal ions bounded onto the interior wall of the apoferritin cage undergo reduction to zero-valence to produce metal magnetic nanoparticles.

Metal–alloy magnetic nanoparticles have also been produced within apoferritin. An example is shown in Figure 8.7, where TEM micrographs of Co:Pt composite nanoparticles formed within apoferritin are shown. The nanoparticles were prepared by using 0.1 M salt solutions of ammonium tetrachloroplatinate ($(HN_4)_2PtCl_4$) and cobalt acetate tetrahydrate ($(CH_3COO)_2Co{\cdot}4H_2O$) as precursors of the metal ions. They were slowly added to the apoferritin solution in 1:1, 2:1 and 3:1 Co:Pt ratios. After allowing time for the metal ions to enter the apoferritin cavity (~30 min) a stoichiometric amount of 0.1 M sodium borohydride was added at pH 8.3 to reduce the metal ions to metal. Nanoparticles containing core sizes corresponding to 1,000 atoms were thus prepared. The resulting solutions were freeze-dried to a powder and TEM pictures were taken before and after annealing the powder at 650°C for 60 min under reducing atmosphere. Spherical morphology is observed with an average particle size of 4.1 nm, which does not increase appreciably upon

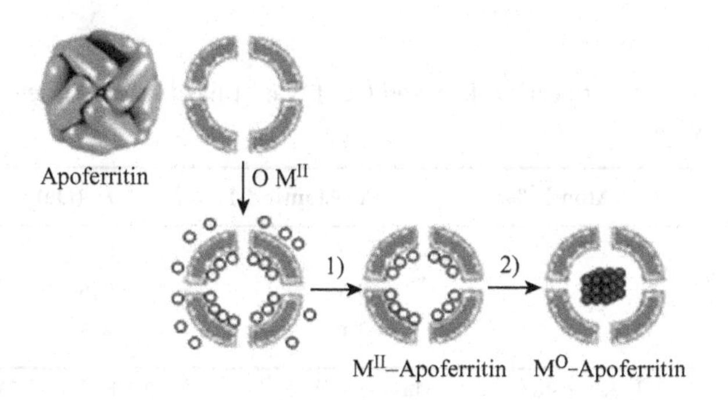

FIGURE 8.6
Schematic representation of apoferritin encapsulated Ni and Co metal nanoparticles. $M^{II} = Ni^{2+}$ or Co^{2+}, M^0 = Co or Ni. (1) Dialysis and chromatography. (2) Addition of $NaBH_4$. (From N. Gálvez, P. Sánchez, J. M. Domínguez-Vera, A. Sorianno-Portillo, M. Clemente-Léon and E. Coronado, Apoferritin-encapsulated Ni and Co superparamagnetic nanoparticles, *J. Mat. Chem.*, 16 (2006) 2757, Copyright Royal Society of Chemistry, reproduced with permission).

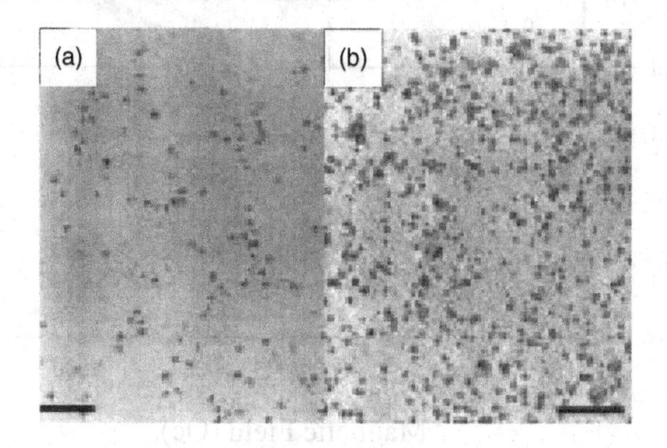

FIGURE 8.7
TEM micrographs of Co: Pt composite material before (a) and after (b) annealing for 60 min in hydrogen at 650°C. Scale bars are 50 nm, and the average particle diameter from multiple micrographs is 4.1 nm, σ = 2.4 nm. (From B. Warne, O. I. Kasyutich, E. L. Mayes, J. A. L. Wiggins and K. K. W. Wong, Self-Assembled Nanoparticulate Co:Pt for Data Storage Applications, *IEEE Transactions on Magnetism* 36 (2000) 3009, Copyright IEEE Magnetics Society, reproduced with permission).

annealing (Figure 8.7). At this temperature, the protein shell appears to carbonize protecting the particles from both oxidizing and sintering, which leads to particle growth.

The composition of the metal particles was determined by Energy Dispersive X-ray (EDX) spectroscopy, as reported in Table 8.2. With increasing Co:Pt ratio, the cobalt content of the particle increases but, for the experimental conditions used, the metal stoichiometry of the resulting nanoparticles does not correspond to that of the precursor salts. In addition, it was observed that the as-prepared nanoparticles were superparamagnetic and exhibited no distinct crystallinity. Upon annealing at 650°C for 60 min in a reducing atmosphere, the nanoparticles became highly crystalline and exhibited a range of magnetic behaviors as indicated in Table 8.2 and Figure 8.8. This behavior can be partially explained by the well-known structural transition of CoPt from fcc (cubic) to fct (uniaxial) phase upon annealing, with the latter exhibiting high uniaxial magnetocrystalline anisotropy of the order of $K_u \approx 5 \times 10^6$ J/m^3.

TABLE 8.2

EDX and Magnetization Data of Ferritin Derived CoPt Nanophases after Annealing at 650°C in Hydrogen Atmosphere

Co:Pt Addition	Co (Atomic %)	Pt (Atomic %)	H_c (Oe)	M_r (emu/g)
1:1	27.2	72.8	SP	SP
2:1	30.4	69.6	246	2.3
3:1	37.0	63.1	315	3.2

Source: From B. Warne, O. I. Kasyutich, E. L. Mayes, J. A. L. Wiggins and K. K. W. Wong, Self-Assembled Nanoparticulate Co:Pt for Data Storage Applications, *IEEE Transactions on Magnetism* 36 (2000) 3009, Copyright IEEE Magnetics Society, Reproduced with Permission.

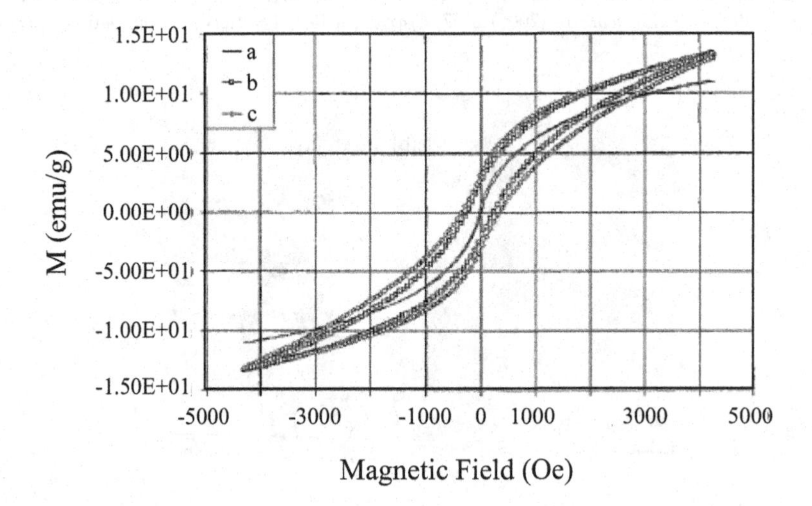

FIGURE 8.8
Room-temperature hysteresis loops of (a) 1:1 CoPt, (b) 2:1 CoPt and (c) 3: CoPt addition samples to apoferritin after annealing in hydrogen at 650°C for 60 minutes. (From B. Warne, O. I. Kasyutich, E. L. Mayes, J. A. L. Wiggins and K. K. W. Wong, Self-Assembled Nanoparticulate Co:Pt for Data Storage Applications, *IEEE Transactions on Magnetism* 36 (2000) 3009, Copyright IEEE Magnetics Society, reproduced with permission).

Example 8.1: Controlled Synthesis and Exchange Anisotropy in Binary Phase Fe/Co Oxides within Apoferritin

In addition to the controlled deposition of inorganic nanoparticles of both iron and cobalt oxides within the apoferritin cage, researchers have attempted the synthesis of mixed iron- and cobalt-oxide nanoparticles within apoferritin. Klem and co-workers (M. T. Klem, D. A. Resnick, K. Gilmore, M. Young, Y. U. Idzerda and T. Douglas, Synthetic control over magnetic moment and exchange bias in all-oxide materials encapsulated within spherical protein cage, *J. Am. Chem. Soc.* 129 (2007) 197) successfully demonstrated that variation in the synthesis conditions can result in the formation of single phase $Co_xFe_{3-x}O_4$ ($0 \le x \le 1$) or segregated, albeit intimately mixed, Fe_3O_4 or $CoFe_2O_4$ and Co_3O_4 phases within the HoSF protein cage.

TABLE 8.3

Coercive Fields and Blocking Temperatures for A Series of Fe and Co Nanomaterials, Biomimetically Synthesized within the Ferritin Protein Cage

Mineral in Ferritin	H_c at 5 K (kG)	T_B (K)
Co_3O_4	0.140	4.2
Fe_3O_4	1.2	30
5% Co in Fe_3O_4	8.2	76
10% Co in Fe_3O_4	7.2	126
33% Co in Fe_3O_4 fast synthesis	8.3	50
33% Co in Fe_3O_4 slow synthesis	7.8	60

Source: From M. T. Klem, D. A. Resnick, K. Gilmore, M. Young, Y. U. Idzerda and T. Douglas, Synthetic Control Over Magnetic Moment and Exchange Bias in All-Oxide Materials Encapsulated within A Spherical Protein Cage, *J. Am. Chem. Soc.* 129 (2007) 197, Copyright American Chemical Society, Reproduced with Permission.

Loading of 1,000 M/protein (M = Fe or Co) was achieved by adding, continuously and simultaneously, solutions of iron ($(NH_4)_2Fe(SO_4)\cdot 6H_2O$), cobalt ($Co(NO_3)_2\cdot 6H_2O$) and oxidant ($H_2O_2$) at a constant rate to an apoferritin solution under an atmosphere of N_2 at pH 8.5 and elevated temperature (65°C), over a defined time period. In the presence of apoferritin, the reaction produced a homogeneous dark brown solution, while control reactions performed in the absence of apoferritin resulted in bulk precipitation of a dark brown solid. The absence of precipitation in the presence of apoferritin indicates that the reaction took place in spatially selective manner, exclusively within the ferritin cage. Co-doping concentrations of 5%, 10% and 33% were used in addition to single component oxide particles of magnetite and cobalt oxide. Metal ion addition for the 33% Co doping was performed in a slow (over 30 min) or a fast (5 min) manner instead of the 15 min addition for all other samples. Spectroscopic analysis supported the notion that Co was incorporated into each nanoparticle in two competing phases, as cobalt oxide, Co_3O_4, or cobalt ferrite, $CoFe_2O_4$. Under fast synthesis conditions (5 min), the nanoparticles were composed of 85% $CoFe_2O_4$ and 15% Co_3O_4, while under slow, extended synthesis time (30 min), the composition corresponded to 25% $CoFe_2O_4$ and 75% Co_3O_4. Their magnetic properties are given in Table 8.3.

The blocking temperatures show a dramatic variation with increasing Co/Fe ratio, as compared to either pure magnetite (T_B = 30 K) or cobalt oxide (T_B = 4.2 K) synthesized in a similar manner within the ferritin cavity. The monotonic increase of the blocking temperature up to 10% Co doping is consistent with the incorporation of Co within the Fe_3O_4 lattice to form cobalt ferrite, $CoFe_2O_4$, while the decrease in T_B at 33% Co is consistent with the formation of segregated phases of magnetite and cobalt oxide. The presence of significant amounts of antiferromagnetic cobalt oxide strongly exchange-coupled to a ferrimagnetic phase can give rise to exchange anisotropy, a concept we first introduced in Section 4.12 in connection with core/shell magnetic nanoparticles containing a ferromagnetic metallic core coated with an antiferromagnetic oxide layer (Co/CoO). As we discussed then, exchange anisotropy arises when the Curie temperature of the ferromagnetic phase exceeds the Néel temperature of the antiferromagnetic phase. It turns out that similar types of interactions are present also in the case of intimately mixed ferr*i*magnetic (rather than ferromagnetic) and antiferromagnetic phases.

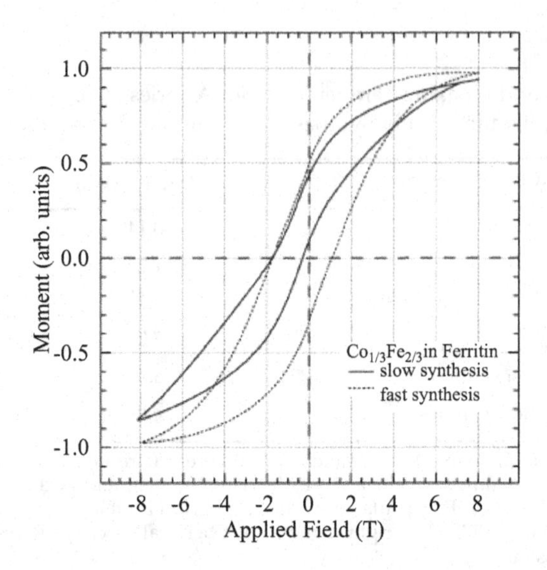

FIGURE 8.9
Hysteresis loops at $T = 2$ K for field-cooled (8 T) mixed oxide material of nominal composition 66% Fe_3O_4/33% Co_3O_4 for slow (solid line) and fast (dashed line) synthesis. The offset between the hysteresis loops is due to the exchange bias effect. (From M. T. Klem, D. A. Resnick, K. Gilmore, M. Young, Y. U. Idzerda and T. Douglas, Synthetic control over magnetic moment and exchange bias in all-oxide materials encapsulated within a spherical protein cage, *J. Am. Chem. Soc.* 129 (2007) 197, Copyright American Chemical Society, reproduced with permission).

A measure of the strength of the exchange anisotropy field can be obtained from measuring the hysteresis loops of samples cooled in an applied magnetic field through the Néel temperature of the antiferromagnetic phase, here Co_3O_4 with $T_N = 30$ K. Exchange coupling between the ferrimagnetic $CoFe_2O_4$ ($T_C = 790$ K), or Fe_3O_4 ($T_C = 858$ K), and antiferromagnetic Co_3O_4 can produce the observed shift in the hysteresis loop of Figure 8.9, due to the fact that the direction of the antiferromagnetic order acquired by Co_3O_4 upon cooling below 30 K is biased, due to the exchange interaction across phase boundaries with the ferrimagnetic phase (see Figure 4.24). The case of slow synthesis, producing 75% of cobalt oxide, shows a large shift in the hysteresis loop (solid line) compared to that of fast synthesis (dashed line), which results in only 15% of the cobalt-oxide phase.

The amount of displacement of the center of the loop to the left of the axes center gives a measure of the exchange anisotropy field, H_{ex}. At 2 K a large shift, corresponding to an exchange field of $H_{ex} \sim 1$ T, is observed for the slow synthesis (30 min) sample, while the measured ratio of exchange to the coercive field, H_{ex}/H_c, is 1.3. For the fast synthesis (5 min) sample, the measured H_{ex}/H_c ratio is reduced to 0.3. This drop in the exchange bias is due to the incorporation of Co into the Fe_3O_4 lattice to form cobalt ferrite, reducing the amount of available Co_3O_4 phase to bias the ferrimagnetic phase. The investigators measured the value of this ratio at various temperatures below 30 K for 10% Co doping and 33% Co doping, slow and fast synthesis, samples. The results are given in Figure 8.10. In all cases, the exchange bias vanishes at $T = 30$ K, the Néel temperature of the antiferromagnetic phase, Co_3O_4.

This example illustrates the power of the biomimetic approach and the use of protein cages as biomolecular templates for the self-assembly of magnetic nanoparticles whose magnetic properties can be fine-tuned between ferrimagnetic and antiferromagnetic properties.

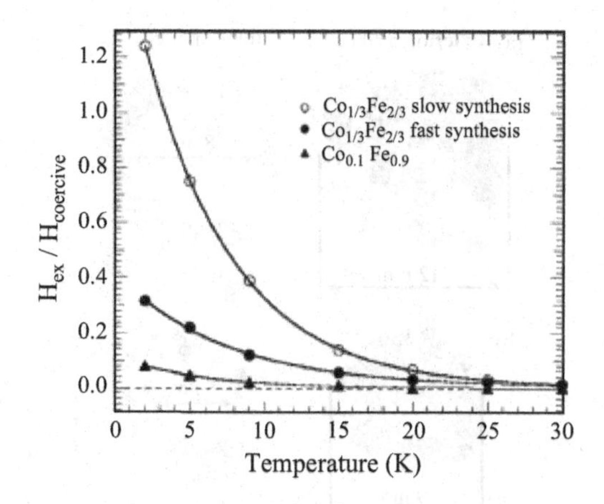

FIGURE 8.10
Temperature dependence of the exchange bias for a slow synthesized 33% Co (open circles), fast synthesized 33% Co (filled circles), and 10% Co (filled triangle) mixed Fe/Co oxide phases synthesized within the ferritin protein cage. (From M. T. Klem, D. A. Resnick, K. Gilmore, M. Young, Y. U. Idzerda and T. Douglas, Synthetic control over magnetic moment and exchange bias in all-oxide materials encapsulated within a spherical protein cage, *J. Am. Chem. Soc.* 129 (2007) 197, Copyright American Chemical Society, reproduced with permission).

8.3 Exploring the Ferritin Superfamily

Mammalian ferritin belongs to a "ferritin superfamily" whereby similar but distinct protein cages occur in bacterial life forms capable of sequestering iron. These afford additional supramolecular templates for materials synthesis. Among these, Dps (Deoxyribonucleic acid (DNA)-binding proteins from starved cells) proteins have been extensively investigated and used in the biomimetic synthesis of magnetic nanoparticles. Dps proteins possess a shell structure made of 12 identical amino acid subunits, as opposed to 24 heteropolymer subunits of the "canonical" ferritins. They are smaller than mammalian ferritins possessing an exterior diameter of ~9 nm and a central cavity of ~5 nm in diameter, as depicted in Figure 8.11. They were first identified in the bacterium *Escherichia coli* in 1992, and since their discovery, homologous structures have been found in other bacteria and archaea. The biological purpose of Dps is the protection of DNA through two distinct mechanisms: (a) the Dps-driven DNA condensation by the formation of a highly ordered and stable Dps-DNA-complex crystalline structure, which protects DNA from various damages and (b) the Fe^{2+} oxidation at a highly conserved ferroxidase center and subsequent biomineralization of iron within the protein cage. Oxidation of iron occurs by hydrogen peroxide according to the reaction of Eq. (8.7), which prevents hydroxyl radical production by the Fenton reaction of Eqs. (7.1) and (7.2), further protecting the DNA from oxidative damage by free radicals.

$$2Fe^{2+} + H_2O_2 + 2H^+ = 2Fe^{3+} + 2H_2O \tag{8.7}$$

The three-dimensional crystal structure of Dps proteins exhibits two types of 3-fold symmetry channels one of which is lined with hydrophilic amino acids that can facilitate

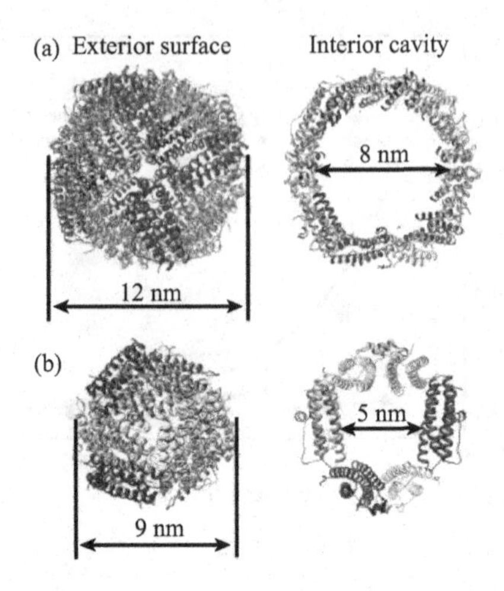

FIGURE 8.11
Ribbon representation of the exterior surface and interior cavity of (a) HuHF, human heavy chain ferritin, and (b) Dps protein from *Listeria innocua*. (From M. Uchida, S. Kang, C. Reichhardt, K. Harlen and T. Douglas, The ferritin superfamily: Supramolecular templates for materials synthesis, *BBA* 1800 (2010) 834, Copyright Elsevier, reproduced with permission).

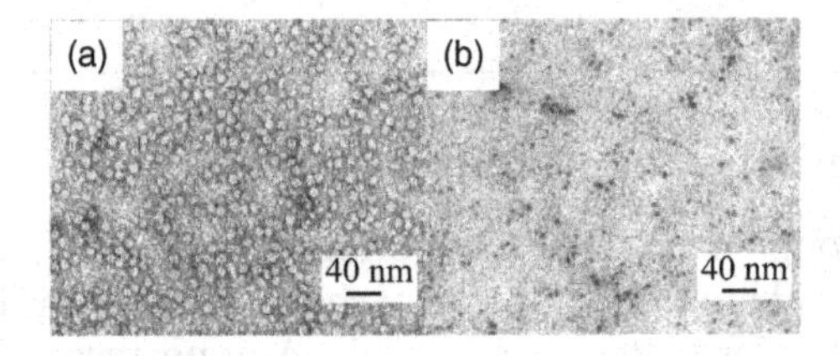

FIGURE 8.12
TEM of the *L. innocua* Dps cage mineralized with γ-Fe_2O_3 (a) stained with uranyl acetate and (b) unstained. (From M. Allen, D. Willits, J. Mosolf, M. Young and T. Douglas, Protein cage constrained synthesis of ferrimagnetic iron oxide nanoparticles, *Adv. Mater.* 14 (2002) 1562, Copyright Wiley-VCH, reproduced with permission).

the entrance of cations to the interior cavity of the protein. As in the case of mammalian ferritin, once the iron ions are inside the cavity, they are electrostatically attracted to the negatively charged surface of the interior wall where the mineralization reactions can be initiated. The electrostatic surface of *L. innocua* Dps is similar to that of canonical ferritins with clusters of glutamic acid residues lining the surface that can facilitate iron cluster nucleation. The ferrihydrite mineral core formed within the Dps cavity can accommodate up to 500 Fe atoms, as opposed to 4,500 Fe atoms in mammalian ferritin. Investigators have used the Dps cage to form mineralized γ-Fe_2O_3 nanoparticles. Treatment of the protein at pH 8.5 and 65°C with 400 Fe^{2+} per protein and stoichiometric amounts of H_2O_2 resulted in the formation of a homogeneous solution with deep brown coloration. TEM micrographs of negatively stained proteins with uranyl acetate shown in Figure 8.12(a) indicate that the mineralized iron-oxide phase is contained within the *L. innocua* Dps cage. The measured iron-oxide mineral particle size of (4.1 ± 1.1) nm diameter of unstained proteins shown in Figure 8.12(b) correlates well with the inner diameter of the Dps protein cage.

Furthermore, the reaction of Co^{2+}, instead of Fe^{2+}, with H_2O_2 at pH 8.5 under an elevated temperature of 65°C results in the formation of Co_3O_4 within the Dps protein cage, while the same reaction carried out at 23°C forms Co(O)OH within the cage, indicating that the temperature at which the reaction is performed can influence the nature of the mineralized phase within the protein. As the formation of many inorganic phases requires reactions that take place at elevated temperatures, often higher than most proteins can tolerate, investigators have sought to isolate protein templates from thermophilic (heat loving) and hyper-thermophilic bacterial and archaeal organisms found in hot springs, acidic soils and near volcano vents.

Such organisms were first discovered in the 1960s by Thomas Brock and have succeeded to expand the temperature range of biomimetic synthesis of magnetic nanoparticles. For example, ferritin has been isolated from the anaerobic marine bacterium *Pyrococcus furiosus*, which lives in thermal springs where temperatures can reach 120°C. The structure of this protein is homologous to other ferritins and synthesis of maghemite within this ferritin cage has been demonstrated. Its wide temperature stability opens up the possibility of biomimetic synthesis at temperatures unattainable with other biomolecular templates. As the number of protein cages isolated from hyper-thermophilic organisms increases, so too will the variety and size of magnetic nanoparticles that can be synthesized using biomimetic approaches.

8.4 Other Protein Cages and Viral Capsids as Constrained Reaction Vehicles

In addition to the ferritins, nature provides us with a large selection of protein cages in the form of viral capsids, which can also function as nanoreactors for the self-assembly of nanomaterials. Increasingly, genetic engineering techniques are being applied to the design of mutant protein cages with electrostatic properties inductive to the self-assembly of magnetic nanomaterials. Figure 8.13 gives examples of space-filling images of viral capsids, protein, and enzyme cages. These include capsids of both plant and bacterial viruses. Non-viral protein cages include the canonical ferritins, Dps proteins, heat shock proteins (Hsp) and the enzyme lumazine synthase. In their biological function, these supramolecular architectures presented in Figure 8.13 serve diverse roles from nucleic acid storage and transport in viruses, chaperon function to prevent protein misfolding and denaturation in heat shock proteins, iron sequestration in the ferritins and enzymatic action in lumazine synthase. However, they all share the same architectural principle in being assembled from a limited number of polypeptide subunits to form robust nanocages; they also share the ability to have their functionality altered through both chemical and genetic means.

The discovery of viruses dates back to 1892 with Dmitri Ivanovsky's description of a non-bacterial pathogen infecting tobacco plants, leading to the discovery of the tobacco mosaic virus in 1898 by Martinus Beijerinck. Since then, over 5,000 viruses have been described in detail; they are the most abundant type of biological entity. Virus particles, known as "virions", consist of three parts: DNA or RNA, long molecules that carry genetic information; a protein coat that protects these genes; and in some cases an envelope of lipids that surrounds the protein coat when they are outside a cell. The shapes of viruses

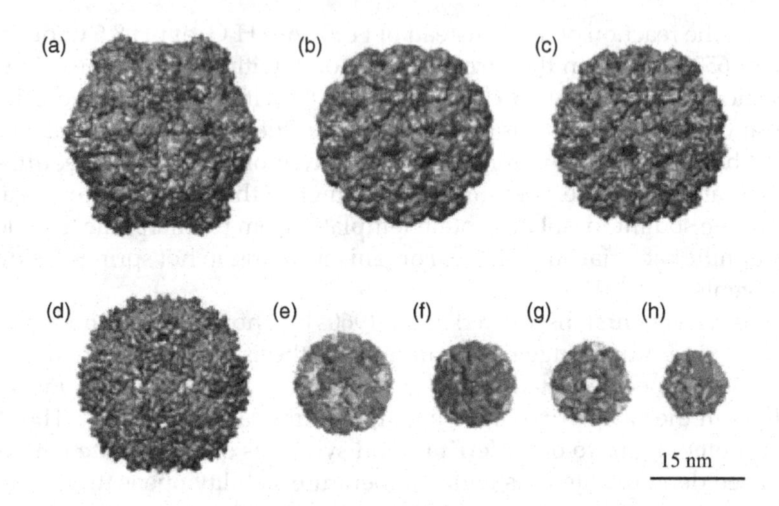

FIGURE 8.13
Space filling images of protein-cage architectures with various (exterior) diameters, *d*. (a) Cowpea mosaic virus, *d* = 31 nm, (b) Brome mosaic virus, *d* = 28 nm, (c) Cowpea chlorotic mottle virus, *d* = 28 nm, (d) MS2 bacteriophage, *d* = 27 nm, (e) lumazine synthase, *d* = 15 nm, (f) ferritin, *d* = 12 nm, (g) small heat shock protein, *d* = 12 nm, (h) Dps protein, *d* = 9 nm. Mosaic viruses are plant viruses that make the leaves of infected plants acquire a specked appearance, while the MS2 bacteriophage is a virus that infects the bacterium *Escherichia coli*. (From M. Uchita, M. T. Klem, M. Allen, P. Suci, M. Flenniken, E. Gillitzter, Z. Varpness, L. O. Liepold, M. Young and T. Douglas, Biological containers: protein cages as multifunctional platforms, *Adv. Mater.* 19 (2007) 1025, Copyright Wiley-VCH, reproduced with permission).

range from simple spherical and icosahedral forms to more complex helical structures. The viral capsids, once depleted of their genomic content, can be engineered to possess negatively charged interior walls and, thus, electrostatically attract Fe^{2+} ions and induce the hydrolytic polymerization of iron within their interior cavities in a way analogous to that of the ferritin protein cage.

The small heat shock protein (Hsp) from the hyper-thermophilic bacterium *Methanococcus jannaschii* (MjHsp) has been isolated and its crystal structure characterized. Like ferritin, it assembles into a 24-subunit cage with large 3-nm-diameter pores at the 3-fold axes (as opposed to 0.4 nm for ferritin), allowing free exchange between the interior and bulk solution. It is stable up to 70°C and in a broad pH range of 5–11. It has been investigated, both in wild-type form and a genetically engineered mutant, for its ability to act as a size-constrained reaction environment for the mineralization of iron, in analogy with ferritin. The replacement of glycine at position 41 on the interior surface of the protein by cysteine generated the HspG41C mutant. Addition of Fe^{2+} to this mutant in air resulted in the formation of a homogeneous rust-colored solution. Transmission electron microscope (TEM) images indicated the presence of small iron-oxide particles with an average diameter of 9 ± 1.2 nm, as depicted in Figure 8.14(a). In contrast, control experiments carried out in the absence of Hsp-templates resulted in rapid, bulk precipitation (Figure 8.14(b)). Both the wild-type and mutant forms of the protein exhibited similar iron mineralization capabilities. The electron diffraction of the bulk precipitate showed *d* spacings that were consistent with the major reflections of lepidocrocite (γ-FeOOH), which is paramagnetic at room temperature and becomes antiferromagnetically ordered at low temperatures with T_N between

FIGURE 8.14
TEM images of (a) iron oxide inside Hsp cages and (b) control sample illustrating the bulk precipitation in the absence of Hsp nanotemplates. The insets show the electron diffraction from each sample. (From M. Flenniken, D. A. Willits, S. Brumfield, M. J. Young and T. Douglas, The small heat shock protein cage from *Methanococcus jannaschii* is a versatile nanoscale platform for genetic and chemical modification, *Nano Letts.* 3 (2003) 1573, Copyright American Chemical Society, reproduced with permission).

50 K and 77 K depending on the degree of crystallinity. In contrast, the TEM of the material mineralized within the MjHsp cage indicated *d* spacings consistent with poorly crystallized ferrihydrite.

Lumazine synthase is a hollow bacterial enzyme composed of 60 subunits that is involved in the synthesis of lumazine, a precursor of riboflavin, also known as vitamin B_2. The structure of the enzyme isolated from *Bacillus subtilis* has been studied to reveal a hollow porous shell with an inner diameter of 7.8 nm and an outer diameter of 14.7 nm. Its ability to mineralize iron has been investigated at various iron loadings: 300, 1,000, 1,500 and 2,000 Fe atoms per enzyme capsid. TEM and EDX analysis confirmed the presence of iron-containing electron dense nanoparticles, while no nanoparticles were observed in control samples prepared in the absence of the enzyme cages. Instead, TEM analysis indicated the presence of micrometer-sized, plate-shaped electron dense particles consistent with the formation of lepidocrocite.

Viral capsids are presently the most extensively studied biological templates in the biomimetic synthesis of inorganic nanoparticles. Due to their well-defined architectures and readily accessible genetic information, they afford unlimited possibilities for structural modification *via* genetic engineering. Their interior space is normally reserved for their genome, where in their native state, the viral capsids act as host containers for nucleic acid storage and transport. However, purified viral coat protein subunits can be easily assembled *in vitro* into empty virions even in the absence of RNA or DNA, and can therefore be utilized as constrained reaction vesicles. In 1998, investigators Douglas and Young (T. Douglas and M. Young, Host-guest encapsulation of materials by assembled virus protein cages, *Nature*, 393 (1998) 152) pioneered the approach of using viral capsids for the synthesis of size-constrained inorganic and organic polymer species. In their experiments, they used Cowpea chlorotic mottle virus (CCMV) as a model system for reversibly gated entrapment of inorganic minerals. CCMV is an icosahedral virus composed of 180 identical subunits forming a protein cage of 18 nm inner and 28 nm outer diameters (Figure 8.13(a). This nanotemplate affords a cavity approximately twice as large as that of apoferritin, defining a new upper limit for crystal growth size of entrapped minerals. The exact size of the capsid depends on the pH of the medium, as it undergoes a pH-dependent reversible swelling behavior. The capsid increases its dimensions by about 10% compared to the non-swollen form when the pH of the medium is raised to ≥6.5. In its swollen (open) state, the CCMV capsid provides 60 open pores of about 2-nm diameter each that allow small external guest molecules to diffuse into the internal space. Such dynamic structural transitions are common in many virions induced by chemical switches, which provide unique molecular gating mechanisms to control the containment and release of entrapped materials.

X-ray structural analysis indicates that highly basic N-termini, from six arginine and three lysine amino acids per subunit (see Table 7.1 for a list of amino acids and their properties), line the interior wall of the CCMV capsid, a total of 1,620 basic amino acids, 9 from each of the 180 subunits, providing a large positive charge to the interior wall of the capsid. The large positive charge is necessary for the native virus to package and condense the anionic viral genome. This is opposite to the case of ferritin, with an interior wall lined with highly acidic C-termini of glutamic acid, providing a negative charge to the interior wall of the protein cage. Thus, the native capsid would exhibit electrostatic attraction for negatively charged species in the bulk solution, unlike ferritin. Indeed, the first material crystallized within the viral capsid was accomplished by the electrostatic attraction and subsequent oligomerization of aqueous molecular tungstate (WO_4^{2-}) ions to form macromolecular complexes of paratungstate ($H_2W_{12}O_{42}^{10-}$). Young and Douglas demonstrated, however, that the N-terminus is not required for the structural integrity of the empty capsid *in vitro*. They proved that it was possible to alter the chemical properties of the protein cage through rational design without disturbing the overall architecture of the capsid. They accomplished this by genetically engineering the N-terminus of the protein to replace the 9 basic residues at the N-terminus with glutamic acid, forming a mutant capsid (subE mutant), with the electrostatic characteristics of the interior wall changed from cationic to anionic. The altered electrostatic character of the interior of the protein favors strong interaction with ferrous and ferric ions, which promotes oxidative hydrolysis and the formation of size-constrained iron-oxide particles within the viral capsid, in a fashion similar to that of apoferritin.

The electrostatically altered viral protein cage catalyzed the rapid oxidation of Fe^{2+}, leading to the formation of iron-oxide magnetic nanoparticles. Specifically, the purified subE mutant proteins were treated with aliquots of Fe^{2+} at pH 6.5 and allowed to oxidize in air. In the presence of the anionic empty subE mutant viral cages, the reaction proceeded to form a homogeneous orange solution. In contrast, under the exact same reaction conditions, but in the presence of wild-type (native) empty cationic viral cages, bulk precipitation of an orange-colored solid was observed, similar to that observed in control experiments in the absence of a viral cage. Single and double mineralization processes, with loading factors of 2,000 and 6,000 Fe atoms per viral protein cage, were carried out resulting in the formation of spherical particles with (8.2 ± 1.6) nm and (24 ± 3.5) nm diameter cores, as imaged by high-angle annular dark field (HAADF) scanning transmission electron microscopy (STEM) shown in Figure 8.15. High-resolution TEM images indicated that the resulting nanoparticles were single-crystalline lepidocrocite.

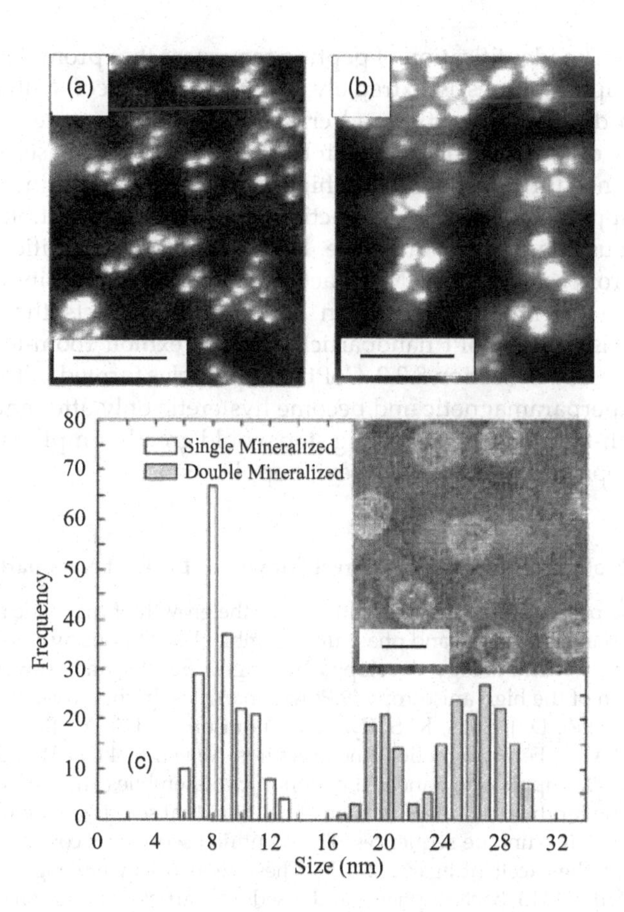

FIGURE 8.15
HAADF STEM images of single (a) and double (b) mineralized CCM-derived protein cages and corresponding size distribution histograms (c). The scale bars in (a) and (b) correspond to 100 nm. The inset (scale bar = 25 nm) shows a bright-field TEM image of uranyl acetate negatively stained specimen. (From T. Douglas, E. Strable, D. Willits, A. Aitouchen, M. Libera and M. Young, Protein engineering of a viral cage for constrained nanomaterials synthesis, *Adv. Mater.* 14 (2002) 415, Copyright Wiley-VCH, reproduced with permission).

8.5 Phage Display Technologies Applied to Magnetic Nanoparticle Synthesis

Biotechnological processes borrowing from protein engineering techniques, such as screening combinatorial peptide libraries using phage display, are currently being explored for the identification of peptide sequences with high specificity toward particular inorganic materials and in some cases with high specificity toward unique crystal faces. Phage display, first described by George P. Smith in 1985, is a laboratory technique for the study of protein–protein, protein–peptide and protein–DNA interactions that uses bacteriophages (viruses that infect bacteria) to connect proteins with the genetic information that encodes them. One of the most common bacteriophages used in phage display is the filamentous

M13 bacteriophage. The identification of peptide sequences that promote the nucleation of specific crystallographic phases is extremely useful in the direct synthesis of magnetic nanoparticles with desired properties and crystallographic structure, such as the hard-magnetic FePt and CoPt nanoparticles in their high anisotropy fct phase, also known as $L1^0$ phase, without the need for post-synthesis high-temperature annealing. Unlike previous examples, in this approach biological interactions control the nucleation of nanoparticles with no isomorphous compliment in nature. For example, the identification of a specific peptide sequence from screening the M13 bacteriophage engineered into the protein cage of MjHsp from the thermophilic archaeon *M. jannaschii* directs the peptide-specific-recognition synthesis of $L1^0$ CoPt nanoparticles, which exhibit room-temperature ferromagnetism. As discussed in Section 8.2.2, CoPt nanoparticles formed within apoferritin are amorphous and superparamagnetic and become hysteretic only after annealing at 650°C. Eliminating the high-temperature annealing step would greatly simplify the production of hard magnetic nanoparticles suitable for device applications.

Example 8.2: Biological Route to the Direct Growth of $L1^0$ FePt Nanoparticles

We illustrate the power of biomimetic synthesis for the growth of magnetic nanoparticles of desired material composition and phase under ambient conditions by a brief discussion of a phage display methodology developed by Angela Belcher and co-workers for the direct production of the high anisotropy FePt nanoparticles. In this work by Belcher *et al.* (B. D. Reiss, C. Mao, D. J. Solis, K. S. Ryan, T. Thomson, and A. M. Belcher, Biological Routes to Metal Alloy Ferromagnetic Nanostructures, *Nano Lett.* 4 (2004) 1127) phage display technology was applied to annealed nanoparticle assemblies and thin films of fct or $L1^0$ FePt, allowing for the rapid identification of peptides that selectively bind to this material from a pool of 10^{10} unique sequences. The identified sequences contained numerous amines, known to be excellent ligands for Pt. These sequences were engineered into the gP8 protein of the M13 bacteriophage and used in particle nucleation experiments. Specifically, 1 mL of engineered phage (10^{12} phage/mL) was mixed with 1 mL of 0.075 M $FeCl_2$ and 1 mL of 0.025 M H_2PtCl_6. After thorough mixing, 1 mL of 0.1 M $NaBH_4$ was added to reduce the metal ions forming the desired nanoparticles. The derived nanoparticles were single crystalline and had an average diameter of 4 ± 0.6 nm, which is larger than the theoretical limit of ~2.5 nm diameter required for thermal stability of $L1^0$ FePt nanoparticles. Control experiments involving wild-type phages that do not express the high $L1^0$ FePt phase affinity peptide-sequence only produced non-crystalline or polycrystalline particles of disordered FePt. Figure 8.16(a) shows a STEM image of the single-crystalline FePt nanoparticles attached to the bacteriophage, while Figure 8.16(b) shows hysteresis loops obtained at 300 K and 5 K.

The linear assemblies observed in Figure 8.16(a) attest to the large magnetic moment of these nanoparticles and the presence of strong magnetostatic interactions, reminiscent of those of bacterial magnetosomes we saw in Figure 7.7(a). The coercivities recorded in Figure 8.16(b) are 1,350 Oe at 5 K and 200 Oe at room temperature. The room-temperature coercivity is rather low for that expected from a sample composed purely of $L1^0$ FePt nanoparticles ($H_c \sim 10$ kOe), indicating that a mixture of phases may be present in addition to the $L1^0$ FePt phase. Alternatively, the coercivity may be reduced due to the fact that the size of these particles is close to their superparamagnetic limit and thus they may experience thermally assisted collective magnetic excitations at room temperature, resulting in a reduction of coercivity. We discussed such processes that occur in single-magnetic-domain particles in Section 4.7. See, in particular, Eq. (4.56). We will consider $L1^0$ FePt nanoparticles further in Chapter 10 where we discuss magnetic recording media, as $L1^0$ FePt is extensively being investigated in the area of high-density magnetic recording.

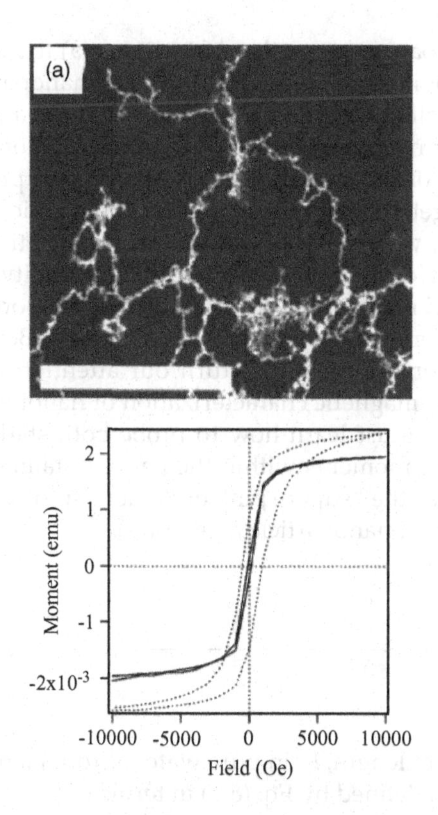

FIGURE 8.16
(a) STEM image of FePt nanoparticles prepared using engineered viral templates. Scale bar 350 nm (b) hysteresis loops of similar particles taken at 300 K (solid line) and 5 K (dashed line) (from B. D. Reiss, C. Mao, D. J. Solis, K. S. Ryan, T. Thomson, and A. M. Belcher, Biological Routes to Metal Alloy Ferromagnetic Nanostructures, *Nano Lett.* 4 (2004) 1127, Copyright American Chemical Society, reproduced with permission).

8.6 Future Promise and Prospects of Biomimetic Synthesis

In this chapter, we have explored protein-cage architectures and their utilization in the controlled crystal growth of various magnetic nanophases. We have taken a glimpse into the power of genetic engineering techniques in tailoring the electrostatic properties of the surfaces of organic templates in order to induce the nucleation and polymerization of a variety of phases, in analogy to the iron biomineralization processes exercised by ferritin *in vivo*. Protein cages are amenable to molecular modifications by both chemical and genetic engineering means. As we have just experienced, the application of genetic engineering techniques in materials science holds great promise for the future of magnetic nanoparticle synthesis and their applications. In addition to designing the interior surfaces of the protein nanotemplates, investigators can also endow the exterior protein surfaces with specific molecule recognition properties that can be used in nanoparticle arraying, or nanopatterning, on flat substrates for applications in the magnetic recording industry, a process coined the "bio-nanoprocess" for the fabrication of nanoelectronic components.

Therefore, the protein coat can serve three functions: (a) to induce the self-assembly of specific magnetic phases within its interior, (b) to keep nanoparticles isolated from each other, preventing uncontrolled coagulation and aggregation, and (c) to direct the formation of ordered arrays of nanoparticles for device applications. Molecular recognition properties on the exterior of the protein coat also hold great promise for applications in nanomedicine such as targeted cancer therapy, bio-imaging, bio-sensor and gene therapy.

In Part IV of this book, we discuss applications of magnetic nanoparticles across the disciplines in great detail; we will then appreciate that the ability to fine-tune the magnetic behavior of the synthesized nanoparticles is of paramount importance in the production of functional nanomaterials for nanotechnological applications. Before we can discuss applications in earnest, however, we must first turn our attention to the experimental techniques used in the detailed magnetic characterization of nanoparticles, which we address in Part III of the book. We must learn how to probe both static and dynamic magnetic phenomena, local magnetic moments within the nanocrystalline lattice, the total particle moment often referred to as the "super-spin" or "macro-spin" of the particle and the collective magnetic properties of nanoparticle assemblies.

Exercises

1. (a) Calculate the Bjerrum length, l_B, in pure water at room temperature. (b) Express the Debye screening length defined by Eq. (8.4) in terms of l_B.
2. Calculate the "double-layer" width for 1 M aqueous solutions of $FeCl_2$ and $FeCl_3$ at room temperature. Compare your values with that of 1 M NaCl solution.
3. What interaction is primarily responsible for the precipitation of iron-oxide or ferrihydrite phases within the interior space of apoferritin rather than in the exterior bulk solution? Explain
4. The formation of magnetoferritin in 1992 is considered to be the seminal experiment for the bio-inspired synthesis of magnetic nanoparticles. Describe, in broad terms, the sequence of fundamental steps undertaken in the synthesis of magnetoferritin.
5. The magnetic moment of magnetoferritin is about 50 times larger than that of ferritin. Explain the origin of such a large difference.
6. Draw similarities/differences in the chemical processes followed in the production of (a) iron oxides and hydroxides as opposed to (b) metal and metal–alloy nanoparticles within apoferritin.
7. Given that viral capsids, unlike ferritin, have interior walls lined with positive charges, explain how it is possible to grow iron-oxide/hydroxide particles within their interior space.

Part III

Magnetometry

9

Macromagnetic vs. Micromagnetic Characterization

> The usual methods of measuring magnetic moments can be divided into three major classes: measurement of a force on a material in a non-uniform magnetic field, measurement of magnetic induction in the vicinity of the sample, and indirect measurements of phenomena which involve the magnetic properties.
>
> **Simon Foner, from Reviews of Scientific Instruments, 30 (1959) 548**

There are many techniques that probe the magnetism of nanomaterials. Most require the application of a magnetic field, and all have a characteristic measuring time, τ_m, to detect and record the magnetic properties of the sample. They can be broadly classified into *macro*magnetic and *micro*magnetic characterization techniques. *Macro*magnetic characterization entails the measurement of the total magnetic moment or the equilibrium static magnetization of the sample, while *micro*magnetic characterization, usually based on the phenomenon of resonance, probes the internal magnetic properties of the material, local magnetic moments and dynamic spin relaxation processes. Techniques that measure the *macro*scopic magnetization of a sample can be further classified into induction and force measurement methods; the former requires the application of a homogeneous and the latter, an inhomogeneous external magnetic field.

The most widespread technique entails the measuring of the magnetic induction in the vicinity of the sample magnetized by a homogeneous external magnetic field. The magnetization of the sample is usually recorded as a function of temperature and/or applied magnetic field strength using a Vibrating Sample Magnetometer (VSM). As the sample vibrates, an electromotive force is induced in a nearby conducting wire loop. The induced voltage is then detected and processed by the associated circuitry. Other specialized methods that constitute indirect measurements of phenomena that involve the magnetic properties, such as Electron Paramagnetic Resonance (EPR), Ferromagnetic Resonance (FMR), AC Susceptibility, Mössbauer Spectroscopy, Nuclear Magnetic Resonance (NMR), polarized neutron scattering, magnetooptical techniques, *etc.*, provide important information on internal magnetism, spin ordering and spin excitation dynamics. Complete characterization of the static and dynamic magnetic properties of nanomaterials necessitates a combination of such techniques in order to collect complimentary information on the m*a*croscopic and m*i*croscopic magnetic properties of the sample, over a wide range of experimental measurement time windows.

In this chapter, we discuss fundamental concepts associated with magnetometry. In the study of nanomaterials, the characteristic measurement time window, τ_m, of the technique used is of paramount importance for the detection and characterization of magnetic ordering phenomena. This is due to thermally excited spin reversals that lead to superparamagnetism, as we have already discussed in Section 4.6. The characterization of dynamic magnetic processes of short spin-relaxation time requires the use of fast spectroscopic methods with short measurement time windows, while static or equilibrium magnetization properties can be investigated over longer measurement times.

DOI: 10.1201/9781315157016-12

9.1 Force Methods

The interaction of a magnetic dipole with a magnetic field was discussed in Sections 1.6.2 and 1.6.3. A magnetic dipole experiences a torque in a homogeneous magnetic field that tends to align it in the direction of the field; and a force in an inhomogeneous magnetic field that pulls it in the direction of the magnetic field gradient. Since the magnetization of a material is defined as the vector sum of all elementary magnetic dipoles per unit volume arising from electronic orbital and spin magnetic moments, the sample as a whole responds in a similar manner to an external magnetic field. The design of magnetometers for susceptibility and magnetic moment measurements is based on this response. Such measurements lead to the *macro*magnetic characterization of the sample, that is, the measurement of its total, static magnetic moment, induced or permanent.

9.1.1 The Faraday Balance

The measurement of the magnetic moment of a sample based on the force it experiences in the presence of an inhomogeneous magnetic field dates back to the 1800's and is attributed to Faraday. Using a sensitive analytical balance with a resolution of 0.01 mg or better one detects the apparent change in weight of a sample due to a vertically applied magnetic force. The sample is placed within the gap of an electromagnet with specially shaped pole pieces, "Faraday pole caps", to produce a constant magnetic field gradient in the vertical direction over the dimensions of the sample. According to Eq. (1.23), the force on a sample with a magnetic moment $\vec{m} = \vec{M}V$, where $\vec{M}$ is the magnetization density assumed constant throughout the material and V the volume of the sample, is given by

$$\begin{aligned} \vec{F}_B &= \vec{\nabla}\left(\vec{m}\cdot\vec{B}\right) = \vec{\nabla}\left(m_x B_x + m_y B_y + m_z B_z\right) \\ &= m_x \vec{\nabla} B_x + m_y \vec{\nabla} B_y + m_z \vec{\nabla} B_z \\ &= m_x \left(\frac{\partial B_x}{\partial x}\hat{x} + \frac{\partial B_x}{\partial y}\hat{y} + \frac{\partial B_x}{\partial z}\hat{z}\right) \\ &\quad + m_y \left(\frac{\partial B_y}{\partial x}\hat{x} + \frac{\partial B_y}{\partial y}\hat{y} + \frac{\partial B_y}{\partial z}\hat{z}\right) \\ &\quad + m_z \left(\frac{\partial B_z}{\partial x}\hat{x} + \frac{\partial B_z}{\partial y}\hat{y} + \frac{\partial B_z}{\partial z}\hat{z}\right) \end{aligned} \tag{9.1}$$

Evidently, for a crystalline sample, the force depends on the magnitude of the magnetization, but, additionally, on the magnetic anisotropy of the material and the spatial variation of the components of the magnetic field along the x-, y- and z-directions, that is, the magnetic field gradient. In the simple case of an isotropic, that is, linear material, the above expression is greatly simplified; the induced magnetization is collinear with the applied field and its magnitude is proportional to the strength of the applied field, as experimentally observed for paramagnetic and diamagnetic samples. If the pole pieces of the electromagnet are so shaped as to produce, at the position of the sample, a constant magnetic field gradient along the vertical direction with negligible field variation along the horizontal directions, the magnetic force can easily be derived.

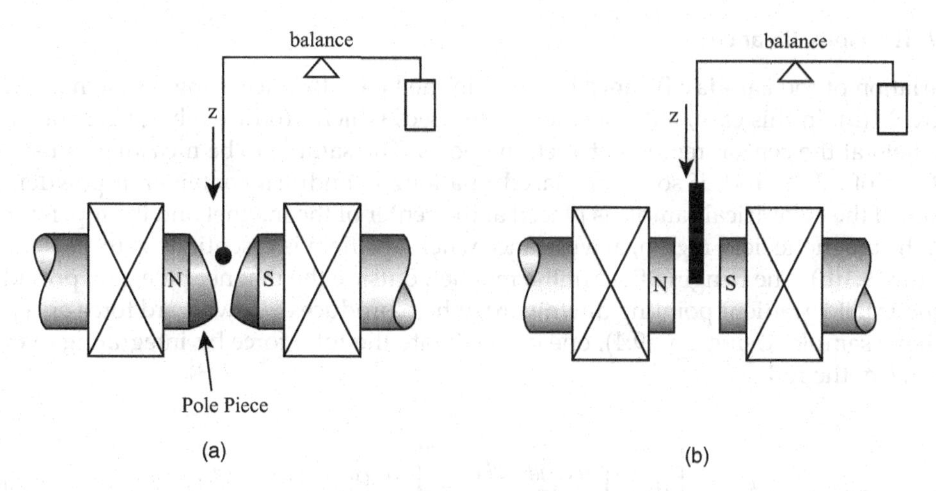

FIGURE 9.1
Schematic representation of the general features of the (a) Faraday Balance and (b) Gouy Balance.

Consider Figure 9.1(a) which gives a schematic representation of the general features of a widely used force magnetometer, the Faraday Balance. The applied magnetic field is along the y-direction and the field gradient along the z-direction (downward). For an isotropic sample, the magnetic moment will be induced along the y-direction ($m_y \neq 0$, $m_x = m_z = 0$) and the only surviving term in Eq. (9.1) is

$$\vec{F}_B = m_y \frac{\partial B_y}{\partial z} \hat{z} \tag{9.2}$$

We can express this force in terms of the magnetic susceptibility, χ, of the sample and the strength of the applied magnetic field, using Eq. (2.39) and $\vec{B} = \mu_0 \vec{H}$, where μ_0 is the permeability of free space. The magnetic force on the sample is thus given by

$$\vec{F}_{mag} = \mu_0 \chi V H_y \frac{\partial H_y}{\partial z} \hat{z} = \frac{\mu_0 \chi V}{2} \frac{\partial H_y^2}{\partial z} \hat{z} \tag{9.3}$$

where V is the volume of the sample. The magnetic force is downward if the susceptibility is positive (paramagnet) producing an apparent increase in the weight of the specimen. The magnetic force would be upward if the susceptibility were negative (diamagnet). One requirement is that the sample dimensions be small so the condition that the magnetic field gradient is constant throughout the sample is satisfied. In principle, the field strength is inversely proportional to the gap width. The inhomogeneity of the field is determined by the shape of the magnet pole pieces. Knowing the volume of the sample and the variation of H_y^2 with z, a characteristic of the Faraday pole cap design, measurement of the magnetic force *via* a sensitive (better than 0.01 mg) analytical or torsion balance allows determination of the susceptibility of weakly diamagnetic ($\chi \approx -0.01 \times 10^{-6}$) to strongly paramagnetic and antiferromagnetic ($\chi \approx 500 \times 10^{-6}$) materials. A variety of high-sensitivity Faraday Balance magnetometers have been developed operating over a wide range of sample temperatures. Typical electromagnets used have maximum fields of about 1 T to 1.5 T and magnetic field gradients of the order of 0.15–0.2 T/cm.

9.1.2 The Gouy Balance

A variation of the Faraday Balance is found in the Gouy Balance, shown schematically in Figure 9.1(b). In this case, flat pole pieces are used, which produce a homogeneous magnetic field at the central region between the poles. The sample to be measured must be in the form of a long rod, if solid, or placed in a long cylindrical container, if powder. The bottom of the cylindrical sample is placed at the center of the magnet and the top, far away from the magnetic field region, at a distance where the field is essentially zero, as depicted in Figure 9.1(b). The fringing field of the magnet, outside the magnet pole gap, provides a magnetic field gradient pointing downward, which produces a downward force on a paramagnetic sample. Using Eq. (9.4), one can estimate the total force by integrating over the length, l, of the rod,

$$F_{mag} = \int_0^l dF_z = \int_0^l \frac{\mu_0 \chi a dz}{2} \frac{\partial H_y^2}{\partial z} = \int_0^{H_y} \frac{\mu_0 \chi a}{2} dH_y^2 = \frac{\mu_0 \chi a}{2} H_y^2 \tag{9.4}$$

Here, a stands for the cross-sectional area of the cylindrical sample with the integration taken over its length along the positive z-axis (downward). As the field at the top of the rod ($z = 0$) is essentially zero, only the strength of the homogeneous field at the center of the pole pieces is needed. A disadvantage of the method is the requirement of samples in the form of long, homogeneous rods, which is difficult to satisfy in general.

The Faraday and Gouy Balances have been adapted to take susceptibility measurements at various sample temperatures, in which case the small size required for the Faraday Balance presents an additional advantage since it is easier to control the temperature of a small sample. The Faraday Balance magnetometer is exceptionally adaptable to high-temperature measurements for the determination of Curie temperatures of materials and other *in situ* magnetic measurements.

Example 9.1: *In situ* Magnetic Measurements of Bimetallic Nanoparticle Formation in Mesoporous Silica Matrices *via* a Faraday Balance Magnetometer

As an example of the use of the Faraday Balance magnetometer in studies of nanomagnetism, we present the interesting study by F. Schweyer-Tihay *et al.* (*Phys. Chem. Chem. Phys.* 8 (2006) 4018) on the nature of metallic nanoparticles obtained from molecular Co_3Ru-carbonyl clusters heated in mesoporous silica matrices.

Mesoporous silica matrices are molecular sieves of porous SiO_2 networks with a variable pore size (2–50 nm), and come in the form of spherical particles, as seen in Figure 9.2(a). They form excellent supports to disperse magnetic nanoparticles used as catalysts for various chemical reactions, as drug- and gene-delivery vehicles and as absorbents in wastewater treatments among other applications. They can be in the form of amorphous silica xerogels or highly organized porous structures like the MCM-41 type matrices, shown in Figure 9.2(b and c), first developed by scientists in Mobil Oil Corporation in the 1990s. The most interesting feature of the MCM-41 matrices is that although they consist of amorphous silica walls, they possess long-range, hexagonal "honeycomb" arrangements forming an ordered framework with straight, uniform cylindrical pores, Figure 9.2(c). The synthesis of the ordered MCM-41 type aluminosilicate matrices extended the naturally occurring microporous aluminosilicate minerals of Zeolite, as in MCM-41 matrices the pore size can be tunable. Their uniform pores can function as space-confined

FIGURE 9.2
(a) Scanning Electron Microscopy of mesoporous silica spheres (from Wikipedia, File: Mesoporous silica SEM.jpg). (b) Transmission Electron Microscopy of MCM-41, mesoporous silica showing a hexagonal or honeycomb order of straight cylindrical pores viewed along and (c) perpendicular to the cylindrical axis (from http://www.nano.gov/timeline).

nanoreactors for the formation of uniform magnetic nanoparticles in a function analogous to that of protein and virus cages, discussed in Chapter 8, in biomimetic magnetic nanoparticle synthesis.

In their study, following well-known techniques from preparations of catalysts, Schweyer-Tihay *et al.* used tetrahydrofuran (THF) solutions of organometallic $NEt_4[Co_3Ru(CO)_{12}]$ clusters and impregnated them into two mesoporous silica matrices: (a) amorphous silica xerogels with pores of 2.0 nm diameter and (b) ordered MCM-41 silica matrices with pores 2.7 nm diameter.

Subsequently, the thermal decomposition of the $NEt_4[Co_3Ru(CO)_{12}]$ clusters within the silica pores was followed by *in situ* high-temperature magnetic susceptibility measurements using a Faraday Balance magnetometer. The magnetometer was equipped with an oven operating under a controlled atmosphere in the temperature range 20–1,000°C at an applied field of up to 1 T (10 kOe). Figure 9.3 gives the magnetic behavior of the clusters

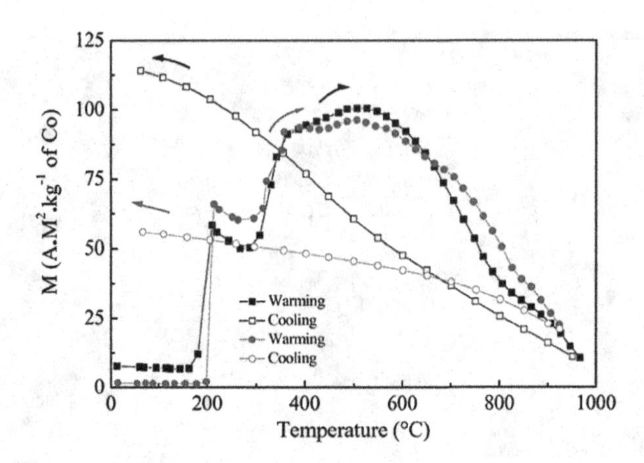

FIGURE 9.3
Magnetic behavior of the cluster impregnated silica matrices under argon as a function of temperature (■, □: xerogel; ●, ○: MCM-41) using a Faraday Balance magnetometer. The arrows indicate the warming and cooling phases of the measurements. (From F. Schweyer-Tihay *et al.*, On the nature of metallic nanoparticles obtained from molecular Co_3Ru-carbonyl clusters in mesoporous silica matrices, *Phys. Chem. Chem. Phys.*, 8 (2006) 4018, Copyright Royal Society of Chemistry, reproduced with permission).

incorporated within the matrices, as the temperature was raised to 1,000°C, at a warming rate of 200°C/h, under inert atmosphere (argon) with the applied field set at 1 T. It was initially thought that the thermal decomposition of the heterometallic carbonyl clusters would lead to the formation of highly dispersed Co_3Ru alloy nanoparticles within the matrices. This alloy has a Curie temperature of 370°C and a room-temperature bulk magnetization of 64.5 Am^2 (kg^{-1} of Co) or 64.5 emu (g^{-1} of Co). However, the magnetic behavior shown in Figure 9.3 did not support this hypothesis. A non-vanishing magnetization is recorded up to 1,000°C, more consistent with the formation of cobalt particles, as cobalt has a Curie point of 1,131°C.

In order to understand the observed behavior, the nanoparticles formed at various stages of heating were analyzed using X-ray diffraction and transmission electron microscopy measurements. The Co_3Ru-carbonyl clusters have an electron count of 60 and are diamagnetic. This is consistent with the observation of zero or very-low magnetization from room temperature to 200°C. Any magnetic signal registered in this temperature range must be due to impurities in the sample. The X-ray diffraction and TEM data indicated that the abrupt increase in magnetization at ~200°C is due to the onset of the thermal decomposition of the carbonyl clusters, which is followed by the nucleation and growth of segregated metallic Co and Ru nanoparticles within the silica pores.

Furthermore, the metallic Co nanoparticles underwent a change in crystallographic structure from the *hcp* to the *fcc* phase as the temperature increased from 200°C to 270°C, producing the observed reduction in magnetization. The further increase of the magnetization at higher temperatures, reaching a broad maximum at 500°C, correlated with the formation of larger cobalt particles. At yet higher temperatures, the magnetization starts to decrease due to the eventual incorporation of Ru atoms within the cobalt nanoparticle structure, leading eventually to the formation of a Co/Ru alloy. The alloy had the same stoichiometry as that of the initial cluster composition of Co_3Ru. Upon cooling back to room temperature, the magnetization curves of the Co_3Ru-silica derived nanocomposites within the xerogels *vs.* the MCM-41 matrices deviate from each other. This is a strong indication that the characteristics of the matrix play an active role in the determination of the magnetic properties of the nanocomposites.

9.1.3 The Alternating Field Gradient Magnetometer

A more recently developed magnetometer using the force method is the Alternating Field Gradient Magnetometer (AFGM or AGM), also known as the Vibrating Reed Magnetometer. The basic design of the AGM is shown in Figure 9.4(a), adapted from the original publication by H. Zijlstra, who was the first to propose it in the 1970s (H. Zijlstra, *Rev. Sci. Instr.* 41 (1970) 1241). It can achieve high sensitivities for the measurement of the magnetic moments of single microscopic particles. The specimen is mounted at the end of a straight thin wire, or reed, and is simultaneously subjected to a static external magnetic field supplied by an electromagnet plus an alternating magnetic field gradient produced by an appropriate coil pair, as depicted in Figure 9.4(a). The alternating force applied to the specimen deflects the wire and causes it to oscillate. The deflection and oscillation of the wire can be observed with an optical microscope. If the frequency of vibration is tuned into a resonant frequency of the reed, the vibration amplitude increases by a quality Q factor of 100. In the original publication, the gradient coil pair was made of copper wire 100 μm in diameter with an outer coil diameter of 0.6 cm and inner diameter of 0.3 cm. The spacing between the two coils was 0.2 cm. The coils contained 350 turns each and were fed by an alternating current with a root-mean-square value of 0.14 A. The amplitude of the alternating field gradient produced by this arrangement was 500 Oe/cm or 5 T/m. The lowest detectable magnetic moment was about 2×10^{-8} emu (erg/Oe) or 2×10^{-11} Am^2. This corresponds to the moment of an iron particle with a mass of 10^{-10} g or, equivalently, a spherical iron particle of about 2 μm diameter.

The original design of the AGM by Zijlstra was further developed by P. J. Flanders in the late 1980s, depicted in Figure 9.4 (b) (P. J. Flanders, *J. Appl. Phys.* 63, 3940 (1988)). In Flanders' much-improved design, the magnetic sample is mounted on the end of a cantilever rod that incorporates a piezoelectric bimorph element. The sample is magnetized by a static, homogeneous magnetic field that can be varied in magnitude. It is simultaneously subjected to a small alternating field gradient, as in the original design. The alternating

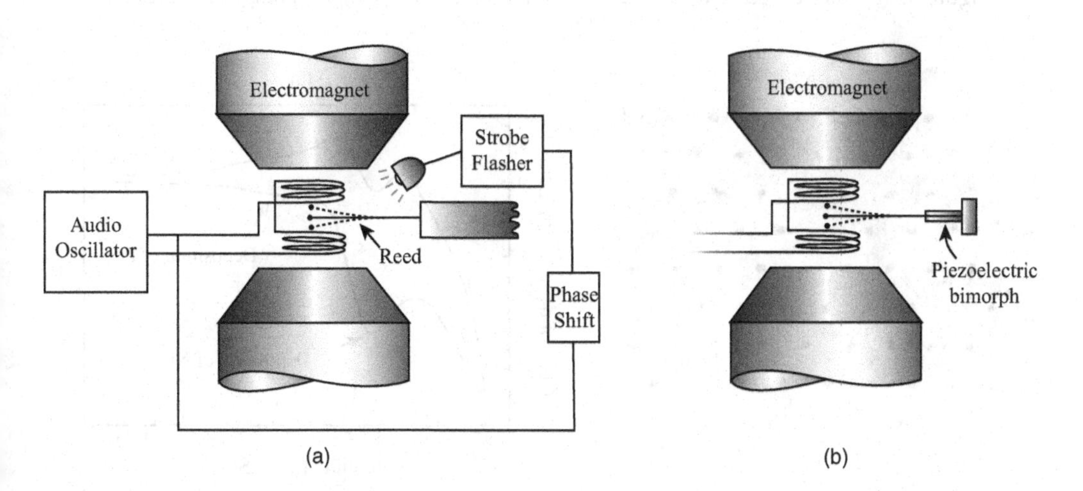

FIGURE 9.4
(a) Principle of operation of the alternating gradient magnetometer; (b) depiction of improved design by Flanders incorporating a piezoelectric bimorph element. ([a] (H. Zijlstra, A vibrating reed magnetometer for microscopic particles, *Rev. Sci. Instr.* 41 (1970) 1241; (Copyright the American Institute of Physics, reproduced with permission). [b] P. J. Flanders, An alternating gradient magnetometer, *J. Appl. Phys.* 63, 3940 (1988)). (Copyright the American Institute of Physics, reproduced with permission).

gradient exerts an oscillating force on the sample, the strength of which is proportional to the magnitude of the field gradient and to the magnetic moment of the sample. The degree by which the cantilever deflects under the influence of this force is measured by the voltage output of the piezoelectric bimorph element. Operating at or near a mechanical resonance frequency of the cantilever, the output signal can be greatly amplified, depending on the mechanical quality Q of the system. The operating frequency is usually in the 100–1,000 Hz range, with mechanical Q values of 25–250. Measurements have been made over a temperature range from 77 to 400 K.

The Flanders magnetometer has a sensitivity of 10^{-8} emu (10^{-11} Am^2). Its sensitivity is limited only by mechanical and acoustic noise in the environment; it is 1,000 times greater than that of a conventional, widely used VSM we introduce in Section 9.2.1. A special feature of the Flanders' magnetometer is that the data collection is very fast. A complete hysteresis loop over ±10 kOe applied field can be made in about 100 s, much quicker than that of a Superconducting Quantum Interference Device (SQUID) magnetometer, which generates point-by-point data, as we will discuss in Section 9.2.2.

Example 9.2: Study of the Magnetic Properties of Ni-Pt Multilayered Nanowire Arrays by AGM Measurements

As an illustration of the use of the AGM magnetometer in the study of nanomagnetic systems, we present the study by Liang and co-workers (H-P Liang *et al.*, *Inorg. Chem.* 44, 3013 (2005)) on highly ordered Ni-Pt multilayered nanowire arrays fabricated in porous anodic aluminum oxide (AAO) templates by the process of pulsed electrodeposition. The investigators prepared AAO templates with pore diameters of 20 and 50 nm. Figure 9.5(a) shows SEM micrograph images of an AAO template with an average pore diameter of 20 nm. The image shows an almost perfect close-packed array of columnar hexagonal cells, each containing a central pore normal to the surface. The inter-pore distance is *ca.* 100 nm and the density of pores is ~1.2×10^{10} pores/cm^2.

Nickel and platinum were electrodeposited in these pores using cyclic voltammetry, Figure 9.5(b), referenced to a saturated calomel electrode (SCE), of a single plating bath

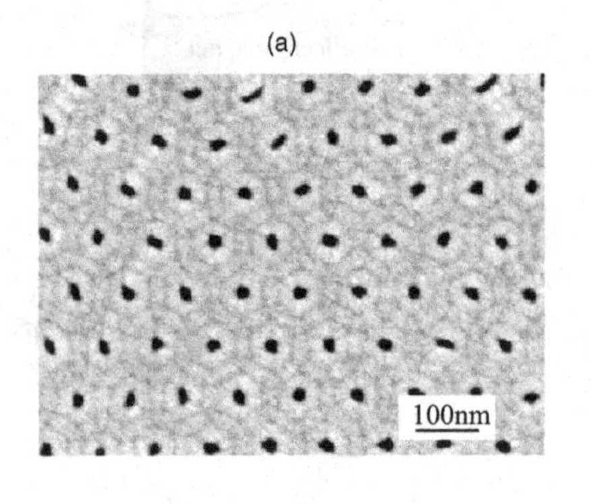

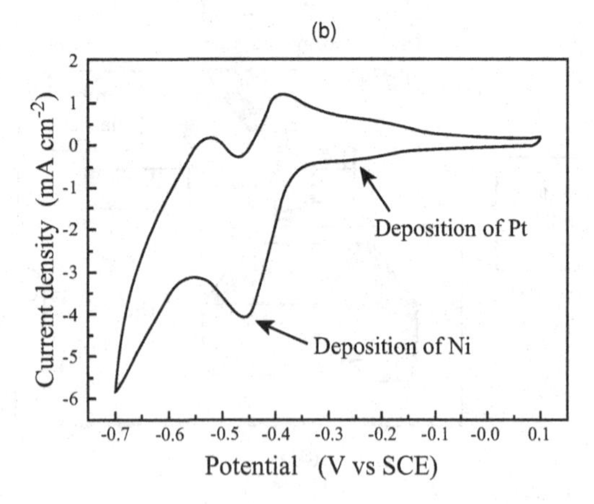

FIGURE 9.5
(a) Top-view SEM images of AAO template with an average diameter of 20 nm. (b) Cyclic voltammogram from a single plating bath containing nickel sulfate (2 M), chloroplatinic acid (2 mM) and boric acid (0.5 M), at room temperature, scan rate = 50 mV/s. (From H-P Liang *et al.*, Ni-Pt multilayered nanowire arrays with enhanced coercivity and high remanence ratio, *Inorg. Chem.*, 44 (2005) 3013, Copyright American Chemical Society, reproduced with permission).

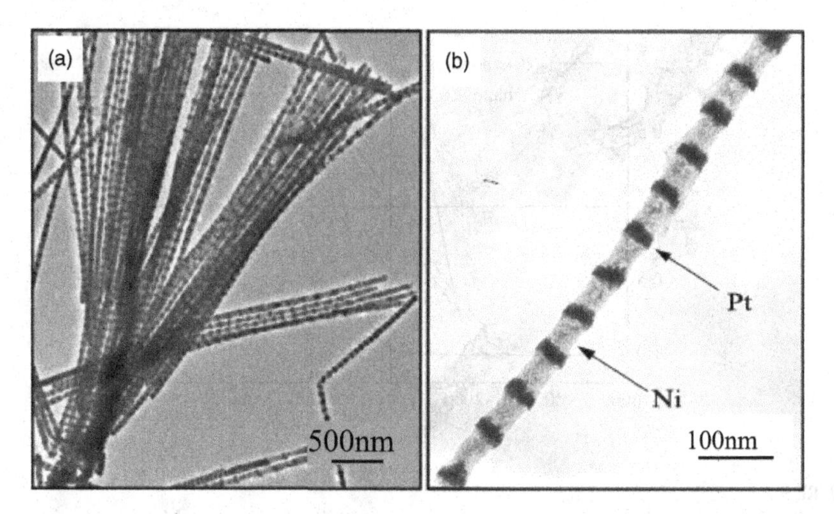

FIGURE 9.6
(a) Low- and (b) high-magnification TEM images of Ni-Pt nanowires liberated from an AAO template with a diameter of about 50 nm. (From H-P Liang *et al.*, Ni-Pt multilayered nanowire arrays with enhanced coercivity and high remanence ratio, *Inorg. Chem.*, 44 (2005) 3013, Copyright American Chemical Society, reproduced with permission).

containing nickel sulfate (2 M), chloroplatinic acid (2 mM) and boric acid (0.5 M). The reduction of Ni^{2+} and deposition of Ni begins at *ca.* −0.33 V and reaches the pronounced peak current at *ca.* −0.46 V. Platinum deposition is evinced by a weak reduction peak at *ca.* −0.26 V due to the fact that the concentration of chloroplatinic acid is very low, 2 mM compared to 2 M for the nickel sulfate. Thus, the majority of the deposit was Ni when a proper deposition potential was chosen, because the concentration of Ni^{2+} was 1,000 times higher than that of $PtCl_6^{2-}$.

In this manner, multilayered Ni-Pt nanowire arrays were fabricated as seen in Figures 9.6(a) and (b), which show typical low- and high-magnification TEM images of Ni-Pt nanowires liberated from an AAO template with about 50 nm pore diameter. In Figure 9.6(b), the alternating dark and bright contrasts correspond to cylindrical Pt and Ni nanoparticles, respectively, within the bamboo-like nanowire. The Pt segments have a length of 18 nm.

Hysteresis loops of nanowire samples embedded within the AAO template were obtained at room temperature using an AGM magnetometer with the magnetic field parallel ($\|$) and perpendicular ($\perp$) to the axes of the nanowires. Figure 9.7 shows the magnetic hysteresis loops obtained for AAO embedded nanowires with a diameter of 20 nm and length of Ni segments of 50 nm. The coercivity with applied field parallel to the axis of the nanowires is much enhanced, $H_c^{\|} = 1{,}169$ Oe, compared to that with the applied field perpendicular to the axis, $H_c^{\perp} = 134$ Oe. These magnetic measurements reveal that the arrays exhibit uniaxial anisotropy with the easy axis along the axis of the nanowires. The coercivity of 1,169 Oe shows remarkable enhancement in comparison to that of bulk Ni of about 0.7 Oe. This enhancement is due to the shape anisotropy of the Ni nanoparticles engineered within the nanowires. By varying the diameter and/or length of the Ni segments (shape anisotropy), the investigators could tailor the hysteresis loop characteristics. The derived parameters of coercivity (H_c) and remnant-to-saturation magnetization ratios (M_r/M_s) for specific values of diameter and length of Ni segments are given in Table 9.1.

With the diameter reduced, the magnitude of the shape anisotropy of the Ni nanoparticles is increased, resulting in an enhancement in coercivity and, therefore, switching field of the multilayered nanowires. Such nanocomposites have important prospective technological applications in chemical sensors, nanoelectrode ensembles and, particularly, in ultra-high-density magnetic recording media.

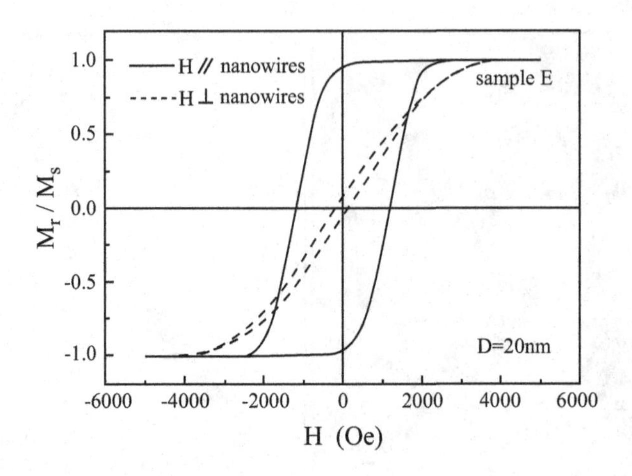

FIGURE 9.7
Magnetization hysteresis loops at room temperature obtained *via* an AGM magnetometer for a sample with 20 nm diameter and Ni length segment of 50 nm embedded in the AAO template. Hysteresis loops for an applied field, parallel and perpendicular to the axis of the nanowires are shown. (From H-P Liang *et al.*, Ni-Pt multilayered nanowire arrays with enhanced coercivity and high remanence ratio, *Inorg. Chem.*, 44 (2005) 3013, Copyright American Chemic Society, reproduced with permission).

TABLE 9.1
Magnetic Properties and Morphologic Features of Ni-Pt Multilayered Nanowire Arrays

Diameter (nm)	Length of Ni Segments (nm)	$H_c^{\parallel}$ (Oe)	M_r/M_s ($\parallel$)	$H_c^{\perp}$ (Oe)	M_r/M_s ($\perp$)
50	50	596	0.57	88	0.08
50	100	682	0.51	56	0.03
50	150	828	0.50	59	0.03
30	50	1,146	0.90	96	0.09
20	50	1,169	0.96	134	0.06

Source: From H-P Liang *et al.*, Ni-Pt multilayered nanowire arrays with enhanced coercivity and high remanence ratio, *Inorg. Chem.*, 44 (2005) 3013, Copyright American Chemical Society, reproduced with permission.

9.2 Induction Methods

Magnetometers based on the induction method make use of Faraday's Law of Induction (Eq. 1.26), which states that an electromotive force (emf) is induced in a conducting wire loop while the magnetic flux through the loop is changing. The induced current is in a direction as to counteract the change in magnetic flux, known as Lenz's Law. If a sample magnetized by a homogeneous magnetic field were to be oscillated in the vicinity of a wire loop, an alternating voltage would be induced in the loop as the sample repeatedly approached and then receded from the loop. The induced signal can be recorded and analyzed. For a given oscillatory motion and geometry of sample *vs.* induction coil arrangement within the magnetometer, one can easily determine the magnetization of the sample from the recorded signal.

9.2.1 The Vibrating Sample Magnetometer

The first design of a versatile Vibrating Sample Magnetometer (VSM) was proposed by American physicist Simon Foner (S. Foner, *Rev. Sci. Instr.*, 30 (1959) 548) and is often referred to as the Foner magnetometer, in his honor. The basic elements of the Foner magnetometer are depicted in Figure 9.8, where the magnetic moment of the sample is measured in continuously swept magnetic fields by the use of an electromagnet.

The sample is attached at the end of a long rod, made of a non-magnetizable material such as a drinking straw, with the other end mounted on a vibrating mechanism, such as a loudspeaker transducer. The sample is located in the region between the pole pieces of an

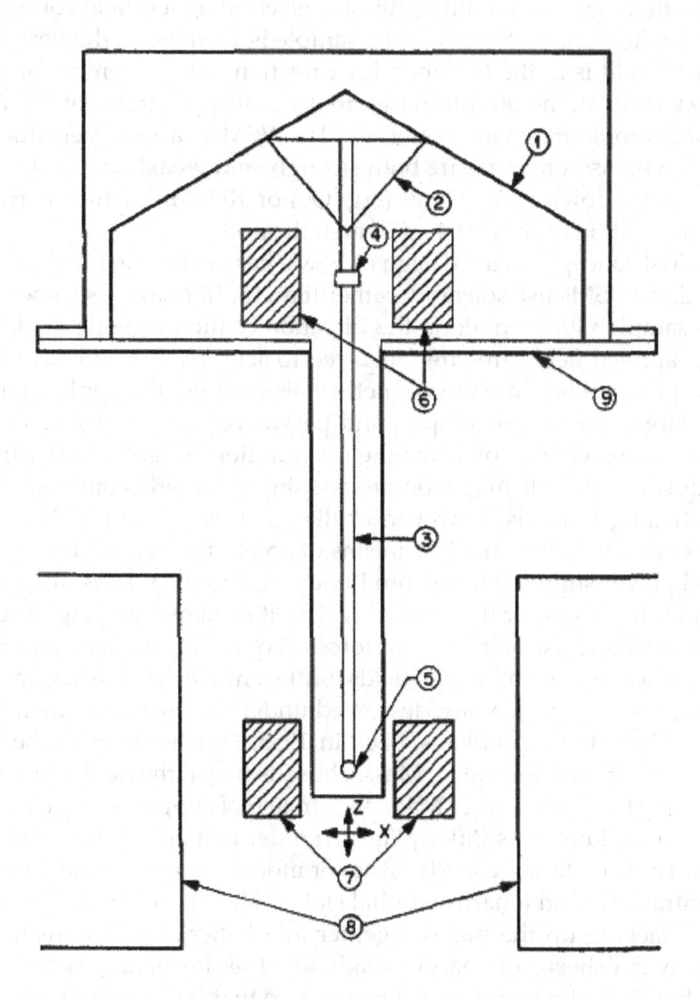

FIGURE 9.8
Basic components of a Foner VSM magnetometer. (1) Loudspeaker transducer, (2) conical paper cup support, (3) drinking straw, (4) reference sample, (5) sample, (6) copper reference detection coils, (7) copper sample detection coils, (8) magnet poles, (9) metal container. (S. Foner, Versatile and sensitive vibrating-sample magnetometer, S. Foner, *Rev. Sci. Instr.*, 30 (1959) 548, Copyright American Institute of Physics, reproduced with permission).

electromagnet where a homogeneous magnetic field exists. The magnetized moving sample induces a small alternating emf in the copper detection pick-up coils placed near it (sample coils). This small alternating emf is amplified, usually with a lock-in amplifier that is sensitive only to signals at the driving vibration frequency. The lock-in amplifier is provided with a reference signal at the frequency of vibration derived from a sensor coupled to the driving mechanism. The vibration frequency of the sample can be varied over a wide range, from below 40 Hz to 100 Hz, depending on the vibrating mechanism, while the amplitude of the vibration can vary from a few mm down to 0.1 mm. The sample's motion is also monitored by placing a small bar magnet as a reference sample at the top of the rod, outside the magnetic field region, at the vicinity of reference pick-up coils, as depicted in Figure 9.8. This reference coil is used to provide a feedback loop to the vibrating mechanism ensuring that the amplitude and the frequency are properly maintained. Figure 9.9 gives a more detailed view of the construction of the Foner magnetometer. The sample must be positioned and vibrated at the center of the gap of the electromagnet, where the magnetic field is homogeneous. With traditional electromagnets fields of up to 2 T (20 kOe) can be applied to the sample. Note that the sample is vibrated in the vertical, z-direction, while the applied field is in the horizontal, y-direction. The placement of multiple detection coils at the vicinity of the sample allows for special applications of the VSM, including the study of anisotropic magnetic materials. The VSM is a very versatile and sensitive instrument. It can be used to measure both strongly and weakly magnetized materials. It can detect a moment down to 10^{-5} emu (erg/Oe) or 10^{-8} Am^2, which corresponds to the saturation magnetization of only 0.04 μg of metallic iron.

Since its original conception, the design of VSMs has evolved in pace with the evolution of magnets. Today's VSMs use solenoid, rather than traditional horseshoe-shaped electromagnets, with sample vibration along the direction of the magnetic field, the axis of the solenoid. Large applied fields are often required to achieve magnetic saturation in highly anisotropic samples. Studies in nanomagnetism also require the application of high magnetic fields, as large surface anisotropies and pervasive canting of surface spins in magnetic nanoparticles prevent complete magnetic saturation, as has been discussed in Section 4.10. The production of high magnetic fields using solenoids constructed with normal (resistive) conducting wires is, however, challenging as it requires large power input. Maintaining a steady magnetic field by means of an electric current has zero efficiency as all the electrical power supplied turns into Joule heat ($P = I^2R$). Thus, the removal of large amounts of heat is necessary, or the magnet will melt from overheating. A second problem arises from the extremely strong Lorentz forces (Eq. (1.17)) that act on current-carrying wires in the presence of large magnetic fields. Sufficient mechanical strength must be provided or the magnet will break apart, destroyed under the force of its own field.

In order to address these problems, back in 1933, Francis Bitter at the Massachusetts Institute of Technology in Cambridge, Massachusetts began the development of high field solenoids that carry his name "Bitter Magnets". In lieu of copper wires to carry the current, he used copper disks, known as Bitter plates, in order to increase the mechanical strength of the magnets. These disks are usually 30 cm or more in diameter and 1 mm in thickness, they have a central hole and a narrow radial slot, as shown in Figure 9.10(a). A magnet is constructed by stacking up the plates together in a helical fashion, insulated from each other by similarly cut sheets, or spacers, made of a thin insulating material. Each copper disk is rotated by 20° with respect to its neighbor, so that the region of overlap provides a conducting path for the current to flow from one disk to the next. The current through the stack of disks is therefore helical and the stack acts as a solenoid. The disks are clamped tightly together and enclosed in a case. Each copper plate and insulating disk has a large

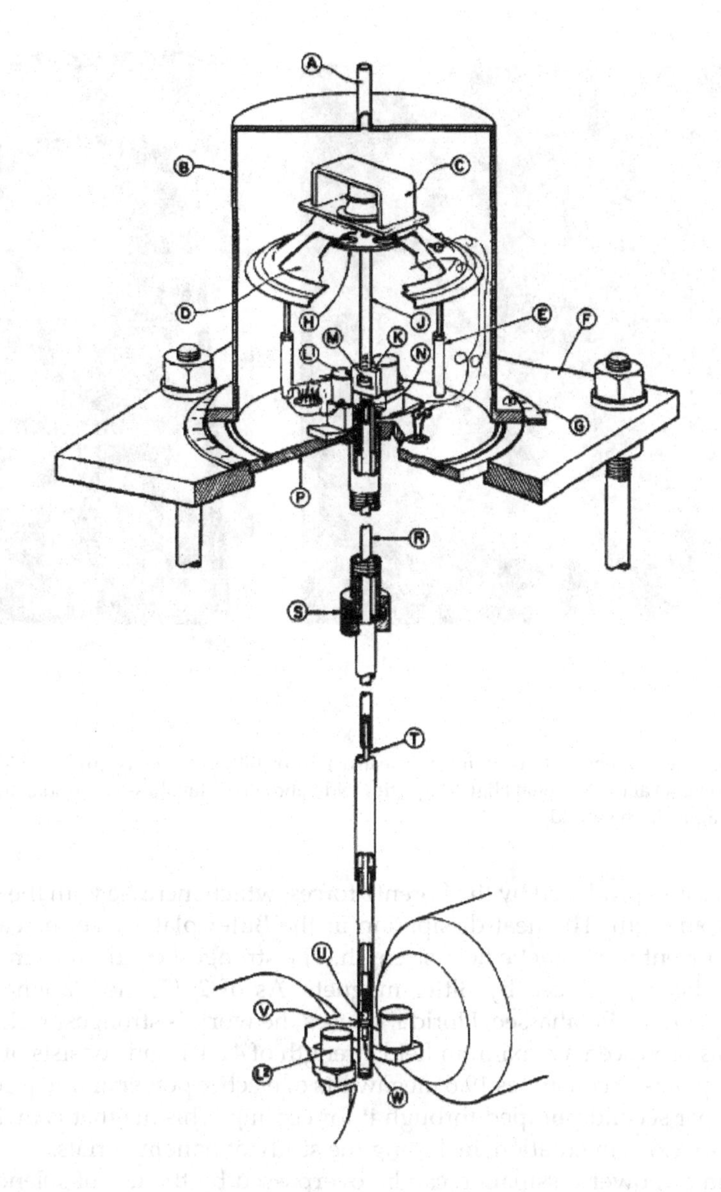

FIGURE 9.9
Detailed construction of the Foner VSM magnetometer. A, evacuation tube; B, removable brass hat; C and D, loudspeaker transducer; E, leveling screws; F, support plate; G, indicator scale; H, plastic element for rigid support; J, threaded plastic rod; K, reference sample holder; M, reference sample; N, threaded plastic connector; P, base plate; R, stainless steel extension rod; S, brass extension-tube; T, solid plastic rod; U, threaded sample holder; W, sample; V, Teflon centering washer; L1, reference pick-up coils; L2, three-dimensional view of sample detection pick-up coils. (From S. Foner, Versatile and sensitive vibrating-sample magnetometer, *Rev. Sci. Instr.*, 30 (1959) 548, Copyright American Institute of Physics, reproduced with permission).

number of small holes (Figure 9.10(a)), which allow forced water, as a coolant, to circulate through the magnet during magnet operation to remove resistive Joule heating created by the large flowing currents. The maximum field produced depends on the maximum current that can be safely passed through the Bitter plates and the size of the bore of the magnet, the central area of the solenoid that defines the available high-field experimental space. The stacked plate design allows the magnet to withstand the enormous outward

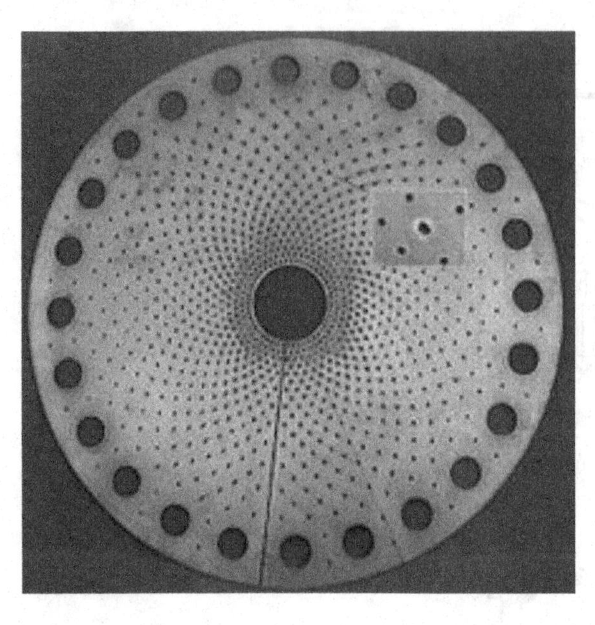

(a)

(b)

FIGURE 9.10
(a) Depiction of a Bitter copper plate used in the construction of Bitter solenoids. (b) The 45-T hybrid Bitter-superconducting magnet at the National High Magnetic Field Laboratory, Tallahassee, Florida, the most powerful DC-field electromagnet in the world.

mechanical pressure produced by the Lorentz forces, which increase with the square of the magnetic field strength. The heat dissipation in the Bitter plates also increases with the square of the current or magnetic field strength. The strongest continuous magnetic fields on Earth have been produced by Bitter magnets. As of 2017, the National High Field Laboratory (NHFL) in Tallahassee, Florida, houses the world's strongest resistive magnet. This system has produced a maximum field strength of 41.4 T and consists of hundreds of separate Bitter plates. It consumes 19.6 megawatts of electric power and requires about 139 liters of water per second pumped through it for cooling. This magnet is mainly used for materials science experimentation, including the study of nanomaterials.

The problem of power dissipation can be overpassed by the use of solenoids made of superconducting wires, since the resistance of the wire is reduced to zero below the critical temperature, T_c, for superconductivity. The strength requirement must, however, still be addressed. Additional considerations arise from the fact that T_c is reduced in the presence of an applied magnetic field and superconductivity is destroyed at large currents. This limits operation of superconducting solenoids below the material's critical current and critical field for superconductivity. Thus, there is a limit as to how large a magnetic field can practically be produced in the laboratory. The best materials for the construction of superconducting magnets are Nb_3Sn, Nb-Zr and Nb-Ti alloys affording fields up to 8 T (80 kOe) or 6.4 MA/m at 4.2 K. These Nb alloys are low-T_c superconductors operated at the boiling temperature of liquid helium, 4.2 K. Combining these traditional low-temperature superconducting materials with the high-T_c cuprous compound superconductors, such as $YBa_2Cu_3O7\text{-}_\delta$ (YBCO) discovered in the 1980's, scientists at the NHFL achieved a maximum field of 32 T in an all superconducting solenoid in 2017.

The strongest continuous manmade magnetic field, 45 T, available to experimenters at NHFL has been produced by a device consisting of a Bitter magnet enclosed within a superconducting solenoid, appropriately called a "hybrid magnet". It has a bore of 32 mm requiring a current of up to 67 kA and a power input of about 20 MW. At the bore, the two magnetic fields add up, reinforcing each other, thus pushing both, electro-magnet and superconducting-magnet technologies to the limit in order to produce the world's highest possible continuous magnetic fields (Figure 9.10(b)).

The VSM has been adapted to measure samples in high and low temperatures by inserting the sample and its supporting rod in a cryogenic Dewar and incorporating heating elements at the place of the sample. Measurements between 4.2 K and 400 K can easily be achieved. Temperatures lower than 4.2 K are also routinely achieved by pumping on liquid helium.

Even higher fields are produced in the United States at the Pulsed Field Facility at Los Alamos Laboratories in New Mexico, using pulsed magnets operating from a couple of seconds down to microseconds. As of 2012, a record of 100-T (non-destructive) pulsed field has been produced at the Los Alamos facility. Compare this to the Earth's magnetic field of 0.0005 T in magnitude. We first mentioned the existence of such powerful magnets in Chapter 7, in connection with our discussion of the magnetic properties of horse spleen ferritin in intense magnetic fields. In Example 7.4, Figure 7.21 presents data on ferritin collected at the world's most powerful magnet at NHFL and the Pulsed Field Facility at Los Alamos Laboratory.

Other pulsed-field facilities exist in Germany, France, China and Japan, dedicated to fundamental studies of the structure of matter at the nanoscale. Self-destructing magnets can reach even higher fields of the order of 1,000 T, but only for a few microseconds, before they explode! Such destructive fields have been produced in Japan. Thus, even though there is no fundamental limit to how strong a magnetic field may be produced in the cosmos, (*e.g.*, some type of neutron stars, known as magnetars, are estimated to produce magnetic fields of the order of 10^{11} T), in the laboratory we are limited in the production of high magnetic fields.

9.2.2 The SQUID Magnetometer

Superconducting quantum interference devices (SQUIDs) are sensitive detectors of magnetic flux. One type of SQUID consists of a superconducting loop interrupted by two insulating barriers, as depicted in Figure 9.11. The configuration of two superconductors (gray shaded area) separated by a thin insulating barrier (white bar) constitutes a "Josephson junction". It is named after the British physicist Brian Josephson, who, in 1962, developed the mathematics of the current and voltage (I–V) characteristics across such barriers. SQUIDs can be fabricated from either low or high critical temperature (T_c) superconductors. Low-T_c SQUIDs operate at the boiling temperature of liquid helium (4.2 K) and are usually made of niobium, which has the highest critical temperature of all elemental superconductors, with $T_c = 9.2$ K at atmospheric pressure. High-T_c SQUIDs can operate at the boiling temperature of liquid nitrogen (77 K) and are made primarily of thin films of YBCO ($YBa_2Cu_3O_{7-x}$) with $T_c = 93$ K. However, they are less sensitive than the conventional low-T_c SQUIDs. The operation of SQUIDs is based on two physical phenomena: (a) flux quantization through a superconducting loop and (b) supercurrent tunneling through the barrier.

Figure 9.11 gives a schematic diagram of what is called a dc-SQUID. An input current, I, enters the superconducting loop of the SQUID at the top and splits into two branches

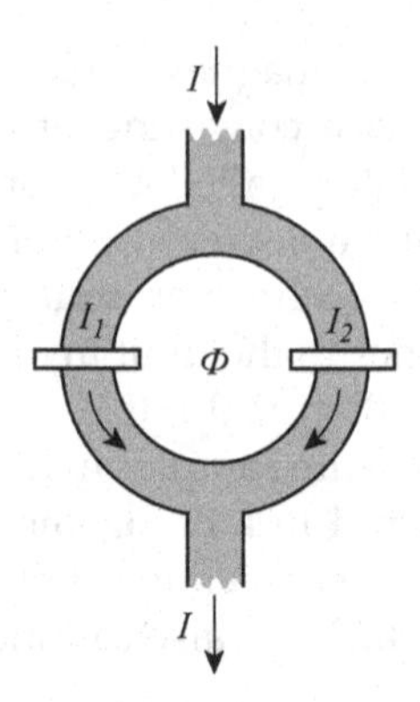

FIGURE 9.11
Diagram of a dc-SQUID. The loop is made of a superconducting metal (gray shaded area), like Nb. The current, I, enters and splits into the two paths. The thin insulating barriers (white bars) on each path are Josephson junctions. Φ represents the magnetic flux threading the dc-SQUID loop.

with currents I_1 and I_2. Φ represents the magnetic flux threading the SQUID loop. In the absence of an external magnetic field, the input current splits equally between the two branches, $I_1 = I_2 = I/2$. Upon the introduction of an external magnetic field, a screening current, I_S, is induced in the loop, which generates a magnetic field of its own opposing the change in flux through the loop, according to the laws of Faraday and Lenz. The currents flowing through the two branches of the SQUID are now different, as the screening current is added to the existing current in one branch and is subtracted from it in the other. This results in currents equal to $I/2 + I_S$ in one branch and $I/2 - I_S$ in the other. In the superconducting state, Cooper pairs of electrons tunnel through a Josephson junction with no voltage drop across the junction. This holds as long as the persistent current through the junction remains below a critical value, I_0, at which the junction becomes resistive. Whenever the current exceeds I_0 in either branch, a voltage V appears across the junction and therefore across the SQUID, which can be recorded with an appropriate electrical circuit.

It is a property of superconducting loops, that the magnetic flux through any area bounded by the loop is quantized. The quantum of magnetic flux, Φ_0, is a fundamental physical constant, $\Phi_0 = h/(2e) = 2.067833758(46) \times 10^{-15}$ T m^2, where h is the Plank's constant, and e is the charge of the electron. The magnetic flux enclosed by the superconducting loop must, thus, always be an integral number of flux quanta. This requirement leads to the following interesting phenomenon: if the introduction of an external field were to increase the flux threading the loop by an amount that exceeded $\Phi_0/2$, half the magnetic flux quantum, the SQUID finds it energetically more favorable to increase the change in flux to Φ_0 rather than to screen it and I_S flows in the opposite direction. Thus, I_S changes direction every time the flux increases by an amount of $n\Phi_0/2$, where n is an integer. As a result, the currents across the Josephson junctions oscillate as a function of the applied flux. If the input or bias current, I_b, is larger than I_0, the critical current for the Josephson junction to switch from the superconducting to the resistive state, the SQUID operates exclusively in the resistive mode, and a voltage V is always detected across the SQUID as indicated in Figure 9.12(a) with I–V characteristics as shown in Figure 9.12(b). An oscillatory voltage across the SQUID is detected, which is a function of the applied magnetic field and has a period of oscillation equal to Φ_0, as shown in Figure 9.12(c). Therefore, in principle, the

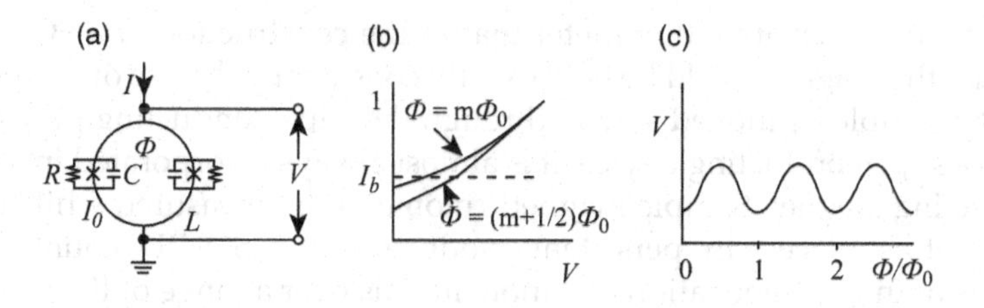

FIGURE 9.12
(a) Schematic of the operation of a dc-SQUID; flux to voltage conversion, (b) *I–V* characteristics of a dc-SQUID, (c) *V vs.* Φ/Φ_0 at constant bias current. The periodic voltage response is due to flux through the SQUID. The periodicity is equal to one flux quantum, Φ_0. (Adapted from R. Kleiner *et al.*, Superconducting quantum interference devices: state of the art and applications, *IEEE Xplore*, 92 (2004) 1534).

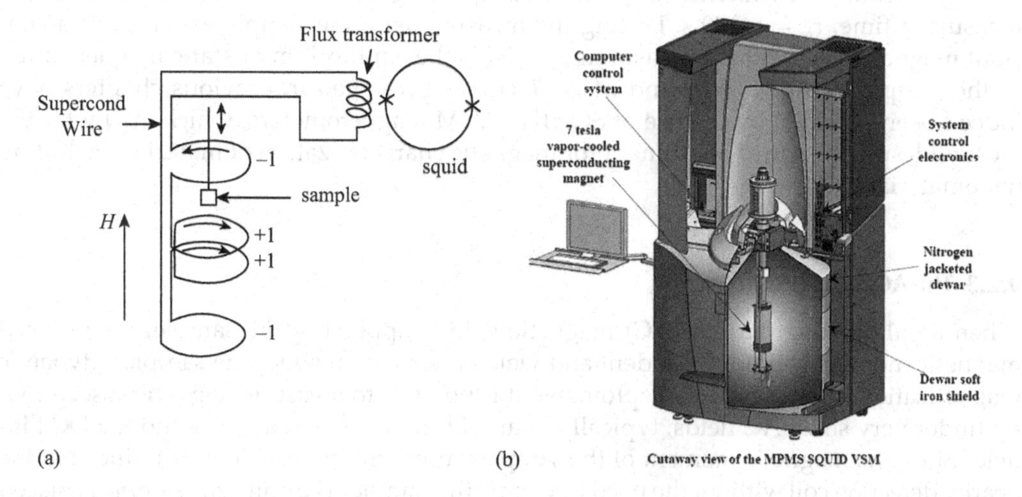

FIGURE 9.13
(a) Schematic representation of the superconducting detection coil coupled to a SQUID. (b) Cutaway view of the Magnetic Properties Measuring System (MPMS) SQUID VSM by Quantum Design Inc.

SQUID is an exquisitely sensitive flux-to-voltage converter. The *I–V* characteristics of Josephson junctions are well explained within a model, which assumes that the Josephson junction is in parallel with a resistance, *R*, and a capacitance, *C*, as depicted in Figure 9.12(a). The screening current is then given by the ratio of the applied flux Φ over the self-inductance, *L*, of the ring ($I_S = \Phi/L$), or the voltage *V* over the resistance *R* across the Josephson junction ($I_S = V/R$). Thus Φ can be estimated as a function of *V*.

SQUIDs are the most sensitive instruments for the detection of small changes in magnetic flux. They can detect magnetic fields as small as 10^{-14} T. Combined with proper electrical circuitry, SQUIDs have been adapted to detect the magnetic signals resulting from the electrical activity of the brain (10^{-13} T) in "magnetoencephalography" and those resulting from the electrical activity of the heart (10^{-10} T) in "magnetocardiography". The advent of SQUIDs ushered the presently very active area of "biomagnetism" with important applications in neuroscience. In SQUID VSMs the copper detection coils of the Foner magnetometer are replaced by superconducting coils that are coupled to a SQUID (Figure 9.13(a)) and the electromagnet is replaced with a superconducting magnet. Sample vibration is along the axis of the solenoid, along the direction of the applied magnetic field. The result is a most sensitive magnetometer, which can detect magnetic moments down to 10^{-9} emu.

The SQUID is such a sensitive flux meter that in the construction of a SQUID-VSM, it is imperative that the magnetic field be held exactly constant by the use of a superconducting shield, and the sample be moved slowly through the superconducting pickup coils. Since the SQUID is a superconducting device, it is almost always incorporated in a system using a superconducting magnet. Sample magnetic moments are measured while the superconducting magnet is placed in persistent mode and the SQUID counts flux quanta. Measurements of the magnetization of nanomaterials over a range of fields, typically −7 T to +7 T, and temperatures, typically $2 < T < 400$ K, are time-consuming and, thus, the systems are usually operated unattended under computer control. Figure 9.13(b) gives a cutaway view of the Magnetic Properties Measuring System (MPMS) SQUID VSM by Quantum Design Corporation, available in many individual investigator laboratories and shared instrumentation facilities.

Magnetization measurements performed using SQUID VSMs have a characteristic measuring time, τ_m = 1–100 s. During the measurement, the sample experiences a constant magnetic field and the measurement yields the equilibrium or static magnetization of the sample. Most *M vs. H* and *M vs. T* curves presented in previous chapters have indeed been obtained by the use of SQUID – VSM magnetometers, which are by far the most widespread type of instrument for magnetic characterization of materials including nanomaterials.

9.2.3 The AC Susceptometer

When an alternating current (AC) magnetic field is applied to the sample, the induced magnetic moment is time-dependent and yields information about the sample's dynamic magnetization. With an AC Susceptometer, it is possible to measure magnetic susceptibility under very small AC fields, typically of amplitude 1–4 Oe, with or without a DC bias field. Since the magnetic moment of the sample varies with time, an emf is induced in the nearby detection coil without the need to vibrate the sample. Typically, the sample is placed at the center of a solenoid carrying an alternating current, resulting in a small AC magnetic field of amplitude H_o and angular frequency ω, $H_{AC} = H_o \cos\omega t$, applied to the sample. For very low frequencies, the M (H) curve obtained is similar to the DC magnetization curve and the susceptibility at each point of the curve is given by the slope, $\chi = dM/dH$, often referred to as the differential susceptibility.

At high frequencies, the induced AC moment lags behind the drive field; it cannot follow the DC magnetization curve. This lag is detected by the susceptometer circuitry. Thus, the AC susceptometer measures two quantities: the magnitude of the susceptibility, χ, and the phase shift, φ, of the induced moment relative to the drive signal. This information is usually given in complex representation, with the susceptibility thought of having an in-phase, or real, component $\chi' = \chi \cos \varphi$ and an out-of-phase, or imaginary, component $\chi'' = \chi \sin \varphi$, where $\chi = \sqrt{\chi'^2 + \chi''^2}$ and $\varphi = \arctan(\chi''/\chi')$. The in-phase component of the susceptibility is related to the time average of the magnetic energy stored in the sample, while the out-of-phase component is related to the energy converted into heat during one cycle of the AC field. The characteristic measuring time is given by the inverse of the frequency of the driving field, $\tau_m = 2\pi/\omega$.

The instrumentation of the AC susceptometer uses the principle of the co-axial mutual inductance technique. The basic design of the measuring circuitry consists of a primary coil placed co-axially around two oppositely wound secondary coils, as shown in Figure 9.14. The primary coil generates the AC magnetic field and the secondary coils

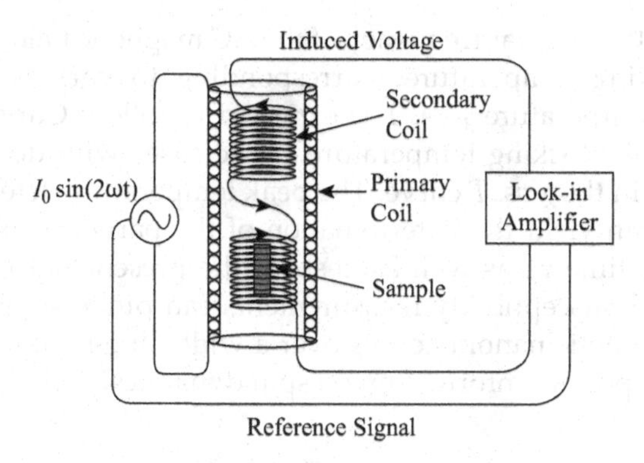

FIGURE 9.14
Coil assembly for AC susceptibility measurements.

sense the induced magnetization of the sample. The two secondary coils have identical dimensions and the same number of loops. When an AC current flows through the primary coil, each secondary coil experiences an induced emf of the same magnitude but of opposite direction. Thus, in the absence of a sample, the net voltage across the two detection coils is expected to be null. However, when a sample is inserted in the middle of one of the secondary coils, a voltage is detected across the two-secondary-coil assembly. The magnitude of the recorded voltage signal is directly proportional to the susceptibility of the sample.

To see how, in principle, this comes about, consider that the sample is introduced in the middle of the bottom coil as shown in Figure 9.14. The emf or voltage induced across each secondary coil is given by

$$V_{top} = -\frac{d\Phi}{dt} = -NA\frac{dB_{top}}{dt} = -NA\mu_0\frac{dH}{dt} \tag{9.5}$$

$$V_{bot} = -\frac{d\Phi}{dt} = -NA\frac{dB_{bot}}{dt} = -NA\mu_0(1+\chi)\frac{dH}{dt} \tag{9.6}$$

N is the number of turns in each coil and A is their cross-sectional area. Here, we have made use of Eqs. (2.37) and (2.38) to relate B to H, assuming for simplicity that the sample fills the entire core of the bottom coil. The net induced voltage across both oppositely wound secondary coils would be

$$V_{net} = V_{bot} - V_{top} = -NA\mu_0\chi\frac{dH}{dt} \tag{9.7}$$

Equation (9.7) can be solved for χ, determining thus the susceptibility of the sample.

The AC susceptometer is also combined with cryogenic equipment and measurements over the temperature range from 4.2 K to 300 K are common. AC susceptometers are easy to construct in the laboratory and many researchers design in-house built units tailored to their field of research. In the area of nanomagnetism, this technique can profitably be utilized in the study of superparamagnetism. This is due to the fact that the blocking temperature, T_B, given by the Néel–Brown theory of Eq. (4.45), depends on the characteristic measurement time, τ_m, of the experiment. Since for AC susceptibility measurements

$\tau_m = 2\pi/\omega$, where ω is the angular frequency of the AC magnetic field, τ_m is tunable; one can easily measure blocking temperatures corresponding to various driving frequencies. Above the blocking temperature χ *vs.* T is expected to follow Curie's Law given by Eq. (2.41), while below the blocking temperature, χ decreases with decreasing temperature. This produces a peak in the χ *vs.* T curve. The peak temperature determines T_B. A series of such measurements can yield the determination of the particle's anisotropy energy and characteristic attempt time τ_0, as well as, test for the presence of interparticle magnetic interactions. Thus, AC susceptibility measurements can probe superparamagnetic relaxation processes in magnetic nanomaterials over a wide range of measurement time windows, contributing important information on spin dynamics.

Example 9.3: Probing Superparamagnetism *via* DC and AC Magnetization Measurements

This example serves to demonstrate the use of DC and AC magnetization measurements in the study of superparamagnetism. Theoretical considerations of superparamagnetism were introduced and discussed extensively in Chapter 4. Superparamagnetic relaxation of uniaxial magnetic nanoparticles is governed by the Néel relation of Eq. (4.45). An important concept associated with superparamagnetism is the "blocking temperature" of an ensemble of magnetic nanoparticles, T_B, defined by Eq. (4.47). We emphasized then that T_B is not a physical quantity associated with the particle, as it depends on the characteristic measuring time, τ_m, of the technique used. Ferritin, which we discussed extensively in Chapter 7, is often referred to as a model superparamagnet. We, thus, use the case of the biomineral magnetic core of horse spleen ferritin to demonstrate how the blocking temperature is determined experimentally, and how its value depends on τ_m.

We start with DC magnetization measurements, which constitute the most common procedure used in the determination of T_B for an ensemble of magnetic nanoparticles. One records the zero-field-cooled (ZFC) and field-cooled (FC) magnetization curves as a function of temperature. Figure 9.15 shows such ZFC/FC curves for horse spleen ferritin

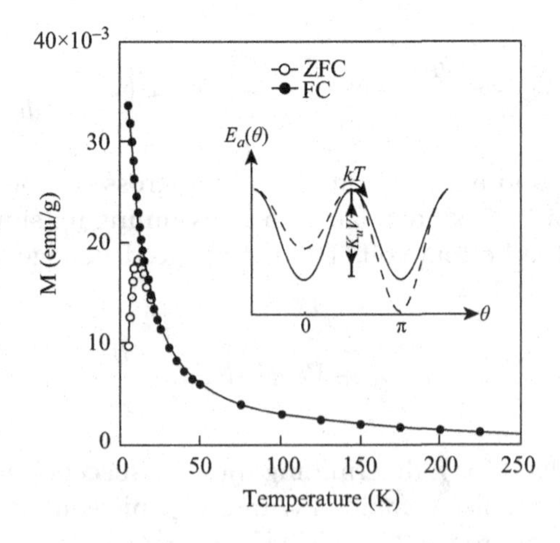

FIGURE 9.15
ZFC/FC magnetization curves for horse spleen ferritin in an applied field of 50 Oe. The inset depicts the magnetic anisotropy energy in the absence (solid line) and in the presence (broken line) of an external magnetic field applied against the anisotropy axis ($\theta = \pi$). (From S. A. Makhlouf, F. T. Parker and A. E. Berkowitz, Magnetic Hysteresis Anomalies in Ferritin, *Phys. Rev. B*, 55, R14717 (1997), Copyright American Physical Society, reproduced with permission).

(HoSF) in the temperature range between approximately 5 K and 225 K, reproduced from the work of Makhlouf *et al.* (S. A. Makhlouf, F. T. Parker and A. E. Berkowitz, Magnetic Hysteresis Anomalies in Ferritin, *Phys. Rev. B*, 55, R14717 (1997)). As we discussed in Section 7.5, the magnetic properties of ferritin arise from uncompensated spins at the surface and defect sites in the interior of the antiferromagnetic ferrihydrite core, resulting in a net moment per ferritin core of ~300 μ_B. When the ferritin sample is cooled down to 5 K in the absence of an externally applied magnetic field, the internal anisotropy field (Eq. 4.17) aligns the particle's magnetic moment along the uniaxial anisotropy axis. It is the property of uniaxial symmetry that moment orientation along or opposite to the direction of the symmetry axis is equally probable; the system exhibits bistability. In terms of energetic considerations, there is equal probability for either of the two potential wells (solid line) shown in the inset of Figure 9.15 to be populated, with half of the particles in the ensemble having their macrospin moment pointing along ($\theta = 0$) and the other half against ($\theta = \pi$) the anisotropy axis. Thus, at $T \rightarrow 0$ one expects zero average bulk magnetization of the ferritin sample, even though each ferritin core particle has a net magnetic moment. Subsequently, an external magnetic field of 50 Oe was applied and the temperature was gradually increased. The applied field tips the bistability of the potential energy wells (broken line) in favor of the direction of the magnetic field, as shown in the inset of Figure 9.15, where the field is applied at an angle $\theta = \pi$ relative to the direction of the anisotropy axis. Temperature contributes thermal energy $U_{thermal} = kT$ that facilitates moment reversals between the two potential wells, over the anisotropy energy barrier $E_a = K_uV$. For $U_{thermal} \approx E_a$ moment transitions into the lower energy well on the right ($\theta = \pi$) exceed those into the higher energy on the left ($\theta = 0$), resulting in a net magnetization along the applied magnetic field direction. The induced magnetization reaches a relatively sharp maximum at ~12 K. At higher temperatures, where $U_{thermal} >> E_a$, thermal energy dominates and the system behaves as a collection of paramagnetic, giant magnetic moments; that is, it exhibits superparamagnetism. The temperature of the sharp maximum in the magnetization curve defines the blocking temperature, T_B. Due to the presence of the protein shell, the magnetic cores are magnetically isolated from each other. Each individual protein core undergoes superparamagnetic relaxation independently, as governed by the Néel relaxation of Eq. (4.45). At T > 100 K the magnetization is rapidly reduced toward zero, as thermal agitation prevents the small field of only 50 Oe to orient the moments. Upon cooling the sample to low temperature in the presence of the field, the magnetization exhibits Curie-like paramagnetism, with the magnetization continuing to increase at low temperatures.

We continue with a discussion of AC susceptibility measurements of ferritin. Figure 9.16(a) shows the results of AC susceptibility studies of horse spleen ferritin taken from the work of Kilcoyne and Cywinski (S. H. Kilcoyne and R. Cywinski, *J. Magn. Mag. Mater.* 140–144, 1466 (1995)). The data show a broad maximum, the location of which depends on the frequency of the AC applied field. At a low frequency $f = 100$ Hz, the maximum is at T ~ 15.2 K and increases to 18.2 K at 33,000 Hz. Since the characteristic measurement time is the inverse of the frequency of the AC field, the maximum, which determines the blocking temperature, changes as expected. We can rewrite Eq. (4.46) as

$$\frac{1}{f} = \tau_0 \exp\left(\frac{K_u V}{kT_B}\right) = \tau_0 \exp\left(\frac{E_a}{kT_B}\right) \tag{9.8}$$

Taking the logarithm on both sides, we obtain

$$\ln\left(\frac{1}{f}\right) = \ln \tau_0 + \left(\frac{E_a}{k}\right)\left(\frac{1}{T_B}\right) \tag{9.9}$$

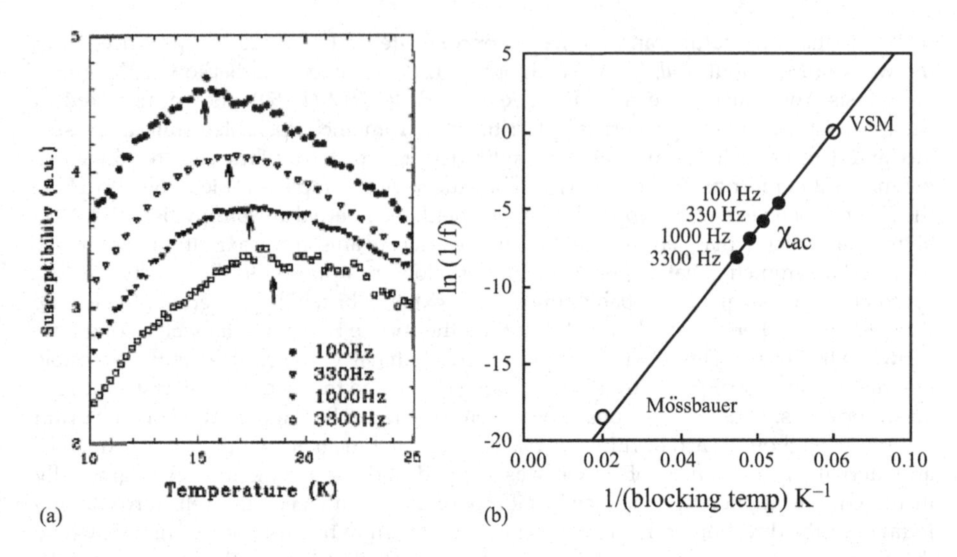

FIGURE 9.16
(a) Frequency-dependent AC susceptibility of ferritin in the vicinity of the susceptibility peak. The arrows indicate the blocking temperature, T_B. (b) The fit of the experimental data of ferritin derived from DC magnetization, AC susceptibility and Mössbauer spectroscopic measurements to Eq. (9.9). (From S. H. Kilcoyne and R. Cywinski, Ferritin: a model superparamagnet, *J. Magn. Mag. Mater.* 140–144, 1466 (1995), Copyright Elsevier, reproduced with permission).

The authors fitted this equation to the AC susceptibility data to extract estimated values for $\tau_0 = 9 \times 10^{-12}$ s and $E_a/k = 318$ K. The fitted line is shown in Figure 9.16(b), extrapolated to higher and lower frequencies, that include the characteristic measurement time of DC magnetization measurements of the order of 1–100 s, corresponding to a frequency of 1–0.01 Hz. Data from Mössbauer measurements, with $\tau_m \sim 10^{-8}$ s or $f \sim 100$ MHz, are also indicated. We introduce Mössbauer spectroscopy and discuss its applications in studies of nanomagnetism in Section 9.3.4.

9.3 Magnetic Resonance Methods

The use of electromagnetic radiation in research for the study of the structure of matter is widespread, including the determination of the magnetic structure of nanomaterials. The resonant absorption of electromagnetic radiation by matter can yield important information about its quantum mechanical structure that determines its electronic and magnetic properties. A number of spectroscopic techniques have been developed that can address specific questions concerning the magnetism and internal spin structure of nanoparticles, nanocomposite and nanostructured materials. Magnetic resonance occurs when the energy levels of a quantized system of electronic and nuclear magnetic moments are Zeeman split by an applied magnetic field and the system absorbs energy at sharply defined frequencies due to transitions between these energy levels. The resonant moment can be that of an isolated ion spin, as in Electron Paramagnetic Resonance (EPR), or a nuclear spin, as in Nuclear Magnetic Resonance (NMR). It could be the ordered magnetization or the total magnetic moment of a ferromagnetic nanoparticle as in Ferro-Magnetic Resonance (FMR). Mössbauer spectroscopy is also a resonance technique where the corresponding energy levels arise from the hyperfine coupling of nuclear and electronic spins. Below, we discuss

(a) the fundamental principles of magnetic resonance techniques and (b) how they have uniquely advanced our understanding of the internal magnetic structure, spin wave excitations and superparamagnetic spin fluctuations in magnetic nanosystems.

9.3.1 Principal Components of EPR/FMR Spectrometers

In EPR/FMR spectroscopies, transitions between Zeeman split levels in the microwave frequency range are used to study electronic/ferromagnetic structure. The basic setup for an EPR/FMR experiment consists of a microwave resonant cavity with an electromagnet, as depicted in Figure 9.17(a). The paramagnetic/ferromagnetic sample is placed in the resonator, which is positioned between the poles of an electromagnet. The source emits microwave electromagnetic radiation, which propagates down the waveguide and into the resonant cavity. A detector is placed at the other end of the waveguide to detect the microwaves. Microwave radiation that is being reflected out of the resonator is channeled into the detector by the circulator.

As we discussed in Section 2.2, Figure 2.7, the unpaired electron-spin moments precess around an applied magnetic field with an angular frequency equal to the angular Larmor

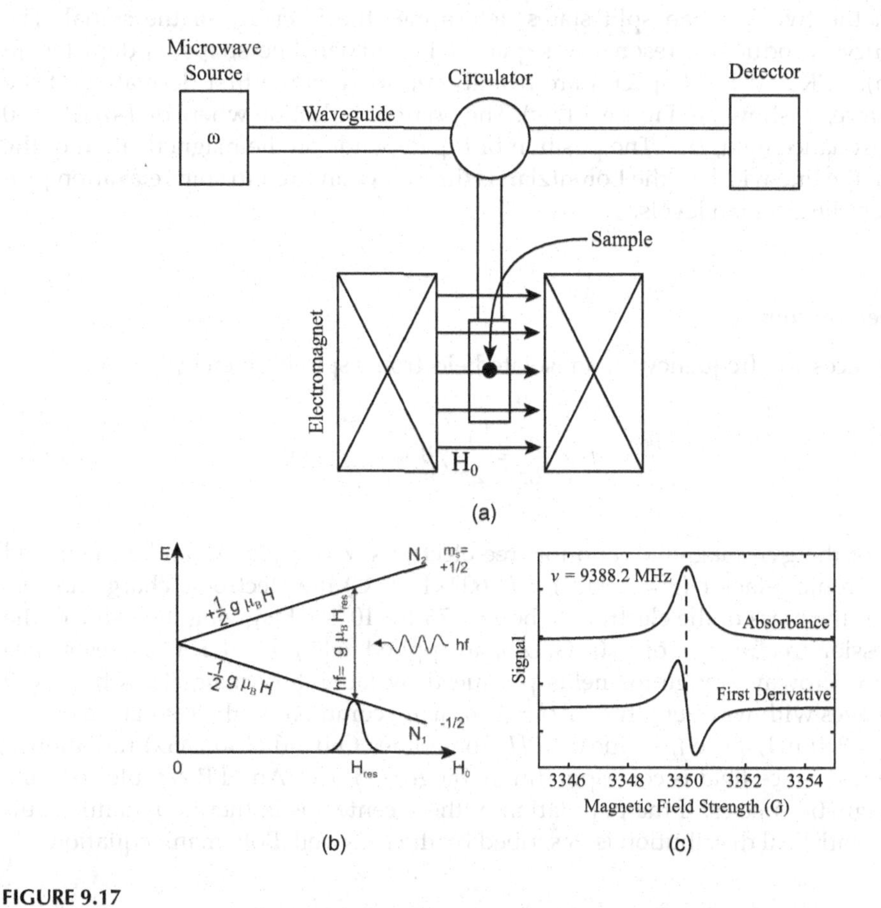

FIGURE 9.17
(a) Basic components of an EPR or FMR spectrometer. (b) Zeeman splitting and microwave absorption at resonance for an S = ½ system. N_1 and N_2 denote the population of the ground and excited spin level, respectively. (c) Definition of the resonance field H_{res}.

precession frequency, ω_L, which is proportional to the strength of the applied magnetic field, as seen from Eq. (2.14). We first discussed the concept of magnetic resonance in Example 2.7 for the case of a spin ½ paramagnet placed in a homogeneous magnetic field, while subjected to a weak alternating magnetic field acting at right angles to the static homogeneous field. When the frequency of the alternating field matches the Larmor precession frequency, the resonance condition is satisfied.

In Figure 9.17, the alternating magnetic field is provided by the microwave cavity. The homogeneous magnetic field, $\vec{H}_0$, Zeeman splits the electronic energy levels, and the microwave radiation can induce magnetic dipole transitions between these Zeeman split electronic spin levels. To search for resonances one must scan over a large range of frequencies. It is, however, difficult to vary the microwave source frequency over large ranges. For this reason, the microwave frequency is kept fixed, while instead the magnitude of the homogeneous magnetic field, H_0, is varied. When the Larmor precession frequency becomes equal to the resonant cavity frequency, the condition for resonance is satisfied, as indicated in Figure 9.17(b) for a spin ½-system. At resonance, absorption increases sharply, which is indicated by a decrease in the microwave intensity at the detector.

In Figure 9.17(b), N_1 and N_2 denote the number of paramagnetic ions in the ground (lower) and excited (upper) spin levels, respectively. The difference in electronic population between the two Zeeman split states determines the intensity of the signal. The absorption curve produces a resonance signal of Lorentzian line shape as depicted in Figure 9.17(b). EPR and FMR spectra are usually reported as the first derivative of the absorption curve, as shown in Figure 9.17(c). The point of inflection where $\mathrm{d}(abs)/\mathrm{d}H_0 = 0$ defines the resonance field, H_{res}. The position of H_{res} depends on the magnetization of the sample, while the line width of the Lorentzian is directly connected to spin relaxation processes between the Zeeman levels.

9.3.2 EPR Spectroscopy

The Larmor precession frequency for an isolated electronic spin is given by

$$f_L = \frac{\omega_L}{2\pi} = \frac{1}{2\pi}\gamma B \tag{9.10}$$

The value of the gyromagnetic ratio for free electrons, $\gamma = (ge/2m_e)$, can be calculated using the electronic g-factor ($g = 2.0023$), e (1.602×10^{-19} C) the electronic charge and m_e (9.109×10^{-31}kg) the mass of the electron, to be $\gamma = 176.1 \times 10^9$ s^{-1} T^{-1} giving the value of the Larmor precession frequency f_L of 28.02 GHz at an applied field $\mu_0 H = 1$ T. Thus, resonance occurs in the microwave range for fields produced by laboratory magnets. X-band (~9 GHz) microwaves with wavelength $\lambda = c/f = 33$ mm are common, so the resonance occurs at about $\mu_0 H = 300$ mT, since f_L is linear in H. Sometimes Q-band (~40 GHz) radiation is used and the resonance field is correspondingly bigger (~1.3 T). An EPR sample contains many paramagnetic species. If the population of these centers is in thermodynamic equilibrium, their statistical distribution is described by the Maxwell–Boltzmann equation

$$\frac{N_2}{N_1} = \exp\left(-\frac{E_{upper} - E_{lower}}{kT}\right) = \exp\left(-\frac{hf}{kT}\right) \tag{9.11}$$

where N_1 is the number of paramagnetic centers occupying the lower energy state, N_2 those occupying the upper energy state, k is the Boltzmann's constant, T is the temperature in Kelvin, h is Plank's constant and $f = \omega/2\pi$ is the microwave frequency. At room temperature, X-band microwave frequencies give $N_1/N_2 \sim 0.998$. This means that the upper energy level has a smaller population than the lower one; transitions from the lower to the higher level (absorption) are more probable than the reverse (stimulated emission), leading to a net absorption of energy. However, the degree of spin polarization at room temperature is low leading to a weak intensity signal. For this reason, most EPR experimental data are collected at lower temperatures, such as the boiling temperature of liquid nitrogen (77 K) or the boiling temperature of liquid helium (4.2 K).

Absorption of radiation is a dynamic process, which tends to equalize the spin population of the two energy levels. This tendency is counterbalanced by the natural tendency of the spin system to regain its thermal equilibrium. The temperature T is defined by the crystal lattice of the system. Thermalization involves the exchange of energy between the spins and the lattice and is known as "spin–lattice relaxation". The line width of the absorption line is inversely proportional to the spin–lattice relaxation time, T_1. The relaxation time gives a measure of the average time the spin spends pointing in a certain direction before it reverses to the opposite direction. If T_1 is very short, the absorption line becomes broad, whereas if T_1 is very long, the line is sharp. The order of magnitude of the relaxation time is provided by the Heisenberg uncertainty principle

$$\Delta E \Delta t \geq \frac{\hbar}{2} \tag{9.12}$$

where ΔE is the uncertainty in absorption energy of the resonant line and Δt is the spin relaxation time. So, if the full width at half maximum (FWHM) of the resonant absorption line in Figure 9.17(b) is of the order of $\Delta H_{res} = 1$ mT, as usually observed, then $\Delta E = g\mu_B \Delta H_{res} \sim 2 \times 10^{-26}$ J, giving an estimate for Δt or equivalently for

$$T_1 = 5 \times 10^{-9}\,\text{s} \tag{9.13}$$

Spin–lattice relaxation is a dissipative process where magnetic energy from the spin reservoir is transferred to the phonon reservoir of the lattice. There is a second non-dissipative relaxation process for magnetically interacting systems, "spin–spin relaxation" with relaxation time denoted by T_2. Spin–spin relaxation arises from dipolar and exchange interactions between spins in proximity and leads to spin dephasing.

The "spin–orbit interaction" is the mechanism by which a spin system couples to the lattice phonons through which thermal equilibrium is attained. Good EPR spectra are obtained with ions where the orbital angular momentum is quenched or absent. These are S-state ions with half-filled shells, such as free radicals ($^2S_{1/2}$), Mn^{2+} or Fe^{3+} ($^6S_{5/2}$), Eu^{2+} and Gd^{3+} ($^8S_{7/2}$). Moreover, the resonant ions should be dilute in the crystal lattice to minimize dipolar and exchange interactions between them, which broaden the resonance line width due to spin–spin relaxation and lead to dephasing of the spins.

Nevertheless, these ions have an intimate interaction with the crystal lattice. The crystal field modifies the electronic states, as we have discussed in Section 2.9. The effect of the crystal fields is to create a "zero-field splitting" of the degenerate energy levels of the ground spin state involved in EPR for states with $S > ½$, which modifies the "effective g-factor" of the lowest energy level, and makes it anisotropic with respect to the crystal axes. To account for these effects, it is common practice to replace the Zeeman Hamiltonian

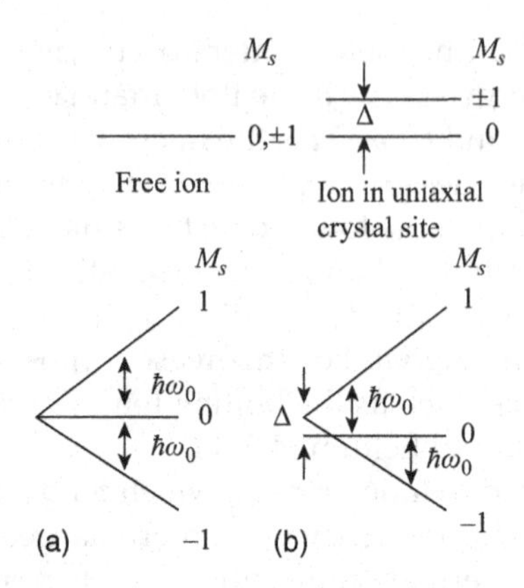

FIGURE 9.18
Zeeman energy levels in the absence (a) and presence (b) of crystal field splitting in the case of uniaxial anisotropy.

of the system with an effective spin Hamiltonian, which describes how the ground state energy levels split in a magnetic field in the presence of zero-field splitting. An effective spin S is chosen so that the magnetic degeneracy is $2S + 1$. Terms in the "Spin Hamiltonian" reflect the crystal symmetry of the resonant ion. Terms are added to the Zeeman term in order to account for the local symmetry. These terms are of the form DS_z^2 for uniaxial symmetry, $E(S_x^2 - S_y^2)$ for an orthorhombic distortion and $D_c(S_x^4 + S_y^4 + S_z^4)$ for cubic symmetry. The crystal field introduces fine structure in the resonance signal that is reflected in the above spin Hamiltonian parameterization.

For example, let us consider the case of $S = 1$ at an ion site having uniaxial symmetry, with the field $\vec{H}_0$ applied along the symmetry crystal axis. The spin Hamiltonian for this case is

$$\mathrm{H}_{\mathrm{spin}} = DS_z^2 - g_{eff}\,\mu_B H_0 S_z \left(\text{Spin Hamiltonian for uniaxial anisotropy}\right) \tag{9.14}$$

The effect of the crystal field is to introduce a fine structure in the EPR spectrum. Figure 9.18 depicts the Zeeman energy levels in the absence and presence of crystal field splitting in the case of uniaxial anisotropy. Microwave transitions must obey the dipole selection rules $\Delta M_s = \pm 1$. Under the action of the crystal field, the degeneracy of the $M_s = 0$, ± 1 azimuthal spin states is partially lifted by a crystal field splitting energy, Δ. The presence of zero-field splitting produces a double EPR resonance absorption peak rather than a single line, as depicted in Figure 9.18(b), where resonance is seen to occur at two different values of H_0 for the allowed transitions between the $M_s = 0$ and $M_s = \pm 1$ states. The magnitude of D (or Δ) is usually reported in wave numbers and it is in the range of 1–10 cm^{-1} (or 1.5–15 K).

There is yet another much weaker interaction that further modifies the splitting of the electronic ground state. This is the "hyperfine interaction" with the nucleus. Nuclei possess quantized angular momentum when their nuclear spin $I \neq 0$. The corresponding nuclear magnetic moment is given by $\mu_n = g_n \mu_N I$, where g_n is the nuclear g-factor, a number of the order of one, and μ_N is the nuclear magneton, previously defined (Eq. 2.20)

$$\mu_N = \frac{e\hbar}{2m_p} = 5.0508x10^{-27}\,Am^2 \tag{9.15}$$

Nuclear moments are three orders of magnitude smaller than electronic moments due to the larger mass of the proton, m_p, in the denominator. An applied magnetic field would separate the $2I + 1$ degenerate nuclear energy levels with $M_I = I, I-1, \ldots, -I$. However, even in the absence of an externally applied field, the unpaired electrons of a magnetic ion create a magnetic field at the nucleus, known as the "hyperfine magnetic field". This field ranges up to 50 T for 3*d* ions and it can be up to 10 times larger for some rare-earth atoms because of the 4*f* electron orbital contribution. These are huge magnetic fields, albeit, in a very small volume. Hyperfine interactions are of the order of 10^{-1} to 10^{-3} K. They give rise to hyperfine structure in EPR, NMR and Mössbauer spectra.

The interactions are represented by a term of the form in the spin Hamiltonian, which couples the nuclear with the electronic spin. The hyperfine coupling constant *A* has units of energy. EPR microwave transitions occur only between levels obeying the dipole selection rules $\Delta M_s = \pm 1$ and $\Delta M_I = 0$. The frequencies required to induce transitions between the nuclear levels lie in the radio-frequency range, MHz, rather than GHz. The EPR resonances then occur at $\hbar\omega_0 = g\mu_B H + AM_I$. Each EPR line splits into $(2I + 1)$ hyperfine absorption lines.

9.3.3 FMR Spectroscopy

EPR is widely used in the study of materials containing non-interacting paramagnetic centers. In the case of ferromagnetic materials, it is the sublattice magnetization that precesses around the direction of the local effective field within the material, $\vec{H}_{eff}$. The precession of the magnetization, $\vec{M}$, is described by the Landau–Lifshitz–Gilbert equation

$$\frac{\partial \vec{M}}{\partial t} = -\gamma\left(\vec{M} \times \vec{H}_{eff}\right) + \frac{G}{\gamma M_s^2}\left(\vec{M} \times \frac{\partial \vec{M}}{\partial t}\right) \tag{9.16}$$

The first term on the right side of Eq. (9.16) corresponds to the precession, as previously described in Example 2.2, Eq. (2.15). The second term introduces a viscous damping, with Gilbert damping constant *G*. It is this term that is responsible for magnetizing the material along the direction of the magnetic field as this term forces the magnetization to spiral down in the direction of the field, as depicted in Figure 9.19. It is the presence of "damping processes", such as spin–lattice relaxation, that allows the orientational magnetic energy of the magnetization to be transformed into lattice energy, heating of the material, and thus eventually the moment to be aligned in the direction of the field. The effective magnetic field, $\vec{H}_{eff}$, includes the static field of the electromagnet, $\vec{H}_0$, the microwave alternating

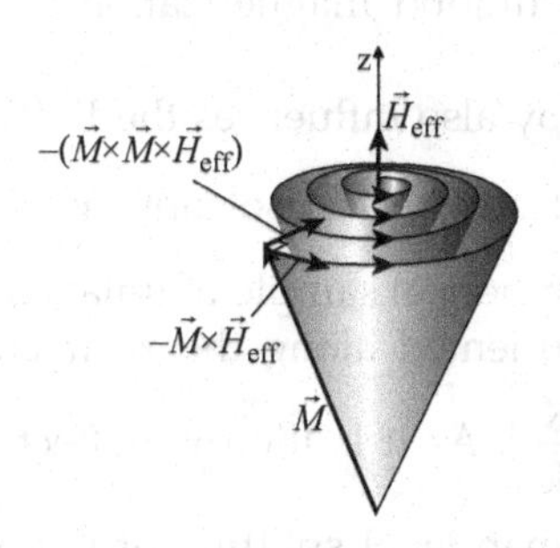

FIGURE 9.19
Precession and alignment of the magnetization in an effective magnetic field, as described by the Landau–Lifshitz–Gilbert equation.

magnetic field component, the demagnetizing field (shape anisotropy) and the magneto-crystalline anisotropy field; γ is the gyro-magnetic ratio. The effective field creates a Zeeman splitting of the energy levels, and the microwave excites magnetic dipole transitions between these split levels.

For the resonant absorption of microwaves to be detected, the microwave radiation must be able to penetrate the material. As we have discussed the three room-temperature elemental ferromagnets, Ni, Fe and Co, are metallic. It is well known that metals reflect electromagnetic waves, allowing penetration only over a thin layer, δ_s, below their surface, an effect known as the "skin depth" effect. The depth of penetration, δ_s, depends on the angular frequency of the electromagnetic radiation, ω, as well as, the conductivity, σ, permeability, μ, and permittivity, ε, of the conductor, according to Eq. (9.17). This poses no difficulty for insulators such as the ferrimagnetic oxides, but for metals, the skin depth of the metal for 10 GHz radiation has to be taken into account. For metallic iron at 10 GHz, the skin depth can be calculated to be of the order of a micrometer, so nanometer-sized particles can definitely be investigated.

$$\delta_s = \frac{1}{\omega}\sqrt{\frac{2}{\varepsilon\mu}}\left[\sqrt{1+\left(\frac{\sigma}{\varepsilon\omega}\right)^2}-1\right]^{-1/2} \quad \text{(Skin depth of a conductor)} \tag{9.17}$$

For ferromagnetic nanoparticles, we have to distinguish clearly between the external field, $\vec{H}_0$, and the demagnetizing field, $\vec{H}_d$, first introduced in Section 3.5.2. The demagnetizing field is given by $\vec{H}_d = -N\vec{M}$ (Eq. (3.54)), where N is the demagnetizing tensor and $\vec{M}$ is the magnetization density. Assuming that the demagnetization tensor is diagonal and the applied field is along the z-direction, the condition for resonance becomes

$$\omega_0^2 = \mu_0^2\gamma^2\left[H_o + (N_x - N_z)M\right]\left[H_0 + (N_y - N_z)M\right] \tag{9.18}$$

known as the Kittel equation for the resonance frequency. Thus, the nanoparticle shape anisotropy will be reflected in the FMR spectra. For a sphere, where $N_x = N_y = N_z = 1/3$, the resonance frequency is reduced to $\omega_0 = \gamma\mu_0 H_0$, as it should be for the case of no anisotropy. For a thin film with $\vec{H}_0$ applied perpendicular to the plane, $N_x = N_y = 0$ and $N_z = 1$. Then, Eq. (9.18) gives the resonance frequency to be $\omega_0 = \gamma\mu_0(H_0 - M)$. For a thin film with $\vec{H}_0$ applied parallel to the plane, in the x-direction, $N_y = N_z = 0$, $N_x = 1$, the resonance frequency is given by $\omega_0 = \gamma\mu_0[H_0(H_0 + M)]^{1/2}$. For the case of a generalized spheroid of revolution, the calculation is more complicated, but the position and structure of the resonance line should contain information on the saturation magnetization and the shape anisotropy of the spheroid nanoparticle.

Magnetocrystalline anisotropy also influences the FMR frequency condition, so to the demagnetizing field, one must add the anisotropy field $H_K = \dfrac{2K}{\mu_0 M_s}$ (Eq. (4.17)). In the absence of shape anisotropy, a spherical sample of uniaxial magnetocrystalline anisotropy, and with the anisotropy axis oriented along the z-direction, would show resonance at $\omega_0 = \gamma\mu_0\left(H_0 + \dfrac{2K}{\mu_0 M_s}\right) = \gamma\left(B + \dfrac{2K}{M_s}\right)$. As it is the total effective field acting on the moment that is responsible for the Zeeman level splitting, it is possible to observe FMR in zero applied field for a single-domain particle, or a single crystal of high anisotropy material magnetized along the z-direction. In an applied field, the presence of the anisotropy field

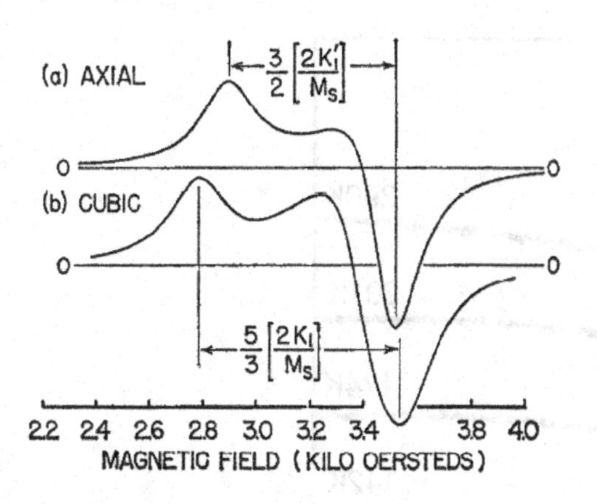

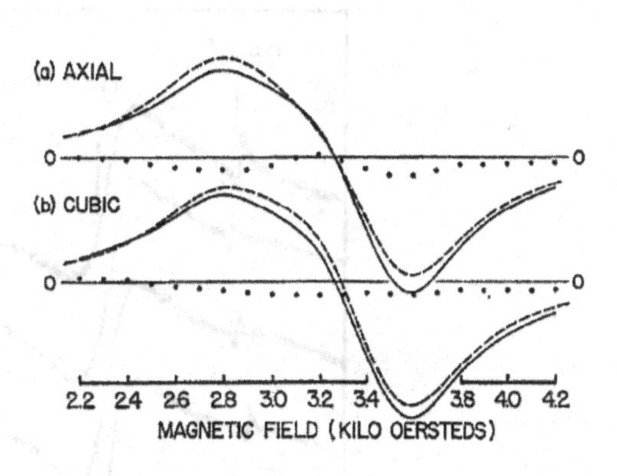

FIGURE 9.20
Computer simulations of FMR fine structure for a random distribution of fine magnetic particles in the case of sharp (left panel) and broad (right panel) resonance absorption lines with (a) axial and (b) cubic anisotropy. (From D. L. Griscom, FMR spectra of lunar fines: some implications of line shape analysis, *Geochimica et Cosmochimica Acta*, 38 (1974) 1509, Copyright Elsevier, reproduced with permission).

has the effect of shifting the resonance to higher frequencies. In an ensemble of randomly oriented uniaxial magnetic nanoparticles, however, one must average over all angles between the applied field and the anisotropy axis. Depending on their relative orientations, the anisotropy field would either add to or subtract from $\vec{H}_0$, producing line broadening, signal asymmetry and fine structure in the resonance lines.

Figure 9.20((a) and (b), left panel) gives computer-simulated FMR spectra of randomly oriented, spherical, single-domain ferromagnetic particles with (a) uniaxial magnetocrystalline anisotropy and (b) cubic magnetocrystalline anisotropy, for the case of small single-crystal line widths ($\Delta B_{res} << 2K/M_s$). Figure 9.20((a) and (b), right panel) gives the corresponding simulated spectra for the case of large Lorentzian line widths ($\Delta B_{res} >> 2K/M_s$). In this case, the fine structure is not observed due to line broadening. In real systems, the most common sources of line broadening are particle size distribution and thermal fluctuations.

Thus, the analysis of FMR spectra can provide a measurement of M_s and K, as well as γ. Detailed spectral characteristics can also yield information on magnetostatic interparticle interactions allowing characterization of assemblages of magnetic nanoparticles. Note that the instantaneous field in the sample is uniform in an FMR experiment, provided the wavelength of the microwaves is much greater than the sample size. At 10 GHz, as we discussed, $\lambda = 3.3$ cm; so, the condition is certainly satisfied for nanometer-sized samples. However, the uniform magnetization assumption throughout the sample is not generally valid. As we know, the spin structure at the surface can deviate significantly from that in the core for nanometer-sized structures.

Example 9.4: FMR Spectroscopy of HoSF

The iron biomineral core of horse spleen ferritin (HoSF) has been investigated by EPR/FMR spectroscopy over a large temperature range. Figure 9.21 shows data by E. Wanjberg *et al.*, (*J. Magn. Res.* 153 (2001) 69) taken in the temperature range from 290 K down to 19 K. The spectra contain sharp spectroscopic features, such as those at g = 4.3 and g = 9, which ride on a broad asymmetrical spectroscopic signal at g = 2.014. The sharp features correspond to EPR spectra of isolated paramagnetic ions, such as single-iron complexes and

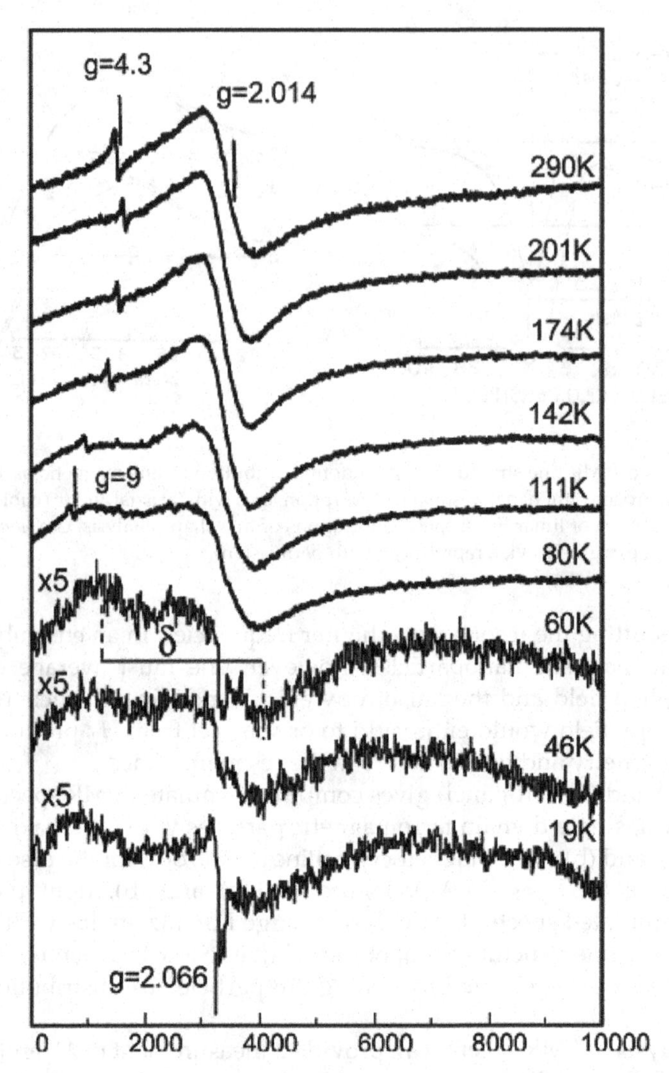

FIGURE 9.21
FMR spectra of horse spleen ferritin as a function of temperature. (From E. Wanjberg, L. J. El-Jaick, M. P. Linhares and D. M. S. Esquivel, Ferromagnetic resonance of horse spleen ferritin: Core blocking and surface ordering temperatures, *J. Magn. Res.* 153 (2001) 69, Copyright Elsevier, reproduced with permission).

oxygen impurities, while the dominant, broad asymmetrical feature is the FMR signal of the ferritin core.

The core absorption signal, visible as a broad signal centered at g = 2.014, becomes increasingly broadened and anisotropic as the temperature is decreased. The anisotropic structure in the spectrum is the effect of the anisotropy field which becomes measurable below the blocking temperature of a particle. The value of δ, peak-to-peak distance in the derivative spectra, and its temperature dependence can be analyzed to obtain the anisotropy field of ferritin, while the intensities of the asymmetric (blocked) and the symmetric (superparamagnetic) absorption as a function of temperature can be used to obtain the blocking temperature. Analysis of the spectra yielded the values of $2K/M = 2.7 \times 10^3$ Oe, $MV = 1.9 \times 10^{-17}$ emu and $T_B = 116$ K.

9.3.4 Mössbauer Spectroscopy

Mössbauer spectroscopy is a nuclear physics technique widely used in the magnetic characterization of iron-containing nanosystems. It makes use of the ^{57}Fe nucleus to probe the internal magnetic field at the iron site through the magnetic hyperfine interactions of the nuclear spin with the unpaired electronic spins. These are the same interactions responsible, in reciprocity, for the hyperfine structure of EPR spectra, in which the nuclear spin perturbs the electronic spin states, as briefly discussed in Section 9.3.2. Mössbauer spectroscopy is based on the phenomenon of the "recoilless nuclear emission and resonant absorption of γ-radiation". The law of conservation of linear momentum requires that for one nucleus to emit a γ-ray and a second nucleus to resonantly absorb it, both nuclei must be embedded in solids, a phenomenon known as the "Mössbauer effect". In addition to the magnetic, Mössbauer spectroscopy gives quantitative information on hyperfine interactions between the nuclear and electronic charge distributions. These are small energies that arise from the electrostatic interaction between the nucleus and its neighboring electrons, in which the electrons around a nucleus perturb the energies of nuclear states. There are three important hyperfine interactions measured by Mössbauer spectroscopy. The first originates from the electron density at the nucleus and gives rise to the "isomer shift", the second is due to the gradient of the electric field at the site of the nucleus and is responsible for the observation of the "nuclear quadrupole splitting", while the third arises from the unpaired electron density at the nucleus and determines the "hyperfine magnetic field" experienced by the ^{57}Fe nucleus. Through these three hyperfine interactions, Mössbauer spectrometry provides unique measurements of electronic, magnetic, and structural properties of materials, including nanomaterials.

A Mössbauer spectrometer consists of simple instrumentation plus a radioactive source that provides the γ-radiation. Figure 9.22 shows the decay scheme of ^{57}Co to its daughter nuclide ^{57}Fe, which is produced at an excited state with a nuclear spin $I = 5/2$ and energy 136 keV above its ground state. Subsequently, the nucleus reaches its ground state either through a direct transition (9%) or by cascade through an intermediary level with an energy of 14.4 keV (91%). Thus, the excited ^{57}Fe nuclei will emit a 14.4 keV gamma ray *via*

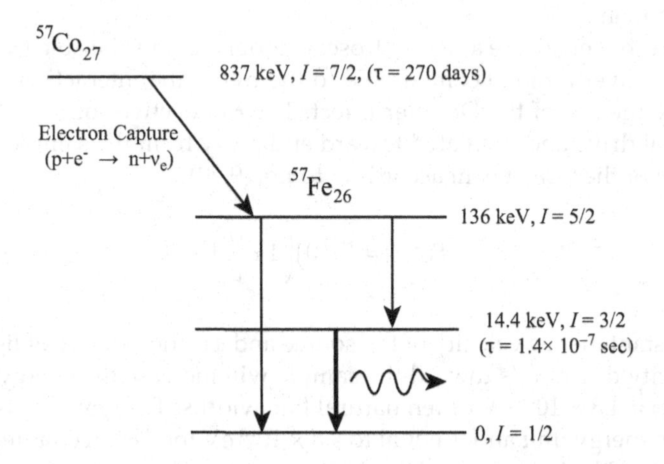

FIGURE 9.22
Energy level scheme of ^{57}Fe derived from the decay of ^{57}Co. Mössbauer spectroscopy involves the 14.4 keV transition.

a magnetic dipole transition from the metastable $I = 3/2$ state to the ground state with $I = 1/2$. The ratio of recoil-free 14.4 keV photons to all the 14.4 keV photons emitted is f, the "recoil-free fraction" of the source. It is this fraction of γ-rays that are used in Mössbauer spectroscopy. The recoil-free fraction is given by

$$f = e^{\left[-E_\gamma^2\langle x^2\rangle/(\hbar c)^2\right]} \tag{9.19}$$

In the above equation, $E\gamma$ is the energy of the 14.4 keV state, $<x^2>$ is the average of the square of the vibrational amplitude of the iron nucleus at the crystallographic lattice of the solid, $\hbar$ is Plank's constant divided by 2π and c is the speed of light. The recoil-free fraction is the statement of the conservation of linear momentum in the quantum statistical processes involved in the emission and absorption of nuclear radiation by nuclei bound to a solid. The Mössbauer effect is simply due to the fact that zero-phonon events have a finite probability to occur. The size of the effect increases at low temperatures where $<x^2>$ decreases. Measurements between 4.2 K and 300 K are common.

The line width of the emitted radiation is limited by the uncertainty principle of Eq. (9.12) with Δt equal to the mean lifetime of the 14.4 keV excited nuclear state of $\tau = 1.4 \times 10^{-7}$ seconds. This is a rather long-lived state, compared to atomic, electronic excited states of mean lifetimes of the order of 10^{-12} seconds, and is, therefore, referred to as a metastable state. This results in the emission of a very "sharp" γ-ray, of well-defined energy. The γ-ray energy distribution is of Lorentzian shape with a natural line width (FWHM) $\Gamma_{nat} = 4.7 \times 10^{-9}$ eV $= 7.52 \times 10^{-28}$ J. Hyperfine interactions cause very small perturbations in the energies of Mössbauer γ rays, of the order of 10^{-9} to 10^{-7} eV, where 1eV $= 1.6 \times 10^{-19}$ J $= 10^4$ K. Compare this to the γ ray itself, which has an energy of 10^5 eV. Thus, the 14.4 keV γ-ray has a fantastic fractional resolution of $\Gamma_{nat}/E\gamma = 10^{-9}/10^5 = 10^{-14}$, the highest ever observed. This extraordinary resolution has allowed the detection of the gravitational frequency shift of a 14.4 keV photon falling on the surface of Earth, in a celebrated experiment by Pound and Rebka at Harvard University in 1959, (R. V. Pound and G. A. Rebka Jr., Gravitational Red Shift in Nuclear Resonance, *Phys. Rev. Letts*. 3(9) 439-441 (1959). It is this extraordinarily high resolution that also makes it possible for small hyperfine perturbations on the ^{57}Fe nuclear energy states to be measured easily, and with high accuracy, using low-cost Mössbauer equipment.

To use the Mössbauer source as a spectroscopic tool we must be able to vary the energy of the γ-ray beam over a range comparable to the hyperfine interaction energies. This is accomplished by the use of the Doppler Effect. The radioactive source is mounted on an electromechanical drive and oscillated toward and away from the sample, thereby modulating the energy of the γ-ray beam according to Eq. (9.20).

$$E_\gamma(\upsilon) = E_\gamma(0)\left(1 \pm \frac{\upsilon}{c}\right) \tag{9.20}$$

where υ is the instantaneous velocity of the source and c is the velocity of light. Moving the source at a velocity of 1 mm/s toward the sample will increase the energy of the photons by 14.4 keV $(\upsilon/c) = 4.8 \times 10^{-8}$ eV or ten natural line widths. The "mm/s" is, thus, a convenient Mössbauer energy unit and is equal to 4.8×10^{-8} eV for ^{57}Fe. A counter placed behind the sample monitors the intensity of the γ-ray beam after it has interacted with the sample. Two geometries are used, the transmission and reflection (or back-scattering) geometry. Figure 9.23 depicts the geometrical arrangement of the basic components in the two

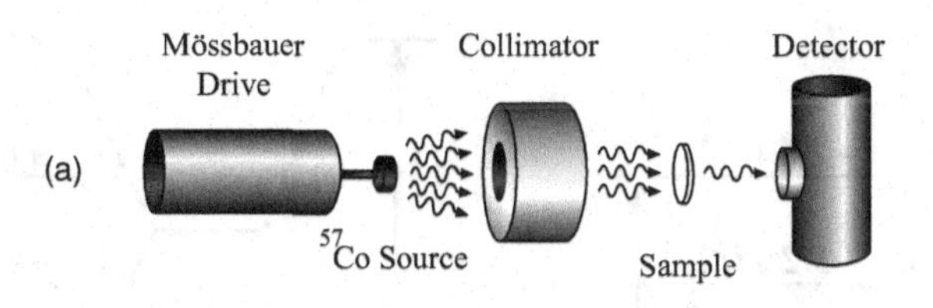

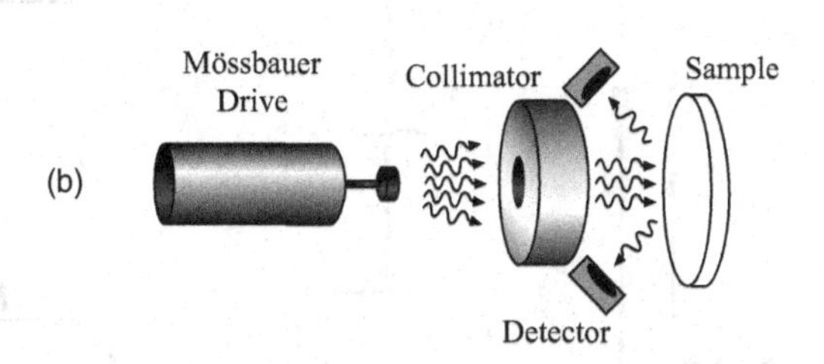

FIGURE 9.23
Experimental arrangements for the (a) transmission and (b) reflection Mössbauer spectrometer.

geometries. The transmission geometry is most generally used for the study of powder, polycrystalline and large single-crystal samples while the reflection or back-scattering geometry is best suited for the study of surfaces and thin-film samples.

Most usually, the source is driven at constant acceleration with the velocity varying linearly from zero up to a maximum value of ~±12 mm/s. In transmission Mössbauer spectroscopy, absorption spectra are reported as the intensity of the γ-ray beam *vs.* the velocity of the source. A proportional counter with associated amplification and discrimination circuitry is employed in the detection and processing of the γ-ray beam intensity. When the Doppler-shifted radiation energy matches the difference in energy between a ground and excited nuclear hyperfine energy level the γ-photon is resonantly absorbed, resulting in the measured Mössbauer absorption spectrum. The nuclear transitions responsible for the observed spectrum are depicted in Figure 9.24.

Only cases corresponding to a single type of hyperfine contribution are depicted. In real systems, all three may contribute simultaneously, giving rise to more complex spectra.

The most generalized Mössbauer Hamiltonian, which includes the possibility of an applied and a hyperfine magnetic field at the nucleus is given by Eq. (9.21),

$$\mathcal{H} = -g_n\mu_N\vec{I}\cdot\vec{B}_{appl} - g_n\mu_N\vec{I}\cdot\vec{B}_{hf} - eQV_{zz}\frac{\left[3I_z^2 - I(I+1)\right] + \eta\left(I_x^2 - I_y^2\right)}{4I(2I-1)} \tag{9.21}$$

In the above expression, the first two terms pertain to both the ground (I = ½) and the excited (I = 3/2) nuclear energy levels while the third term applies only to the excited energy level, which has a non-vanishing quadrupole moment, Q; g_n is the nuclear g-factor, V_{zz}, is the principal component of the electric field gradient assumed diagonal, η is the asymmetry parameter of the electric field gradient ($\eta = \frac{V_{yy} - V_{xx}}{V_{zz}}$), and e is the electronic charge. The major contribution to the magnetic hyperfine field comes from the "Fermi-contact interaction" of the nucleus with the unpaired electron density at its site. Unpaired

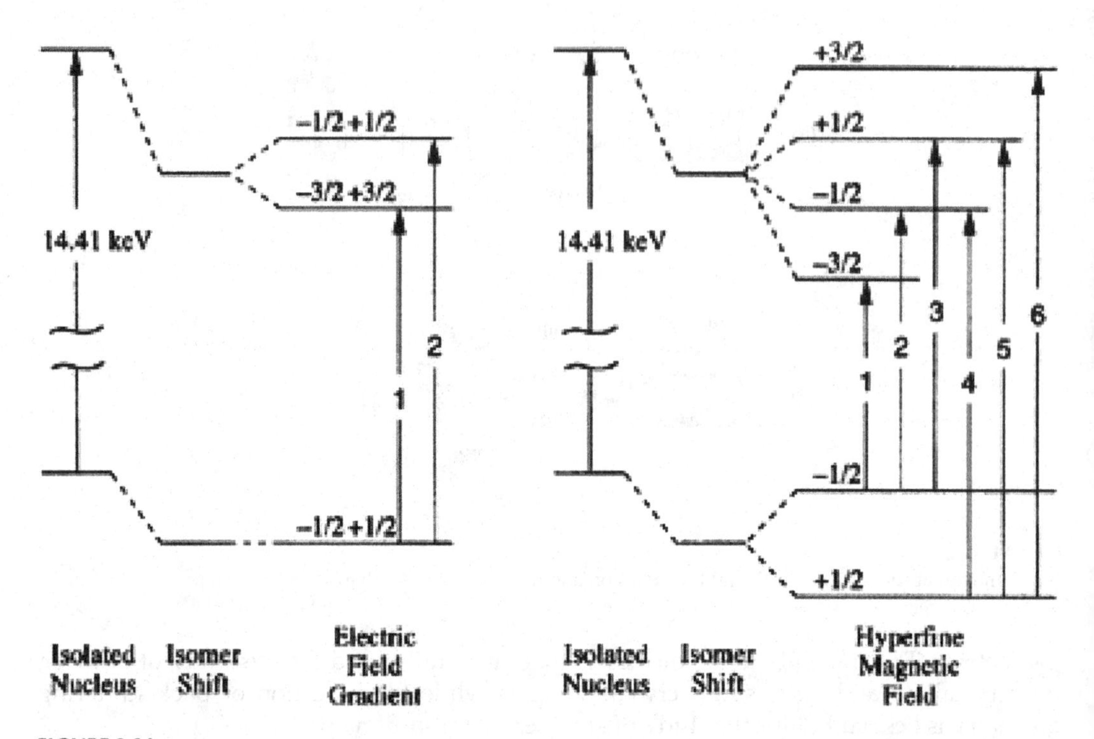

FIGURE 9.24
Nuclear transitions giving rise to Mössbauer spectra. Energy level diagrams for ^{57}Fe in the presence of (a) an isomer shift resulting in a single line absorption spectrum, (b) an electric field gradient giving rise to a quadrupole doublet, and (c) a hyperfine magnetic field giving rise to a six-line magnetic spectrum. Transitions corresponding to $\Delta m_I = 0, \pm 1$ are allowed.

electrons are in 3*d* shells, which have no electron density at the nucleus, but they polarize the 1*s*, 2*s* and 3*s* core electron shells which do have some charge density there. The core polarization contribution is about ~11 T/μ_B in iron. A further contribution comes from 4*s* conduction electrons. For non-S state ions, there are also orbital and dipolar contributions produced by unquenched orbital angular momentum and non-spherical atomic spin distribution. The magnetic hyperfine field in magnetically ordered solids faithfully follows the ordered moment, and it falls to zero at the Curie or Néel temperature.

Applied fields can only be a small perturbation to the hyperfine field, as they are an order of magnitude smaller than the hyperfine magnetic fields in conventional Mössbauer experimental setups using superconducting solenoids. However, important additional information can be obtained due to induced spin polarizations and modification of transition probabilities between energy levels, affecting the relative intensities of spectral absorption lines. In a ferromagnetic system, the applied field subtracts from the hyperfine field due to the strong negative fields produced by the Fermi-contact interaction.

A thin film of bcc Fe (α-Fe) metal foil enriched in ^{57}Fe (only 2.2% of natural iron is ^{57}Fe) is used to calibrate the Mössbauer spectrometer. Figure 9.25 shows the Mössbauer spectrum of a 6 μm metal iron foil. Data were acquired at 300 K in transmission geometry with a constant acceleration spectrometer. The points are the experimental data. The solid line is a fit to the data for six independent Lorentzian absorption lines with unconstrained centers, widths and depths. A ^{57}Co source in Rh matrix was used, but the zero of the velocity scale is the centroid of the α-Fe spectrum itself. Isomer shifts are, therefore, reported

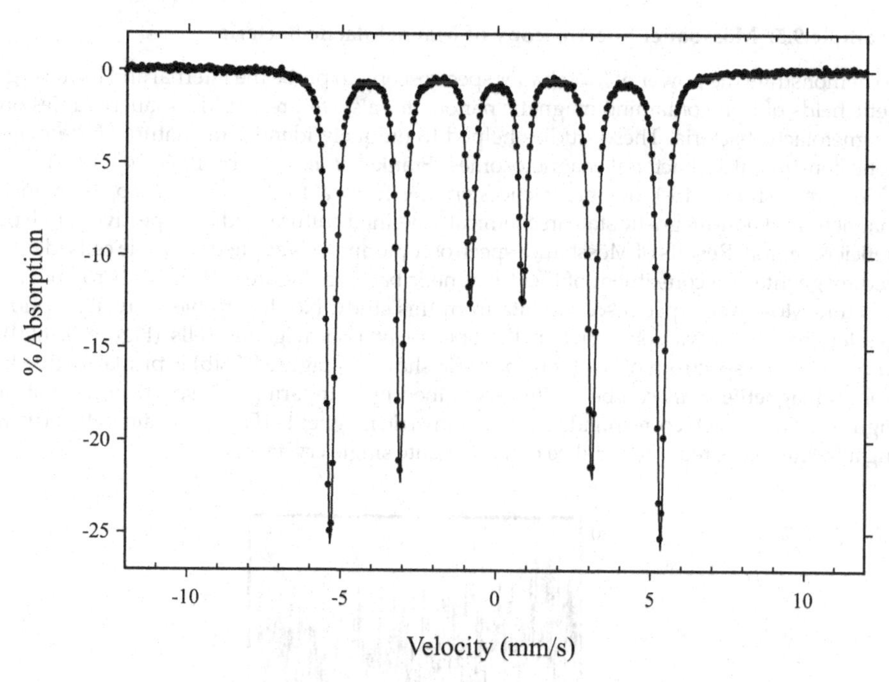

FIGURE 9.25
Calibration spectrum of 6 μm metal iron foil at room temperature.

relative to the centroid of this spectrum. The separation between peaks 1 and 6 is 10.62 mm/s. This splitting corresponds to 330 kOe, the internal field of metallic iron. It is used as a calibration standard (31.073 kOe/mm/s).

The line positions are related to the splitting of the nuclear energy levels in Figure 9.24, but the relative intensities, measured by the area under the absorption lines, depend on the angle θ between the Mössbauer gamma-ray propagation direction and the nuclear spin moment. The outer, middle and inner line intensities are in the ratio of 3: x: 1: 1: x: 3, where $x = 4\sin^2\theta/(1+\cos^2\theta)$. That is, the outer and inner lines are always of the same relative intensity but the middle lines can vary in relative intensity between $x = 0$ (for $\theta = 0°$) and $x = 4$ (for $\theta = 90°$). For the case of polycrystalline samples with no externally applied magnetic field, one must consider the average of $\sin^2\theta$ over all possible directions of the nuclear spin moment in the full solid angle according to

$$\left\langle \sin^2\theta \right\rangle = \frac{1}{4\pi}\int_0^{2\pi}\int_0^{\pi}\left(\sin^2\theta\right)\sin\theta d\theta d\varphi \tag{9.22}$$

and similarly for $\cos^2\theta$ in order to determine the value of x. As you can easily determine for yourself using Eq. (9.22), the most usual case of powder samples would correspond to the average value of $x = 2$ and magnetic Mössbauer spectra with relative line intensities 3 : 2: 1 : 1 : 2 : 3. However, in single crystals or under applied fields, the relative line intensities can give important information about preferential moment orientation and magnetic ordering (see Example 9.6).

Example 9.5: Mössbauer Spectroscopy of Magnetotactic Bacteria

To demonstrate the power of Mössbauer spectroscopy to probe the internal and local magnetic fields of iron-containing magnetic nanomaterials, we present Mössbauer studies on magnetotactic bacteria. These studies helped to uniquely identify the nature of the magnetic constituent in bacterial magnetosomes. Frankel *et al.* (*Science*, 1979. 203: pp. 1355–1356) grew strain MS-1 magnetotactic *spirillum* from a freshwater swamp, both in a magnetic and non-magnetic state in chemically defined culture media, respectively, rich or deficient in iron. Results of Mössbauer spectroscopic analysis applied to whole cells identified magnetite as a constituent of these magnetic bacteria. Figure 9.26 presents room temperature Mössbauer spectroscopic data from this study. No discernible γ-ray absorption greater than 0.2% was observed in the spectra of non-magnetic cells (Figure 9.26(a)). In contrast, the spectrum of magnetic bacteria shown in Figure 9.26(b) is primarily due to iron in magnetite. This can be easily ascertained by comparing this spectrum to that of Figure 9.26(c), which corresponds to stoichiometric magnetite (Fe_3O_4), obtained by crushing a freshly prepared interior slice of a magnetite single crystal.

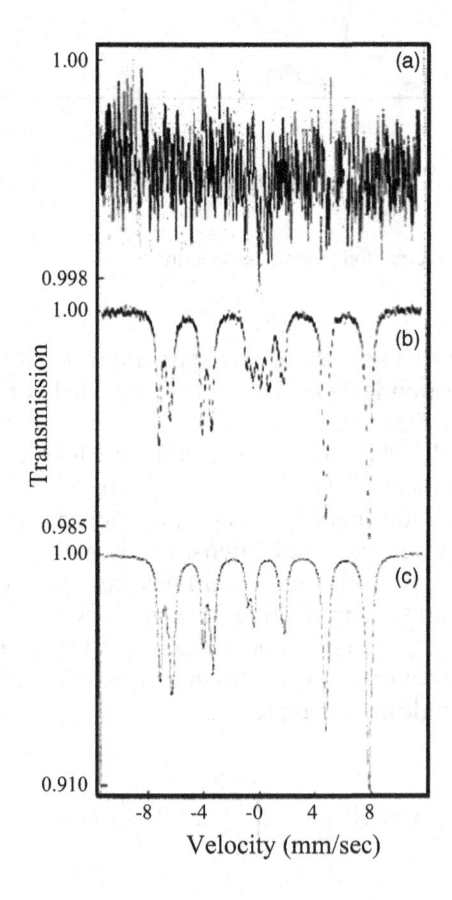

FIGURE 9.26
Mössbauer spectrum at room temperature of (a) 350 mg of freeze-dried non-magnetic cells of strain MS-1 grown in chemically defined medium deficient in iron, (b) 350 mg of freeze-dried magnetic cells of strain MS-1 grown in chemically defined medium containing iron and (c) stoichiometric magnetite (Fe_3O_4), obtained by crushing a freshly prepared interior slice of a magnetite single crystal. Solid lines through the spectra tracings are theoretical least-squares fits to the data, based on Lorentzian line shapes. (From R. B. Frankel, R. P. Blakemore and R. S. Wolfe, *Science*, 203 (1979) 1355, Copyright American Association for the Advancement of Science, reproduced with permission).

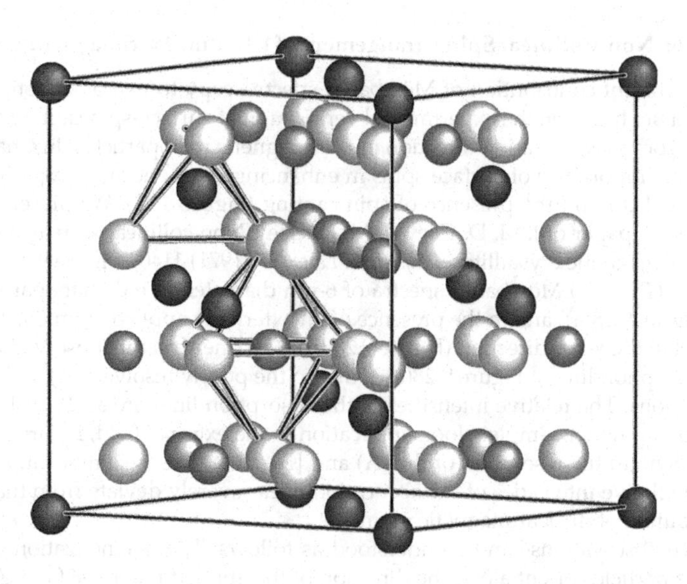

FIGURE 9.27
Crystallographic structure of magnetite. White color balls depict oxygen ions, black color balls depict iron ions in tetrahedral (A) coordination sites and gray color balls depict iron ions in octahedral [B] coordination sites.

Magnetite contains both ferric ($3d^5$, Fe^{3+}, $S = 5/2$) and ferrous ($3d^6$, Fe^{2+}, $S = 2$) iron ions. In the bulk, magnetite has an inverse spinel crystallographic structure of antiparallel spin sublattices $(A\uparrow)[B\downarrow][B\downarrow]O_4$, where (A) represents tetrahedral and [B] octahedral iron coordination to oxygen ions, as depicted in Figure 9.27, and the vertical arrows depict the relative iron ion spin orientation. A strong antiferromagnetic interaction between (A) and [B] sublattices results in an overall ferrimagnetic order, rendering strongly magnetized particles. The electronic structure of Fe_3O_4 can be depicted as $(Fe^{3+})[Fe^{3+}][Fe^{2+}]O_4^{2-}$. At room temperature, there is fast electronic hopping of the sixth $3d$ electron of the ferrous ion, $[Fe^{3+} \leftrightarrow Fe^{2+}]$, within the [B] sublattice, and the ^{57}Fe nucleus records an average valence of $Fe^{2.5+}$. In stoichiometric bulk magnetite, room-temperature Mössbauer spectra exhibit the superposition of two, partially resolved, magnetic sub-spectra one for Fe^{3+} in (A) sites with isomer shift $\delta_{(A)} = 0.27$ mm/s and hyperfine field $H_{hf(A)} = 491$ kOe, and the other for $Fe^{2.5+}$ in [B] sites with isomer shift $\delta_{[B]} = 0.66$ mm/s and hyperfine field $H_{hf[B]} = 453$ kOe with an intensity ratio $(Fe^{3+})/[Fe^{2.5+}] = 0.5$ (Figure 9.26(c)). The intensity is given by the area under the absorption spectrum; the ratio of intensities measures the relative abundance of the two iron species, assuming that the recoil-free fraction is similar at the tetrahedral and octahedral iron subsites.

There are two significant differences between the spectrum of the magnetotactic bacterial cells in Figure 9.26(b) and that of stoichiometric magnetite in Figure 9.26(c), as indicated by the presence of a quadrupole doublet at the center of the spectrum in Figure 9.26(b) and the different relative intensities of the two lowest velocity absorption lines. The central quadrupole is due to superparamagnetic magnetite due to smaller, nascent magnetosomes. The local magnetic field reverses direction as the particle's magnetization fluctuates between opposite directions of the easy axis of magnetization at a frequency larger than the Larmor precession frequency of the nuclear spin in the internal magnetic field within the material ($\tau_s << \tau_m$); the ^{57}Fe nucleus registers an average magnetic field of zero resulting in a non-magnetic, quadrupolar spectrum. The difference in (A)/[B] subsite intensity ratio can arise from the introduction of vacancies in the [B] sublattice or from oxidation of Fe_3O_4 to γ-Fe_2O_3 on the surface of the particle. Maghemite, γ-Fe_2O_3, has the same crystallographic structure as magnetite, but contains only high-spin ferric ions ($3d^5$, Fe^{3+}, $S = 5/2$).

Example 9.6: Non-Collinear Spin Arrangement of Ultrafine Ferrimagnetic Crystallites.

Another important contribution of Mössbauer spectroscopy to micromagnetic studies in nanomagnetism has been the experimental verification of surface-spin canting away from the particle core magnetization direction in ferrimagnetic nanoparticles. In Chapter 4, we discussed the importance of surface spins in enhancing magnetic anisotropy in nanoparticles (Figure 4.20) and the presence of spin canting (Figure 4.22). We present here data from the classic paper of J. M. D. Coey (J. M. D. Coey, Non-collinear spin arrangement in ultrafine ferrimagnetic crystallites, *Phys. Rev. Letts*, 27 (1971) 1140). Figure 9.28 shows low temperature (T = 5 K) Mössbauer spectra of 6-nm diameter γ-Fe_2O_3 nanoparticles in the absence (Figure 9.28(a)) and in the presence of an externally applied magnetic field (B_{app} = 5 T) parallel to the γ-ray direction (Figure 9.28(b)). The fine structure observed in the high velocity absorption line of Figure 9.28(a) is due to the poorly resolved (A) and [B] sublattices of Fe^{3+} ions. The relative intensities of the absorption lines are as 3 : 2 : 1 : 1 : 2 :3 as expected for a powder sample. Upon application of the external field, Figure 9.28(b), two things happen: (a) the resolution of the (A) and [B] sublattices is significantly improved, and (b) the relative intensities of the absorption lines severely deviate from that of a random distribution of nuclear magnetic moments.

The above observations can be understood as follows: The magnetization or "macrospin" of the particles orient along the direction of the applied magnetic field. As a result, the antiferromagnetically coupled (A) and [B] sublattices orient, respectively, along and opposite to the applied field direction, as indicated in Figure 9.28(b). The applied field of 5 T adds to the internal field of the (A) sublattice, and subtracts from that of the [B] sublattice, drastically increasing the resolution of the two sublattices. If all iron moments were oriented along or opposite to the applied field, the intensity of the Δm = 0 lines (lines 2 and 5) should vanish. This is indeed observed in the case of bulk γ-Fe_2O_3. The persistence of Δm = 0 absorption line intensity, albeit drastically diminished, in the nanoparticles indicates the presence of canted spins away from the sublattice magnetization direction. These spins are associated with the surface of the nanoparticles, where the long-range magnetic order is abruptly interrupted. While in the bulk, the surface-to-volume ratio is much too small for surface spins to be detected, the large surface-to-volume ratio in nanosystems

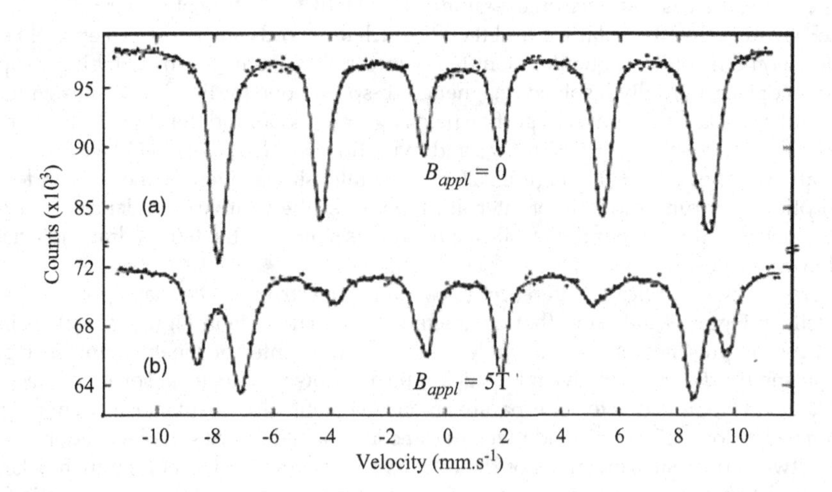

FIGURE 9.28
Mössbauer spectra of ultra-fine (6 nm) crystallites of γ-Fe_2O_3 at T = 5 K (a) without and (b) with an applied field of 5 T parallel to the γ-ray direction. The solid lines are least-square fits of the experimental data to twelve Lorentzian lines as shown. (From J. M. D. Coey, Non-collinear spin arrangement in ultrafine ferrimagnetic crystallites, *Phys. Rev. Letts*, 27 (1971) 1140, Copyright American Physical Society, reproduced with permission).

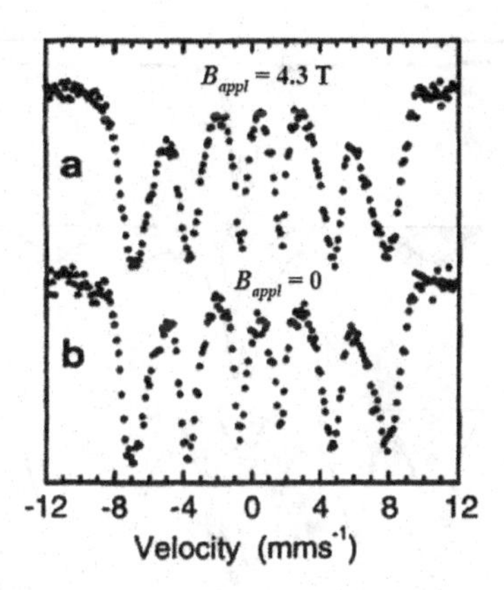

FIGURE 9.29
Mössbauer spectra of 4-nm iron particles after exposure to O_2 gas flow for 26 hours. The spectra were obtained at 5 K in a magnetic field of 4.3 T applied along the γ-ray propagation direction (a) and in zero applied field (b). (From F. Bødker, S. Mørup and S. Linderoth, Surface Effects in Metallic Nanoparticles, *Phys. Rev. Letts*, 72, (1994), Copyright American Physical Society, reproduced with permission).

allows surface spins not only to be detected, but also to play a significant role in the overall magnetic behavior of the nanoparticles, as we have extensively discussed in Chapter 4.

Figure 9.29, reproduced from another classic paper by Bødker, Mørup and Linderoth (F. Bødker, S. Mørup and S. Linderoth, Surface Effects in Metallic Nanoparticles, *Phys. Rev. Letts*, 72, (1994)) informs on similar studies of iron oxide nanoparticles in zero and in applied magnetic fields. In this study, the investigators prepared 4-nm-diameter metallic nanoparticles on carbon supports, which they studied *in situ* before and after controlled oxidation by oxygen gas. After exposure to an O_2-gas flow for 26 hours, all the iron was oxidized. The figure shows the Mössbauer spectra of the oxidized sample at T = 5 K in a magnetic field of B_{appl} = 4.3 T applied parallel to the gamma-ray direction (Figure 9.29(a)) and that in zero field (Figure 9.29(b)).

Surprisingly, in the 4.3 T spectrum, the Δm = 0, lines 2 and 5, have the same relative intensity as in the zero-field spectrum. The only effect of the applied field is a broadening of the lines. The widths of lines 1 and 6 increase by about 0.35 mm/s; otherwise, the spectra remain unaffected. This indicates the presence of extensive spin canting throughout the nanoparticle. Thus, a further reduction of particle size from 6 nm diameter to 4 nm diameter drastically affects the internal spin structure. How can one understand this?

According to Table 4.3, the characteristic exchange length scale, l_{ex}, for magnetite is estimated to be 4.9 nm. Maghemite, sharing the same crystallographic structure as magnetite, is expected to share the same exchange length scale. Thus, we can understand the drastically different behavior of the 6-nm *vs.* the 4-nm γ-Fe_2O_3 nanoparticles by the presence of non-collinear magnetic structure in the latter, presumably due to extensive spin canting permeating throughout the particle volume, as the diameter passes from above to below the characteristic exchange length scale for maghemite.

Example 9.7: Mössbauer Spectroscopy of Ferritin

To demonstrate dynamic magnetic characterization of magnetic nanoparticles we present Mössbauer studies of horse spleen ferritin carried out by the author (G. C. Papaefthymiou, Particle Nanomagnetism, *Nano Today*, 4, 438 (2009)). Figure 9.30 shows the full spectral profile of reconstituted horse spleen ferritin to 1,500 ^{57}Fe/protein in the temperature range from 4.2 to 80 K. As temperature increases above 4.2 K, the spectra first broaden due to the

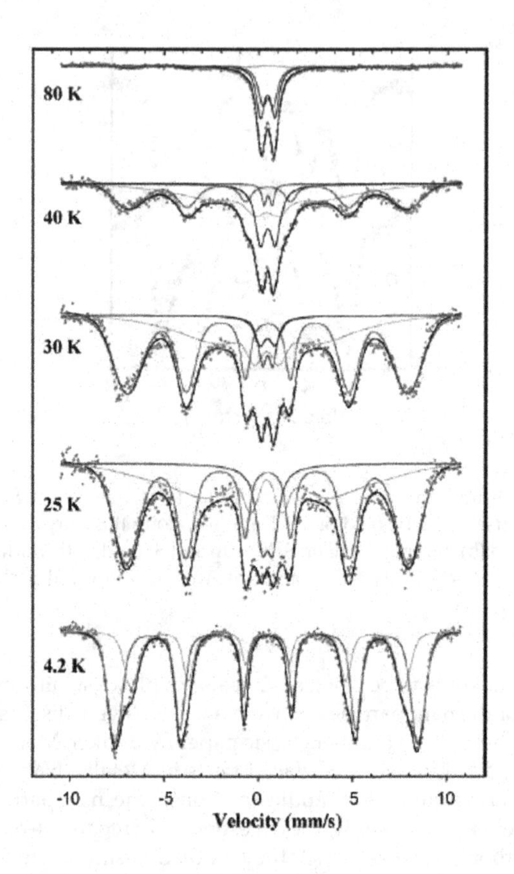

FIGURE 9.30
Mössbauer spectra of lyophilized, *in vitro* reconstituted HoSF (1,500 ^{57}Fe/protein) at various temperatures. Solid lines (black) through the experimental points are least-square fits to a superposition of iron subsites: purple, interior core sites; green, surface sites; red and blue, superparamagnetic sites. (From G. C. Papaefthymiou, Nanoparticle Magnetism, *Nano Today*, 4, 438 (2009), Copyright Elsevier, reproduced with permission). A color version of this image is available at https://www.routledge.com/9781439818466

onset of dynamic spin relaxation processes, prior to entering the superparamagnetic regime with full magnetization reversals at thermal energies above the superparamagnetic energy barrier K_uV (See Eq. (4.45) and Figure 4.15). Simultaneously, the magnitudes of the observed hyperfine fields, as measured by the overall splitting of the outer absorption lines, decrease prior to their collapse at the superparamagnetic blocking temperature (Eq. 4.47) and the appearance of a quadrupole doublet. The reduction in the magnitude of the magnetic hyperfine field is due to the precession of the particle's magnetization vector or "macrospin" about its anisotropy axis, at temperatures insufficient to induce spin reversals, according to the Collective Magnetic Excitations Model discussed in Section 4.7, which successfully describes spin dynamics below the blocking temperature, where the magnetization vector is trapped or blocked within a single potential well in Figure 4.15. The hyperfine field follows the dynamics of the sublattice magnetization of Eq. (4.55) and is described by a similar equation of the form

$$H_{hf} = H_{hf}^{0}\left(1 - \frac{kT}{2K_uV}\right) \tag{9.23}$$

where H_{hf}^{0} is the saturation hyperfine magnetic field at $T \to 0$, K_u is the uniaxial magnetic anisotropy density, V the volume of the particle, and k is Boltzmann's constant.

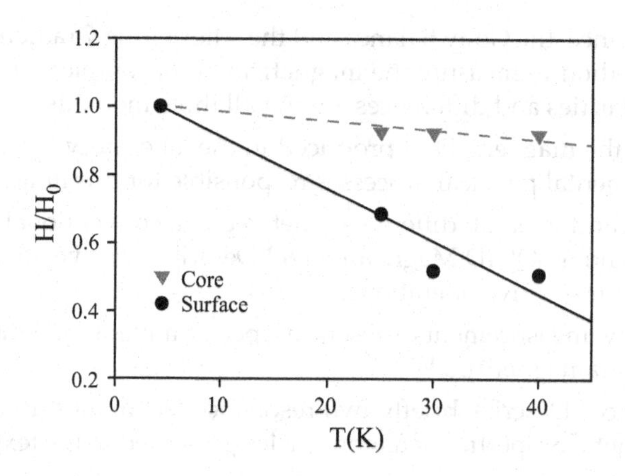

FIGURE 9.31
Temperature dependence of the reduced hyperfine magnetic fields at the interior, core (▾) and surface (●) iron sites. The steeper reduction of surface fields is indicative of a more complex potential energy landscape at the surface. (From G. C. Papaefthymiou, Nanoparticle Magnetism, *Nano Today*, 4, 438 (2009), Copyright Elsevier, reproduced with permission).

The ferritin core contains antiferromagnetically coupled octahedral Fe^{3+} sites in the form of ferrihydrite, as extensively discussed in Section 7.4.2. Unlike magnetite, ferrihydrite contains only octahedral iron coordination sites. The fine structure observed on individual Lorentzian absorption lines, best resolved at 4.2 K in Figure 9.30 (bottom spectrum) cannot, therefore, be associated with tetrahedral *vs.* octahedral sites. It is associated with core *vs.* surface iron sites. The core iron sites produce larger hyperfine fields at 4.2 K and smaller quadrupole splitting at 80 K, while surface atoms that experience a lower degree of coordination compared to the core, exhibit smaller hyperfine fields at 4.2 K and larger quadrupole splitting at 80 K. At intermediate temperatures, the spectra undergo dynamic spin fluctuations, giving broadened, featureless absorption spectra. The area under the absorption lines corresponding to surface *vs.* core iron sites shows characteristically different temperature dependence as shown in Figure 9.31.

For the interior iron sites of the core of the particle, the temperature dependence of the reduced hyperfine magnetic fields is consistent with the Collective Magnetic Excitations model of Eq. (9.23). A maximum of 15% diminution in the hyperfine field is expected through this process before the spectrum collapses to a paramagnetic doublet due to superparamagnetism. The precipitous collapse of the reduced hyperfine field at surface sites indicates the presence of additional surface spin excitation modes, such as spin-wave excitations, where surface anisotropies introduce spin canting and a greater complexity in the potential energy landscape at the surface.

Exercises

1 Gadolinium is a strongly paramagnetic metal with susceptibility $\chi = 4.8 \times 10^{-1}$. Determine the magnetic force exerted on a spherical specimen of 4-mm diameter, when placed in a Faraday Balance. Assume that the magnetic field strength set in the electromagnet is 1 T and the Faraday pole pieces produce a field gradient of 0.15 T/cm.

2 The Faraday Balance, the Gouy Balance and the Alternating Gradient Magnetometer all use the force method to measure the magnetization of samples. Describe in your own words, the similarities and differences among all three methods.

3 The strength of the magnetic field produced in the laboratory is limited. Give the reasons and fundamental physical processes responsible for this limitation.

4 What are the fundamental differences between a conventional Vibrating Sample Magnetometer and a SQUID Magnetometer? Describe relative advantages and disadvantages in their respective operations.

5 AC susceptibility measurements present a special advantage when it comes to the study of superparamagnetism. Why?

6 What is resonance? Describe briefly two resonance techniques used in the determination of the magnetic properties of nanoparticles presented in the text. Contrast similarities and differences.

7 What are hyperfine interactions? What is their energy range? What property of the first excited state of the ^{57}Fe nucleus gives Mössbauer spectroscopy its fantastic resolution?

8 The equation of motion of the magnetization $\vec{M}$ is described by the Landau–Lifshitz–Gilbert equation, Eq. (9.16). What is the physical meaning of each term on the right-hand side of the equation?

9 Carry out the integration indicated in Eq. (9.22) to determine the average value of $\langle \sin^2\theta \rangle$ and $\langle \cos^2\theta \rangle$ over all angles (in three dimensions) encountered in the case of randomly oriented powder samples.

10 In magnetic nanoparticles, the establishment of parallel and antiparallel magnetic sublattices breaks down as the size of the particle approaches the characteristic length scale of the material. How was this manifested by Mössbauer spectroscopy for the case of nanoscale maghemite?

Part IV

Applications across the Disciplines

The novel magnetic behavior of matter at the nanoscale is not only of basic scientific interest but also of direct utility in industry. In the late 20th century, a worldwide, well-developed semiconductor-electronics-based technology has led to a variety of devices that have changed our lives. Now, in the dawn of the 21st century, a new worldwide effort is underway—the development of nanotechnology. Nanotechnology was first foreseen by Richard Feynman in his now-famous 1959 lecture "There is Plenty of Room at the Bottom", where he predicted that the manipulation of matter at the nanoscale will have an enormous number of technical applications. This has indeed come to fruition for the case of nanomagnetism with vast applications currently being explored in science and engineering, health and medicine and other aspects of society. The giant industries of information technology, automotive and biomedical innovations are actively working with nanomagnets. Nanomagnets are poised to play an increasingly important role in modern technology, based on the new physical concepts involving nanoscale magnetic materials. Applications of nanomagnets can already be found in such diverse fields as medical imaging and drug delivery, sensors and computing, energy and environmental remediation.

Below the critical volume for domain wall formation, magnetic nanoparticles contain only a single magnetic domain giving rise to large magnetic moments per unit volume, as well as, large magnetic anisotropies. These properties are advantageous in nanotechnological applications and allow simple theoretical modeling of the processes involved in response to the application of external magnetic fields, as many applications require. There are already numerous areas of nanoelectronics, magnetic separation methodologies and magnetic resonance image enhancement, to name a few, where nanomagnetism is already being utilized profitably. Emerging new technologies relying on multifunctional properties, such as a combination of magnetic and ferroelectric properties or the possibility to operate a magnetic state by the application of an electric field at the nanoscale using multiferroic materials are under intense investigation. It is anticipated that magnetic nanocomposites and multilayered materials will become increasingly important in sensors and magnetic memory devices. Metallic nanomagnets, thin magnetic films and multilayers will continue to play a pivotal role in magneto-transport applications, such as those based on the giant-magnetoresistance effect, where their resistance is drastically changed upon the application of an external magnetic field, in sensor technologies.

Historically, some of the earliest engineering applications of nanomagnetism utilize colloidal magnetic nanoparticles suspended in hydrocarbon fluids for vacuum sealing, such as the formation of dynamic O-ring seals, bearings and dampers in various manufactured goods from lasers and computer disk drives to computed tomography scanners, cars, *etc.* In dynamic O-ring seals, a combination of permanent magnets and pole pieces create intense magnetic field gradient regions that retain the ferrofluid in the form of distinct rings around a rotating shaft in order to provide a hermetic seal. Seals capable of maintaining a vacuum of 10^{-8} Torr and a fluid-shaft/interface speed up to 10 m/s are easily attained. Early applications of magnetic fluids can also be found in the Bitter method of magnetic imaging as applied to studies of permanent magnet materials, geological specimens, quality control of magnetic recording media and the identification of defects in massive steel structures. In such applications, the ferrofluid, usually containing nanoparticles of magnetite, is spread over the surface of the object to be studied. The particles are attracted by the magnetic field gradient at regions of intense magnetic field and thus congregate at magnetic domain walls and magnetic defects. The final pattern can be imaged under an optical or electron microscope. This allows very small, less than 100 nm, magnetic features to be resolved in multidomain ferromagnets or in magnetic recording media, where the nanoparticles of the ferrofluid aggregate at magnetic transition regions. The Bitter method is named after Francis Bitter who introduced the technique in 1931, by demonstrating the imaging of the magnetic wall structure on the surface of a piece of Co and FeSi.

More recently, dispersions of magnetic nanoparticles on various supports have extensively been utilized in high technology, such as the burgeoning magnetic recording industry; while in permanent magnet research, the "exchange-spring nanocomposite magnet" concept leads in materials design for future permanent magnets in large-scale applications, such as electrical vehicles and wind turbines. Even more recently, there have been intense investigations in the use of magnetic nanoparticles for biomedical applications. These include a broad spectrum of both *ex vivo* applications in biotechnology and *in vivo* applications in biomedicine. *Ex vivo* applications include cell sorting and cell purification by magnetic separation, protein and DNA fragment separation, molecular biology investigations, enzyme immobilization and others. *In vivo* applications include both the diagnosis and treatment of disease. These applications take advantage of the fact that the saturation magnetization and coercivity at the nanoscale can be tailored for specific applications through size, shape and surface modification, while for biomedicine, the surface of magnetic nanoparticles can be rendered biocompatible and functionalized by the attachment of surfactants and cell-receptor recognition peptides, and that they can be guided or modulated by external magnetic fields.

In Part IV of this book, we discuss applications of nanomagnetism in select areas of nanotechnology and nanomedicine. As before, we do so by including illustrative examples from the literature, which reflect the evolution and current state of nanomagnetic technology in these fields. The number of applications is large and growing at a rapid pace, prohibiting complete coverage of the field. Our discussion of selective examples, however, will serve to illustrate the evolution and power of nanomagnetism in shaping today's nanotechnological advancements with profound societal impact on telecommunications, information storage, human health and disease.

10

Magnetic Recording Media

Why cannot we write the entire 24 volumes of Encyclopedia Britannica on the head of a pin?

Richard Feynman, the annual meeting of the American Physical Society, December 29, 1959, California Institute of Technology

We start with the application of nanomagnetism to magnetic recording, the rapid development of magnetic storage media during the 20th century and the continuing quest for higher-density magnetic media in the 21st century. Magnetic recording systems probably represent the fastest developing area of high technology in the world today, primarily driven by economic forces and largely enabled by the recent extraordinary developments in high-sensitivity magnetoresistive sensors, an area in which nanomagnetism has played a huge role in ushering the new generation of magnetoresistive read-heads and the dawn of "spintronics".

10.1 The Principles of Magnetic Recording

The earliest publication known on magnetic recording is credited to American engineer Oberlin Smith, who in 1888 suggested the possibility of the magnetic recording of sound using steel dust suspended on cotton threads; these dust particles were to be magnetized in accordance with the alternating current of a microphone source. In 1898, the Danish engineer Valdemar Poulsen demonstrated magnetic wire recording in his invention of the "telegraphone". In 1928, German engineer Fritz Pfleumer invented the magnetic tape. Magnetic wire and magnetic tape recording involve a magnetizable medium that moves past a stationary recording head, as shown in Figure 10.1(a). The recording head is comprised of a coil wrapped around a gapped inductor made of a soft magnetic material. A soft magnetic material has low coercivity and high magnetic saturation. Soft bulk ferrites, such as MnZn and NiZn ferrites have been used, as well as thin films of Permaloy, a metallic Fe-Ni magnetic alloy characterized by high magnetic permeability. When current passes through the coil a magnetic field appears in the gap of the inductor, as shown. The recording medium moves with velocity $\vec{v}$, relative to the head, and in close proximity to the gap. The head field magnetizes the medium according to the current in the coil and thus the time-varying electrical signal in the coil is converted into a spatially varying magnetic pattern along a track on the surface of the medium. The original sound recording devices used "analog recording", where the recording head created magnetization patterns along the wire length whose amplitude and frequency replicated the recorded sound. Some high fidelity sound recording is still performed by analog recording. However, today's

DOI: 10.1201/9781315157016-14

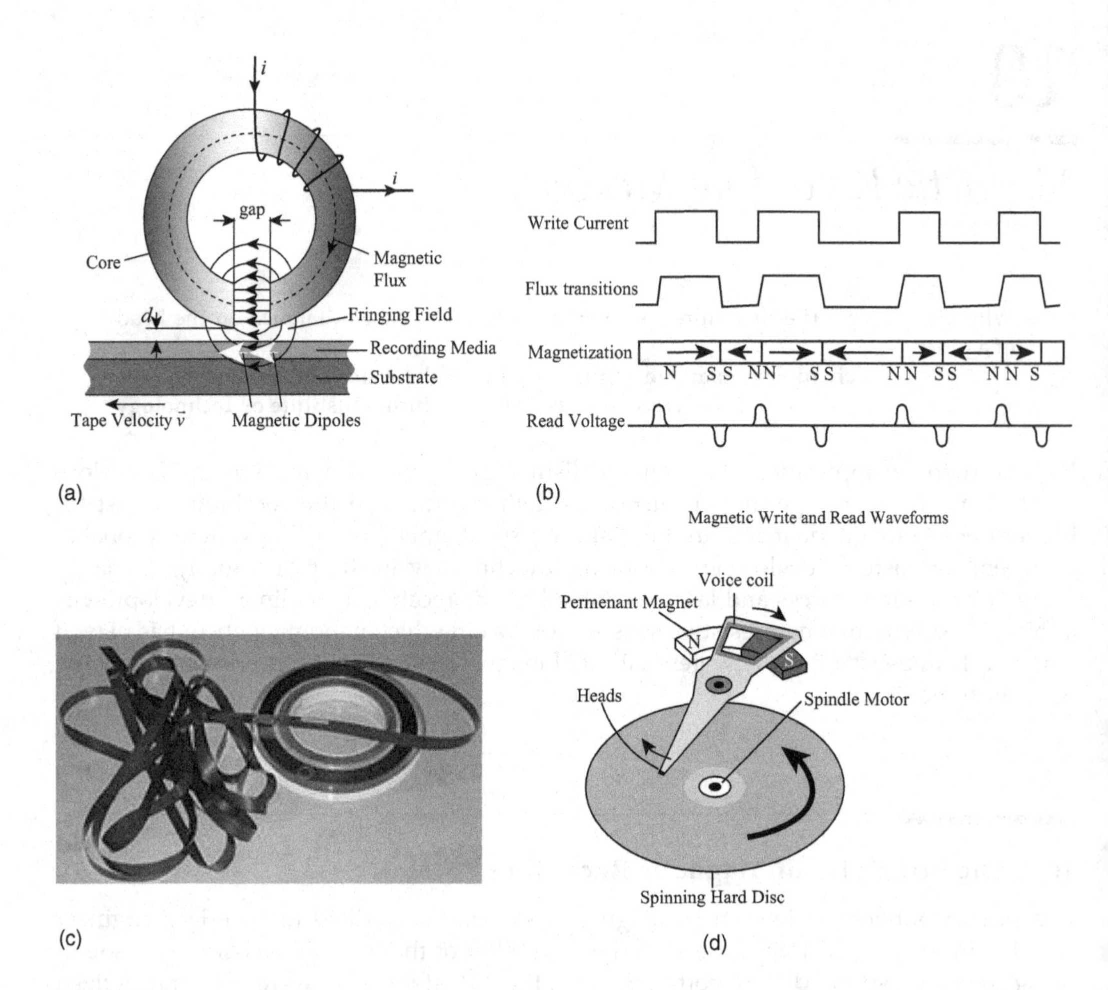

FIGURE 10.1
Principles of magnetic recording (a) depiction of magnetic tape write-/read-head arrangement, (b) schematic representation of write/read electrical signals, (c) magnetic tape and (d) hard disk geometry.

information technology storage media use exclusively digital recording, where binary coded information is recorded in bit form "0" or "1" represented on the medium by regions of opposite magnetic polarity, as indicated in Figure 10.1(b). The ring gap determines the minimum bit size of a recorded area on the disk. Thus, repeated reversals of current produce a sequence of transitions between regions of opposite magnetization on the recording medium.

In the read-back process, the fringing fields associated with magnetic reversals at the medium's surface induce electric currents in the passive coil of the inductive head according to the Faraday effect as the surface is moved with velocity, $\vec{v}$, once again past the gap of the head; the larger the magnitude of the velocity, the larger the amplitude of the induced current in the coil. The spatially varying magnetic pattern on the magnetic storage medium is thus converted back into a time-varying electrical signal, detected as an induced emf or voltage in the coil and processed by the electrical circuitry. With the exception of long-term storage of information, the flexible magnetic tape (Figure 10.1(c)) is increasingly being replaced by hard disk devices (Figure 10.1(d)), schematically depicted. The same principles apply to magnetic recording on the surface of a hard disk, except that the magnetic tracks are now circular, concentric with the disk and that the head in this case is not stationary, but can be moved radially across the disk. The head itself is attached to a slider, an aerodynamically shaped block that allows the head to maintain a consistent flying height above the disk. In turn, the slider is connected to a suspension arm that is controlled by an actuator which can move the head to any track on the disk, from the inner to the outer diameter. This allows access to recorded information much faster than in the case of the magnetic tape, where information can only be processed serially. The geometry of the head gap is a compromise between what works best for reading and what works best for writing. When it's time to write, a pattern of electrical pulses representing the data pass through the coil in the writing element of the recording head, producing a related pattern of magnetic fields at a gap in the head nearest the disk. These magnetic fields alter the magnetic orientations of bit regions on the disk itself, so the bits now represent the data.

The first hard-disk computer was introduced by IBM (International Business Machines) corporation in 1956, the IBM 305 RAMAC or "Random Access Memory Accounting" machine with a storage capacity of 5 million bits of information or 5 MB. It consisted of fifty, 24-in diameter disks, which had a storage density of 2 Kbits/in^2. This can be compared to today's desk-top personal computers, which have a capacity of 2 TB, use 3.5-in disks with a storage density of 500 Gbits/in^2. The design of the IBM RAMAC was motivated by the need for real-time accounting in business. The access arm moved in and out the disk under servo control to select a recording track. The depiction of the process of magnetic recording in Figure 10.1 corresponds to "parallel magnetic recording", where the magnetization of the medium lies in the plane of the tape or disk. Today's hard-disk devices utilize "perpendicular magnetic recording" media, where the magnetization is perpendicular to the plane of the disk, as discussed later.

10.2 Particulate Magnetic Recording Media

Particulate magnetic media are composed of single-magnetic-domain particles dispersed on non-magnetic polymeric supports. Special requirements must be met on particle size and shape, magnetization and coercivity. The particles must have large remanent magnetization in order to produce a strong enough signal for the recorded information to be easily read on playback. Their coercivity must be sufficiently large to guarantee long-term stability and protection against inadvertent erasure, but sufficiently low to enable easy writing and re-writing of information at modest magnetic field strengths. Ideally, their hysteresis loop must be "square", that is, with the remanent magnetization similar to the saturation magnetization. The elementary magnetic grains within the recording medium must be physically small and magnetically isolated from one another, but not so small as to exhibit superparamagnetism. These prerequisites were first met by two types of media: (a) particulate dispersions of maghemite (γ-Fe_2O_3), Co-modified γ-Fe_2O_3, chromium oxide (CrO_2) or barium ferrite ($BaO{\cdot}6Fe_2O_3$), and (b) granular thin films of Fe or Co-Fe alloys. Within the recording medium, the particles or grains must be well oriented and uniformly dispersed at a high packing fraction. A typical dispersion of γ-Fe_2O_3 particles used as magnetic tape medium would contain needle-like (acicular) particles with a length of ~300 nm and diameter of ~60 nm (axial ratio of 5:1) packed at a volume fraction of 40% in a flexible organic layer. The large axial ratio imparts shape anisotropy to the magnetic nanoparticles

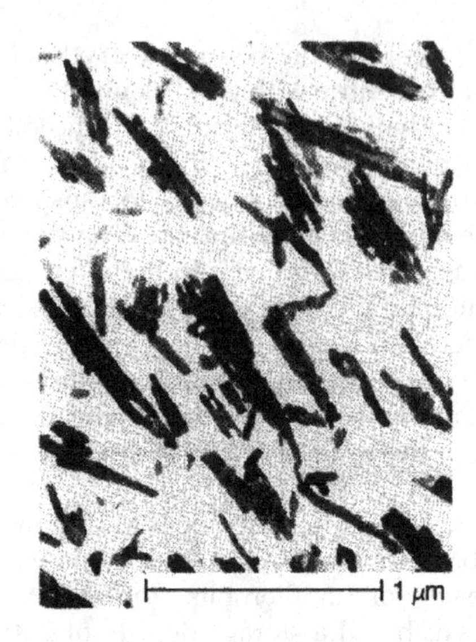

FIGURE 10.2
TEM of commercially available acicular particles of Co-coated γ-Fe_2O_3 suitable for many audio- and video-tape applications.

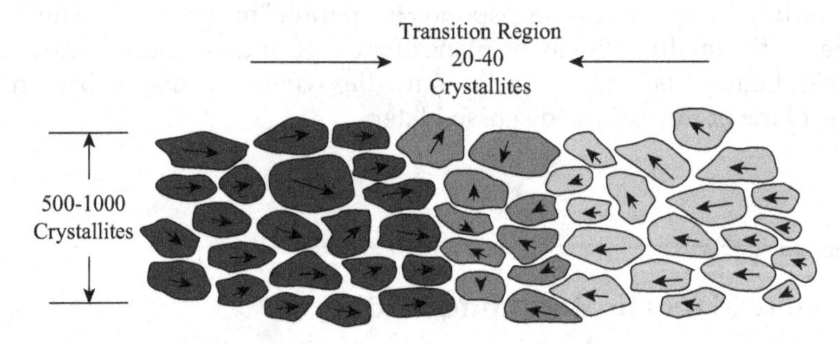

FIGURE 10.3
Schematic diagram of a typical particulate medium.

(see Eq. (4.32)). A TEM micrograph of typical acicular particles of Co-coated γ-Fe_2O_3 with coercivity in the range of 700–750 Oe is shown in Figure 10.2. The recording of information is achieved by switching the magnetization of the single-domain particles between their two stable directions, that is, along and opposite to the particle's easy axis of magnetization. Figure 10.3 gives a schematic diagram of a typical particulate medium, indicating the transition region between two oppositely magnetized regions in a recording track.

The small arrows within the grains represent the magnetization directions of each individual particle, while the large arrows drawn above the grains give the average magnetization of the regions. As implied from the figure the recorded signal is the result of a statistical averaging over many particles, not all of them necessarily perfectly aligned within the track. Table 10.1 gives some physical properties of particulate media used for magnetic recording. The larger coercivity barium hexaferrite is used in cases of permanently recorded information such as credit cards and identity cards, where there is no intention of erasing and re-writing information.

TABLE 10.1

Approximate Physical Properties of Particulates used in Recording Media

Material	Saturation Magnetization (mT)	Coercivity (kA/m)	(Oe)	Average Particle Size (μm)	Particle Shape
γ-Fe_2O_3	440	30	375	0.5 × 0.1	Needle
Co Modified γ-Fe_2O_3	460	60	750	0.5 × 0.1	Needle
CrO_2	600	70	875	0.4 × 0.05	Needle
Fe	2,100	125	1,362	0.15 × 0.05	Needle
$BaO{\cdot}6Fe_2O_3$	460	200	2,500	0.14 × 0.05	Disk

10.3 Granular Magnetic Recording Media

In the early 1980s, hard disks were developed containing polycrystalline Co-based thin films deposited on the disk by DC magnetron sputtering. Magneton sputtering was introduced in Chapter 5 (see Figures 5.12 and 5.13). The technology was further developed in the 1990s with the introduction of CoPtCr ternary alloys deposited by high-pressure sputtering, which caused physical voids between grains in order to minimize magnetostatic interactions between them. Further grain-isolation techniques were subsequently introduced by using various quaternary Co-based alloys CoXYZ (X = Pt, Ni…., Y = Cr, Ta, B, P…, Z = Nb, Mo…). The quality of these granular films as magnetic recording media is primarily determined by their crystallographic and microstructural properties, which are controlled by deposition conditions, such as deposition rate and temperature, type of sputter gas, etc. Furthermore, these granular media are deposited on elaborate substrates, which, by epitaxial growth, can affect their microstructure in the film, that is, their crystallographic orientation, grain size and grain size distribution, by influencing particle nucleation and growth during the sputtering process. The choice of cobalt alloys is dictated by the large intrinsic uniaxial magnetocrystalline anisotropy of cobalt along the c-axis (Figure 3.12(c)) with $K_{mc} = 4.5 \times 10^5\ J/m^3$, which imparts good thermal stability to the magnetic grains for long-term information storage. A thin carbon overcoat and a lubricant layer are added to protect these metallic media from oxidation and physical damage.

Figure 10.4 gives a TEM micrograph of a CoPtCrB medium together with a schematic of the transition region between two oppositely magnetized regions. Under high-temperature deposition conditions, the non-magnetic Cr migrates at the grain boundaries and acts to magnetically isolate the Co-rich grains from each other, so that each one can act independently under the action of the writing field. Note that the magnetic grains have variable sizes and irregular shapes. As the grains are single domain, that is their size is below the critical size for magnetic-wall formation, the transition boundary has to go around the grains, as depicted in the inset, meandering around grains of opposite polarity. There are many grains across the width of the track and therefore the exact location of the transition along the recording track is not very well defined. This randomness in (a) particle size, (b) easy-axis orientation and (c) transition width introduces medium noise in the recording system. For this reason, a group of grains is used to store a bit of information and

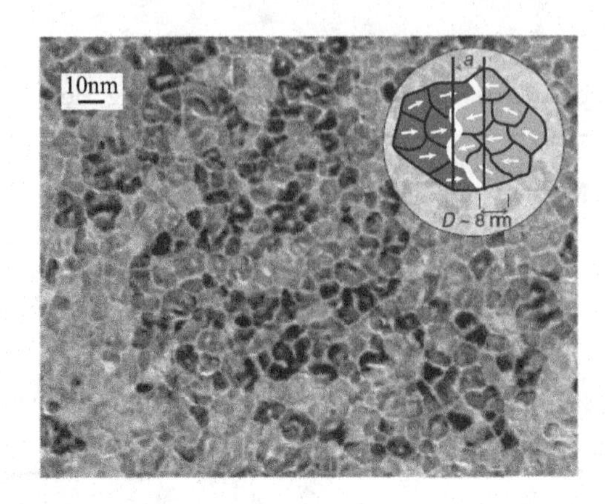

FIGURE 10.4
TEM image of CoCrPtB recording media. The amorphous non-magnetic Cr-rich boundaries, which magnetically separate the Co-rich grains, are visible as light gray areas. The inset schematically shows a magnetic transition meandering between the grains. (From A. Moses *et al.* Magnetic recording: advancing into the future, *J. Phys. D: Appl. Phys.* **35** (2002) R157–R167). Copyright Institute of Physics, reproduced with permission)

determine the transition region. The signal-to-noise ratio (*SNR*) is proportional to the square root of N (Eq. [10.1]), where N is the number of grains in a bit.

$$SNR \propto \sqrt{N} \tag{10.1}$$

The *SNR* is a key indicator of the recording performance of the medium, as it gives a measure of how reliably the bits can be read. According to Eq. (10.1), *SNR* can be improved by increasing the number of grains in a bit. Doing so, however, tends to keep the areal density of magnetic information storage low, unless smaller grain sizes were to be used. In addition, narrower grain-size distributions and smaller variations in magnetic properties are essential for good media *SNR*. For example, media with crystalline anisotropy fields, H_K, that vary widely from grain to grain provide poor *SNR*.

Whereas the magnetization reversal process in isolated particles is reasonably well understood within the classic Stoner and Wohlfarth model of coherent spin rotations discussed in Section 4.3, it is considerably more demanding to quantitatively model the same process in a macroscopic sample containing a distribution of fine particle sizes, a distribution of anisotropy axis directions and the possibility of the existence of significant magnetostatic interparticle interactions. These factors affect the observed response of the magnetic medium to an applied magnetic field. All constituent single-domain particles contribute to the observed hysteresis loop, resulting in a distribution of switching fields rather than the unique switching field of a single isolated particle. In addition, the orientational texture of the assembly affects the observed response, since the coercivity depends on the angle between the applied field and the anisotropy axis (see Figure 4.10). Due to the above, the coercivity, or the median switching field, of a magnetic material does not completely characterize its properties for recording purposes. The breadth of the switching field distribution (*SFD*) is also crucial. The *SFD* is defined *via* the derivative of the magnetization with respect to the applied field, dM/dH, at the vicinity of the coercive field, as demonstrated in Figure 10.5, which depicts a characteristic hysteresis loop for a recording medium and

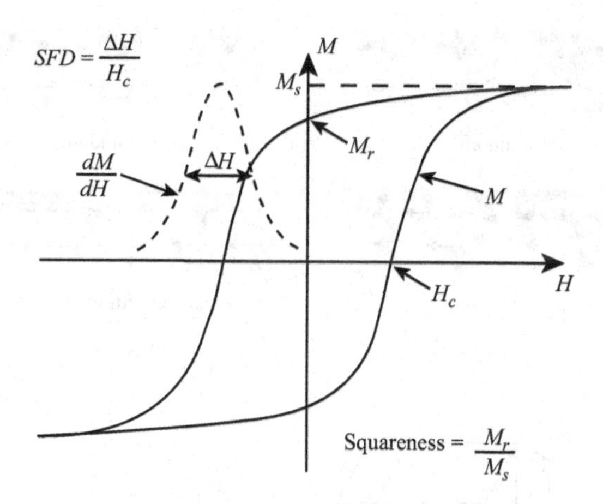

FIGURE 10.5
Typical hysteresis loop for a recording medium and definition of the term *SFD* and squareness.

defines *SFD* and squareness. Specifically, the full width at half maximum of the peak of the derivative curve, ΔH, divided by the median coercive field, H_c, gives the value of the *SFD*,

$$SFD = \Delta H / H_c \tag{10.2}$$

A narrow *SFD* allows for the writing of sharp, well-defined magnetic transitions and therefore facilitates the recording of information at high densities. A broad *SFD* diffuses the transitions and leads to a variety of problems, such as the inadvertent erasure or overwriting of old information with new.

However, the requirement of large N is due partially to the presence of a particle size distribution and the randomness in easy-axis orientation in traditional media with non-zero probability for the anisotropy axis of some grains to point out of the plane of the track, even lie in the direction perpendicular to the track and therefore not contributing to the recorded signal. One could obtain the same value of *SNR* with a smaller number of grains by using grains of a narrower size distribution and by enhancing the component of magnetization in the plane of the film and along the recording track direction, thus improving the performance of the recording medium. Figure 10.6(a) gives a schematic representation of the process. Deposition techniques have been developed where grain magnetization orientation within the plane has been enhanced by using a number of underlayers. For example, a Cr alloy deposited on a hard disc at about 250°C can grow with a (200) texture. When a Co alloy is subsequently deposited on this Cr layer the Co alloy will acquire a (110) texture, as shown in Figure 10.6(b). Since the c-axis is the easy axis of a Co crystal, the grains with (100) texture will have a larger component of their magnetization within the plane of the substrate than grains whose c-axis tilts away from the disk surface. These are commonly referred to as "oriented longitudinal recording media". However, a certain amount of randomness still remains.

The magnetic recording medium stores data based on the two stable magnetic configurations corresponding to M_r and $-M_r$ on the hysteresis loop (Figure 10.5). The stability of the medium is determined by the magnetic anisotropy energy of the material, K_uV, which gives the energy barrier, E_B, to be surmounted for reversal of the magnetization from one stable state to the other (see Figure 4.15). Here, K_u is the uniaxial magnetic anisotropy and

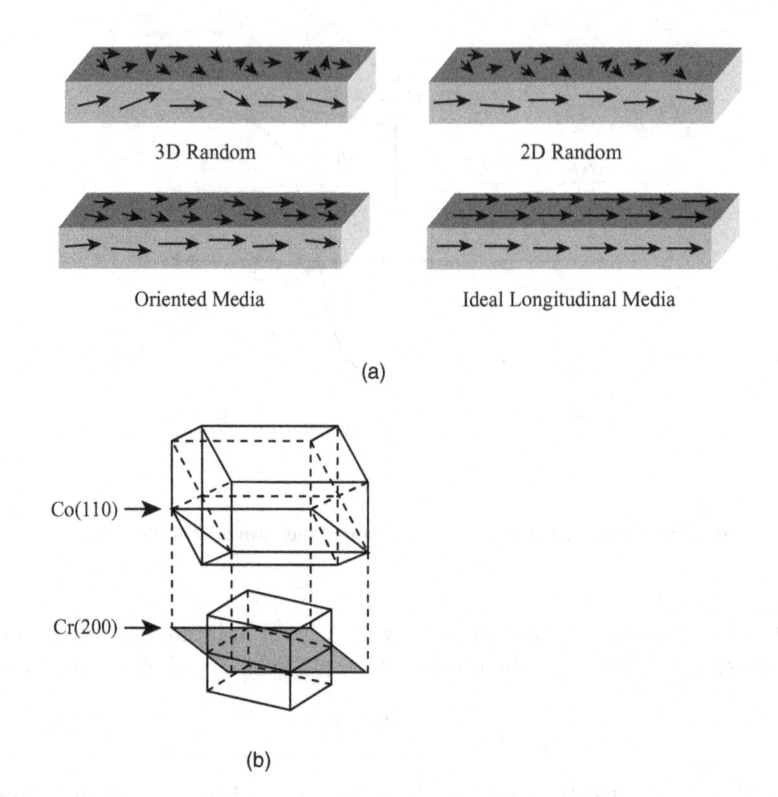

FIGURE 10.6
(a) Schematic representation of the reduction in the randomness of the magnetization direction of thin-film longitudinal recording media. (b) Depiction of epitaxial growth of Co(110) on Cr(200) planes.

V the volume of the grains, on average. The greater the amount of energy that is required to randomize the electronic spins either thermally or magnetically, the less likely that the information stored in the medium will be erased due to stray magnetic fields or excess heat. The recording is often done on the time scale of 10^{-8} s, and desired storage times are on the order of 10^{+8} s. This difference, of 16 orders of magnitude, dictates the desired range of anisotropy energies for the magnetic storage media. In order to ensure the stability of the recorded information, but also allow the writing and re-writing of information to occur under available recording-head field strengths, it is estimated that the magnetic anisotropy energy must be more than 60-fold higher than thermal energies at operating hard-disk-drive temperatures of about 340 K or $K_uV/k_BT > 60$. Therefore, the search for better-performing materials to make up hard disk media is centered upon the tailoring of the anisotropy energy, and thus, the coercivity of the candidate materials, since the coercivity is directly related to the anisotropy, $H_c = 2K_u/\mu_0 M_s$ (Eq. 4.31), at least in the case of non-interacting particles undergoing coherent spin rotation.

As mentioned earlier, one bit of information is the statistical average over many grains. This is required in order to obtain good *SNR*. Let us consider the process of increasing the areal density of bits on a magnetic track of a hard disk drive (HDD), a process that has relentlessly been pursued by the magnetic recording industry in the miniaturization of electronic devices. Figure 10.7 gives a schematic of such a track and defines the parameters to be considered. B is the length of a bit in the tack, W is the width of the track, t the thickness of the medium and d is the height above the track at which the recording head flies. In order to increase the areal density, we must reduce the bit length B and the track width W. However, in order to maintain the same *SNR*, we must maintain the same number of grains

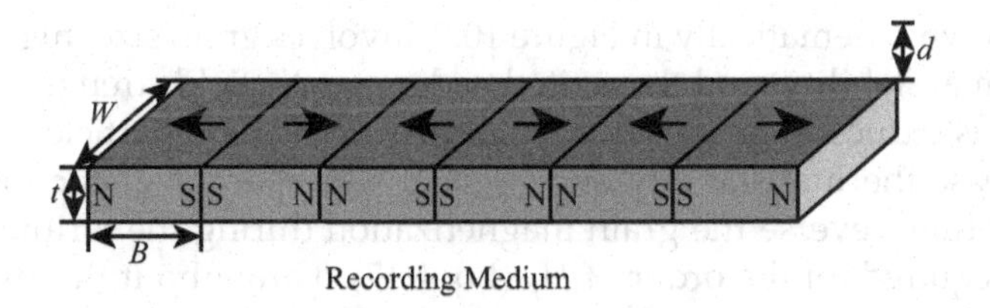

FIGURE 10.7
Schematic representation of the parameters of a magnetic recording track. *B* is the bit length; *W*, the track width; *t*, the track thickness; and *d*, the recording head-to-track distance.

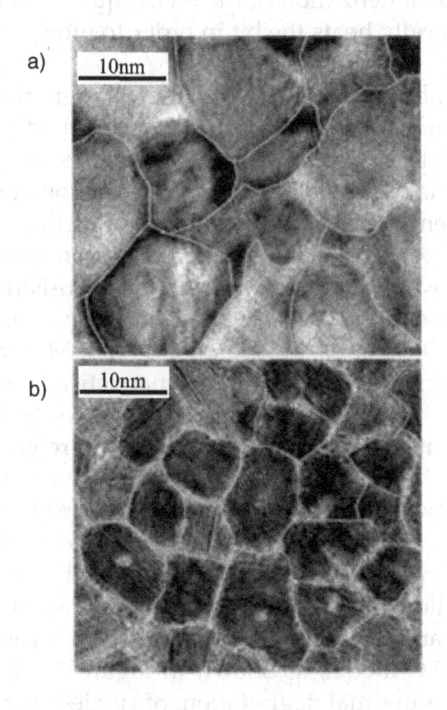

FIGURE 10.8
TEM images for two different CoPtCrB hard disk media with different areal magnetic recording densities. (a) TEM of 13 nm diameter grain media with an areal data density of about 10 Gbits/in^2. (b) TEM of 8.5 nm diameter grain media corresponding to 35 Gbits/in^2 areal density. (From Eric E. Fullerton, David T. Margulies, Andreas Moser and Kentaro Takano, Advanced magnetic recording media for high-density data storage, *Solid State Technology*, vol. 44, no. 9, Sept. 2001, p. 87. Copyright PennWell Publishing Corp., reproduced with permission).

per bit. This would necessarily require a reduction of the individual grain volume *V*. Indeed, the grain size of magnetic recording media has steadily decreased over the years. As an example, Figure 10.8 shows TEM images for two different CoPtCrB disk media. The medium imaged at the top supports a data density of about 10 Gbits/in^2 with an average grain diameter of about 13 nm, while the one imaged at the bottom corresponds to an areal density of 35 Gbits/in^2 with a grain diameter of 8.5 nm. In such high-density recording media, each bit contains about 100 grains.

As the volume of the grains is reduced in the scaling process, the magnetization of the grains may become unstable due to thermal fluctuations, that is, the grains may become superparamagnetic, and the stored information may be lost. This is known as the "superparamagnetic limit" in the scaling to higher bit area densities. The "superparamagnetic trilemma", as shown schematically in Figure 10.9, involves grain size (high density), media anisotropy (thermal stability) and the write-head magnetic field strength (ease of writing); as the grain size is reduced, the anisotropy field H_K of the nanoparticle must be increased in order to increase thermal stability and prevent superparamagnetic loss. Therefore, in order to purposefully reverse the grain magnetization during the writing process, larger head fields are required, of the order of H_K (Eq. 4.17). Throughout the 1990s, as the grain size diminished, the write-head fields were increased by successively replacing the inductive write-head Ni-Fe alloys with a saturation magnetization of 1.0 T with Co-Ni-Fe alloys with saturation magnetization of 2.1 T and finally Co-Fe with saturation magnetization of 2.4 T. High anisotropy media, where the writing fields produced by the writing head are insufficient to reverse the magnetization of the bits, require "thermally assisted magnetic recording" where a laser locally heats the bit in order to momentarily lower its coercivity during the writing process.

Traditional engineering, however, requires that all design parameters of the magnetic recording system, that is, the recording head, the recording media and the spacing between media and recording head, be scaled to smaller dimensions simultaneously as areal density increases. The "scaling laws" therefore require a reduction of the recording layer thickness t as well, or equivalently the magnetic thickness of the medium defined as $M_r\,t$. Reducing the magnetic thickness also introduces thermal instability. In early 2001, a new design of thin-film media was introduced, known as "Anti-Ferromagnetically Coupled" (AFC) media, in order to extend the use of parallel magnetic recording media beyond their superparamagnetic limit. Developed independently at IBM and Fujitsu Research, AFC media consist of two magnetic layers antiferromagnetically coupled through a non-magnetic Ru layer, about 0.6 nm thick. Figure 10.10 shows a schematic drawing of a transition region between two bits in an AFC medium. In the remanent state, the layers are magnetized in antiparallel directions, resulting in an effective magnetic thickness of M_rt_{eff}, given by the difference between the magnetic thicknesses M_rt of the two layers, M_rt_{eff}, = $M_rt(top) - M_rt(bottom)$. The reduction in effective magnetic thickness leads to a sharper transition, while the enhanced total physical thickness of the structure improves thermal stability. The AFC media allow independent optimization of stability and M_rt_{eff} by adjusting the physical thickness and magnetic properties of each layer. As a result, the thermal stability is enhanced in AFC media, as shown in Figure 10.11, which compares signal amplitude loss as a result of thermal degradation, of single-layer media with AFC media. As it can be seen, the AFC media suffer from a smaller amplitude loss for smaller M_rt_{eff} than single-layer media with $M_rt = M_rt_{eff}$, allowing scaling to smaller grain volumes. AFC media first appeared in the market in 2001 using 2.5-in HDDs and offered storage capacities of up to 30 GB.

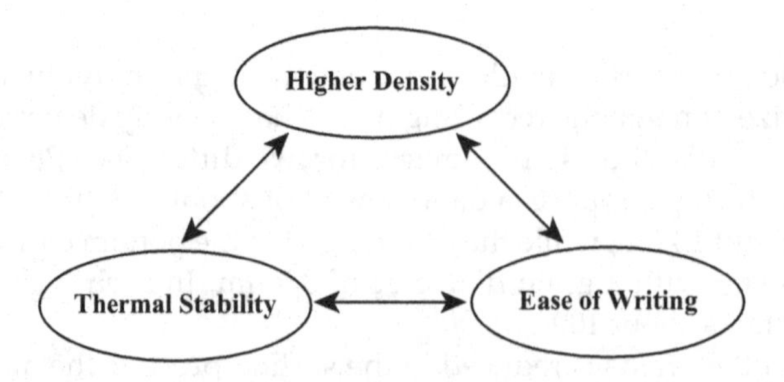

FIGURE 10.9
Schematic depiction of the superparamagnetic trilemma in high-density magnetic recording media.

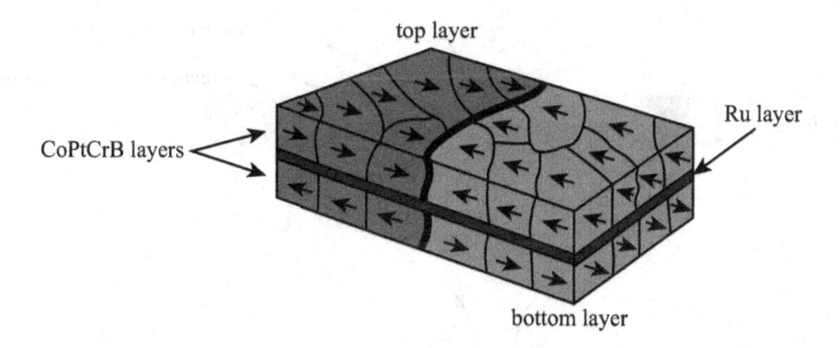

FIGURE 10.10
Schematic representation of a magnetic transition in AFC (anti-ferromagnetically coupled) media.

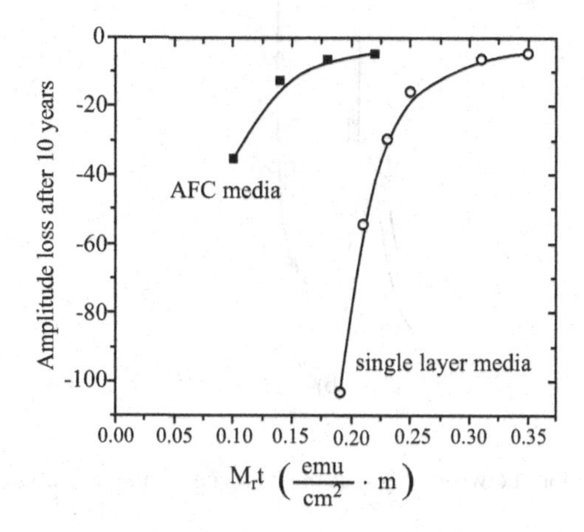

FIGURE 10.11
Comparison of amplitude loss as a result of thermal degradation of single-layer media and AFC media. The AFC media exhibits less amplitude loss for smaller M_rt_{eff} than single-layer media. (From A. Moser *et al.* Magnetic recording: advancing into the future, *J. Phys. D: Appl. Phys.* **35** (2002) R157–R167), Copyright Institute of Physics, reproduced with permission).

10.4 The Transition from Parallel to Perpendicular Magnetic Recording Media

Since 2006, the most significant development in increasing areal density of magnetic recording has come from replacing "parallel magnetic recording" with "perpendicular magnetic recording". While in parallel magnetic recording media, the magnetization direction of the bits is in the plane of the track, in perpendicular magnetic recording media the magnetization of the bits is normal to the plane of the track. To demonstrate the principle behind the increase in areal densities by switching from parallel to perpendicular recording, let us consider the difference in magnetic interactions within longitudinal *vs.* perpendicular magnetic recording media, as depicted in Figure 10.12. Note that in the case of parallel recording, the magnetic field lines, emanating from the magnetic poles of the bit, point in the opposite direction to the magnetization of an adjacent bit. This indicates that

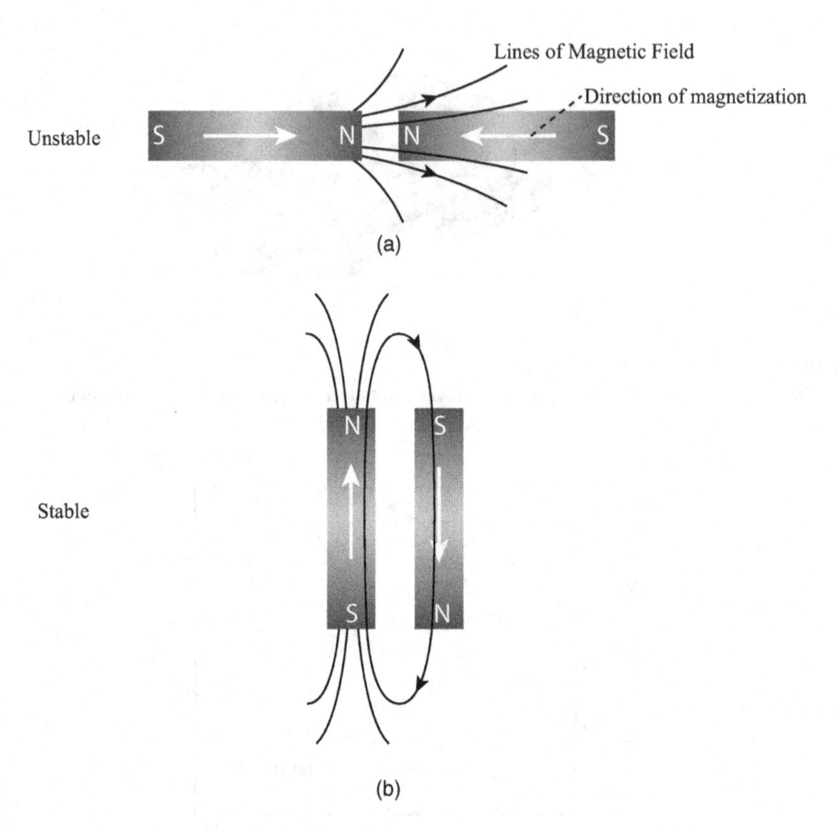

FIGURE 10.12
Difference in magnetic interactions between magnetic bits of reversed magnetization in (a) parallel recording and (b) perpendicular recording.

parallel recording has a weakness of unstable magnetization in a region where the magnetization is reversed. In contrast, in the case of perpendicular recording, the magnetic field lines emanating from a bit are in the same direction as the magnetization direction of adjacent bits, contributing to the stability of the recorded information and allowing for greater proximity between bits. In perpendicular recording media, the thermal stability of magnetic grains is about five times higher than in parallel magnetic recording media. Note also that magnetic bits are larger for parallel magnetic recording as compared to perpendicular magnetic recording, further allowing for higher density under the perpendicular magnetic recording geometry.

Since the bit magnetization is perpendicular to the track, the writing field must also be perpendicular to the track. This required a major re-design in magnetic write-heads and the introduction of a "soft magnetic underlayer" as indicated in Figure 10.13, which depicts the major differences between the principles of parallel and perpendicular recording modes of magnetic information storage used today. We note that these systems contain a head composed of separate read and write elements, which flies in close proximity to the granular recording medium. The inductive write element records the data in horizontal (Figure 10.13[a, b]) or perpendicular (Figure 10.13[c, d]) magnetization patterns. Note that the inductive ring element in Figure 10.13(a) contains symmetric poles (P1 and P2) in the case of parallel recording, with the induced field in the gap lying in the plane of the track. In contrast, in Figure 10.13(c) of perpendicular recording, the inductive write element

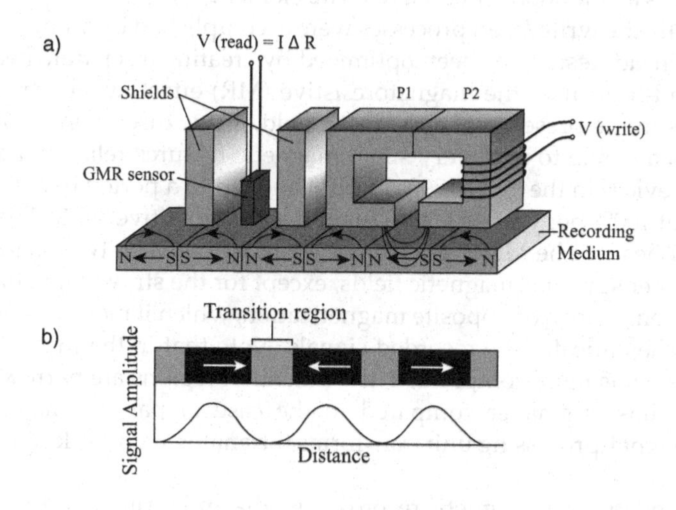

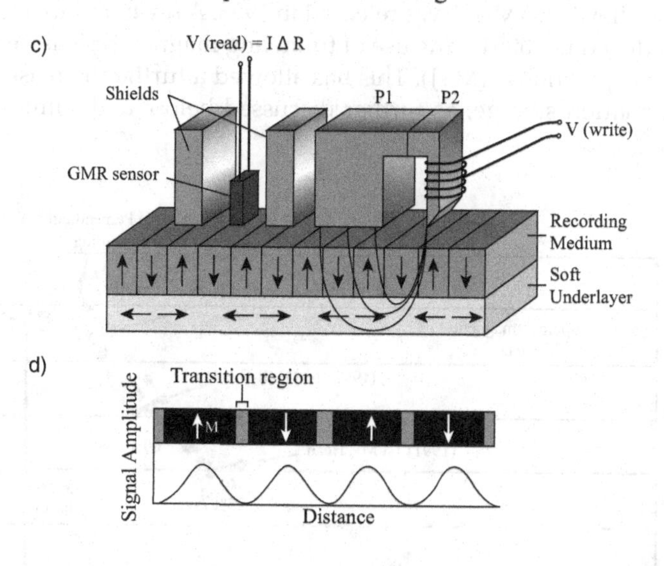

FIGURE 10.13
(a) Schematic representation of current parallel magnetic recording scheme depicting a symmetrical "ring" type inductive write element, GMR read-head and shield. (b) Schematic representation of the resulting bit geometry in parallel recording and associated signal. (c) Schematic representation of today's perpendicular magnetic recording scheme depicting an asymmetrical "monopole" type inductive write element, return pole, GMR read-head, shield and soft magnetic underlayer. (d) Schematic representation of the resulting bit geometry in perpendicular recording and associated signal.

contains very asymmetric poles (P1 and P2), with P2 being very narrow. This results in a strong magnetic field gradient at the vicinity of P2, as indicated by the converging magnetic field lines. The perpendicular magnetic recording medium is deposited on a magnetically soft underlayer that pulls the magnetic flux below the magnetic recording layer. Thus, the magnetic field immediately below P2 is strong and points precisely perpendicular to the magnetic track. It is this field that is responsible for the writing of magnetic information under the perpendicular magnetic recording geometry. For this reason, this write-head is often referred to as a "monopole inductive write element".

While originally the write/read processes were accomplished by a single inductive ring, improvement in head design has been optimized by creating a separate head for reading. The separate read-head uses the magnetoresistive (MR) effect, where the resistance of a material changes in the presence of a magnetic field, as we discuss in greater detail later. These MR heads are able to read very small magnetic features reliably. They were introduced in HDD devices in the early 1990s by IBM and led to a period of rapid areal density increases of about 100% per year. In 2000, "giant magnetoresistive" (GMR) heads started to replace MR read-heads. The MR read element is placed between two magnetic shields to protect it from other spurious magnetic fields, except for the stray magnetic field from the transitions between regions of opposite magnetization, which it measures. Figures 10.13(c) and (d) depict schematically, the recorded signals. Note that in the case of perpendicular recording, the signal is more compressed, the transition regions are narrower and the size of the magnetic bits is smaller compared to the case of parallel magnetic recording. Subsequently, a signal-processing unit transforms the analog read-back signal into a stream of data bits.

The above innovations in magnetic recording media and write-/read-head design have resulted in a continuous and steady increase in magnetic recording density over the last decades, ever since IBM's RAMAC was released in 1956. A further innovation in read-head sensors was introduced in 2005 by the use of tunneling magnetoresistance sensors, known as "magnetic tunnel junctions" (MTJ). This has allowed a further increase in areal density of magnetic information storage, as further discussed below and summarized in Figure 10.14.

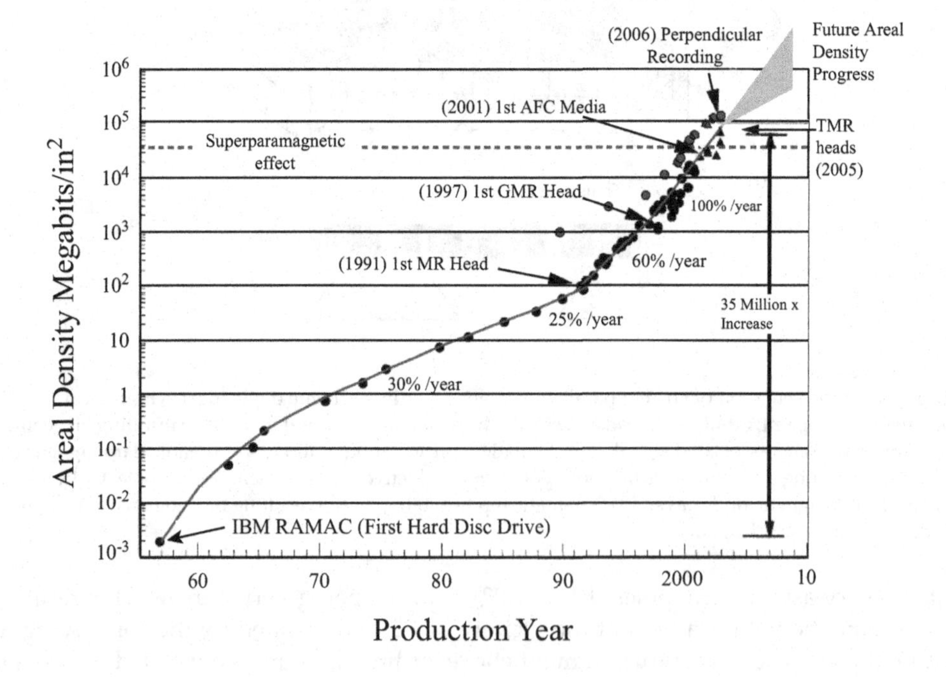

FIGURE 10.14
Evolution of magnetic recording media, areal magnetic storage densities and read-/write-head innovations since 1956.

10.5 Bit-Patterned Magnetic Recording Media

The ultimate high areal density should be achieved by the possibility of utilizing a single-magnetic-domain particle to store a bit of information, while maintaining a good signal-to-noise ratio. Researchers are investigating avenues to achieve this goal in the study of "bit-patterned media" for extending future storage densities to very high values (1Tb/in^2) without the need for high write-fields. A patterned recording medium consists of a regular array of magnetic elements, each of which has uniaxial magnetic anisotropy. The easy axis can be oriented parallel or perpendicular to the substrate. The grains within each patterned element are coupled so that the entire element behaves as a single-magnetic domain. The major advantages of such a scheme are (a) transition noise is eliminated because the bits are now defined by the physical location of the elements and not by the meandering boundary between two oppositely magnetized regions of a thin film, and (b) very high data densities can be obtained because the stability criterion now refers to the volume and anisotropy of the entire magnetic element, not to the individual grains of which it is composed. Furthermore, the elements can be as small as a few nanometers, deposited with less than 50 nm periodicity implying exceedingly high densities.

Implementation of bit-patterned media, however, will require the development of new schemes for addressing the array elements and for the detection of magnetic fields on a very small scale. In addition, the mass production of bit-patterned media is exceedingly challenging. Fabrication of large-area arrays of elements with dimensions on the sub-50-nm scale cannot be done using conventional "optical lithography" used extensively in the microfabrication industry to produce geometrical patterns on a substrate for the production of integrated circuits. Nanolithography techniques are needed, such as electron-beam and X-ray lithographies. These must be combined with magnetic materials deposition processes, such as electrodeposition or chemical vapor deposition. Electrodeposition is well-suited to making high aspect ratio particles. Prototype arrays of Ni or Co alloys such as NiFe, NiCo and CoPt, as well as, multilayers such as Ni/Cu, and granular alloys such as Co-Cu have been produced by electrodeposition. Chemical vapor deposition has been used to produce prototype arrays of flat, tapered or conical particles consisting of pure metals, alloys such as NiFe, and multilayer structures. Self-assembly methods have also been applied in which magnetic particles are synthesized chemically and assembled as close-packed monolayers on a surface. Figure 10.15 illustrates such examples from the literature as reviewed by C. Ross (C. Ross, Patterned Magnetic Recording Media, *Annu. Rev. Mater. Res.* 2001, 31:203–35). Although these methods are relatively slow, they allow tall structures to be made in precise locations on a substrate.

Figures 10.15(a–g) give examples of bit patterning using various nanolithographic and chemical vapor deposition techniques, while Figures 10.15(i) and (j) give an example of self-assembled FePt nanoparticles on a substrate. Self-assembly is a particularly promising approach that can be easily scaled-up to mass production. Magnetic nanoparticles can be synthesized and assembled on a surface to form a regular array, or superlattice as we discussed in Chapter 6 (see Figure 6.10), without the need for nanolithography. This can enable the rapid production of very fine-scale structures over large areas. Structures with periods of the order of 10 nm can be made, a feature size that is beyond that of most lithography techniques. However, it is critical to control the size distribution of the particles to obtain uniform magnetic properties and a well-ordered self-assembly. Even if the particles were monodispersed in size, the tendency of magnetic particles to agglomerate can impede separation and self-assembly. Metal particles also tend to oxidize. Oxidation,

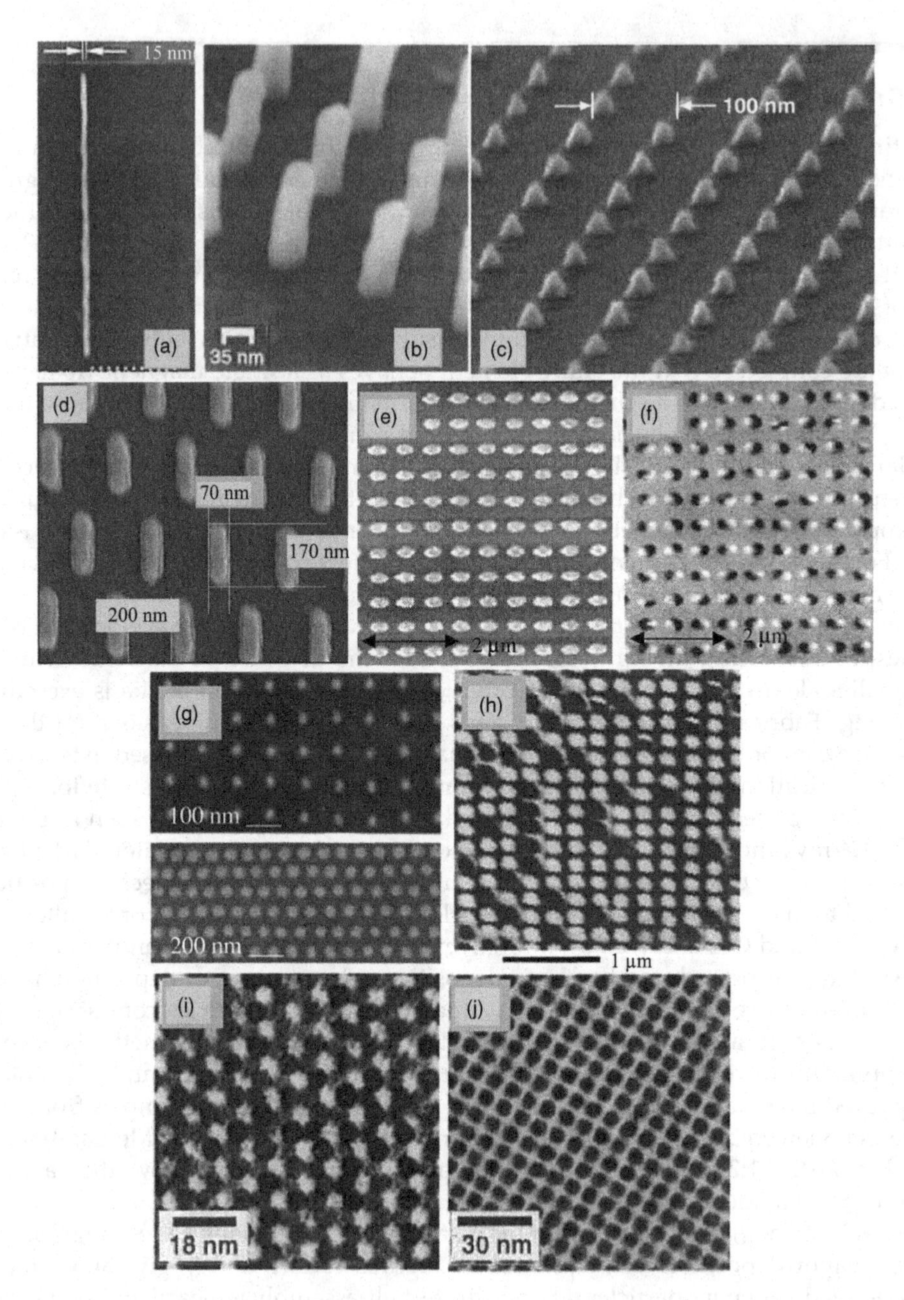

FIGURE 10.15
Examples of bit-patterned media consisting of regular arrays of magnetic nanoparticles (a) Evaporated Ni bar, 15 nm wide, made using electron-beam lithography. (b) Electrodeposited Ni pillars, 35nm diameter, 100-nm period, made using electron-beam lithography. (c) Evaporated Ni-alloy pyramids, 30-nm diameter, 100-nm period, made using achromatic interference lithography. (d) Etched ellipses of a Co/Cu multilayer, 70 × 170 nm, 10-nm thick, made using interference lithography. (e) Topological, and (f) corresponding magnetic image of 140 × 250-nm evaporated Co ellipses made using electron-beam lithography. In-plane dipole images (dark and light poles) from the ellipses can clearly be seen in (f). (g) Arrays of Fe pillars grown by CVD in two separate geometries, and (h) a magnetic force micrograph of another array showing that pillars are magnetized as dipoles with vertical moment up or down (dark or light). (i, j) FePt particles, 6 nm in diameter, self-assembled on a substrate to form regular arrays with a nearest neighbor spacing of ~4 nm. (From C. Ross, Patterned Magnetic Recording Media, *Annu. Rev. Mater. Res.*, 2001, 31:203–35). (Copyright Annual Reviews, reproduced with permission).

agglomeration, and self-assembly can be controlled by coating the particles with a passivating layer of precise thickness and chemical functionality to protect the particles and to ensure uniform separation between the particles once assembled onto a surface. We note, however, that even though these particles would typically form a close-packed array with excellent short-range order, their long-range order may be poor. The domain size or correlation length of the structure can be increased by improving the monodispersity of the particles and by careful growth and annealing of the self-assembled monolayer.

Example 10.1: Ferromagnetic FePt Nanocrystal Superlattices for Magnetic Recording

The FePt nanoparticles shown in Figures 10.15(i) and (j) were synthesized by reduction of platinum acetylacetonate and decomposition of iron pentacarbonyl in the presence of oleic acid and oleyl amine stabilizers, as reported by S. Sun *et al.* (S. Sun, C. B. Murray, Dieter Weller, Liesl Folks and A. Moser, Monodisperse FePt Nanoparticles and Ferromagnetic FePt Nanocrystal Superlattices, *Science* 287 (2000) 1989). The FePt particle composition was readily controlled, and the size was tunable from 3- to 10-nanometer diameter with a standard deviation of less than 5%. These nanoparticles self-assemble into three-dimensional superlattices. Thermal annealing converted the internal particle structure from a chemically disordered face-centered-cubic (fcc) phase of low coercivity to the chemically ordered face-centered-tetragonal (fct) phase of high coercivity and transformed the nanoparticle superlattices into ferromagnetic nanocrystal assemblies. The fcc → fct phase transition common in FePt (and CoPt) alloys shown in Figure 10.16 is under intense investigation due to the high coercivity of the fct phase, also known as $L1_0$ phase, with promising applications in high-density perpendicular magnetic recording of the future. In addition to an observed change in crystallographic lattice constants (from $a = c$ to $a \neq c$) in the ordered fct phase, all iron atoms reside on the same crystallographic plane with the Pt atoms occupying crystallographic sites on an adjacent plane, in contrast to the disordered fcc phase, where Fe and Pt atoms occupy the various crystallographic sites randomly.

This transition can be followed *via* the evolution of the XRD pattern shown in Figure 10.17, top panel. The transition to the fct phase is complete at the annealing temperature of 600°C. While the as-synthesized fcc assemblies are superparamagnetic at ambient temperature the annealed fct assemblies are stable nanoscale ferromagnets at room temperature. This makes them an important class of materials in permanent magnetic applications because of their large uniaxial magnetocrystalline anisotropy ($K_u > 7 \times 10^6$ J/m^3) and good chemical stability. Indeed, such assemblies have been demonstrated to be chemically and mechanically robust and shown to support high-density magnetization reversal transitions, as shown in Figure 10.17, bottom panel. This figure shows the read-back sensor voltage signals from written data tracks recorded at various linear densities from 500 to 5,000 flux changes per millimeter (fc/mm). The recording medium is an approximately 120-nm-thick assembly of 4-nm FePt nanocrystals with an in-plane room temperature coercivity of $H_c = 1{,}800$ Oe.

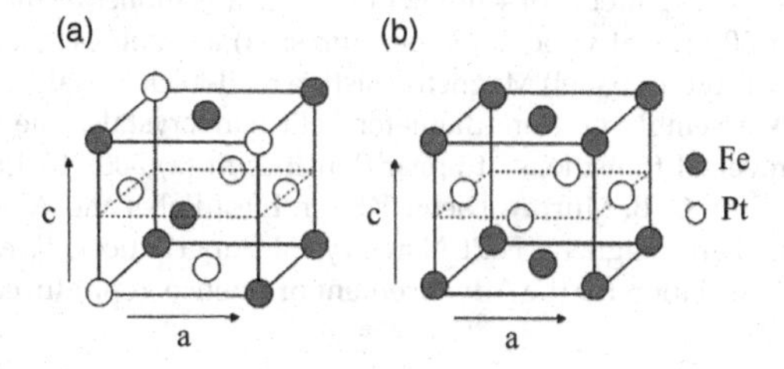

FIGURE 10.16
Schematic depiction of the crystallographic phase transition of FePt alloys from the disordered (a) fcc phase, where Fe and Pt atoms are randomly distributed on the various crystallographic sites, to the highly ordered (b) fct crystallographic phase, where Fe and Pt atoms are located on separate crystallographic planes.

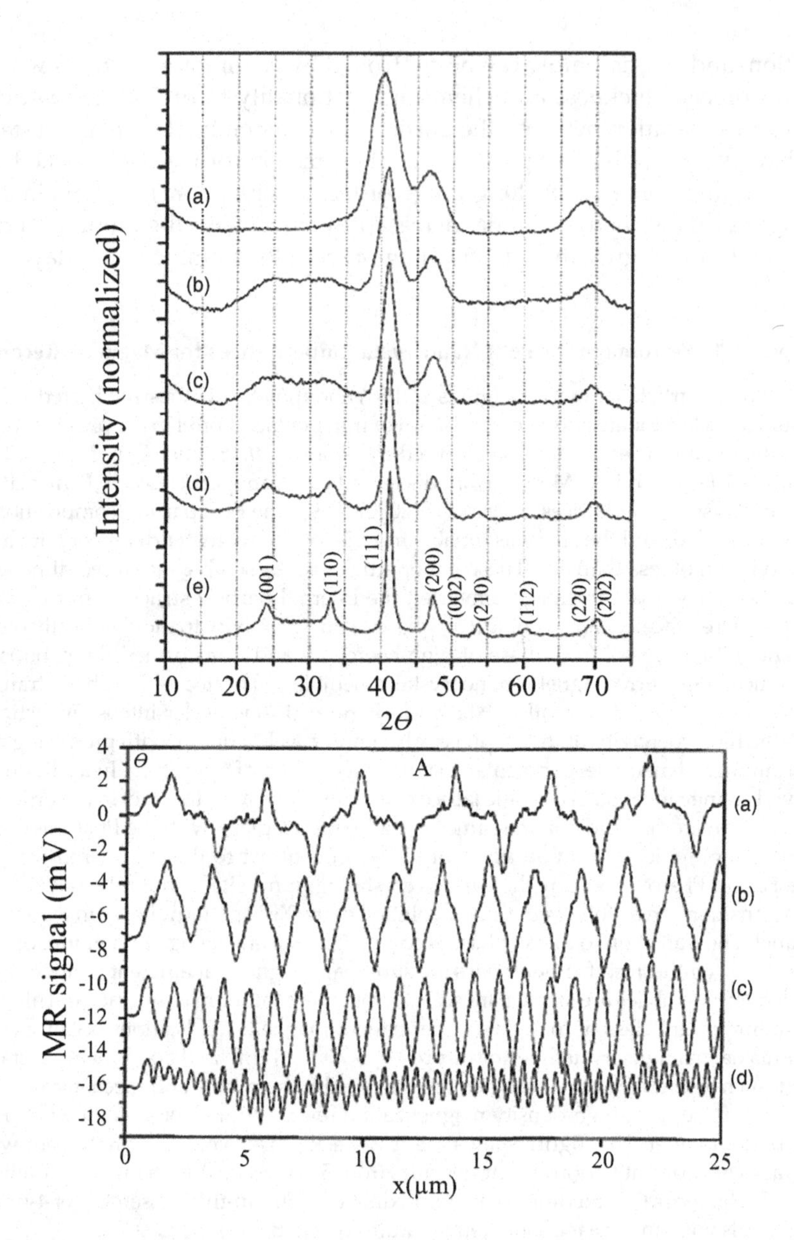

FIGURE 10.17
(Top panel) Evolution of XRD patterns of 4-nm FePt particle assemblies as they are annealed under atmospheric N_2 gas for 30 min. at various temperatures: (a) as-synthesized, (b) 450°C, (c) 500°C, (d) 550°C and (e) 600°C. (Bottom panel) Magnetoresistive read-back signals from written bit transitions in a 120-nm-thick assembly of 4-nm-diameter FePt nanocrystals. The individual line scans reveal magnetization reversal transitions at linear densities of (a) 500, (b) 1,040, (c) 2,140 and (d) 5,000 fc/mm. (From S. Sun, C. B. Murray, Dieter Weller, Liesl Folks and A. Moser, Monodisperse FePt Nanoparticles and Ferromagnetic FePt Nanocrystal Superlattices, *Science* 287 (2000) 1989). (Copyright American Association for the Advancement of Science, reproduced with permission).

10.6 GMR, TMR and the Dawn of Spintronics

Giant leaps forward in the development of high-density magnetic recording media were enabled through the introduction of magnetoresistive read-head elements that can sense very small magnetic fields, as discussed earlier (Figure 10.14). The principles of operation that govern the ability of the heads to detect small magnetic fields lie on the interdependence between magnetization and electron transport. This interdependence links the value of the resistance, R, of a ferromagnetic conductor to the direction of the current flow relative to the direction of the magnetization, $\vec{M}$, as first observed in 1856 by William Thomson. This variation in resistance, known as asymmetric magnetoresistance (AMR), is given by

$$\frac{\Delta R}{R} = \frac{R_{max} - R_{min}}{R_{min}} \tag{10.3}$$

and is of the order of only 1% effect. Even though the observed effect was very small, IBM introduced magnetoresistive read-heads based on AMR in 1991, resulting in a major technological step forward, as seen in Figure 10.14, increasing the growth of areal storage density from 25% per year to 60% per year.

Magnetoresistive read-heads sense the field through its influence on the magnetization direction of the head. The fundamental physics involved deal with the conductivity of the ferromagnetic metals Fe, Ni and Co and their alloys. In these ferromagnetic materials, both 4*s* and 3*d* electron bands contribute to the density of states at the Fermi level (E_F) as seen in Figure 3.9. The itinerant electron theory of ferromagnetism, discussed in Section 3.3.2, indicates that there is an imbalance in "spin-up" and "spin-down" 3*d* electrons at the Fermi level due to the strong exchange interaction. It is indeed this imbalance that is responsible for the observation of a ferromagnetic moment in these metals (Figure 3.10). It is, primarily, the 4*s* electron band that contributes the conduction electrons. Thus, the resistivity of the material depends primarily on s electron scattering, a process that is dominated by the spin conserving *s*-to-*d* electronic transitions, as the conduction electrons move through the lattice of the ferromagnet. The spin imbalance of 3*d* electrons at E_F results in strongly spin-dependent scattering probabilities. The 4*s* electrons can undergo many scattering events while they retain their spin orientation, before their spin flips to the opposite direction. Thus, in the limit where spin-flipping can be ignored, conduction takes place through two parallel spin-channels of "spin-up" and "spin-down" electrons that have very different resistivities. This is due to very different mean free paths λ_{up} and λ_{down}, the mean distance an electron travels between scattering events, for "spin-up" and "spin-down" 4*s* electrons. In usual thin films, these mean free paths scale from a few nanometers to a few tens of nanometers. The discovery of GMR in 1988, by the 2007 Nobel Prize in Physics recipients Albert Fert and Peter Grünberg, became possible when magnetic multilayers of individual layer thicknesses of the order of the mean free paths were built, so evidence for "spin-dependent" electronic transport could be observed.

Figure 10.18 illustrates schematically the principle of GMR for a simple three-layer arrangement consisting of two identical ferromagnetic metal layers (FML1 and FML2) separated by a non-magnetic metal layer (NML). Assuming that $\lambda_{up}^{FML} >> \lambda_{down}^{FML}$ and that the thicknesses t_{FML} and t_{NML} of the individual layers are designed to obey the relationships $\lambda_{up}^{FML} > t_{FML} > \lambda_{down}^{FML}$ and $t_{NML} << \lambda_{NML}$ the GMR effect can be observed.

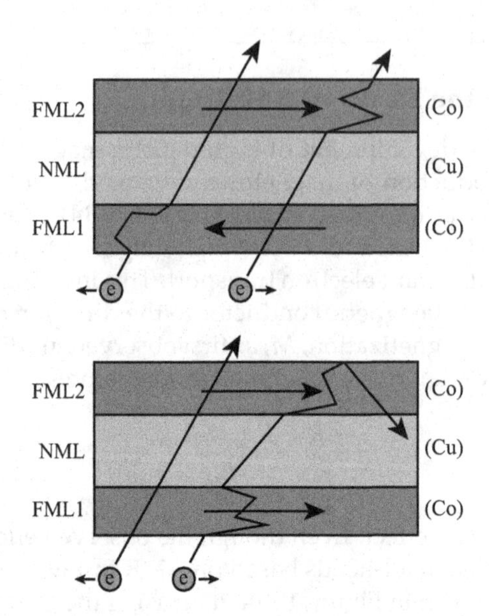

FIGURE 10.18
Schematic representation of the GMR effect for a simple 3-layer arrangement of two ferromagnetic metal layers (Co) separated by a non-magnetic metal layer (Cu).

Let us consider the two situations whereby the two ferromagnetic layers are magnetized either parallel (P) bottom scheme, or antiparallel (AP) top scheme, to each other. In the case of P-magnetization configuration, the spin-up electrons (those with their spin antiparallel to the magnetization) can travel through the trilayer nearly unscattered, providing high conductivity, and therefore, low resistance. In the opposite case of AP-magnetization configuration, both spin-up and spin-down electrons scatter, in one or the other FML, giving rise to high resistance. In analogy to Eq. (10.3), the size of the effect is measured by

$$\frac{\Delta R}{R} = \frac{R_{AP} - R_P}{R_P} \tag{10.4}$$

It can be of the order of 100% or more for multilayers with a large number of FML/NML periodic arrangements. Thus, it was named giant magnetoresistance; the GMR effect is an outstanding example of how in nanomagnetism, artificially structured materials at the nanoscale can uncover fundamental physical phenomena that can bring about new functionalities and technological innovations.

A particularly useful device based on the phenomenon of GMR is the spin-valve sensor. In its simplest form, it consists of a trilayer arrangement such as the one depicted in Figure 10.19, in which one of the FML, say the bottom layer, has its magnetization pinned to an underlayer consisting of an antiferromagnetic material. Then, a strong coupling is provided by the exchange-bias effect at the interface of the ferromagnetic layer with the antiferromagnetic underlayer. Exchange bias was discussed in Section 4.12. The antiferromagnetic layer has no net magnetic moment, and thus, it is not sensitive to an external magnetic field. However, it can have a large magnetic anisotropy, which transferred to the ferromagnetic layer through interfacial exchange interactions can strongly stabilize the magnetic orientation of the adjacent ferromagnetic layer. In contrast, the magnetization of

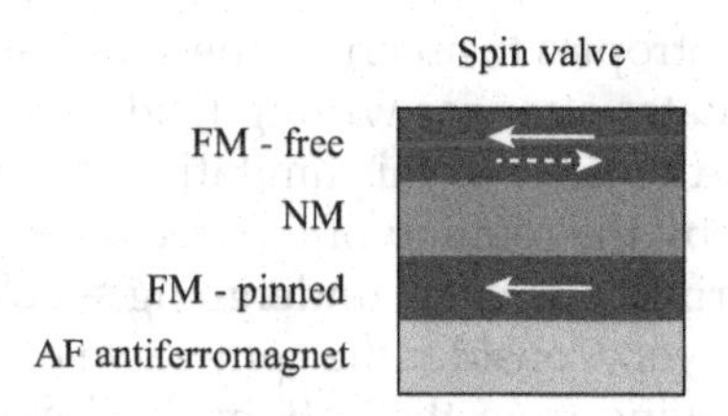

FIGURE 10.19
Schematic representation of the operation of a spin-valve.

the top layer is free to rotate under the action of an external magnetic field, thus resulting in reversible P or AP relative magnetic orientation configurations of the two FMLs in the absence or presence of an external magnetic field. The rotation of the top layer magnetization to form a P-configuration "opens" the flow of electrons; its rotation to form an AP-configuration "closes" the flow of electrons, thus operating as a valve. Standard spin valves have magnetoresistance values of 5–6%. They were introduced in read-head elements in 1997 by IBM increasing the areal density growth of magnetic recording from 60% to 100% (Figure 10.14).

If one were to replace the metallic, non-magnetic layer situated between the two FMLs in Figure 10.18 with a thin, insulating, non-magnetic layer, one would expect that, at least classically, all conduction through the trilayer arrangement will cease, as the insulator layer will act as a barrier to electron flow. However, quantum mechanically, there is a non-zero probability that the electrons tunnel through the thin insulating barrier. Indeed, if the metal layer in Figure 10.18 were to be replaced by a (~1–2 nm) thin, insulating non-magnetic oxide such as alumina (Al_2O_3) one would create a "magnetic tunnel junction" (MTJ). The concept of spin-dependent electron tunneling between ferromagnetic films separated by an insulating thin film was initially proposed by Jullière in 1975. However, practical realization had to await the development of nanofabrication technologies for thin film deposition and nanopatterning that allowed to finely controlling the thickness of the deposited films. In 1995, the first MTJs were produced using amorphous Al_2O_3 as the insulating layer. The resulting "tunnel magnetoresistance" (TMR) reached a limit of 70% at room temperature. Much higher effects were later obtained by replacing the amorphous alumina with single-crystal MgO insulating barriers, taking advantage of symmetry-dependent tunneling effects. The MTJ is a spin-valve device with magnetoresistance values that can reach up to two orders of magnitude higher than standard GMR spin-valves, with dimensions scaled down to well below 100 nm. Indeed, TMR sensors were introduced in read-head technologies in 2005 by Seagate (Figure 10.14). More recently, research has been carried out in developing a novel "magnetic random access memory" (MRAM) in which the binary information "0" and "1" is recorded on the two opposite orientations of the magnetization of a spin-valve free layer along its easy axis of magnetization. In 2006, the first 4-MB MRAM based on spin-valves was commercialized by Freescale Corporation.

The inter-dependence of the development of "spintronics" and "nanomagnetism" is obvious. Practical applications of GMR and TMR could not be realized without the development of techniques in the realm of nanomagnetism to control the thickness and magnetic anisotropy of deposited ferromagnetic films by design at the nanoscale and the exchange-bias interfacial magnetic interactions. Researchers are also investigating concepts of writing magnetic information on spin-valve-based memories by eliminating the need to apply an external magnetic field. As we have discussed earlier, the use of smaller and smaller particles in magnetic recording necessitates the introduction of higher and higher particle magnetic anisotropies to ascertain thermal stability against superparamagnetic relaxation. This necessitates stronger writing fields, which are hard to achieve over smaller and smaller spaces. A remedy to this limitation is being sought in the concept of "spin-transfer torque" whereby the rotation of the free layer magnetization is carried out by the passage of "spin-polarized currents" undergoing *s-d* electron scattering in the rotationally free-magnetization layer. Consider the spin-valve of Figure 10.19 in AP-magnetic orientation and a current flowing from the bottom fixed-magnetization layer to the top free-magnetization layer. If the fixed-magnetization layer is thick enough the current will immerge from the bottom ferromagnetic layer polarized along the fixed magnetization of the bottom layer due to the strong *s-d* exchange interactions. Subsequent *s-d* exchange in the free-magnetization layer tends to align the magnetization of the electrons in the direction of the magnetization in the top layer. The principle of conservation of angular momentum would dictate that the angular momentum lost by the electrons be transferred to the magnetization of the top ferromagnetic layer, which unlike the bottom layer is free to rotate from the AP to the P-magnetic configuration. In other words, the top layer feels a "torque" that tends to rotate its magnetization along the opposite direction of its anisotropy axis. Thus, as GMR enabled the control of electronic transport through magnetization, spin-transfer torque now allows magnetization to be controlled by electronic transport.

Exercises

1. What advantages do hard disk magnetic recording media offer compared to magnetic tape storage media?
2. The early magnetic recording processes utilized particulate magnetic media deposited on a polymeric surface. What special requirements had to be met by the elemental magnetic particulates within the recording medium?
3. Contrast granular, thin-film media to particulate media. The quality of the recording medium is characterized by the signal-to-noise ratio (SNR). Describe strategies by which the SNR can be increased leading to better granular magnetic recording media.
4. The hysteresis loop of a recording medium gives a measure of its quality. What characteristics of the hysteresis loop must be optimized to obtain high-quality magnetic recording?
5. In the quest of achieving ever higher magnetic storage densities, engineers have stumbled upon the "superparamagnetic limit" and the "superparamagnetic trilemma". Define these concepts; explain how they arise and how they can be circumvented.
6. Describe how anti-ferromagnetically coupled media extended the use of parallel magnetic recording beyond the "superparamagnetic limit".
7. Contrast parallel to perpendicular magnetic recording media. Which geometry exhibits higher stability at high magnetic recording densities? Explain.
8. Define the concepts of antisymmetric magnetoresistance, giant magnetoresistance, and tunneling magnetoresistance, which are broadly used in magnetic-read-head sensors in modern magnetic recording devices.
9. Describe the principles of operation of a "spin-valve".
10. What is "spin-transfer torque"? What are its potential applications?

11

Permanent Magnets

It is proposed to make permanent magnets of composite materials consisting of two suitably dispersed ferromagnetic and mutually exchange-coupled phases, one of which is hard magnetic in order to provide a high coercive field, while the other may be soft magnetic just providing a high magnetic saturation.

E. Kneller and R. Hawig, *IEEE Transactions on Magnetics*, 27 (1991) 3588

11.1 Introduction to Permanent Magnetism

A "permanent magnet" is an object made from a permanently magnetized material, which creates its own persistent magnetic field around it. The earliest permanent magnets known to humankind since antiquity are naturally magnetized pieces of the mineral magnetite, known as loadstones, which attract pieces of iron. The earliest practical device made of a permanent magnet is the magnetic compass used for navigation since the 12th and 13th centuries AD. In today's highly technological society, permanent magnets can be found in a myriad of devices, from magnetic recording systems discussed in the previous chapter, to car motors and wind-mill-based energy generation stations.

We know that spontaneously magnetized materials consist of randomly oriented magnetic domains to minimize magnetostatic energy. The magnetic domains can be oriented by the application of an external magnetic field, thus, creating a magnet. A magnet made of material with a high anisotropy field, H_A (or H_K, Eq. (4.17)), given by $H_A = \frac{2K}{\mu_o M_s}$, where K is the magnetocrystalline anisotropy and M_s is the saturation magnetization, can resist self-demagnetization by its internal demagnetizing field, $\vec{H}_d = -N\vec{M}$, where N is the demagnetization factor dependent on shape (Eq. (3.54)). Such a magnet can retain its magnetization over time, thus creating a "permanent magnet". Materials appropriate for the construction of permanent magnets are known as "hard magnetic materials". Permanent magnets can be contrasted to electromagnets, which lose most of their magnetization when the current in the coil wrapped around a material of low anisotropy field, known as "soft magnetic material", is reduced to zero. The evolution of permanent magnets in modern technology has been in step with the discovery of materials with increasingly higher magnetocrystalline anisotropies.

The ideal hysteresis loop for a permanent magnet must be wide and square, as shown in Figure 11.1. Saturation necessitates the application of an external magnetic field with a magnitude of the order of the saturation magnetization. As a review of the magnetization process, the figure also indicates the initial magnetization curve I and defines the parameters of a hysteresis loop, M_s, M_r and H_c. The hysteresis loop defines "technical magnetism". It informs on an intrinsic magnetic property of the material, namely the saturation magnetization, M_s, which is the spontaneous magnetization that exists within a magnetic

DOI: 10.1201/9781315157016-15

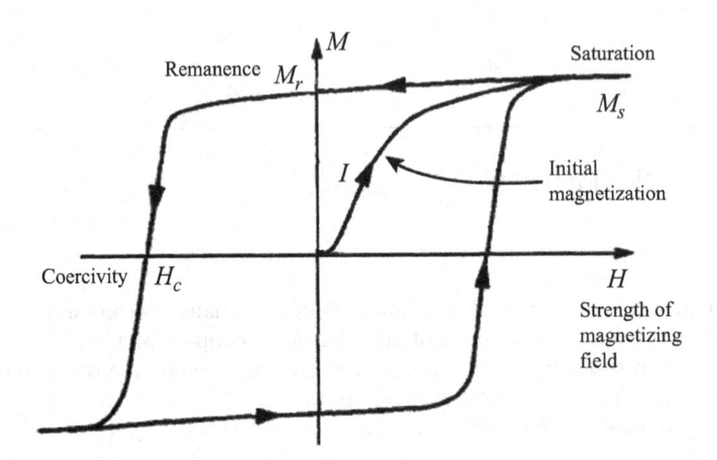

FIGURE 11.1
Schematic representation of a somewhat idealized (square) hysteresis loop for a permanent magnet material including the initial magnetization curve, *I*. The parameters defined by the hysteresis loop saturation magnetization (M_s), remanence (M_r) and coercivity (H_c) are indicated in the figure.

domain of a ferromagnet, and two extrinsic properties, the remanence, M_r, which is a measure of the remaining or persisting magnetization after the applied field has been removed, and the coercivity, H_c, which is a measure of the strength of an opposite magnetic field that brings the magnetization back to zero. The extrinsic parameters depend on a number of external factors including the sample's micro- or nanostructure, shape, thermal history or the rate at which the field is swept in order to trace the loop. In real systems, however, it is very difficult to obtain perfect square loops.

In the study of permanent magnets, the hysteresis loop is presented either as the dependence of the magnitude of the magnetization $\vec{M}$ (in units of kA/m or emu/cm^3), or that of the magnetic induction $\vec{B}$, also referred to as the magnetic flux density (in Tesla or Gauss), on the strength of applied field $\vec{H}$ (in kA/m or Oe). For soft magnetic materials, the *M-H* and *B-H* loops have rather similar shapes, but for hard magnetic materials, suitable for permanent magnet applications the two loops are quite different, as indicated schematically in Figure 11.2.

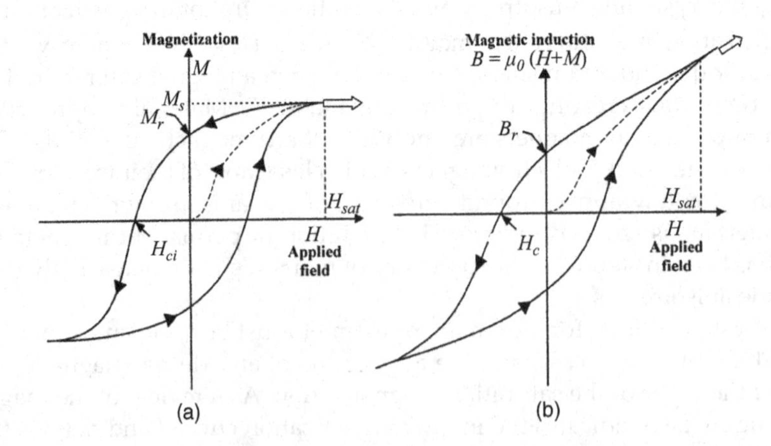

FIGURE 11.2
Schematic representation of realistic hysteresis loops of a permanent magnet material: (a) *M-H* hysteresis loop and (b) *B-H* hysteresis loop. Note the lack of saturation in the *B-H* loop and the definition of H_{ci} in conjunction with the *M-H* loop.

While the $M(H)$ curve reaches saturation, beyond which a further increase in the applied field has no effect on the value of the magnetization (Figure 11.2(a)), the magnetic induction $B(H)$ does not saturate (Figure 11.2(b)). Also, the coercivity H_c of the B-H loop is smaller than that of the M-H loop, which is often indicated as H_{ci} for "intrinsic coercivity", Figure 11.2(a). For simplicity, however, most authors do not adhere to this practice of distinguishing these two concepts of coercivity, using the simpler notation of H_c for either case. In this book, we also adopt this simplification.

11.2 Maximum Energy Product $(BH)_{max}$

In the study of permanent magnets, it is the second quadrant of the hysteresis loop that is of relevance, because its characteristics determine the magnet's suitability for a particular application. The "figure of merit" for a permanent magnet, which defines the magnet's quality or stored magnetostatic energy, is the maximum energy product $(BH)_{max}$, measured in kJ/m³ or MG·Oe (Mega-Gauss Oersted).

In order to understand the meaning of $(BH)_{max}$, let us consider the scheme presented in Figure 11.3. A donut-shaped permanent magnet of length l_m is shown with a small air gap of length l_g. The magnet is fitted with pole pieces made of soft magnetic material that guide the magnetic flux into the gap. The main function of a permanent magnet in a device is to provide an external field $\vec{H}$ in a magnetic circuit, such as the field in this gap. The donut-shaped magnet of length l_m and cross-sectional area A_m provides a field H_g in a small volume, $V_g = A_g l_g$, where A_g is the cross-sectional area of the pole pieces at the air gap. The shape of the magnet assures that all magnetic flux resides in the magnet or is highly localized within the gap, with negligible fringing fields. The magnetic field lines for $\vec{B}$ are continuous, that is, they close upon themselves, as required by the fact that the divergence of the magnetic field is everywhere zero ($\vec{\nabla} \cdot \vec{B} = 0$, no magnetic monopoles exist). Thus,

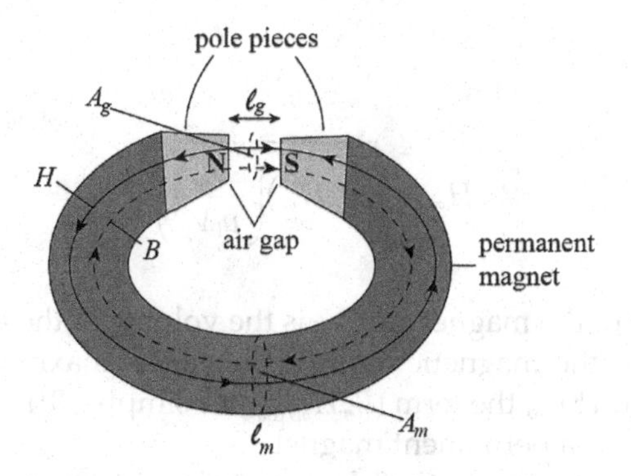

FIGURE 11.3
Schematic illustration of an open magnetic circuit comprised of a ring, made of a permanent magnet material with length l_m and cross-sectional area A_m, two pole pieces, made of a soft magnetic material, and an empty air gap of length l_g and cross-sectional area A_g.

according to the integral form of Maxwell's equation for magnetostatics, the surface integral of $\vec{B}$ over any closed surface always vanishes

$$\oint \vec{B} \cdot d\vec{a} = 0 \tag{11.1}$$

Then, it follows that the flux of the magnetic field as you go around the loop is conserved, and, in particular, it is the same inside the magnet and in the air gap.

$$B_m A_m = B_g A_g \tag{11.2}$$

B_m and B_g are the magnetic induction or magnetic field flux densities inside the magnet and in the gap, respectively.

The field in the gap, H_g, is the demagnetizing field that emanates from the North pole of the magnet and terminates at the South pole. Thus, the magnetic field lines for $\vec{H}$ are discontinuous; they reverse directions at the poles. $\vec{H}$ and $\vec{B}$ are parallel within the gap but antiparallel within the magnet. Ampère's Law for $\vec{H}$ in the absence of any electrical currents (Eq. (2.34)), then gives

$$\oint \vec{H} \cdot d\vec{l} = 0 \tag{11.3}$$

Equation (11.3) then implies that

$$H_m l_m = H_g l_g \tag{11.4}$$

Multiplying Eq. (11.2) with Eq. (11.4), we obtain

$$B_m A_m H_m l_m = B_g A_g H_g l_g \tag{11.5}$$

Using the fact that $\vec{B}_g = \mu_0 \vec{H}_g$ in the gap and rearranging, we observe that

$$H_g^{\,2} = (B_m H_m)\left(\frac{A_m l_m}{\mu_0 A_g l_g}\right) = (B_m H_m)\left(\frac{V_m}{\mu_0 V_g}\right)$$

and

$$H_g = \sqrt{(B_m H_m)\left(\frac{V_m}{\mu_0 V_g}\right)} \tag{11.6}$$

where V_m is the volume of the magnet and V_g is the volume of the air-gap. Therefore, for a given magnet geometry, the magnetic field in the gap is maximized when the product $(B_m H_m)$ is at a maximum. Thus, the term $(B_m H_m)_{max}$, or simply $(BH)_{max}$, is used in determining the "figure of merit" of a permanent magnet.

The magnetic induction or magnetic flux density inside the magnetic material is given by

$$B_m = \mu_0 (H_m + M) = \mu_0 (H_d + M) = \mu_0 (-NM + M) = \mu_0 (1 - N) M \tag{11.7}$$

where $H_m = H_d$, the demagnetizing field, and N is the demagnetizing factor determined by the shape of the magnet, as discussed in Example 3.3. The values of N fall in the range of $0 \leq N \leq 1$.

The condition

$$\frac{\partial(B_m H_m)}{\partial N} = 0 \tag{11.8}$$

gives the value of N that maximizes the energy product. It, thus, determines the shape of an optimum magnet. It is left as an exercise for you to prove that this condition is satisfied for $N = \frac{1}{2}$. The corresponding geometrical shape of the magnet is that of an elongated ellipsoid of revolution, which can be approximated to that of a squat cylinder with a height-to-diameter ratio of 0.3, as illustrated in Figure 11.4.

Furthermore, for an ideal permanent magnet, the $M(H)$ hysteresis loop should be a perfect square, as shown in Figure 11.5(a), with $M_s = M_r$ and $H_c = M_s$. In the second quadrant of the hysteresis loop, the magnetization remains constant at M_r (Figure 11.5(a)). The corresponding values for B_m and H_m (or H_d) for optimized magnet geometry with $N = \frac{1}{2}$ would therefore be $B_m = \frac{\mu_0}{2} M_r$, and $H_m = \frac{1}{2} M_r$ giving an energy product for such an ideal magnet of

$$BH = \frac{1}{4} \mu_0 M_r^2 \tag{11.9}$$

In real magnets, the hysteresis loop is never absolutely square (Figure 11.2(a)). Thus, the above value of Eq. 11.9 is taken to indicate the upper limit of a possible maximum energy product,

$$(BH)_{max} \leq \frac{1}{4} \mu_0 M_r^2 \tag{11.10}$$

In most cases, the condition $H_c \geq \frac{1}{2} M_s$ leads to a large energy product.

Understanding $(BH)_{max}$ has been crucial in the development of low-mass permanent magnets with high-energy products and high stability against de-magnetizing. In the first half of the 20th century, magnets were manufactured in the shape of horseshoe or long cylinders. As new hard materials with high magnetic anisotropy were discovered, squat cylinders eventually replaced these shapes.

The "figure of merit" equals twice the maximum possible magnetostatic energy density available from the magnet, and it is represented graphically as the largest-area rectangle

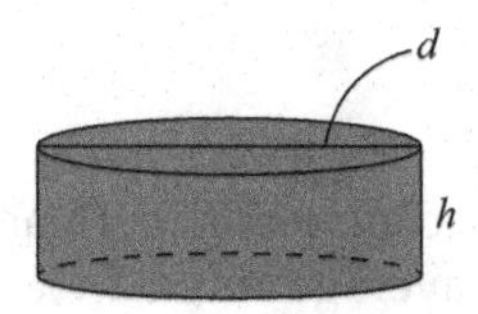

FIGURE 11.4
Illustration of a squat cylinder with a height-to-diameter ratio of 0.3. This shape maximizes the energy product, $B_m H_m$.

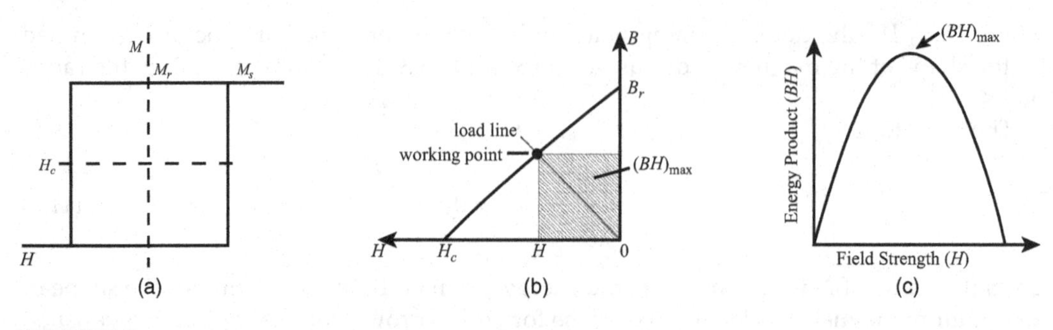

FIGURE 11.5
(a) Schematic representation of the *M-H* loop as a perfect square for an idealized permanent magnet. (b) Corresponding representation of the 2nd quadrant of the *B-H* loop for this idealized perfect permanent magnet indicating (i) $(BH)_{max}$, as the largest-area rectangle inscribed inside the *B-H* loop, (ii) the load line and (iii) the working point of the magnet. (c) Plot of parabolic *BH vs. H* curve. The maximum of this curve defines $(BH)_{max}$.

that can be inserted in the second quadrant of the *B-H* loop, as shown in Figure 11.5(b). This linear relationship of *B* as a function of *H* corresponds to the case of an ideal permanent magnet with the square *M*(*H*) hysteresis loop of Figure 11.5(a), where the magnetization remains constant and equal to M_s throughout the second quadrant. Then,

$$B(H) = \mu_0\left[M(H) + H\right] = \mu_0\left[M_s + H\right] = \mu_0 H + \mu_0 M_s \tag{11.11}$$

B is linear in *H* with a slope of μ_0 and a *y*-axis intercept at $B_r = \mu_0 M_s$. The diagonal line drawn from the origin to the opposite vertex of the $(BH)_{max}$ rectangle is known as the "operating line of the magnet" and the point it intersects the *B*-curve is known as the "operating point". These parameters are of relevance in the design of permanent magnets used in dynamic operations within magnetic circuits. A plot of the product *BH* as a function of the field strength *H* follows a parabolic curve as shown in Figure 11.5(c). The maximum of the parabola gives an alternate definition of $(BH)_{max}$.

11.3 Evolution of Permanent Magnet Materials

Figure 11.6 depicts the evolution of permanent magnets in the 20th century. It indicates the increase in the value of $(BH)_{max}$ over time. It also includes schematics of the shapes and relative sizes of magnets made of different materials corresponding to the same energy product, or magnet capacity. Table 11.1 gives the intrinsic magnetic properties of various hard magnetic materials. It also includes the magnetic properties of the three elemental ferromagnets, Fe, Co and Ni, for comparison.

11.3.1 Steel-Based Magnets

At the beginning of the 20th century, commercial permanent magnets were based on steels, which are iron alloys containing on average about 2.1% carbon by weight. They were used in the hardened state obtained by quenching, that is, rapidly cooling the steel sample from an initial high temperature. In steels, high coercivity was accompanied with physical hardness, which is where the term "hard" magnetic material originates. The superior

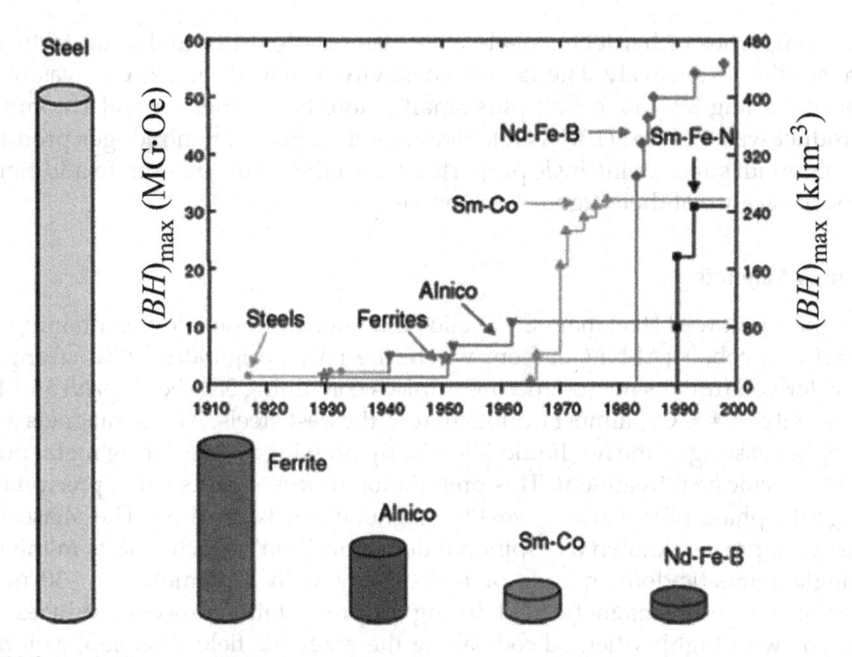

FIGURE 11.6
Evolution of permanent magnet materials and development of maximum energy product, $(BH)_{max}$ in the 20th century. As higher magnetic anisotropy materials were discovered, the size of magnets of comparable magnetostatic energy content was successively decreased.

TABLE 11.1
Some Intrinsic Magnetic Properties of Hard Magnetic Materials Used in the Production of Permanent Magnets

Material	T_C (K)	M_s (×10⁶ A/m)	K_1 (×10⁶ J/m³)	A (×10⁻¹² J/m)	δ_w (nm)	R_{SD} (nm)	Crystallographic Structure
Fe	1,043	1.71	0.048	8.3	40	6	bcc
Co	1,388	1.40	0.45	10.3	14	34	hcp
Ni	631	0.49	−0.005	3.4	82	16	fcc
$L1_0$-CoPt	840	0.80	4.9	10.0	4.5	310	Tetragonal
$L1_0$-FePt	750	1.14	6.6	10.0	3.9		Tetragonal
MnBi	630	0.62	1.2				Hexagonal
$BaFe_{12}O_{19}$	723	0.38	0.33	6.3	14	290	Hexagonal
$SrFe_{12}O_{19}$	733	0.37	0.35				Hexagonal
$SmCo_5$	1003	0.84	17	22	3.6	425	Hexagonal
Sm_2Co_{17}	1190	1.03	3.3	14	6.5*	250	Rhombohedral
$Nd_2Fe_{14}B$	585	1.28	5.0	7.7	3.9		Tetragonal

* Calculated from Eq. (3.70) $\left(\delta_w = \pi\sqrt{\frac{A}{K_1}}\right)$.

For comparison the magnetic properties of Fe, Co and Ni are also included. (T_C, Curie temperature, M_s, saturation magnetization, K_1, first magnetocrystalline anisotropy constant, A, exchange stiffness of crystallographic lattice, δ_w, Bloch magnetic domain wall thickness, R_{SD}, radius of single magnetic domain particle formation).

mechanical properties of hardened steels were also of interest in tool manufacturing, so they were studied extensively. The largest coercivity achieved, of 230 Oe, was obtained with steel containing 30–40% cobalt plus small amounts of tungsten and chromium. Its energy product was 8 kJ/m^3 (1 MG·Oe). Steel-based magnets are no longer produced as new materials with superior intrinsic properties have entered the market. In addition, they are expensive because of their high cobalt content.

11.3.2 Alnico Magnets

Alnico magnets followed steel magnets. Alnico is a family of iron alloys containing aluminum, nickel and cobalt (Al-Ni-Co, from where the name originates) with strong shape anisotropy derived from its microstructure. Alnicos containing 58% Fe, 30% Ni and 12% Al have a coercivity of 400 Oe, almost double that of the best steels. Alnico magnets are produced by either casting of the hot liquid alloy, or by pressing and sintering metal powders followed by specific heat treatment. This preparation process results in the precipitation of a ferromagnetic phase of Fe-Co in a weakly magnetic matrix of Al-Ni. The phase separation occurs by a process known as "spinodal decomposition" which results in the formation of single-magnetic-domain rods of Fe-Co alloy with a diameter of ~30 nm. The application of an external magnetic field during the precipitation process produces "anisotropic Alnico" with highly oriented rods along the magnetic field direction, as indicated schematically in Figure 11.7.

Thus, the source of coercivity in alnico is due to its microstructure and derives from the high shape-anisotropy of the magnetic nanorods. Alnicos, therefore, are considered to be the original "nanostructured permanent magnets".

11.3.3 Ferrite Magnets

Ferrite (or ceramic) magnets are made of $BaFe_{12}O_{19}$ or $SrFe_{12}O_{19}$, which are referred to as hexaferrites due to their hexagonal crystallographic structure that endows them with high intrinsic coercivity; as opposed to cubic ferrites, which have a spinel crystallographic structure and low coercivities. The hexaferrites are made by solid-state reaction synthesis. Mixing $BaCO_3$ with Fe_2O_3 and firing the mixture at 1,200°C, for instance, make barium hexaferrite. In these compounds, the hexagonal *c*-axis is the easy axis of magnetization.

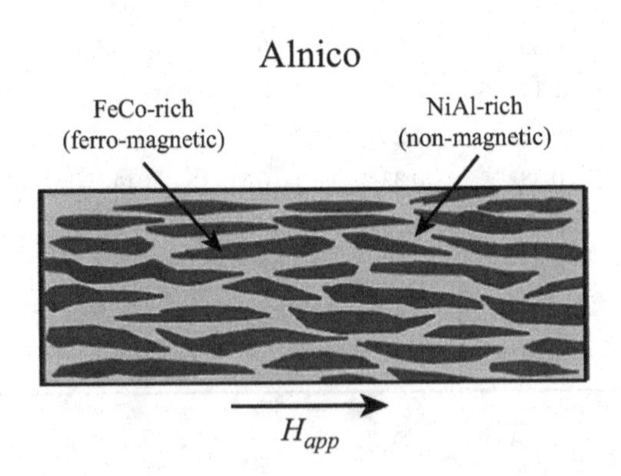

FIGURE 11.7
Illustration describing the microstructure of anisotropic Alnico. Long rods of ferromagnetic Fe-Co alloy with a diameter of ~30 nm are precipitated within a non-magnetic matrix of Al-Ni. The application of an external magnetic field $\vec{H}_{app}$ during the precipitation process aligns the nanorods in the direction of the field.

The magnetocrystalline anisotropy constant for $BaFe_{12}O_{19}$ has the value of $K = 3.3 \times 10^5$ J/m^3, while that for $SrFe_{12}O_{19}$ is a bit higher (Table 11.1).

Morphologically, $BaFe_{12}O_{19}$ forms platelets, disk-like particles, with the *c*-axis normal to the plate. The large magnetocrystalline anisotropy leads to much larger coercivity compared with the steel-based magnets, higher energy density and smaller magnet volume. To produce a ferrite magnet, the as-synthesized $BaFe_{12}O_{19}$ powder is ball-milled to reduce the particle size and is then pressed dry in a die and sintered at 1,200°C. This process produces "isotropic magnets" where the plate-like particles of $BaFe_{12}O_{19}$ are randomly oriented. Non-interacting, randomly oriented nanoparticles are known to produce a remanence of $M_r = \frac{1}{2} M_s$, resulting in a non-square $M - H$ hysteresis loop (see discussion of Figure 4.11). Partial orientation can be achieved by pressing, as the platelets have the tendency to pack together with their surfaces parallel to each other. The application of a magnetic field during a wet pressing process results in complete *c*-axis orientation in the direction of the applied field and the production of an "anisotropic magnet" with enhanced coercivity. The highest energy product $(BH)_{max}$ achieved for commercial ferrite magnets is ~28 kJ/m^3 or 3.5 MG·Oe. The hexaferrites have important technical advantages, as they are inexpensive, chemically inert and easy to process. They also possess relatively high Curie temperatures (Table 11.1) with outstanding performance at high temperatures.

11.3.4 Rare-Earth–Based Magnets

Many rare-earth elements are ferromagnetic with very strong anisotropy arising from the orbital angular momentum of the 4*f* electronic shell and spin–orbit interactions. Unfortunately, they all have Curie temperatures below room temperature. However, they form "intermetallic compounds" with all three room temperature transition metal ferromagnets, Fe, Co and Ni, with Curie temperatures well above room temperature. These intermetallic compounds retain the strong magnetic anisotropy of the rare-earth element, which is intrinsic to the atom and does not depend on the atom's interactions with its surroundings. Thus, they are good candidates for permanent magnet applications.

Let us define the term "intermetallic compound". When two metals of similar size are combined to form an alloy, either a "solid solution" or an "ordered superstructure" of the two metals is produced. The degree of order depends on the thermal treatment of the alloy. The ferromagnetic 3*d* elements Fe, Co and Ni form a variety of alloys with each other and with other metals. However, when metals of different sizes are combined, they form "intermetallic compounds" of well-defined composition, rather than "solid solutions". The atomic radii of 3*d* transition elements are ~0.125 nm, while the radii of the (4*f*) rare-earth elements are ~0.180 nm. Thus, the 4*f* elements occupy a volume three times that of the 3*d* elements, a fact that hinders the formation of solid solutions or ordered superstructures of the two metals, leading to intermetallic formation. An intermetallic compound is made up of two or more elements that produce a new phase with its own composition, crystal structure, and properties. On the other hand, interstitial compounds are formed when a small atom with an atomic radius less than 0.1 nm (like boron, carbon or nitrogen) enters an interstitial site in a 3*d* metal alloy. Hydrogen, on the other hand, can also enter the interstitial sites of rare-earth elements and their compounds. The addition of such interstitial atoms produces dilation of the crystal lattice with striking changes in magnetic properties of the alloys, where the interatomic exchange is particularly sensitive to interatomic distance. Such processes have been widely exploited in the search of new materials for permanent magnet applications.

As seen in Figure 11.6, the most significant increases in the energy product came in the 1960s and 1980s with the discovery of the intermetallic rare-earth compounds Sm-Co and

Nd-Fe-B, respectively. The intermetallic alloy $SmCo_5$ typically contains 36% samarium by weight with the balance being cobalt. It has a hexagonal crystal structure and a large uniaxial anisotropy constant $K = 17.2 \times 10^6$ J/m³, with the easy axis of magnetization along the *c*-axis of the unit shell. The $SmCo_5$-based magnets contain particles of the alloys of about 10 μm in size, each particle consisting of a single crystal of $SmCo_5$. The particles are aligned in a magnetic field, that is, physically rotated, so the easy axes of all the particles are parallel to each other and to the applied field and then compressed in a die. The production of "anisotropic magnets" by magnetic field alignment increases M_r by a factor of two, relative to non-oriented "isotropic magnets" and, therefore, the energy product by a factor of four. $SmCo_5$-based magnets attained values of $(BH)_{max}$ ranging from 16 to 25 MG·Oe, which is approximately 128–200 kJ/m³ (Figure 11.6).

Another intermetallic compound of samarium and cobalt has also been used in the production of permanent magnets: Sm_2Co_{17}. This notation is actually a code name for a much more complex composition. The actual composition consists of two atoms of rare-earth samarium and 13–17 atoms of transition metals, such as Co, Fe and Cu. However, the transition metal content is always rich in cobalt. Other elements like Zr and Hf may also be added in small quantities to achieve a better heat treatment response. The alloy generally contains about 25% samarium by weight. In magnets based on Sm_2Co_{17}, the microstructure consists of bands of $SmCo_5$ separating regions of Sm_2Co_{17}, at a very fine scale. The maximum energy products of Sm_2Co_{17} type magnets range from 20 to 32 MG·Oe, or about 160–260 kJ/m³ (Figure 11.6).

The high cost of Sm-Co-based magnets motivated continued research to identify new compounds for stronger and more affordable magnets. In 1984, the new compound $Fe_{14}Nd_2B$ was discovered with a tetragonal crystal structure, very strong uniaxial magnetic anisotropy and a Curie temperature above room temperature. This compound has large iron content. Iron is much cheaper than cobalt. As a result, Fe-Nd-B-based magnets (also referred to as neodymium magnets) quickly replaced Sm-Co magnets in many applications. This was facilitated by the fact that many processing techniques used in the manufacture of Sm-Co magnets were applicable to the production of the new Fe-Nd-B magnets. Fe-Nd-B-based magnets reached an energy product of 50 MG·Oe or 398 kJ/m³. One disadvantage of Fe-Nd-B magnets is the fact that they have a relatively low Curie temperature of 312°C. This makes their magnetic properties sensitive to temperature variations close to room temperature, rendering them inappropriate for energy applications at high temperatures. Some of the Fe may be replaced by Co, which raises the Curie temperature and improves the temperature stability of the magnet. Some of the Nd may be replaced by heavy rare-earth elements like Dy, which increases the coercivity, but lowers the saturation magnetization because Dy couples antiferromagnetically to Fe. In addition, such atomic substitutions increase the cost of the magnets and make their production dependent on expensive heavy rare-earth elements that are in short supply.

Another approach to increasing the coercivity, applicable to all permanent magnet alloy materials including the ferrites, $BaFe_{12}O_{19}$ and $SrFe_{12}O_{19}$, is to reduce the particle size to micrometer size and mix them with a polymer to produce "bonded magnets". This allows the production of mountable or flexible magnets (refrigerator magnets). However, the packing fraction of magnetic material is about 70%, resulting in reduced saturation magnetization and energy product. Most of the Fe-Nd-B-based magnets are manufactured in China, the world's largest rare-earth element producer. The neodymium-based magnets are also prone to corrosion that necessitates an additional step of coating the magnets to protect them from oxidation. Despite these shortcomings, the rare-earth–based magnets are superior to Alnico and ferrite magnets; they would replace them completely, except for

TABLE 11.2

Selected Magnetic Properties of Some Permanent-Magnet Materials or Nanocomposites

Material	B_r (T)	$\mu_0 H_c$ (T)	$(BH)_{max}$ ($\times 10^3$J/m^3)
Alnico (isotropic)	0.6–1.0	0.1–0.2	10–30
Alnico (anisotropic)	1.3–1.4	0.3–0.4	40–80
$BaFe_{12}O_{19}$ (sintered)	0.39	0.3	28
$SrFe_{12}O_{19}$	0.2–0.4	0.13–0.37	10–40
$SmCo_5$ (metal-bonded)	0.92	1.88	175
$SmCo_5$ (polymer-bonded)	0.58	1.00	60
$Sm_2Co_{17}/SmCo_5$ (sintered)	1.08	1.00	225
$Nd_2Fe_{14}B$ (sintered)	1.0–1.4	1.0–2.5	200–440
$Nd_2Fe_{14}B$ (polymer-bonded	0.6–0.7	0.75–1.5	50–100
$Nd_2Fe_{14}B/\alpha$-FeCo (nanocomposite)	1.4–1.6	1.0–1.8	320–480
MnBi (Nanostructured powder)	0.25	1.0–2.1	14

their cost. Table 11.2 gives the maximum energy product and $B(H)$ hysteresis loop parameters for some of the permanent magnet materials and nanocomposites discussed.

Returning to Figure 11.6 we observe that the overall increase in $(BH)_{max}$during the last century has been dramatic; increasing from 1 kJ/m^3 to over 400 kJ/m^3. Equation (11.6) would imply that massive, huge magnets (with large V_m) are necessary to produce strong magnetic fields. This dramatic increase in $(BH)_{max}$ has allowed the miniaturization of permanent magnets that are found today in many household appliances, cordless power tools, computers, cell phones and other consumer electronic gadgets, enabling the current revolution of nanotechnology.

As mentioned earlier, Table 11.1 lists some properties of compounds with high intrinsic magnetic anisotropies that have been exploited in the evolution of permanent magnet materials. For comparison, the intrinsic magnetic properties of the three elemental ferromagnets have also been included. Even though, it is the spontaneous magnetization that ultimately governs the properties of a magnet, the intermetallic compounds that modern magnets are based on have inferior saturation magnetization than the elemental ferromagnets or some of their alloys, such as the alloy $Fe_{65}Co_{35}$ which is known to have the highest recorded spontaneous magnetization of any magnetic material with $M_s = 1.95$ MA/m. This is due to the fact that high coercivity requires a high anisotropy field, $H_K = 2K/\mu_o M_s$; the fact that the saturation magnetization, M_s, appears in the denominator indicates that materials with high anisotropy fields tend to have low saturation magnetization, in addition to having high magnetic anisotropy density. Thus, the search for better and cheaper permanent magnet materials is still on.

11.4 Future Permanent Magnet Materials

In the absence of discovering new magnetic alloys with higher coercivity and saturation magnetization, high-quality permanent magnets of the future will require engineering of their micro-/nanostructure in order to stabilize the magnetically fully saturated state against the creation of reverse polarization domains and self-demagnetization. One must design the microstructure of the magnet in a way that it resists the nucleation of reverse domains and prevents domain walls from moving easily to expand areas of reversed magnetization. The latter is known as "magnetic wall pinning" by the introduction of defects in the crystallographic structure. Therefore, control of coercivity implies control of microstructure at the nanoscale. Nanoscopic tailoring will play a fundamental role in the design of next-generation permanent magnet materials.

As discussed in Chapter 4, single-magnetic-domain particles exhibit higher coercivities than their multi-domain counterparts. This is due to the fact that magnetic saturation can only be achieved *via* the energetically demanding process of Stoner and Wohlfarth type coherent rotation of magnetic moments, rather than through the easy process of wall movement. Coherent magnetization reversals result in large coercivities that can approach the magnetic anisotropy field, H_K. This coercivity enhancement achieved by creating isolated magnetic domains with the critical single-domain radius R_{SD} of Eq. (4.5) is motivating the nanostructural engineering of novel magnetic materials. In fact, for some hard magnetic materials, the critical radius for single-domain formation is quite large, $R_{SD} > 100$ nm (Table 11.1). Furthermore, at the nanoscale, additional contributions to the anisotropy from surface, strain and shape effects (see Chapter 4) can be introduced to augment the magnetocrystalline anisotropy of the material. Contributions of surface anisotropy are important only at the nanoscale due to the large surface-to-volume ratio in nanostructures. Strain effects are often found in inter-granular regions or at the interfaces between two phases, due to lattice mismatch between their constituent materials. The crystal field can be modified by strain at the nanoscale, altering the magnetocrystalline anisotropy and the direction of the easy axis of magnetization. Another type of anisotropy that originates at interface regions arises from quantum mechanical exchange across the boundary between two phases, already discussed in Section 4.12. This exchange anisotropy, also known as the exchange-bias effect, manifests itself as a lateral shift of the hysteresis loop to the left (Figure 4.23), resulting in increased coercivity.

A system of non-interacting, single-domain particles with their easy axis directions randomly distributed possesses a remanence half its saturation magnetization, $M_r = \frac{1}{2} M_s$ (Figure 4.11). Crystallographic or easy axis alignment of the nanoparticles and the engineering of interparticle and interfacial magnetic interactions can achieve increased remanence. Thus, in addition to material composition, nanostructuring methodologies offer additional degrees of freedom for the control and tailoring of both the intrinsic (K, M_s) and extrinsic (H_c, M_r) properties of a magnet.

11.4.1 Exchange-Coupled Hard/Soft Magnetic Phases: The Exchange-Spring Magnet

As seen in Table 11.1, most hard magnetic materials have rather low saturation magnetization. In order to increase the saturation magnetization, the concept of magnetic coupling across interfaces has been evoked in the proposition of the "nanocomposite magnet", in which a hard magnetic phase providing high coercivity is combined with a soft magnetic phase providing high saturation magnetization. The combination of two, or more, magnetic phases at the nanoscale could create a material with properties superior to those of either phase. As an illustration consider core/shell nanoparticles (Figure 11.8(a)) with a hard ferromagnetic core, such as a binary (rare-earth)/(transition-metal) intermetallic compound, $SmCo_5$ or Sm_2Co_{17}, and a soft ferromagnetic shell, such as Fe or Co. In such nanocomposites, the anisotropy comes primarily from the aspherical 4*f* electronic orbitals

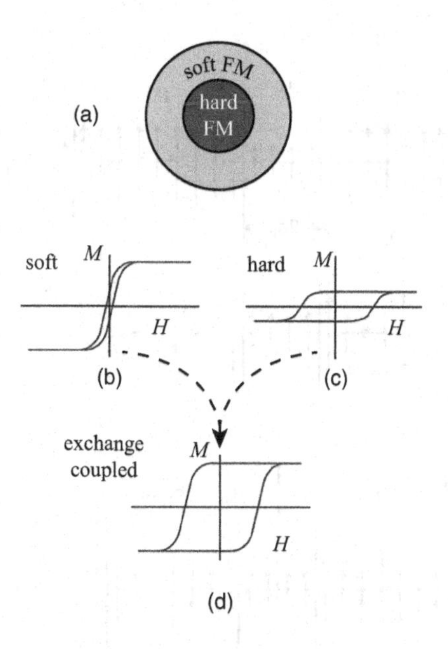

FIGURE 11.8
Demonstration of the principle of the exchange-coupled nanocomposite magnet. A hard-core/soft-shell magnetic nanoparticle with strong interfacial exchange could exhibit a single-phase like hysteresis loop with large coercivity and saturation magnetization leading to an enhanced energy product. See text.

of the rare-earth atoms that exhibit strong spin–orbit coupling, while the high magnetic saturation is contributed by the transition metal. A strong magnetic exchange coupling across the interface, governed by the intimate contact between the core and shell material, can stabilize the magnetically soft shell against demagnetization. Such coupling can result in "cooperative magnetic switching", where the soft and hard magnetic phases reverse in unison. If this were the case, the nanocomposite would behave as a single magnetic phase with high coercivity and relatively high magnetization. The soft magnetic material, on its own, has a narrow hysteresis loop with high magnetic saturation but very small coercivity (Figure 11.8(b)). In contrast, the hard magnetic material has large coercivity but suppressed saturation magnetization (Figure 11.8(c)). The exchange-coupled composite system should maximize the energy product by rendering a broad, square hysteresis loop with both high magnetic saturation, due to the presence of the soft magnetic phase, and high coercivity, due to the presence of the hard magnetic phase (Figure 11.8(d)).

The proposition to construct such magnets was advanced in 1991 by Kneller and Hawig based on theoretical considerations (Eckart F. Kneller and Reinhard Hawig, The Exchange-Spring Magnet: A New Material Principle for Permanent Magnets, *IEEE Transactions on Magnetics*, 27 (1991) 3588). Figure 11.9 gives a schematic of a one-dimensional model of the micromagnetic structure of the exchange-coupled composite material used by these authors as a basis for the calculation of the critical dimensions of the two phases, which would produce single-phase hysteresis loops with enhanced energy product.

The figure depicts soft magnetic phases, m, with high saturation magnetization (M_m, long arrows) and low coercivity sandwiched between hard magnetic phases, k, of high coercivity but low saturation magnetization (M_k, short arrows). Figure 11.9(a) corresponds to the saturated remanence state. The strong exchange coupling between the two materials at the interface region keeps their magnetization directions aligned. With increasing field applied in the direction opposite to the magnetization direction, the magnetic moments in

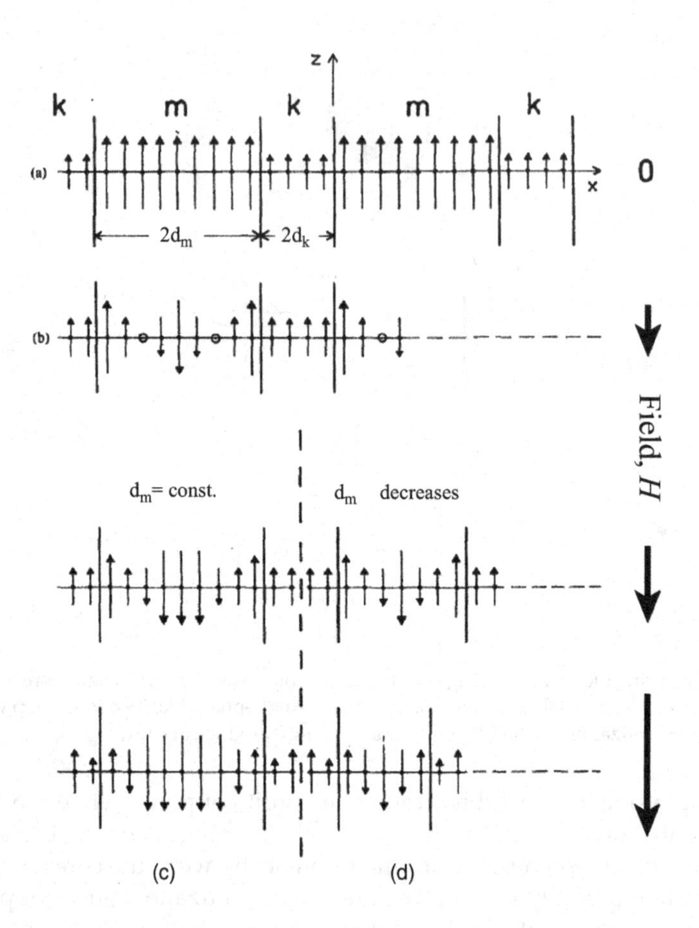

FIGURE 11.9
Schematic representation of a one-dimensional model of the micromagnetic structure of an exchange-coupled composite material. The soft magnetic phase (*m*) is depicted with long arrows (high magnetization M_m) and the hard magnetic phase (*k*) with short arrows (low magnetization M_k). (a) Magnetic moment configuration at magnetic saturation. (b) The onset of magnetization reversal in the middle of the soft phase upon application of a magnetic field in the direction opposite to the magnetization. (c) Demagnetization in an increasing magnetic field at a constant overcritical width $d_m >> d_{mc}$. (d) Demagnetization at decreasing width $d_m \rightarrow d_{mc}$. (Adapted from Eckart F. Kneller and Reinhard Hawig, The Exchange-Spring Magnet: A New Material Principle for Permanent Magnets, *IEEE Transactions on Magnetics*, 27 (1991) 3588, Copyright IEEE, with permission).

the central region of the soft magnetic phase are the first to start rotating into the direction of the applied field, while those next to the hard phase remain pinned in the direction of magnetization of the hard phase, due to the strong interfacial coupling interactions. In the case where the dimensions of the soft magnetic phase, d_m, are relatively large, of the order of the magnetic-domain-wall width of the soft phase, $\delta_m = \pi\sqrt{A_m/K_m}$ (Eq. (3.70), two 180° walls are shown to form reversibly within the soft phase (Figure 11.9 (b and c)). As long as the applied field is less than the coercive field of the hard phase the process of magnetization rotation in the soft phase should be reversible. With further increasing the strength of the applied magnetic field these walls move toward the hard phase, eventually penetrating within the hard phase to produce irreversible magnetization rotation. For smaller d_m, of the order of the magnetic-domain-wall width of the hard phase, $\delta_k = \pi\sqrt{A_k/K_k}$, where $\delta_m >> \delta_k$ due to the fact that $K_m << K_k$, one can reach a critical dimension for best permanent magnet behavior as $d_m \rightarrow \delta_k$ (Figure 11.9(d)).

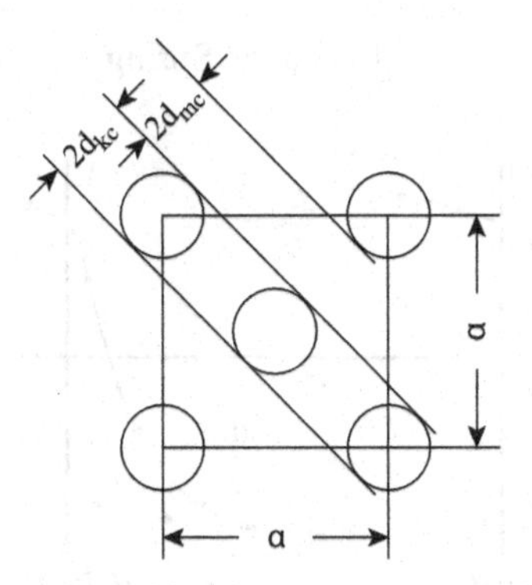

FIGURE 11.10
Schematic model of an optimal microstructure. The hard magnetic *k*-phase forms precipitates (spheres) within an fcc space lattice of the soft *m*-phase with equal critical dimensions $d_{mc} = d_{kc}$. (Adapted from Eckart F. Kneller and Reinhard Hawig, The Exchange-Spring Magnet: A New Material Principle for Permanent Magnets, *IEEE Transactions on Magnetics*, 27 (1991) 3588, Copyright IEEE, with permission).

Figure 11.10 gives the schematic of a favorable microstructure, where spherical precipitates of a hard phase are depicted on an fcc lattice of a soft phase, with equal critical dimensions $d_{mc} = d_{kc}$. For such a configuration, the volume fraction of the hard phase, $v_k = \frac{V_k}{V}$, where V_k is the volume of the hard phase and V is the total volume of the composite, is given by $v_k = \frac{\pi}{24\sqrt{2}} \approx 0.09$, indicating that the nanocomposite magnet may contain as little as 10% of the hard phase, which contains the expensive rare-earth elements. This still holds for the case of hard phase precipitates in a bcc lattice, where v_k is calculated to be of the same order of magnitude, $v_k = \frac{\pi\sqrt{3}}{64} \approx 0.09$.

Figures 11.11(a) and (b) depict the demagnetization curves of such exchange-spring magnets corresponding to optimized critical dimensions for the soft phase, $d_m = d_{mc}$, and for non-optimized $d_m >> d_{mc}$, respectively, indicating the reversible and irreversible regions. For comparison, the demagnetization curves expected for a conventional magnet and that of a mixture of two independent, uncoupled phases are also shown in Figures 11.11(c) and (d), respectively. In the figures, the continuous lines with the arrows indicate reversible regions of the demagnetization curve, while the dashed lines indicate irreversible ones. The process of reversibility, an important feature of exchange-spring magnets, is further elaborated in the next section.

11.4.2 Magnetization Reversal in Exchange-Spring Magnets

The theoretical predictions of Kneller and Hawig agree with experimental observations. The mechanism of action responsible for cooperative magnetization switching in such exchange-coupled magnets can be easily illustrated by considering bilayer or multilayer thin films as model experimental systems. These models are composed of alternate layers of the hard and soft phases; their exchange-spring behavior can be understood in terms of various composition parameters, such as (a) individual intrinsic magnetic properties, (b) dimensions, (c) relative volume fractions and (d) relative geometric arrangement of the soft and hard phases. Not surprising, as predicted by theory, the most important parameter for cooperative switching behavior turns out to be the dimension of the soft phase. When a soft layer is sandwiched between two hard layers, there is a critical soft-layer

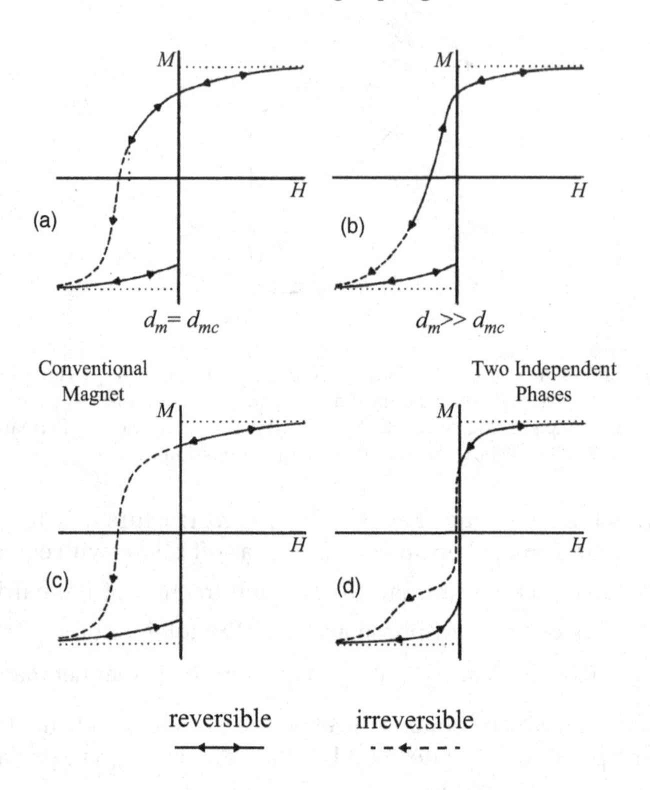

FIGURE 11.11
Schematic depiction of theoretically predicted demagnetization curves for an exchange-spring nanocomposite magnet with (a) optimized critical dimension of the soft phase, $d_m = d_{mc}$, (b) non-optimized dimensions, $d_m >> d_{mc}$. (c) Demagnetization curve of a conventional magnet. (d) Demagnetization curve of a mixture of uncoupled soft and hard phases reversing independently and producing a constricted hysteresis loop. Reversible regions are indicated by a solid line with superimposed arrows, while irreversible regions are indicated by a broken line. (Adapted from Eckart F. Kneller and Reinhard Hawig, The Exchange-Spring Magnet: A New Material Principle for Permanent Magnets, *IEEE Transactions on Magnetics*, 27 (1991) 3588, Copyright IEEE, with permission).

thickness below which the soft layer atomic moments are rigidly coupled to those of the hard layer and the two phases reverse simultaneously, presenting a single-phase-like hysteresis loop. For thicker soft layers, the soft phase reverses at significantly lower fields and the hysteresis loop becomes constricted. In agreement with theory, the critical soft layer thickness found experimentally is roughly equal to $2\delta_k$, twice the width of the domain wall of the hard phase.

We now illustrate the process of magnetization reversal in nanocomposite exchange-spring magnets by considering a hard/soft ferromagnetic bilayer of Sm-Co/Fe, depicted schematically in Figure 11.12. In the absence of an external field and for a thickness of the soft layer below the critical thickness, $t_c = 2\delta_k$, the magnetization of the soft layer, M_m, is ferromagnetically coupled to that of the hard layer, M_k, through exchange interactions at the interface, the strength of which can be represented by an exchange field, H_{ex}. Upon application of an increasing external magnetic field, H_{app}, in a direction opposite to the magnetization direction within the bilayer, the ferromagnetic coupling at the interface remains unperturbed until the strength of the applied field exceeds H_{ex}. As previously

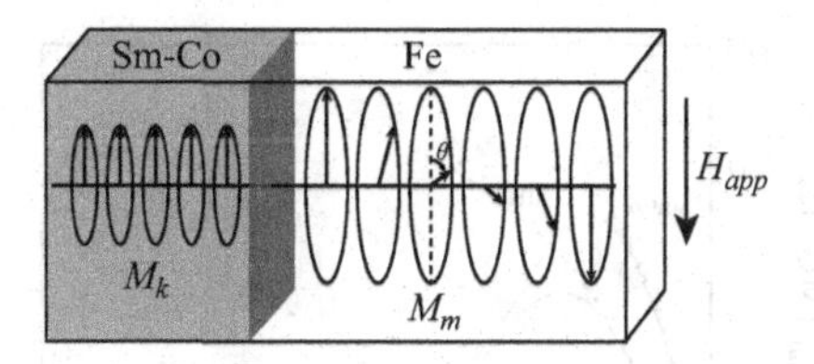

FIGURE 11.12
Schematic representation of a Sm-Co/Fe hard/soft magnetic bilayer demonstrating the mechanism of the restoring torque responsible for the reversible regions of the demagnetization curves of Figure 11.11.

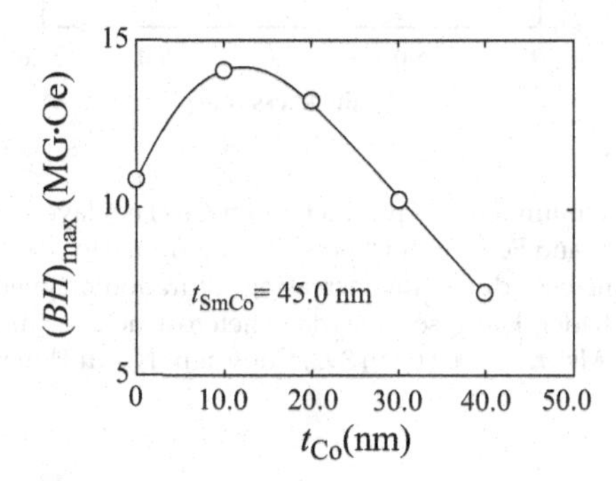

FIGURE 11.13
Maximum energy product for a case of $[Sm\text{-}Co/Co]_{10}$ multilayers, with a fixed Sm-Co hard-phase layer thickness at 45 nm, as a function of a variable Co soft-phase layer thickness. The solid line is a guide to the eye. (Adapted from E. E. Fullerton, J. S. Jiang and S. D. Bader, Hard/soft magnetic heterostructures: model exchange-spring magnets, *Journal of Magnetism and Magnetic Materials*, 200 (1999) 392, Copyright North Holland (Elsevier), with permission).

noted in our discussion of Figure 11.9, the magnetic moments of atoms of the soft phase located farther from the interface are the first to start rotating into the direction of the applied magnetic field, while those lying at the interface remain pinned in the direction of magnetization within the hard phase. For $H_{app} < H_{ex}$, the atomic magnetic moments in the soft layer exhibit continuous rotation, reminiscent of a Bloch wall, with the angle of rotation, θ, decreasing with decreasing distance from the interface. Such systems produce reversible demagnetization curves, due to the fact that moments in the soft layers rotate back into alignment with those within the hard phase upon removal of the externally applied magnetic field. Thus, the moments experience a restoring torque, in analogy with the restoring force experienced by a body in the mechanical analog of the mass-spring system. It is the visualization of this reversible process that inspired the name "exchange-spring magnet" in analogy to the elastic motion of mechanical springs.

Figure 11.13 shows the maximum energy product for the case of $[Sm\text{-}Co/Co]_{10}$ multilayers, with a fixed Sm-Co hard-phase layer thickness at 45 nm, as a function of a variable Co soft-phase layer thickness. Initially, $(BH)_{max}$ increases above that of the hard phase, reaches a maximum at a soft-layer thickness of about 12.5 nm and decreases with further increase of the soft-layer thickness. This maximum occurs at $t_{Co} \sim 12$ nm, corresponding to a critical thickness for the soft layer of the order of $2\delta_k$ (see Table 11.1). Although a single Sm_2Co_{17} layer has a rather large coercivity of $H_c = 3$ T, its low saturation magnetization yields a low energy product, $(BH)_{max} = 11$ MG·Oe. The energy product increases by almost 30% to reach the value of about 14 MG·Oe at optimum soft layer thickness in $[Sm\text{-}Co/Co]_{10}$ multilayers,

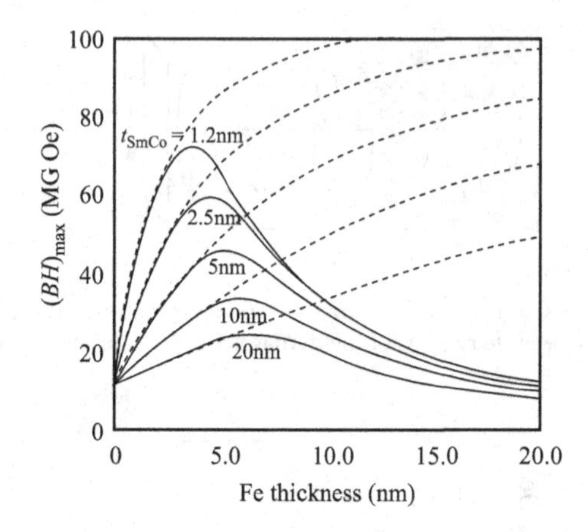

FIGURE 11.14
Numerical calculations for the maximum energy product of Sm-Co/Fe bilayers with the Sm-Co layer thickness varying between 1.2 nm and 20 nm, and Fe-layer thickness varying from 0 to 20 nm. The dashed curves depict the ideal maximum energy product increase due to the increasing saturation magnetization of the bilayer. (From E. E. Fullerton, J. S. Jiang and S. D. Bader, Hard/soft magnetic heterostructures: model exchange-spring magnets, *Journal of Magnetism and Magnetic Materials* 200 (1999) 392, Copyright North Holland (Elsevier), reproduced with permission).

aided by the larger saturation magnetization of the soft layer. This large percent increase in the energy product demonstrates the value of the exchange-spring nanostructuring, where nanomagnetism strategies can play a very important role in the development of next-generation permanent magnets. Even though the saturation magnetization of the multilayers keeps increasing with further increase of the soft-layer thickness, the coercivity decreases due to the softening of the magnetic properties for Co far away from the interface, degrading the value of $(BH)_{max}$.

Numerical calculations for the maximum energy product of Sm-Co/Fe bilayers with a Sm-Co layer thickness varying between 1.2 and 20 nm, and Fe-layer thickness varying from 0 to 20 nm are shown in Figure 11.14. The calculated $(BH)_{max}$ increases with decreasing hard-layer thickness. It initially increases with increasing soft-layer thickness, up to a maximum, and then decreases with further increase of soft-layer thickness. Also shown as dashed curves is the ideal maximum energy product according to Eq. (11.9), due to the increasing saturation magnetization of the bilayer. The calculations show that for bilayers with thin constituent layers one can achieve a maximum energy product above that of Nd-Fe-B magnets, which have $(BH)_{max} \sim 55$ MG·Oe (Figure 11.6).

Example 11.1: Hysteresis Loops in Sm-Co/Fe Bilayers

Detailed hysteresis loops of exchange-spring-coupled bilayers of Sm-Co/Fe have been investigated by Fullerton, Jiang and Bader (E.E. Fullerton, J. S. Jiang and S. D. Bader, Hard/soft magnetic heterostructures: model exchange-spring magnets, *Journal of Magnetism and Magnetic Materials* 200 (1999) 392). Bilayers are the simplest structures with which the exchange-spring principle can be demonstrated. These investigators prepared Sm-Co/Fe bilayers on single-crystal MgO (110) substrates coated with 20-nm Cr (211) buffer layer. This produced an oriented Sm-Co layer with uniaxial in-plane anisotropy. The magnetically soft Fe layer was polycrystalline with a (110) texture. Figure 11.15 shows the hysteresis loops for the reduced magnetization (M/M_s) with an applied field along the easy axis

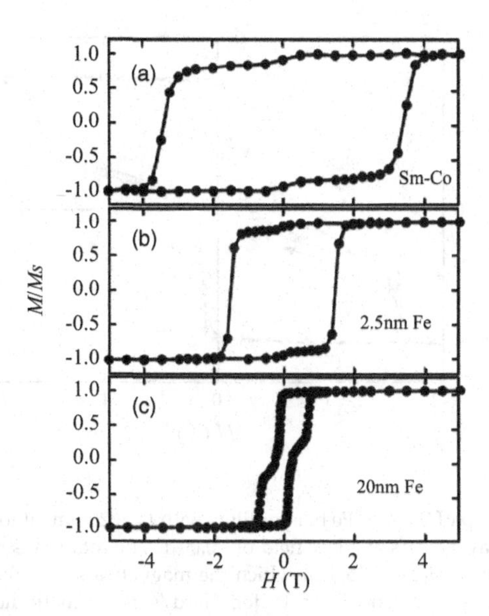

FIGURE 11.15
Room-temperature (T = 300 K) hysteresis loops for (a) a single Sm-Co film, (b) a Sm-Co/Fe bilayer with an iron thickness of 2.5 nm, less than the critical thickness for simultaneous switching and (c) an Sm-Co/Fe bilayer with an iron thickness of 20 nm, larger than the critical thickness for simultaneous switching. (From E.E. Fullerton, J. S. Jiang and S. D. Bader, Hard/soft magnetic heterostructures: model exchange-spring magnets, *Journal of Magnetism and Magnetic Materials* 200 (1999) 392, Copyright North Holland (Elsevier), reproduced with permission).

direction of magnetization. For a single Sm-Co layer, a single-phase-like square hysteresis loop is observed (Figure 11.15(a)) with a coercivity of about 3.5 T. The bilayer with a 2.5 nm thick Fe layer also shows a single-phase hysteresis loop, indicating the presence of cooperative switching, where both the soft and hard layers reverse simultaneously as a single unit (Figure 11.15(b)). This is to be expected as the thickness of 2.5 nm is below its critical value for simultaneous switching. The observed behavior is the manifestation of the strong magnetic coupling of the soft Fe layer to the hard Sm-Co layer. The coercivity, however, is reduced by about 50% to 1.7 T. Figure 11.15(c) shows the room temperature hysteresis loop for a bilayer with an Fe-layer thickness of 20 nm. This thickness exceeds twice the Bloch magnetic wall width of the hard phase given in Table 11.1, that is, this is above the critical value for simultaneous switching. Indeed, the two phases do not switch simultaneously as a unit, resulting in a complex non-square, constricted hysteresis loop.

Figure 11.16 shows the low temperature (T = 25 K) hysteresis loop of the reduced magnetization for the Sm-Co/Fe (20 nm) bilayer of Figure 11.15(c) up to an applied field of 2 T. The fact that the two layers are strongly coupled only at the interface would mean that the orientation of the magnetization in the soft layer (M_m) should be fully reversible as long as the applied field does not exceed the switching field of the hard layer. This behavior is clearly demonstrated in this hysteresis loop of the bilayer with 20-nm thick soft layer. The iron layer starts to switch at a reverse applied field of only −0.09 T, where a sharp drop in magnetization is observed, reflecting the low coercivity of the soft layer. This is then followed by an asymptotic approach to saturation upon increasing the strength of the reverse field, until the hard layer switches irreversibly at H_{irr}. The switching of the iron layer is completely reversible up to applied fields of about −1.5 T. This characteristic shape and reversible exchange-spring behavior have been observed in other hard/soft bilayers by various investigators. These include bilayers of SmCo/NiFe, SmCo/CoZr, $CoFe_2O_4$/(Mn, Zn)Fe_2O_4.

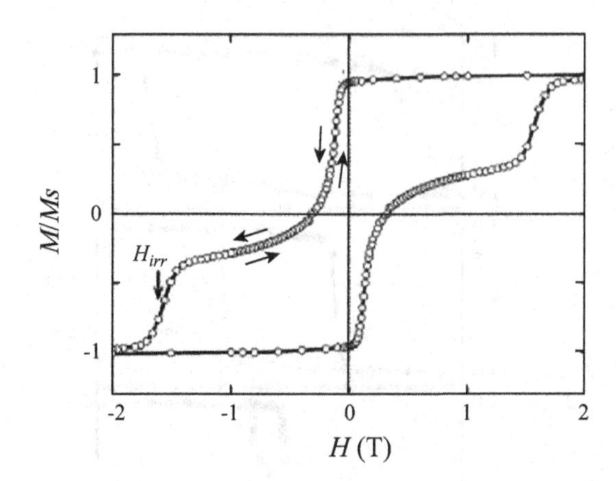

FIGURE 11.16
Constricted hysteresis loop of Sm-Co/Fe bilayer films with t_{Fe} = 20 nm, at low temperature (T = 25 K). Rotation of the soft layer starts at a low field of −0.09 T. The rotation is reversible, as indicated by the arrows, up to a field H_{irr} ~ −1.5 T at which the magnetization of the hard phase reverses. (From E.E. Fullerton, J. S. Jiang and S. D. Bader, Hard/soft magnetic heterostructures: model exchange-spring magnets, *Journal of Magnetism and Magnetic Materials* 200 (1999) 392, Copyright North Holland (Elsevier), reproduced with permission).

Example 11.2: FePt-Based Exchange-Coupled Nanocomposite Magnets.

Another hard magnetic material that is under intense investigation for exchange-spring magnet applications is the face-center-tetragonal (fct) phase of the alloy FePt, also denoted as $L1_0$-FePt phase, which possesses high magnetocrystalline anisotropy density of the order of $K = 5 \times 10^6$ J/m^3 (Table 11.1). Sun *et al.* (H. Zeng, J. Li, J. P. Liu, Z. L. Wang and S. Sun, Exchange-coupled nanocomposite magnets by nanoparticle self-assembly, *Nature*, 40 (2002) 395) used nanoparticle self-assembly and adjuvant thermal treatment to synthesize an exchange-spring nanocomposite magnet material containing FePt as the hard phase and Fe_3Pt as the soft phase with an energy product of 20.1 MG·Oe. This value exceeds the theoretical limit of 13 MG·Oe for single-phase isotropic FePt by over 50%.

The investigators dispersed FePt and Fe_3O_4 nanoparticles, of selected concentrations and sizes, in hexane and mixed them *via* ultrasonic agitation. Subsequently, simply allowing the hexane to evaporate produced three-dimensional binary assemblies of the nanoparticles. The size of the FePt–nanoparticles was kept at 4-nm diameter and that of the Fe_3O_4–nanoparticles was varied from 4 to 12 nm diameter. Figure 11.17 shows transmission electron microscopy (TEM) images of Fe_3O_4/FePt binary assemblies formed with different sizes and a fixed mass ratio of 1:10. For the 4 nm/4 nm assembly (Figure 11.17(a)), Fe_3O_4 and FePt nanoparticles occupy random sites in a hexagonal lattice; but for the 8 nm/4 nm assembly (Figure 11.17(b)), a local ordering immerges, whereby each big particle (Fe_3O_4) is surrounded by 6–8 small particles (FePt). The micrographs of the 12 nm/4 nm assembly (Figure 1.17(c)), however, tell a different story; a clear phase separation is observed with the 12 nm and 4 nm particles forming their own particle lattice arrays. These ordering structures depend mainly on the particle size ratio.

The binary assemblies were subsequently annealed for one hour at 650°C under a reducing atmosphere composed of argon gas mixed with 0.5% hydrogen gas. This thermal treatment converted the Fe_3O_4/FePt binary assembly into a FePt /Fe_3Pt binary assembly *via* the following processes that take place during annealing. First, the iron oxide is reduced to iron; second, the structure of FePt is converted from the disordered face-centered-cubic (fcc) to the ordered face-centered-tetragonal (fct) phase (see Figure 10.16); and third, the organic stabilizers around each particle are desorbed, allowing the nanoparticles to sinter.

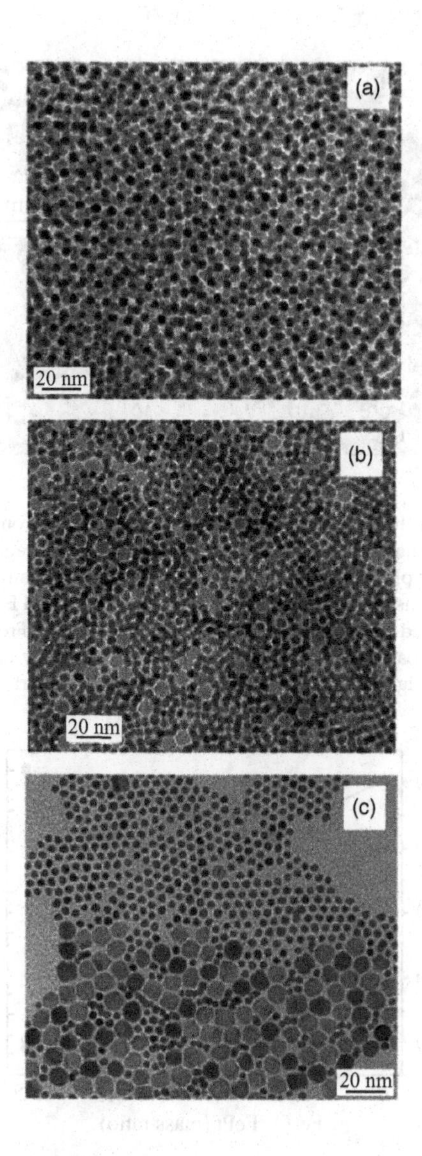

FIGURE 11.17
Transmission electron microscope images showing binary nanoparticle assemblies. (a) Fe_3O_4 (4 nm)/FePt (4 nm) assembly, (b) Fe_3O_4 (8 nm)/FePt (4 nm) assembly, and (c) Fe_3O_4 (12 nm)/FePt (4 nm) assembly. The binary assemblies contained Fe_3O_4 and FePt nanoparticles with a mass ratio of Fe_3O_4: FePt of 1:10, formed by solvent evaporation of the mixed nanoparticle dispersions on amorphous carbon-coated TEM grids. All images were acquired using a Philips CM12 microscope at 120 kV. (From H. Zeng, J. Li, J. P. Liu, Z. L. Wang and S. Sun, Exchange-coupled nanocomposite magnets by nanoparticle self-assembly, *Nature*, 40 (2002) 395, Copyright Nature Research, reproduced with permission).

The transformation of FePt to its ordered fct phase provides a hard magnetic phase contributing a large coercivity to the nanocomposite. Partial inter-diffusion between Fe and FePt during sintering creates Fe_3Pt, which has fcc structure with low coercivity and high magnetic saturation, providing the soft magnetic phase to the nanocomposite. Figure 11.18 is a high-resolution transmission-electron-microscope (HRTEM) image of the sintered sample obtained from an original 4 nm (Fe_3O_4): 4 nm (FePt) binary assembly. The coalesced particle is seen to consist of two distinct phases with dimensions of the order of 5 nm each.

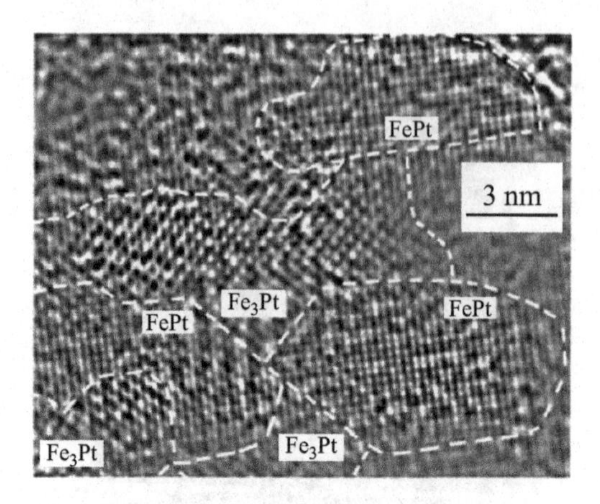

FIGURE 11.18
Structural characteristics of the FePt/Fe_3Pt nanocomposite obtained from an Fe_3O_4 (4 nm): FePt (4 nm) assembly, after annealing. The figure shows a typical HRTEM image of a sintered FePt/Fe_3Pt particle. The FePt and Fe_3Pt phases are in coexistence as different domains within the particle, with each domain having a dimension of about 5 nm, showing a modulated FePt/Fe_3Pt spatial distribution. The image was acquired using a Jeol 4000EX HRTEM at 400 kV. (From H. Zeng, J. Li, J. P. Liu, Z. L. Wang and S. Sun, Exchange-coupled nanocomposite magnets by nanoparticle self-assembly, *Nature*, 40 (2002) 395, Copyright Nature Research, reproduced with permission).

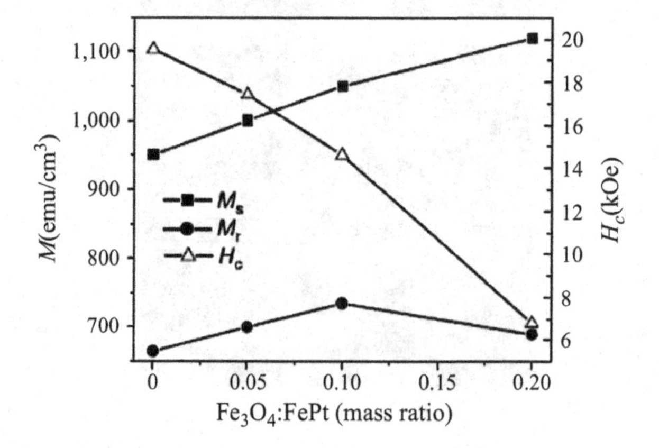

FIGURE 11.19
Saturation magnetization M_s, remanence M_r and coercivity H_c of the FePt/Fe_3Pt nanocomposites from the annealed Fe_3O_4 (4 nm): FePt (4 nm) assemblies as a function of Fe_3O_4:FePt mass ratio. All the data were collected at room temperature on a superconducting quantum interference device (SQUID) magnetometer with an applied magnetic field up to 70 kOe. (From H. Zeng, J. Li, J. P. Liu, Z. L. Wang and S. Sun, Exchange-coupled nanocomposite magnets by nanoparticle self-assembly, *Nature*, 40 (2002) 395, Copyright Nature Research, reproduced with permission).

Elemental analysis from spatially resolved energy dispersive spectroscopy (EDS) confirmed that the two regions correspond to compositions with Fe: Pt ratios close to 1:1 and 3:1, respectively, with the Fe_3Pt phase uniformly dispersed into the FePt matrix.

The magnetic properties of the FePt/Fe_3Pt nanocomposites vary with different initial mass ratios of Fe_3O_4 and FePt. Figure 11.19 shows the saturation magnetization (M_s), remanence (M_r) and coercivity (H_c) of the FePt/Fe_3Pt composites from the Fe_3O_4 (4 nm):FePt (4 nm) assemblies as a function of the initial mass ratio of Fe_3O_4:FePt. The figure shows that

M_s (filled squares) increases monotonically from 950 emu/cm^3 for pure FePt to 1,110 emu/cm^3 for the 1:5 mass ratio, while H_c (open triangles) decreases sharply from 19 kOe to 6.8 kOe. The observed M_s of the pure FePt assembly is about 15% lower than the value reported for bulk FePt (1,100 emu/cm^3). This is expected due to finite-size effects arising from non-collinearity of surface moments or spins with the interior magnetic order of the particle, as discussed in Section 4.11. The surface spins are somewhat disordered and tend to lie perpendicular to the direction of the magnetocrystalline axis of symmetry, an effect that is also known as spin canting.

The remanence (filled circles) shows a maximum value of 740 emu/cm^3 at the 1:10 mass ratio, which represents a 17% increase from pure FePt. The values of the ratio M_r/M_s for all samples are greater than 0.6.

Figure 11.20(a) shows the hysteresis loop obtained for the composite derived from the 4 nm:4 nm assembly with 1:10 mass ratio, shown in Figure 1.17(a). Although the sample consists of both magnetically hard and soft phases, the hysteresis behavior is similar to the one expected from a single-phase material. This indicates that the two phases are exchange-coupled at the interface between the FePt and Fe_3Pt regions, obliging their magnetic moments to undergo cooperative switching. This is in accord with Kneller's and Hawig's

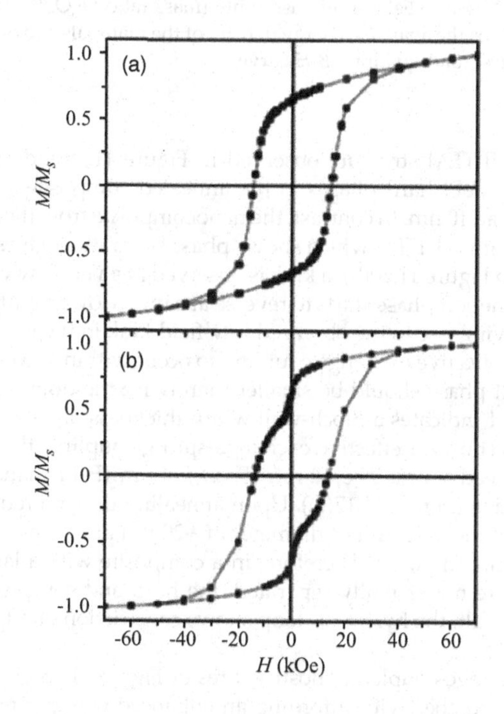

FIGURE 11.20
Typical hysteresis loops of two FePt-based nanocomposites. The composites were made from the annealed Fe_3O_4:FePt assemblies with particle mass ratio kept constant at Fe_3O_4:FePt of 1:10. (a) FePt/Fe_3Pt nanocomposite from Fe_3O_4 (4 nm):FePt (4 nm) assembly. The loop shows single-phase-like behavior, indicating effective exchange coupling between FePt and Fe_3Pt. (b) A nanocomposite from Fe_3O_4 (12 nm):FePt (4 nm) assembly. Owing to the phase separation in the 12 nm:4 nm assembly as illustrated in Figure 11.17(c), the annealed sample contained large body-centered-cubic (bcc) Fe grains as confirmed by HRTEM and EDS, rendering a nanocomposite with the hysteresis showing two-phase behavior. The kink at a low field is related to the magnetization reversal of the soft Fe phase. (From Hao Zeng, Jing Li, J. P. Liu, Zhong L. Wang & Shouheng Sun, Exchange-coupled nanocomposite magnets by nanoparticle self-assembly, *Nature*, 40 (2002) 395, Copyright Nature Research, reproduced with permission).

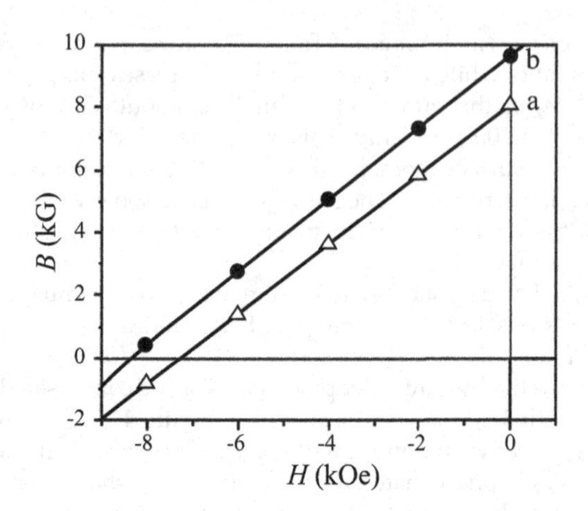

FIGURE 11.21
Second-quadrant *B-H* curves for (a) an annealed 4 nm FePt nanoparticle assembly and (b) an annealed hard/soft exchange-coupled FePt/Fe_3Pt nanocomposite. The composite was obtained from the annealed Fe_3O_4 (4 nm):FePt (4 nm) assembly (mass ratio Fe_3O_4:FePt of 1:10). The energy product, $(BH)_{max}$, describes the available energy density of the materials and is defined as the maximum *BH* product of the second-quadrant *B-H* curve.

calculations, as the HRTEM structure observed in Figure 11.18 indicates spatially well-mixed Fe_3Pt soft and FePt hard phases in the annealed composite with the soft grains limited to sizes less than 10 nm. In contrast, the nanocomposite from the Fe_3O_4 (12 nm):FePt (4 nm) assembly of Figure 1.17(c), which shows phase separation, gives a constricted hysteresis loop, shown in Figure 11.20(b); a kink is observed at a very low value of the applied field indicating that the soft phase starts to reverse at a much lower applied field compared to the hard phase, giving rise to the observed low-field kink in the hysteresis loop. As we discussed earlier, for effective exchange coupling to occur within a two-phase magnet, the dimension of the soft phase should be smaller than twice the domain-wall width of the hard phase. Table 11.1 indicates a Bloch-wall width thickness of 3.9 nm for fct-FePt (also known as $L1_0$-FePt). Thus, for effective exchange-spring coupling, the soft phase dimensions should not exceed 8 nm. The 12-nm (Fe_3O_4): 4-nm (FePt) nanoparticle assembly shows phase separation (Figure 11.17(c)). Upon annealing under a reducing atmosphere, large nanoparticles of metallic iron of diameter of ~20 nm are formed that constitute the soft phase in this nanocomposite. Therefore, in a composite with a large amount of soft phase, even if they are not spatially separated, the hard and soft phases cannot switch cooperatively. As a result, the hysteresis loop shows constriction or a two-phase behavior (Figure 11.20(b)).

The optimally exchange-coupled nanostructures of Figure 11.18 yield both high remanent magnetization and coercivity, affording an enhanced energy product $(BH)_{max}$ compared to a single-phase material consisting of the hard fct FePt phase alone, as shown in Figure 11.21. In this figure, curve (a) illustrates the *B-H* demagnetization curve of an annealed 4-nm FePt nanoparticle assembly, while curve (b) corresponds to the nanocomposite obtained from annealing a Fe_3O_4 (4 nm):FePt (4 nm) assembly with a 1:10 mass ratio (Figure 11.18(a)). According to the experimental data of Figure 11.21, the measured $(BH)_{max}$ for the curve (a) is 14.7 MG·Oe, while that of curve (b) reaches 20.1 MG·Oe, exceeding the value for single-phase FePt assembly by 37%. This $(BH)_{max}$ enhancement clearly indicates effective exchange coupling between the hard and soft phases.

The above examples illustrate that by engineering the nanoscale dimension and spatial distribution of the hard and soft phases, an enhanced energy product can be achieved. They also illustrate the great potential that nanomagnetic tailoring holds for the future fabrication of high-performance exchange-spring magnets. Such approaches can be extended to multi-component systems for fundamental studies in nanomagnetic engineering and applications to functional nanocomposites and devices.

Exercises

1 Define the "energy product" of a permanent magnet. Prove that the energy product is maximized when the demagnetization factor $N = \frac{1}{2}$. What magnet shape does this correspond to?

2 Give examples of technological devices that utilize permanent magnets.

3 What is the intrinsic coercivity, H_{ci}, of a permanent magnet? How does it differ from the coercivity, H_c?

4 Give graphical definitions of $(BH)_{max}$.

5 Figure 11.10 gives the schematic of a microstructure, where spherical precipitates of a hard phase are depicted on an fcc lattice of a soft phase, with equal critical dimensions $d_{mc} = d_{kc}$. Prove that for such a configuration, the volume fraction of the hard phase, $v_k = \frac{V_k}{V}$, where V_k is the volume of the hard phase and V is the total volume of the composite, is given by $v_k = \frac{\pi}{24\sqrt{2}} \approx 0.09$. Furthermore, prove that for the case of hard phase precipitates in a bcc lattice, v_k is calculated to be, $v_k = \frac{\pi\sqrt{3}}{64} \approx 0.09$, which is of the same order of magnitude.

6 What is the fundamental principle and promise of the exchange-spring magnet design? Which parameter needs to be optimized to obtain maximum energy product?

7 Prove that the experimental data presented in Figure 11.21 indicates an increase in $(BH)_{max}$ from 14.7 MG·Oe for single-phase FePt nanoparticle assembly to 20.1 MG·Oe for the FePt/Fe3Pt nanocomposite, as stated in the text.

12

Biomedical Applications of Nanomagnetism

> Physical laws underlying the intravascular magnetic guidance of a novel drug carrier are discussed. The drug carrier is a magnetically responsive drug-bearing microsphere .... of an albumin matrix in which a prototype drug ... and ultrafine Fe_3O_4 particles are entrapped. An *in vitro* analogue of the human circulatory system is used .... Retention of the microspheres by a magnetic field is shown to vary with the linear velocity of the viscous suspending medium and to be dependent on the magnitude of the applied magnetic force. This system permits extracorporeal control over the distribution of intravascular soluble chemotherapeutic agents and allows their concentration at specific body sites.
>
> **From "Magnetic Guidance of Drug-Carrying Microspheres" by A. Senyei, K. Widder and G. Czerlinski, *J. Appl. Phys.* 49 (1978) 3578**

The novel and rapidly developing fields of "nanomedicine" and "medical nanorobotics" trace their origins back to the birth of nanoscience and nanotechnology in the 1970s, as the above quotation from an *in vitro* magnetically guided drug-delivery study indicates. The application of nanotechnology to biomedical science and medicine was both obvious and unavoidable. We already explored the many physical, chemical and biomimetic routes to obtaining magnetic nanoparticles of well-defined composition and narrow size distribution. For the first time, nanotechnology allowed for the controllable manipulation of matter in the nanometer scale, sizes ranging from 1 to 100 nm. These dimensions are smaller than the size of biological cells (10–100 μm); they are comparable to the size of proteins (5–50 nm), genes (2 nm wide and 10–100 nm long) and viruses (20–450 nm). This meant that nanoparticles had the ability to function at the cellular and molecular level of biological interactions. Coated with biological molecules to be rendered biocompatible, nanoparticles could interact with and bind to a variety of biological entities, such as cells, proteins, viruses, nucleic acids or DNA fragments *in vitro* or *in vivo*. Through such interactive processes, nanoparticles promise to revolutionize the diagnosis and treatment of disease.

Since magnetic fields do penetrate within human tissue, *in vivo* administered magnetic nanoparticles can be acted upon by an external magnetic field through "action-at-a-distance" interaction. Taking advantage of the particle's magnetic moment, one could manipulate the particle, and therefore, any biological or pharmaceutical entity bound to it, *via* an extracorporeal magnetic field in a variety of ways. By applying a static magnetic field gradient, one could exert a force on the magnetic nanoparticles, and thus manipulate their intravascular transport, distribution, retention and immobilization at specific tissue sites. By applying a time-varying magnetic field, one could resonantly excite the magnetic nanoparticles resulting in the transfer of energy from the magnetic field source to the nanoparticles, effectively heating the tissue where the nanoparticles are deposited. Alternatively, this inherent thermal energy of the nanoparticles controlled remotely by the alternating magnetic field can be utilized to induce the controlled release of drugs attached to the magnetic carriers. Furthermore, the presence of magnetic nanoparticles in the vicinity of tissue being imaged by nuclear magnetic resonance (NMR) affects

DOI: 10.1201/9781315157016-16

proton-relaxation processes, endowing the magnetic nanoparticles image-contrast enhancing properties.

In this chapter, we discuss the physical principles that lead to the various applications of nanomagnetism in biotechnology and medicine. To do so, we must first address the issue of rendering the magnetic nanoparticles biocompatible to ensure their long lifetime within the bloodstream, before the body's defense mechanisms remove them from circulation. We must also consider various ways to functionalize these nanoparticles at the molecular level, so that they can perform a variety of tasks within the human body, aiding the diagnosis and treatment of disease at the cellular and molecular level, that is, to transform them into "nanorobots" within the human circulatory system. The study of biological and cellular processes at the nanoscale is a strong driving force behind the development of bio-nanotechnology.

12.1 Biocompatibility and Functionalization of Magnetic Nanoparticles

Invariably, applications of magnetic nanoparticles to biotechnology and nanomedicine require nanoparticles in the form of a ferrofluid. The production of colloidal ferrofluids that are stable against aggregation in both a biological medium and a magnetic field is essential. The stability of a magnetic colloidal suspension results from the equilibrium between attractive and repulsive forces. Van der Waals forces induce strong short-range isotropic attraction; electrostatic forces induce long-range isotropic repulsion; while magnetic dipolar forces induce anisotropic interactions, which are found to be globally attractive if the anisotropic interparticle potential is integrated over all directions. Finally, encapsulation of the magnetic nanoparticle within an organic or inorganic coat introduces steric repulsion. Colloidal stabilization of magnetic particles can be achieved by playing on one or both of the two repulsive forces: electrostatic and steric.

Monodispersed in size, shape and composition, high-moment superparamagnetic iron-oxide nanoparticles (SPIONs) are ideal for biomedical applications. However, as we have seen in Section 6.9, high monodispersity has been achieved for nanoparticles stabilized by surfactants like oleic acid, making them hydrophobic, that is, soluble in organic solvents, rather than in aqueous media. For biological applications, these nanoparticles must be rendered hydrophilic, that is, soluble in water, *via* surfactant addition or surfactant exchange. Surfactant addition is achieved through the adsorption of amphiphilic molecules; the hydrophobic segment of the molecule forms a double-layer structure with the original surfactant (*i.e.*, oleic acid) adsorbed on the surface of the nanoparticle, while hydrophilic groups are exposed to the outside imparting water solubility, as depicted schematically in Figure 12.1(a). Surfactant exchange is the direct replacement of the original surfactant with a new "bifunctional surfactant", as depicted in Figure 12.1(b).

The bifunctional ligand exchange strategy provides biocompatibility and better colloidal stability in physiological conditions. These bifunctional surfactants are capable of binding to the nanoparticle surface tightly *via* a strong chemical bond; while the other end of the surfactant, having a polar character, allows the nanoparticles to be dispersed in water. The case of surface coating the nanoparticles with dopamine terminated polyethylene glycol (PEG) ligands is depicted in Figure 12.2. Stability is rendered by the presence of strong interactions between the bi-dentate binding group at the iron oxide surface and the

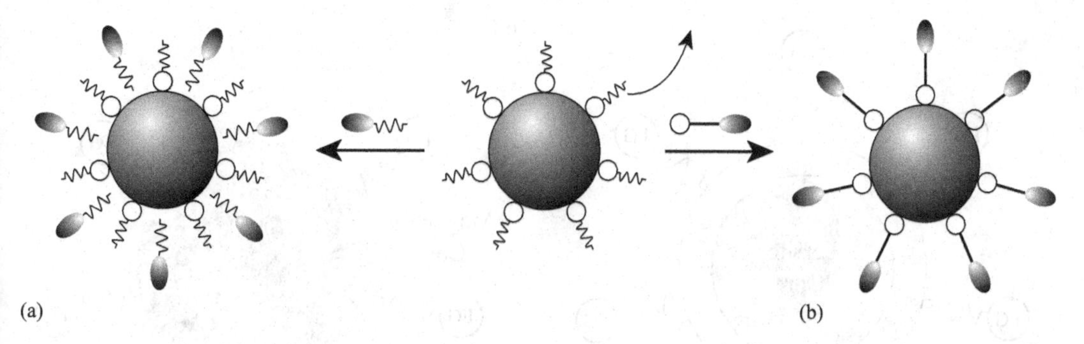

FIGURE 12.1
Scheme of rendering a hydrophobic magnetic nanoparticle hydrophilic by (a) surfactant addition and (b) surfactant exchange.

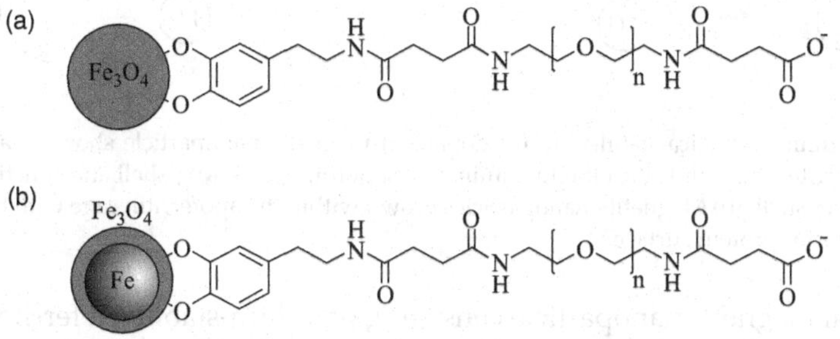

FIGURE 12.2
Examples of water-soluble magnetic nanoparticles. (a) Magnetite nanoparticles and (b) Fe/Fe_3O_4 core/shell nanoparticles terminated with dopamine and coated with bifunctional PEG ligands.

carboxyl group on the other end of the surfactant. In addition, such bifunctional surfactants provide for the immobilization of biomolecules onto the magnetic nanoparticles. Carboxyl- or amine-terminated PEG ligands can readily attach to various molecules, for added functionality to the nanoparticles.

Magnetic nanoparticles have been rendered hydrophilic and biocompatible by a surface coat consisting of various polymers in addition to PEG such as polyvinyl alcohol (PVA), proteins (albumin, apoferritin), phospholipids or polysaccharides (dextran), as well as, a shell of inorganic coating material (silica, gold, or gadolinium). For example, chemical surface modification of iron oxide or ferrite nanoparticles by silica coating has been shown to prevent particle aggregation in aqueous solutions and to improve their chemical stability. Under physiological conditions, silica-coated nanoparticles are negatively charged providing coulomb repulsion of the magnetic nanoparticles in addition to the steric repulsion due to the thickness of the silica shell. Furthermore, the presence of surface silanol groups (Si-O-H, analogous to the hydroxy groups, C-O-H, contained in alcohols) can lead to an easy reaction with various coupling agents to covalently attach specific ligands to silica-coated magnetic particles. These ligands, called functional groups, can interact with cell receptors on the surface of the cell membrane, binding the magnetic nanoparticle to the target cell. In addition, silica can endow the nanoparticles with the ability to carry a payload of pharmaceuticals, chemotherapeutic agents, radioactive materials or genes if, for example, the particles are further coated with a layer of mesoporous silica (MS), rather than just solid silica (SiO_2). The porosity of the mesoporous silica can provide a large volume for loading such therapeutic agents onto the nanoparticle. Figure 12.3(a) depicts schematically, a

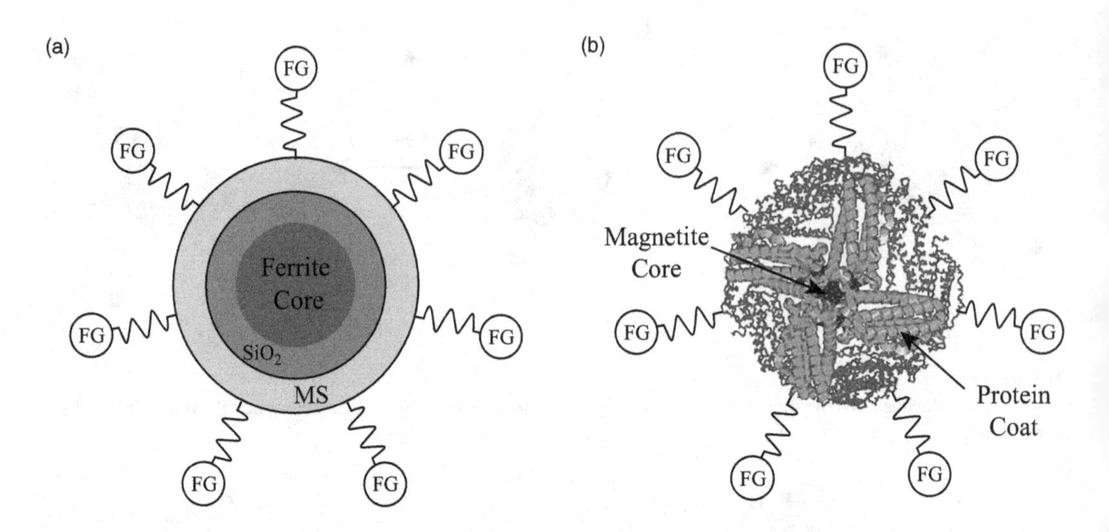

FIGURE 12.3
(a) Schematic diagram of a silica-stabilized, functionalized magnetic nanoparticle showing a core/shell/shell structure with a ferrite core, solid silica (SiO_2) within a mesoporous silica (MS) shell and functional groups (FG) attached to the outer shell. (b) Magnetite nanoparticles grown within the apoferritin cage with functional ligands attached to the exterior protein surface.

multifunctional magnetic nanoparticle consisting of a silica-stabilized ferrite nanoparticle, forming a core/shell/shell nanoarchitecture with attached functional groups (FG). The functional groups may be carboxyl (-COOH), amine (-NH2, -NHR, -NR2), aldehyde (-CHO), thiol (-SH) groups, antibodies, peptides, biotin, avidin, streptavidin and other molecules, which act as attachment points for the coupling of cytotoxic drugs or target antibodies to the magnetic nanoparticle complex.

As discussed in Chapter 8, under biomimetic nanoparticle synthesis, the interior surface of the apoferritin nanotemplate can be modified through genetic engineering to induce the nucleation and growth of a variety of material phases, such as Fe_3O_4, within the protein cage to form magnetoferritin. In this instance, it is the protein shell that imparts the nanoparticles with hydrophilic properties and biocompatibility. In addition, functional groups or targeting agents can be attached to the exterior surface of the protein shell for functionalization of the magnetic nanoparticle, as depicted in Figure 12.3(b).

In the following sections, we discuss some promising, important applications of functionalized magnetic nanoparticles both *in vitro* and *in vivo*. We start with *in vitro* applications to biotechnology and continue with *in vivo* applications in medicine. In the latter, problems of toxicity and particle "opsonization", leading to their removal from circulation by the immune system, make the applications even more challenging.

12.2 *In Vitro* Applications

12.2.1 Magnetic Separation

Biological samples, such as blood samples, are exceptionally complex because they invariably contain multiple components. In disease diagnosis, it is necessary to isolate the specific target entity, such as an infectious agent, from the rest of the components in order to speed up the screening process and enhance the detection limit. Using functionalized magnetic nanoparticles to tag the target entity and employing a magnetic field gradient, one can apply a force on the magnetic nanoparticle and drive it, along with its attached entity, out of solution. Therefore, surface modification of the magnetic particles with specific molecules, or functional groups, which only bind to the target entity, but not other non-target components, is the only prerequisite.

As we have previously discussed a homogeneous magnetic field can only exert a torque on the magnetic moment of a nanoparticle; a magnetic field gradient is needed in order to apply a force. Upon application of an inhomogeneous magnetic field the nanoparticles will be magnetically aligned in the direction of the field and pulled into the region of highest magnetic field strength, that is, the direction of the magnetic field gradient. This phenomenon is known as "magnetophoresis", which refers to the motion of magnetic components relative to their non-magnetic surrounding medium under a non-homogeneous magnetic field. Figure 12.4 depicts the concept behind the magnetic separation technique in biotechnology. While the magnet is held against the tube, the non-magnetic fluid can be decanted or pipetted away. This technique has aided the detection of various biological components such as bacteria, viruses and parasites, as well as, rare circulating tumor cells in the blood for early diagnosis of disease.

Due to the importance of magnetic separation, many separator designs have been developed. These can be broadly classified into two major types: those based on (a) high-gradient magnetic separation (HGMS) and (b) low-gradient magnetic separation (LGMS) according to the magnitude of the magnetic field gradient employed in the separation process. Separators using HGMS consist of a column, which is loosely packed with randomly entangled magnetically susceptible steel wires, and placed between the poles of an electromagnet, as schematically depicted in Figure 12.5(a). When the electromagnet is turned on, the magnetically susceptible wires within the column de-homogenize the magnetic field,

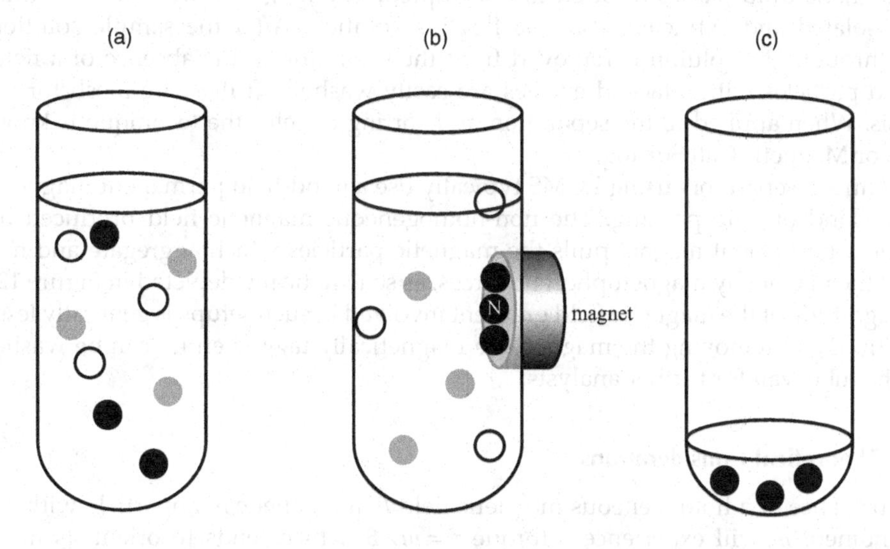

FIGURE 12.4
Concept of magnetic separation. (a) A magnetically tagged biological entity (black spheres) in a complex biological sample. (b) Attachment of a permanent magnet on the side of the tube attracts the magnetically tagged entities. (c) Concentrated targeted entities after decanting or pipetting away the non-magnetic solution while the magnet is held in place.

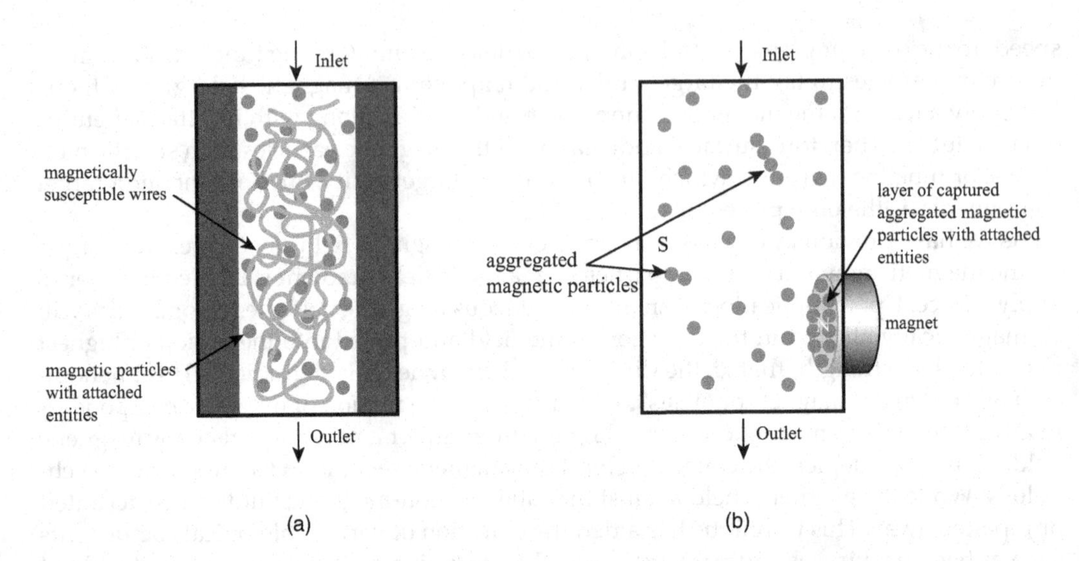

FIGURE 12.5
(a) Schematic of an HGMS separator. The column is packed with entangled magnetizable wires to produce strong and localized magnetic field gradients in their vicinity. Magnetic particles flowing through will be acted upon by an exceptionally strong magnetophoretic force and be captured or immobilized on the wires. (b) Schematic of LGMS separator. A permanent magnet creates a magnetic field gradient across the solution containing the particle suspension. The magnetic particles aggregate and migrate toward the region adjacent to the magnet and are separated from the solution.

producing a high magnetic field gradient (~10^3–10^4 T/m) close to their vicinity. As the biological sample containing the suspension of magnetic particles is passed through the column, entities attached to magnetic particles are captured on the wires while the rest non-magnetic fluid passes through uninterrupted. The magnetically labeled entities are hence isolated and extracted from the flowing solution. After the sample solution has flown through, the column is removed from the separator. In the absence of a field, the retained particles with attached entities are easily washed off the wire mesh for further analysis. When applied to the separation and sorting of cells, the technique is known as MACS or Magnetic Cell Sorting.

In contrast, separators using LGMS typically use a hand-held permanent magnet without any kind of wire packing. The non-homogeneous magnetic field produced by the hand-held permanent magnet pulls the magnetic particles which aggregate and migrate toward the magnet by magnetophoretic forces, as schematically depicted in Figure 12.5(b). The magnitude of the magnetic field gradient involved in such setups is generally less than 100 T/m. Upon removing the magnet, the magnetically tagged entity can be washed off from the tube wall for further analysis.

12.2.2 Theoretical Considerations

In the presence of a homogeneous magnetic field $\vec{B}$, a magnetic nanoparticle with a magnetic moment $\vec{m}$ will experience a torque $\vec{\tau} = \vec{m} \times \vec{B}$, which tends to orient its magnetic moment in the direction of the field, where the orientational magnetic energy of the particle, $U_B = -\vec{m} \cdot \vec{B}$, is minimized. If the magnetic field is inhomogeneous, in addition to the torque, the particle will also experience a magnetic force $\vec{F}_m = -\vec{\nabla}(U_B)$ or $\vec{F}_m = \vec{\nabla}\left(\vec{m} \cdot \vec{B}\right)$. Under the action of the torque, magnetic saturation of the superparamagnetic nanoparticle suspension is quickly reached. The nanoparticle magnetic moment is rendered parallel to the magnetic field and the expression for the force is simplified to

$$\vec{F}_m = m\vec{\nabla}(B) \tag{12.1}$$

indicating that the force is in the direction of the gradient of the scalar field, that is the magnitude of the magnetic field.

For a particle of volume V_p and magnetization density M, the magnitude of its magnetic moment is given by

$$m = V_p M \tag{12.2}$$

The magnetic susceptibility χ of the medium due to the dispersed superparamagnetic nanoparticles, ignoring the much weaker diamagnetic contribution of water and other non-magnetic entities in the complex biological sample, is given by

$$\chi = \frac{M}{H} \tag{12.3}$$

Combining the above three equations, and using $\vec{B} = \mu_0 \vec{H}$, where μ_0 is the permeability of free space, we can express the magnetic force as

$$\vec{F}_m = \frac{V_p}{\mu_0}\chi B\vec{\nabla}B = V_p \chi \vec{\nabla}\left(\frac{1}{2\mu_0}B^2\right) = \frac{4\pi}{3}R^3 \chi \vec{\nabla}\left(\frac{1}{2\mu_0}B^2\right) \tag{12.4}$$

where R is the radius of the particle, reiterating that the force is in the direction in which magnetostatic field energy density increases most steeply, for $\chi > 0$, as is the case here. This is the force that attracts iron filings when brought near the pole of a permanent magnet; and it is the same force that is in action in biomedical applications of magnetic separation.

In order to achieve separation of the targeted entity from the complex biological fluid, this magnetic force must overcome the hydrodynamic drag force acting on the magnetic particle in the flowing fluid, which is given by

$$\vec{F}_d = -6\pi\eta R\vec{v} \tag{12.5}$$

This drag force is a frictional force acting on the interface between the flowing fluid and the particle, η is the dynamic viscosity of the medium, R is the radius of the particle and $\vec{v}$ is the velocity of the magnetic nanoparticle relative to the flowing non-magnetic fluid (*i.e.*, water, for which $\eta = 8.9 \times 10^{-4}$ N s/m^2). Ignoring thermal fluctuations and buoyancy, one can get an estimate of the velocity of the magnetic nanoparticle relative to the non-magnetic fluid, referred to as the "magnetophoretic velocity", by equating $\vec{F}_m$ to $-\vec{F}_d$ to get

$$\vec{v} = \frac{R^2\chi}{9\mu_0\eta}\vec{\nabla}(B^2) \tag{12.6}$$

Thus, the magnetophoretic velocity of the particle is proportional to the gradient of B^2, the magnetic susceptibility of the medium and the square of the particle radius; and inversely proportional to the viscosity of the medium. The quantity

$$\xi = \frac{R^2\chi}{9\eta} \tag{12.7}$$

defines the "magnetophoretic mobility" of the particle, a parameter that describes how easy it is to manipulate the magnetic nanoparticle in a particular medium.

The situation for an immuno-magnetically labeled cell is more complex as the magnetic force on the cell depends on the number of magnetic particles attached to the cell membrane, while the viscous drag force is due to the complete cell–nanoparticles complex. Generally, five parameters will influence the magnetophoretic mobility of an immuno-magnetically labeled cell: (a) the antibody binding capacity (ABC), that is, the number of primary antibodies binding to the cell, (b) the secondary antibody binding amplification factor (ψ), (c) the number n of magnetic particles bound to one antibody, (d) the particle–magnetic field interaction parameter of the magnetic particles ($V_p\chi$) and (e) the cell diameter (D_c). This value depends primarily on the number of antigen molecules per cell, and also on other variables such as the strength of antibody binding, steric hindrance, binding affinities, and non-specific binding events. The amplification factor ψ is obtained by binding a secondary antibody with multiple magnetic particle interaction sites to the primary antibody and n is the number of magnetic particles that conjugate to the (second) antibody. In analogy to Eq. (12.7), the magnetophoretic mobility for the immuno-magnetically labeled cell can be written as

$$\xi_{cell} = \frac{ABC\psi n V_p \chi}{6\pi\eta D_c} \tag{12.8}$$

where again D_c is the diameter of the cell. Figure 12.6 gives a pictorial representation of the various parameters that appear in Eq. (12.8).

One can adjust these parameters to minimize the separation time. High-gradient magnetic separators are very efficient and can process large amounts of fluids in a short time. In addition to cells, other biological species can be separated such as proteins, DNA, viruses, bacteria, *etc*. In such cases, the antigen–antibody interactions are replaced by other lock and key interactions known in biology, such as that between biotin and avidin or streptavidin.

Both, functionalized magnetic micro-beads, containing many nanoparticles encapsulated within a polymer matrix, and individually functionalized magnetic nanoparticles have been used. According to Eq. (12.4), the magnetic force on a micro-bead is large due to the large radius of the micro-bead, as compared to that of an individual nanoparticle. However, the surface-to-volume ratio associated with the nanoparticles is much larger compared to the micro-beads, enabling a larger number of attachments to occur between nanoparticles and targeted entities, for a given nanoparticle concentration in the complex biological sample. This increased binding capacity results in greater separation efficiency, as the smaller-sized nanoparticles afford absorptive areas of 100–1,000 times larger compared with those of micro-beads.

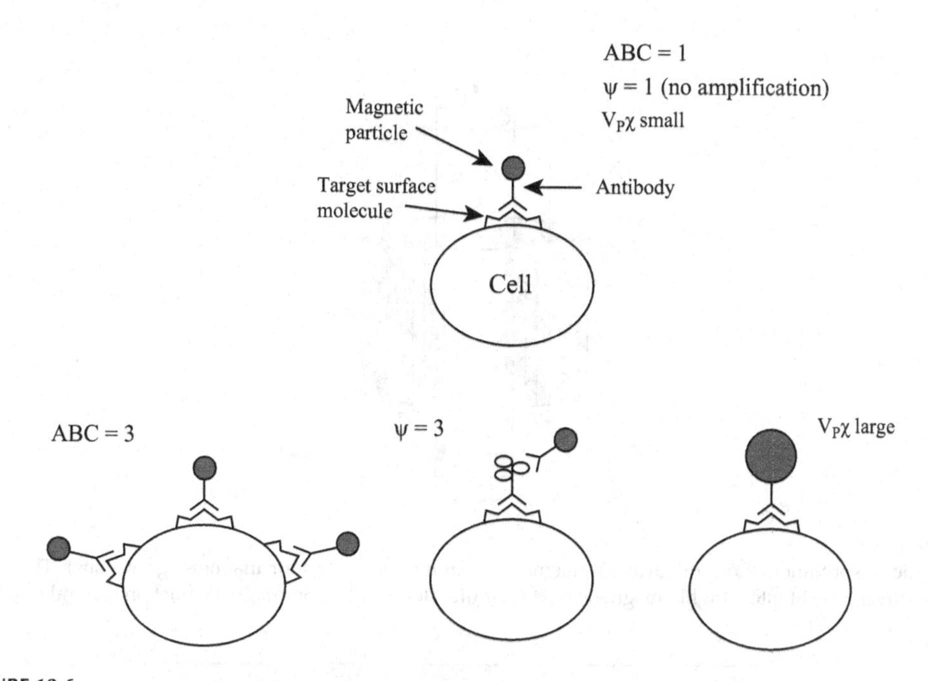

FIGURE 12.6
Comparison of immuno-magnetically labeled cells with different values of antibody binding capacities (ABC), secondary antibody amplification (ψ), or particle–magnetic field interaction parameter ($V_p\chi$). An increase in any of these parameters will increase the amount of magnetic material bound to the cell and therefore increases the magnetophoretic mobility.

12.2.3 Continuous Flow and Microfluidic Magnetic Separators

Other designs of magnetic particle separation systems with a continuous flow have also been developed, taking advantage of laminar flow patterns or using "microfluidics". Figure 12.7 is a schematic diagram of a continuous flow magnetic quadrupole separator.

Here, four magnets embrace the sample carrying tube in an arrangement to induce a maximum magnetic field gradient toward the outer side of the tube, perpendicular to the fluid flow. The sample mixture enters the system at the top; as the sample is carried along the channel by the flow of the fluid, those components that interact more strongly with the field gradient are carried transversely across the channel thickness. A division of the flow at the channel outlet using a stream splitter completes the separation into two fractions. The radial particle separation velocity is induced by the field gradient according to Eq. (12.6). In this case, separation is achieved within a mobile phase without the use of a stationary phase. The input sample stream is split into (a) a highly magnetic and (b) a nonmagnetic streamflow.

Figure 12.8 gives a schematic of a magnetic dipole separator. The magnetic field gradient is directed toward the magnet. A complex biological sample containing different entities labeled with magnetic nanoparticles of various sizes is introduced into the device on the left. As the fluid flow moves to the right, different-sized magnetic nanoparticles experience different magnitudes of magnetic gradient force according to Eq. (12.4). Entities labeled with the larger nanoparticles experience a greater deflection toward the magnet, compared to those attached to smaller nanoparticles. This results in a laminar fluid flow that allows fractionation of the fluid at the receiving slots on the right. Different types of

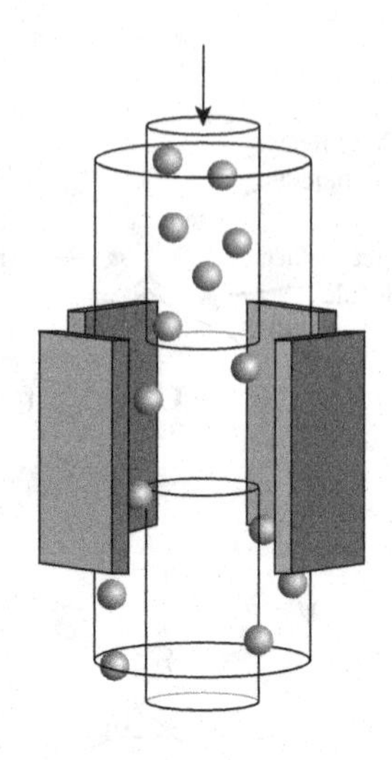

FIGURE 12.7
Schematic of continuous flow quadrupole magnetic separator, utilizing four magnets (gray slabs). The input sample stream is split into a highly magnetic fraction (outer flow) and a non-magnetic fraction (central flow).

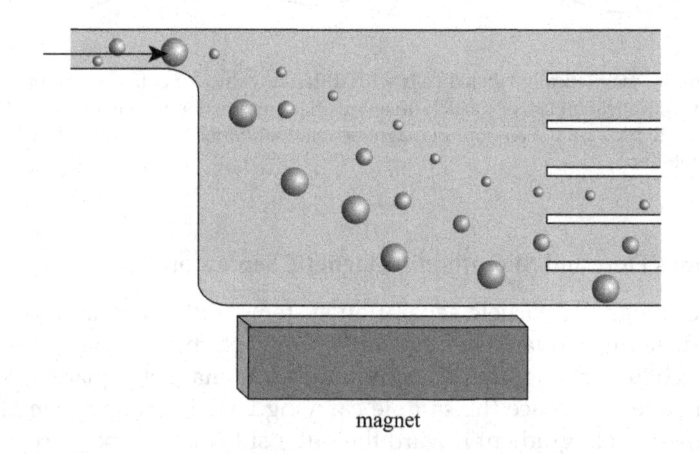

FIGURE 12.8
Schematic of dipole magnetic separator. Different-sized magnetic particles with their attached entities are deflected into different outlet ports, leading to fractionation of the input biological sample into separate components enriched in a particular biological entity of the complex fluid.

magnetic particles with their attached entities will be deflected into different outlet ports, effectively fractionating the input biological sample into separate components enriched in a particular biological entity of the complex fluid.

In recent years, various designs of microfluidic magnetic separators have been introduced that are complex and miniaturized. Such designs are based on the magnetic force generated by an array of magnetized stripes that are positioned at an angle with respect to the hydrodynamic flow direction. These stripes create a series of magnetic field gradients that deflect the magnetic nanoparticles, or magnetic beads, from their original flow direction. Figure 12.9(a) is a schematic diagram of such a microfluidic chip. The magnetic stripes or wires are shown at an oblique angle relative to the direction of the fluid flow. The vector sum of the hydrodynamic and magnetic forces creates a resultant force vector, which

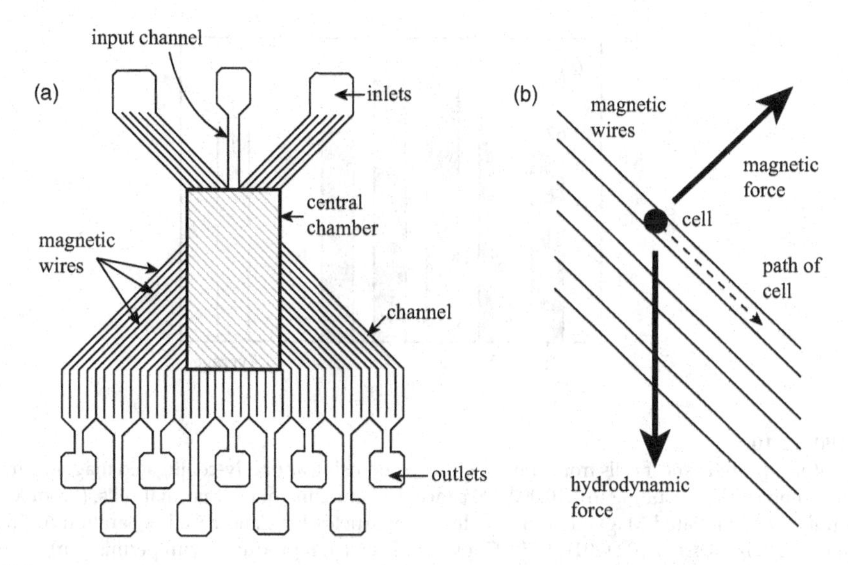

FIGURE 12.9
(a) Schematic of a microfluidic magnetic separator. (b) Depiction of the hydrodynamic and magnetic forces acting on the magnetically tagged cells.

moves the magnetic bead with their attached entities out of the main flow stream, as illustrated in Figure 12.9(b). On the other end of the chip, a large number of outlets allow for the collection of fractionated samples. The non-magnetic fraction is collected at the central outlet, while the most magnetic fraction is collected at the outer outlets.

The impact of such microfluidic devices, also called "lab-on-a-chip" or miniaturized analysis systems, cannot be overestimated. Microfluidics provides attractive solutions for many problems in chemical and biological analysis, especially for in-field use or point-of-care testing.

Example 12.1: Cancer Cell Separation from Fresh Whole Blood

Cancer is one of the biggest public health concerns worldwide. Research has shown that circulating tumor cells can be found in patients' blood long before the primary tumor is detected and are the first messengers of impending metastatic cancer. Only a few such cancer cells may be present in whole blood, however, in the background of billions of normal white and red blood cells. Detection of these circulating tumor cells can play an important role in the early diagnosis of disease. Due to their extremely low concentration, however, they cannot be detected. A milliliter of human blood contains about 10^7 white blood cells, 10^{10} red blood cells and 10^8 platelets, while the number of tumor cells in cancer patients may be <50. This means that one needs to detect less than 50 tumor cells in the presence of 10 billion blood cells.

Despite these astounding numbers, immunomagnetic separation of circulating tumor cells from whole blood has been demonstrated by the use of polymer-coated iron-oxide nanoparticles functionalized with antibodies against human epithelial growth factor receptor2 (anti-HER2) using a low-gradient separator, (Xu *et al.*, Antibody Conjugated Magnetic Iron Oxide Nanoparticles for Cancer Cell Separation in Fresh Whole Blood, *Biomaterials*, 32 (2011) 9758). HER2 is a cell membrane protein that is overexpressed in several types of human cancer cells. One milliliter of fresh human blood placed in a test tube was spiked with HER2 overexpressing human breast cancer cell line SK-BR3. A 0.05 mL antibody functionalized iron-oxide nanoparticle solution, at a concentration expressed as

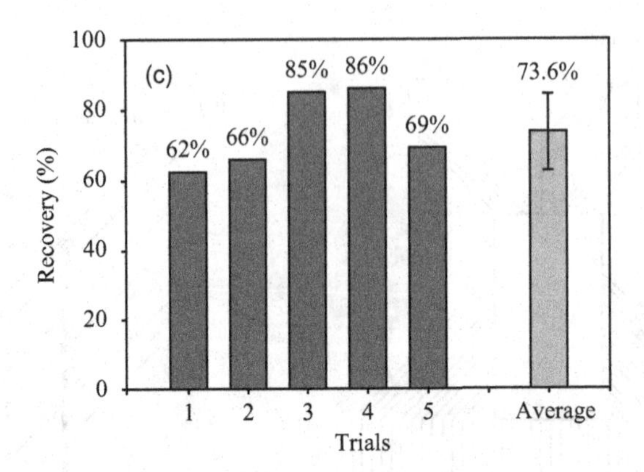

FIGURE 12.10
Separator captured cancer cells from spiked whole blood using antibody-conjugated magnetic iron-oxide nanoparticles, resulting in 1:10,000,000 cancer cell enrichment over normal cells. (From Xu *et al.* Antibody Conjugated Magnetic Iron Oxide Nanoparticles for Cancer Cell Separation in Fresh Whole Blood, *Biomaterials*, 32 (2011) 9758, Copyright Elsevier, reproduced with permission).

2 mg Fe/mL, was added. The nanoparticles attached preferentially to the cancerous cells due to their high level of HER2 surface receptors compared to the normal cell surfaces. The tube was subsequently introduced to a SuperMag™ low-gradient magnetic separator with a gradient of 100 T/m for one hour to allow magnetic isolation of the cancerous cells. The cancerous cells attached to the antibody-functionalized magnetic nanoparticles stuck to the wall of the tube, where the field was stronger, allowing for supernatant fluid removal with a pipette. The results showed that the process exhibited an enrichment factor (cancer cells over normal cells) of 1:10,000,000. Figure 12.10 from the study of Xu *et al.* indicates the percentage of iron-oxide-nanoparticle-antibody captured cancerous cells from spiked whole blood. It can be seen that on average, 73.6% of the magnetic nanoparticles were captured by the separator, with some trials indicating percentages as high as 86%.

Example 12.2: Extracorporeal Blood Cleansing for Sepsis Therapy

Sepsis is a serious condition with 30–50% mortality rate. It results from the presence of microbial pathogens in the bloodstream that triggers systemic inflammation. Sepsis often overcomes the most powerful antibiotic therapies and can lead to multi-organ systems failure, "septic shock" and death. Sepsis afflicts 18 million people worldwide every year, and can lead to death even in state-of-the-art hospital intensive care units; its incidence is increasing because of the emergence of antibiotic-resistant microorganisms. The condition is characterized by a high load of pathogens in the blood. Due to the multiple types of pathogens present, their identification is not feasible, obviating targeted antibiotic therapy. Thus, removal of the pathogens from the bloodstream is highly desirable in the absence of identification. An extracorporeal blood-cleansing therapy system that can rapidly remove microorganisms and endotoxins from blood using microfluidic magnetic separation principles has been demonstrated by Kang *et al.* (Kang *et al.*, An extracorporeal blood-cleansing device for sepsis therapy, *Nature Medicine*, 20 (2014) 1211).

The authors describe an extracorporeal microfluidic device that incorporates a flow channel design inspired by the microarchitecture of the spleen, which can be used to cleanse pathogens from the flowing blood of patients with sepsis. The pathogen capturing agents are composed of magnetic nanobeads coated with a genetically engineered version of human Mannan Binding Lectin (MBL) protein that binds to a wide variety of pathogens,

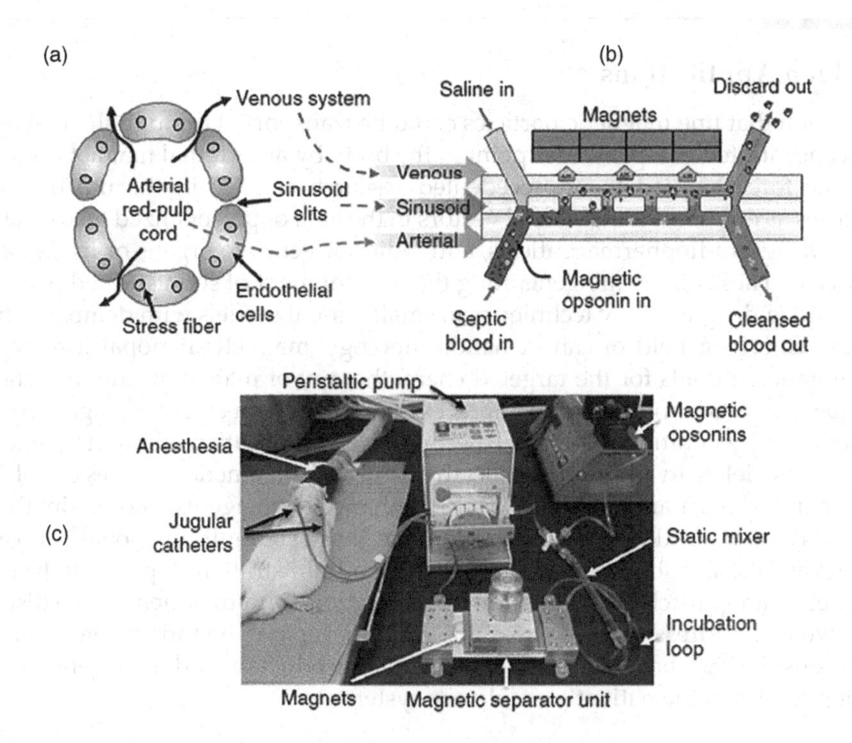

FIGURE 12.11
An extracorporeal blood-cleansing setup based on microfluidic magnetic separator principles that mimic the function of the human spleen (see text). (From Kang *et al.*, An extracorporeal blood-cleansing device for sepsis therapy, *Nature Medicine*, 20 (2014) 1211, Copyright American Association for Cancer Research, reproduced with permission).

referred to as magnetic "opsonins", as their mode of action is similar to opsonization employed by the immune system to engulf foreign bodies, bacteria and dead or dying cells for easy recognition by phagocytes and removal from the bloodstream. We discuss this process in greater detail later. A schematic of the extracorporeal setup and microfluidic device comprising this blood-cleansing device is shown in Figure 12.11.

This microfluidic magnetic separator mimics the function of the human spleen (Figure 12.11(a)) by incorporating a high flow vascular arterial channel perfused with contaminated whole blood and interconnected by open slits to a parallel low-or intermittent-flow venous sinusoid channel perfused with isotonic sterile saline (Figure 12.11(b)). Magnetic opsonins are added to the flowing septic (contaminated) blood, are passed through an incubation loop to promote binding to the pathogens before entering the arterial channel of the microfluidic device. Stationary magnets positioned directly above the sinusoid channel pull the magnetic opsonins and bound pathogens through the open slits, into the saline-filled channel and into a discard collection vial. The cleansed blood is then returned back to the patient. It was observed that the device efficiently removed multiple bacteria, fungi and endotoxins from whole blood flowing through a single microfluidic unit at up to 1.25 L/h *in vitro*. In experiments that used rats, (Figure 12.11(c)), infected with *Staphylococcus aureus* or *Escherichia coli* bacteria, the device cleared >90% of bacteria from blood, reduced pathogen and immune cell infiltration in multiple organs and decreased inflammatory cytokine levels. In a model of endotoxemic shock, this extracorporeal therapy increased survival rates after a 5-h treatment.

12.3 *In Vivo* Applications

The realization that fine magnetic particles could be transported through the vascular system and concentrated at a particular point in the body by an external magnet *via* the magnetophoretic force has led to unprecedented research activity in the use of magnetic nanoparticles as non-viral translational vectors in the area of personalized medicine for the delivery of drugs, radiopharmaceuticals, antibodies or genes to organs or tissues affected by the disease. These efforts are generating the development of sophisticated *in vivo* magnetically guided drug-delivery techniques in small animal models with demonstrable success. In the emerging field of cancer nanotechnology, magnetic nanoparticles represent promising nanomaterials for the targeted chemotherapy of malignant tumors. The particles' unique properties and ability to interact remotely with an external magnetic field are also exploited for generating local hyperthermia *via* the application of low-frequency alternating magnetic fields. In addition, the fact that superparamagnetic particles can also function as magnetic resonance imaging (MRI) enhancement agents, aiding in the early detection of disease, has led to the realization that these are multifunctional nanosystems that can act as "theranostic" agents, since a single administration of properly functionalized magnetic nanoparticles can aid both in the diagnosis and treatment of the disease.

Before we can address each of these applications separately and in combination, we must first consider the challenges presented by the introduction and management of magnetic nanoparticles in the patient's circulatory system.

12.3.1 Avoiding the Mononuclear Phagocytic System

Nanoparticles introduced into the circulatory system are quickly recognized as foreign entities by the immune system that triggers a process known as "opsonization" for their removal from circulation. Opsonization is the molecular mechanism used by the immune system to chemically modify the surface of invading foreign entities for enhanced recognition and removal by the "phagocytic system". It is a general process employed for the removal of any unwanted entities be it cell debris, due to cell death and regeneration in healthy individuals, or bacterial or viral pathogens in the presence of disease. Through this process, magnetic nanoparticles introduced into the bloodstream by intravenous injection are rapidly coated by "opsonins", that is, blood plasma components such as antibodies, complement proteins or circulating proteins. This renders the particles recognizable by the body's major defense system, the "mononuclear phagocytic system" (MPS), also known as the "reticuloendothelial system" (RES). The MPS/RES is a diffuse system of specialized cells that are phagocytic, that is, they engulf and internalize any unwanted material, thus effectively removing it from circulation. It is a network of cells and tissues found throughout the body, especially in the blood, general connective tissue, spleen, liver, lungs, bone marrow, and lymph nodes. In particular, the "macrophage" cells of the liver, known as Kupffer cells, and to a lesser extent, the macrophages of the spleen and circulation, play a critical role in the removal of "opsonized particles".

Figure 12.12(a) gives a schematic of the fate of magnetic nanoparticles after intravenous injection; while Figure 12.12(b) compares blood residence time for various nanoparticles as a function of their size and their ultimate clearance pathways. Thus, through the process of "opsonization" and "phagocytosis", they rapidly accumulate primarily in the liver, spleen, lymph nodes and bone marrow. In addition, due to the high blood supply associated with sites of inflammation and cancerous tumors, an increased level of particle accumulation is

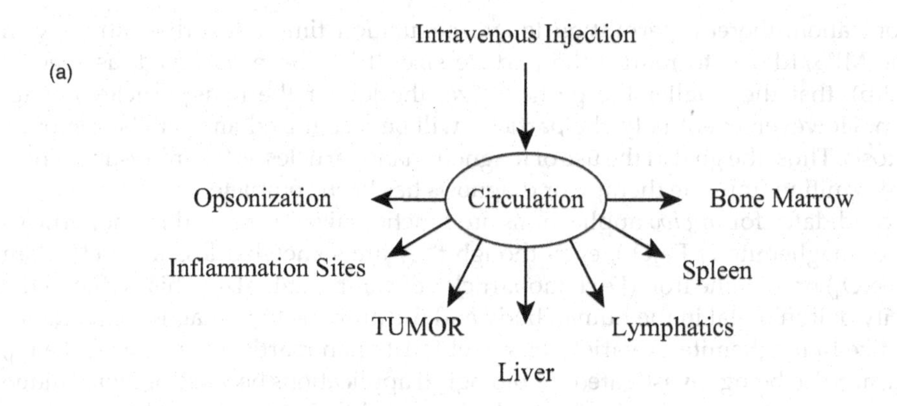

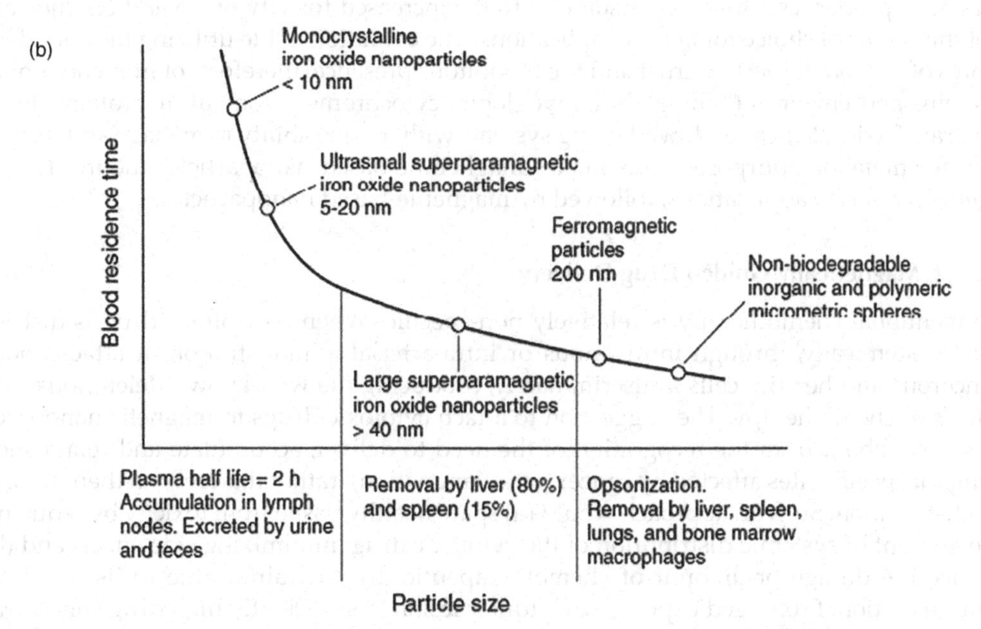

FIGURE 12.12
(a) Ultimate biodistribution of magnetic nanoparticles after intravenous injection into the circulatory system due to opsonization and (b) qualitative diagram showing the evolution of blood residence time with particle size.

also associated with such sites. Most solid tumors possess pathophysiological characteristics that are not observed in normal tissue or organs; that is, extensive angiogenesis leading to hyper vasculature of defective vascular architecture and increased production of permeability mediators. This results in the so-called enhanced permeability and retention (EPR) effect, which has been observed to be universal for solid tumors. This enhanced vascular permeability is what sustains an adequate supply of nutrients and oxygen needed for rapid tumor growth. Thus, by these natural biodistribution processes, one expects increased accumulation of nanoparticles at the tumor sites. Colloidal nanoparticles, even as large as 300 nm, are able to extravasate into the tumor interstitium due to the hyper-permeable vasculature of solid tumors. This is known as "passive nanoparticle targeting" of tumors.

Surface modification of the nanoparticles to render them biocompatible can mask their presence in the bloodstream and avoid immediate recognition by the immune system. The role of the dense brushes of polymers or protein coats shown in Figures 12.1 and 12.3 is to inhibit opsonization, thereby permitting longer circulation times. A further strategy in avoiding the MPS/RES is to reduce the particle size. It has been observed, as seen in Figure 12.12(b), that the smaller the particle size, the longer the nanoparticles remain in circulation. However, eventually, the particles will be recognized and will be removed by phagocytosis. Thus, the goal in the use of magnetic nanoparticles as "theranostic agents" is to delay recognition until the therapy or diagnosis has been completed.

The best candidates for *in vivo* applications are functionalized iron-oxide nanoparticles in the form of maghemite (γ-Fe_2O_3), even though they are somewhat less magnetic than magnetite (Fe_3O_4) or metallic iron (Fe) nanoparticles of comparable size. This is due to the higher toxicity of iron metal in the human body and the propensity of magnetite nanoparticles to oxidize to maghemite. A variety of spinel ferrite nanoparticles, such as $CoFe_2O_4$, $NiFe_2O_4$, *etc.*, are also being investigated for biological applications because of their unique magnetic properties. However, again due to the increased toxicity of Co and Ni, they are not the agents of choice for *in vivo* applications. The evolution of life utilizing the abundant stores of iron on the earth's crust and the ubiquitous presence, therefore, of iron-containing proteins and enzymes (hemoglobin, myoglobin, cytochromes, iron-sulfur proteins, ferritin, transferrin, *etc.*) has endowed living systems with unique ability to manage and recycle iron for metabolic purposes. This makes maghemite-based nanoparticles the most ideal agents for *in vivo* applications, followed by magnetite-based nanoparticles.

12.3.2 Magnetically Guided Drug Delivery

Conventional chemotherapy is relatively non-specific. When a cytotoxic drug is distributed systemically, through intravenous or intra-arterial administration, it attacks both cancerous and healthy cells indiscriminately, producing the well-known deleterious side effects of chemotherapy. The suggestion to attach cytotoxic drugs to magnetic nanoparticles came about from the recognition of the need to deliver, accumulate and retain such drugs at specific sites affected by cancer (or inflammation), rather than rely on their natural biodistribution. Such an approach would (a) spare healthy tissue from toxicity by reducing the amount of systemic distribution of the cytotoxic drug, minimizing side effects and (b) reduce the dosage or amount of chemotherapeutic drug required, due to its localized administration. Prolonged exposure of a tumor lesion to sufficiently high drug concentrations is a prerequisite for therapeutic efficacy, which can be enabled with selective targeting and localization.

The physical principle behind magnetic drug targeting is the same as that of magnetic separation. The functionalized ferrofluids carrying the cytotoxic drug are introduced into the bloodstream *via* intravenous or intra-arterial injection at a vein/artery close to the targeted tumor. A strong magnet is placed outside the body over the location of the tumor. Figure 12.13(a) depicts a potential physical arrangement of patient and magnet, while Figure 12.13(b) gives a schematic of the physical process of particle deposition in the tumor. As the circulating ferrofluids pass by the vicinity of the tumor, they experience a magnetic force of the form given by Eq. (12.4), due to the presence of the inhomogeneous magnetic field produced by the magnet. The magnetic force exerted onto the magnetic nanoparticles must be strong enough to overcome the hydrodynamic drag force of Eq. 12.5, due to the blood flow in the artery or vein. This suggests that magnetic targeting would be more successful for slow arterial blood-flow velocities; especially if the target is close to the magnet and if larger microspheres are used to carry the cytotoxic drug. For this reason, both functionalized nanoparticles and porous biocompatible polymers (or block copolymers) in which magnetic nanoparticles are precipitated within the pores have been used as drug

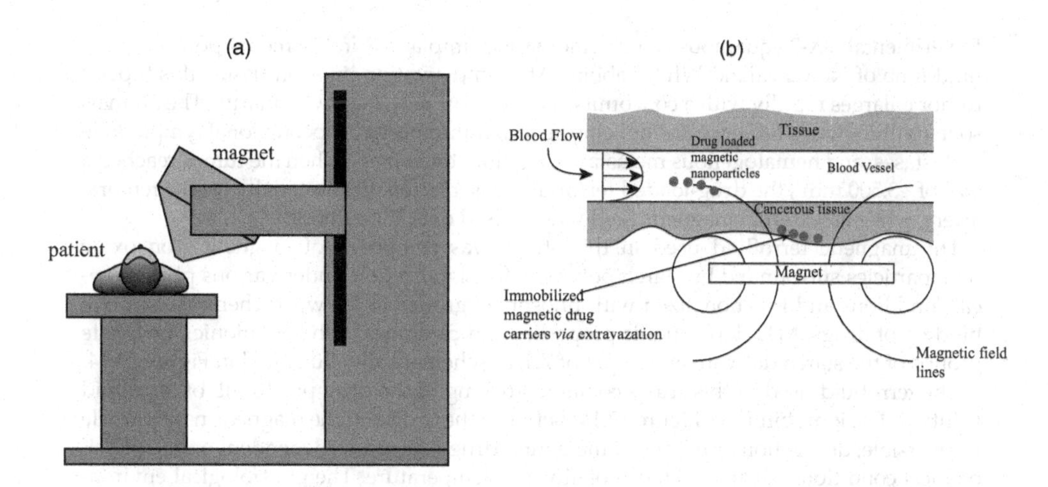

FIGURE 12.13
(a) Potential physical arrangement of magnet and patient in magnetically guided drug delivery. (b) Functionalized magnetic nanoparticles targeted to a specific location in the body extravasate from the blood vessel under the action of a magnetic force due to the magnetic field gradient produced by the external magnet.

carriers. Model studies based on magnetite particles indicate that a magnetic field of the order of 0.2 T with field gradients of ~8 T/m would be required for femoral arteries, while an increased magnetic field gradient of over 100 T/m will be needed for carotid arteries, where blood-flow velocities are higher. The magnetic field must penetrate deep enough into the body so that it reaches the ferrofluid-carrying blood vessel. Blood vessel walls are quite permeable, and thus, the magnetic nanoparticles or microspheres carrying the cytotoxic drugs are expected to extravasate into the tumor cells under the pull of the magnetophoretic force. Figure 12.13(b) gives a depiction of the process of magnetic nanoparticle extravasation to a targeted organ.

Many investigations in small animal models have validated this concept. As magnetic fields can penetrate only a short distance within the body, the technique cannot be applied to tumor sites located deep inside the body, although proposals by scientists to implant magnets inside the body at the tumor site are under investigation. A most natural application of the technique would be on brain and sarcoma tumors, which may be located close to the surface, and thus amenable to guidance and localization by an external magnet.

Example 12.3: Magnetic Guidance for Frug Delivery in Cancer Treatment

To demonstrate the efficacy of the technique of magnetically guided drug delivery, we present a small animal study by Alexiou and collaborators (C. Alexiou *et al.* Locoregional Cancer Treatment with Magnetic Drug Targeting, *Cancer Research*, 60 (2000) 6641). The authors treated squamous cell carcinoma in rabbits with ferrofluids bound to mitoxantrone (MTX), an anticancer agent, which acts against a variety of hematologic and solid tumors by impairing DNA replication, transcription and repair, and may cause adverse effects such as nausea, fever, anemia, immunosuppression, and cardio-toxicity. The drug-bearing ferrofluid was concentrated at the tumor area with an external magnetic field.

Experimental VX-2 squamous cell carcinoma was implanted in the media portion of the hind limb of New Zealand White rabbits. After implantation into soft tissue, this type of tumor enlarges rapidly with a concomitant increase of peripheral vascularity. The animals soon (within two to three weeks) develop central tumor necrosis, locoregional lymph node metastases, and hematogenous metastases (*e.g.*, into the lungs). When the tumor reached a size of ~3,500 mm^3, the drug-loaded ferrofluid was injected intra-arterially *via* the femoral artery, while an external magnetic field was focused onto the tumor.

The magnetic ferrofluid used in this study was composed of magnetic iron oxide nanoparticles surrounded by starch polymers for stabilization under various physiological conditions and functionalized with phosphate groups to allow for chemo-absorptive binding of drugs. MTX has cationic properties and combines with the anionic phosphate groups of the starch derivatives at a pH of 7.4, as schematically illustrated in Figure 12.14.

The ferrofluid used in this study contained 6.5 mg of the drug per 10 mL of ferrofluid solution. The ionic binding (Figure 12.14) between the drug and the magnetic nanoparticle is reversible; desorption or release of the bound drug is possible, dependent on the physiological conditions, such as pH, osmolality and temperature. The physiological environment can be varied by changing the blood electrolyte concentration according to specific needs. In this study, 100% desorption of the drug took place within 60 minutes, as seen in Figure 12.15, after delivery to the tumor.

The drug must be desorbed from the surface of the nanoparticle in order to be able to act freely once concentrated at the tumor by the magnetic field. In this study, an electromagnet with a specially designed pole piece to produce a strong magnetic field of 1.7 T in the region of the tumor surface was used. The authors compared the outcomes of tumor volume evolution and appearance of metastasis under various conditions of treatment with MTX. Figure 12.16 depicts the results obtained for (a) treatment with the drug-loaded ferrofluids under magnetic guidance and (b) classical systemic chemotherapy treatment, as compared with the control (no treatment).

It is observed that with magnetically guided drug delivery, using only 50% of the systemic dose, tumor volume was decreased to 1% its original size, with no deleterious effects such as alopecia (loss of hair), within 15 days after treatment. Chemotherapy treatment with the anticancer-drug-loaded ferrofluids in the absence of guidance, at 50% of the systemic dose, failed to decrease the tumor volume and prevent metastasis. At 75–100%

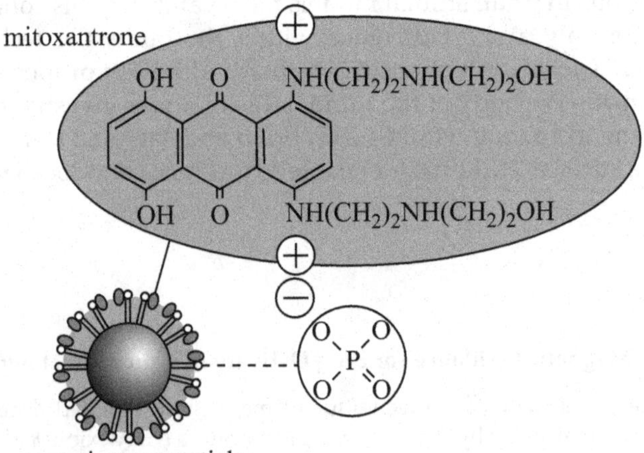

FIGURE 12.14
Schematic representation of a magnetic nanoparticle loaded with the anticancer drug mitoxantrone *via* ionic binding. (From C. Alexiou *et al.* Locoregional Cancer Treatment with Magnetic Drug Targeting, *Cancer Research*, 60 (2000) 6641, Copyright American Association for Cancer Research, reproduced with permission).

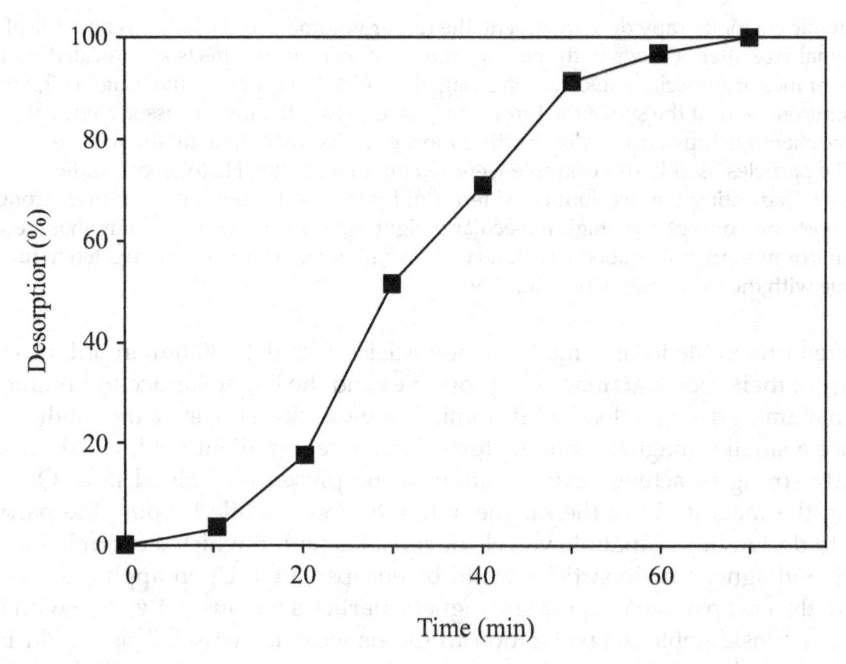

FIGURE 12.15
Percentage of drug desorbed as a function of time. (From Alexiou *et al*. Locoregional Cancer Treatment with Magnetic Drug Targeting, *Cancer Research*, 60 (2000) 6641, Copyright American Association for Cancer Research, reproduced with permission).

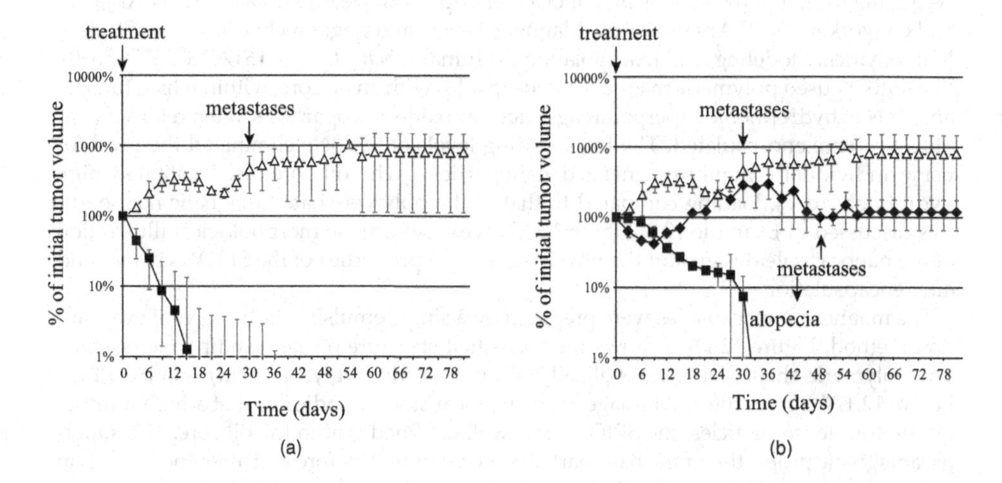

FIGURE 12.16
Progression of disease with (a) magnetically guided chemotherapy with the application of MTX at 50% of the regular systemic dose (■), compared with the control (Δ) (no treatment); and (b) without magnetically guided chemotherapy with the application of MTX at 20–50% of the regular systemic dose (♦) and 75–100% of the regular systemic dose (■), compared with the control (Δ), both after a singular treatment at t = 0. (Adapted from C. Alexiou *et al*. Locoregional Cancer Treatment with Magnetic Drug Targeting, *Cancer Research*, 60 (2000) 6641, Copyright American Association for Cancer Research, with permission).

systemic chemotherapy dose treatment, the tumor volume was reduced to about 9% of its original size after 30 days with the appearance of deleterious effects as indicated by the appearance of alopecia. Thus, by using magnetic guidance to deliver the drug in high concentrations only at the site of the tumor, one can decrease the toxicity associated with systemic chemotherapy and render the drug more effective within the tumor volume.

The particles used in this example were 100 nm in diameter. Histological studies by the authors indicating the distribution of ferrofluid particles throughout the tumor strongly support the concept that high-molecular-weight substances such as chemotherapeutic agents or monoclonal antibodies loaded on ferrofluids can be effectively targeted to tumor tissue with the aid of magnetic guidance.

It is often preferable to use smaller nanoparticles, less than 30 nm, in order to take full advantage of their superparamagnetic properties and the larger surface-to-volume ratio in order to maximize drug payload while minimizing toxicity. However, the smaller particles experience a smaller magnetophoretic force for a given gradient field, which may not be sufficiently strong to achieve extravasation in the presence of blood flow. One path to increasing the magnitude of the magnetophoretic force, while keeping the particle size small, is to design biocompatible vessels or capsules within which a controllable number of superparamagnetic nanoparticles could be encapsulated. Upon application of a magnetic field, the encapsulated superparamagnetic particles will instantly align with the field imparting a considerable magnetization to the nanocarrier vessel. This would induce a strong magnetophoretic force on the nanocarrier, or, nanocapsule (NC). The design of polymeric nanocapsules as carriers of chemotherapeutic drugs with encapsulated superparamagnetic nanoparticles is under intense investigation. Such an approach is presented in the following example.

Example 12.4: Polymeric Magnetic Nanocarriers for Drug Targeting

As an illustration of the promising nanocarrier approach, we present a study by Al-Jamal and co-workers (K. T. Al-Jamal *et al.*, Magnetic Drug Targeting: Preclinical *In Vivo* Studies, Mathematical Modeling, and Extrapolation to Humans, *Nano Letters*, 16 (2016) 5652–5660). The authors used polymeric magnetic nanocapsules with an oil core, within which tunable amounts of hydrophobic superparamagnetic-iron-oxide-nanoparticles, often referred to as SPIONs, were encapsulated. The high loading levels of SPIONs eliminated the need for using individually highly magnetized nanoparticles. The oil core also facilitated high hydrophobic drug loading compared to that of the polymer-coated magnetic nanoparticles discussed in Example 12.3. Figure 12.17 gives a schematic morphological illustration of the nanocapsule design and the physico-chemical properties of the SPIONs before and after encapsulation.

The magnetic nanocapsules were prepared by a single emulsification - solvent evaporation method. Figure 12.17(a) shows the core–shell structure of the resulting nanocarrier. The polymeric shell was functionalized with poly(lactic-co-glycolic acid) (PLGA–PEG). Figure 12.17(b) gives the TEM image of the nanocapsules; it indicates that a high number of electron-dense particles, the SPIONs, are well-confined within the oil core. The superparamagnetic properties of the nanoparticles were studied, before and after encapsulation within the nanocarrier, by SQUID magnetization and Mössbauer spectroscopic studies, shown in Figures 12.7(c) and (d), respectively. The results indicate that the superparamagnetic properties of the nanoparticles were not altered upon encapsulation. In particular, the Mössbauer signature obtained indicates the presence of fairly dispersed, non-aggregated magnetic nanoparticles, with strong superparamagnetic properties. This design affords a many-fold increase in the magnitude of the magnetophoretic forces acting on the nanocapsules as a whole, compared to those exerted on individual SPIONs, as the magnetic force on the nanocapsule is directly proportional to the cumulative iron-oxide-nanoparticle volume within the magnetic core of the nanocapsule. The nanocapsules had a size of ~135 nm

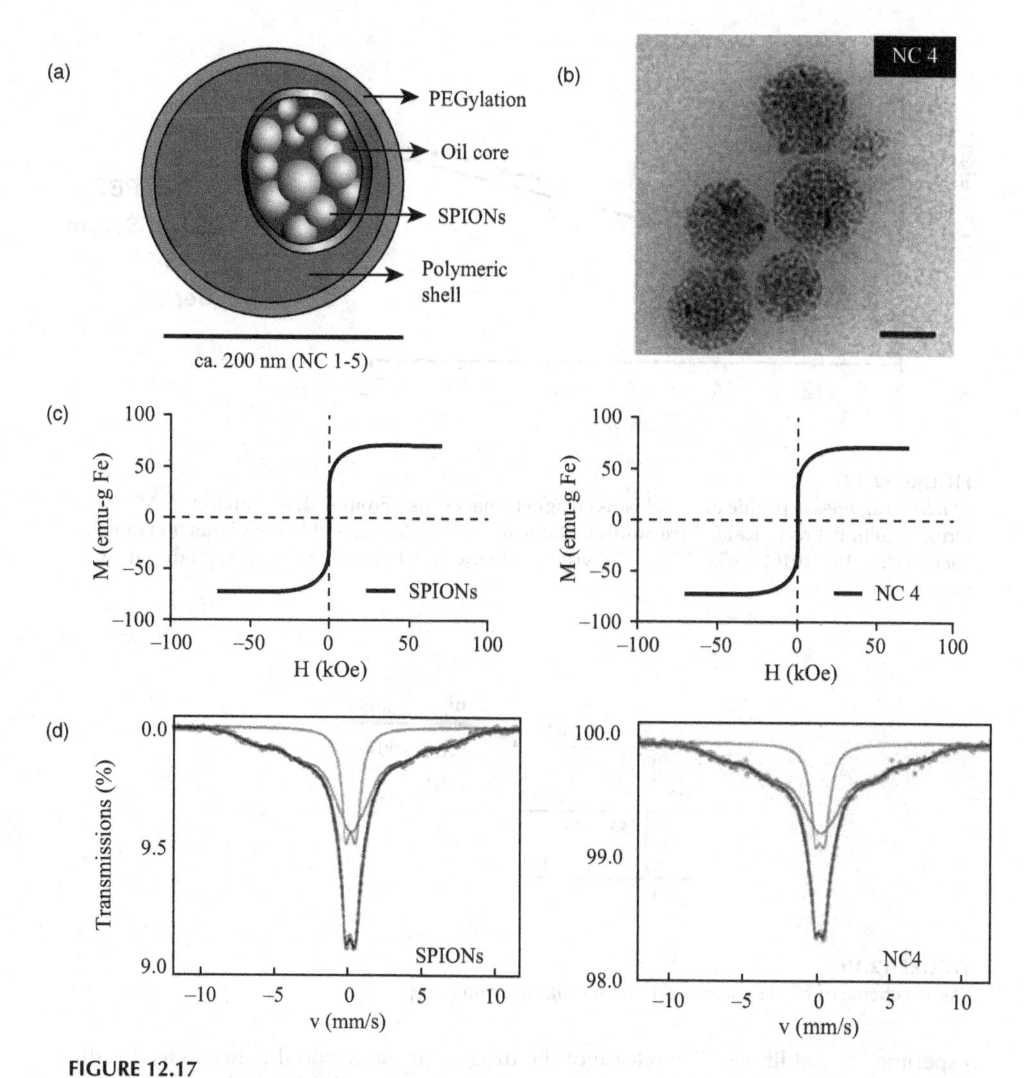

FIGURE 12.17
(a) *Circa* 200-nm hydrodynamic diameter nanocapsule (NC) architecture containing hydrophobic SPIONs. (b) TEM micrograph of the nanocapsules with encapsulated SPIONs (bar 100 nm). (c) SQUID magnetometry of SPIONs before (left) and after (right) encapsulation. (d) Mössbauer spectroscopic signatures of SPIONs before (left) and after (right) encapsulation. The superparamagnetic properties were not compromised upon encapsulation. (From K. T. Al-Jamal *et al.*, Magnetic Drug Targeting: Preclinical *In Vivo* Studies, Mathematical Modeling, and Extrapolation to Humans, *Nano Lett.*, 16 (2016) 5652–5660, Copyright American Chemical Society, reproduced with permission).

diameter, while the SPIONs had a corresponding size of ~10 nm obtained from TEM measurements (Figure 12.17(b)); the hydrodynamic size of the nanocapsules was in the range of 203–218 nm and no significant changes were observed over a three-month storage period. Nanocells with varying amounts of encapsulated magnetic nanoparticles were prepared, containing from 0% up to *ca.* 7% w/w SPION/nanocell.

In addition, nanocells were the carriers of the cytotoxic anticancer drug docetaxel (DTX). The nanocells allow for the slow, controlled release of the drug. The release of DTX from the magnetic nanocells was studied *in vitro* in the presence or absence of serum under sink conditions at 37°C. The release profile was compared to that of DTX dissolved in dimethylsulfoxide (DMSO) dialyzed against phosphate-buffered saline (PBS) solution and used as the 100% release control. Figure 12.18 gives further details on the release profile and

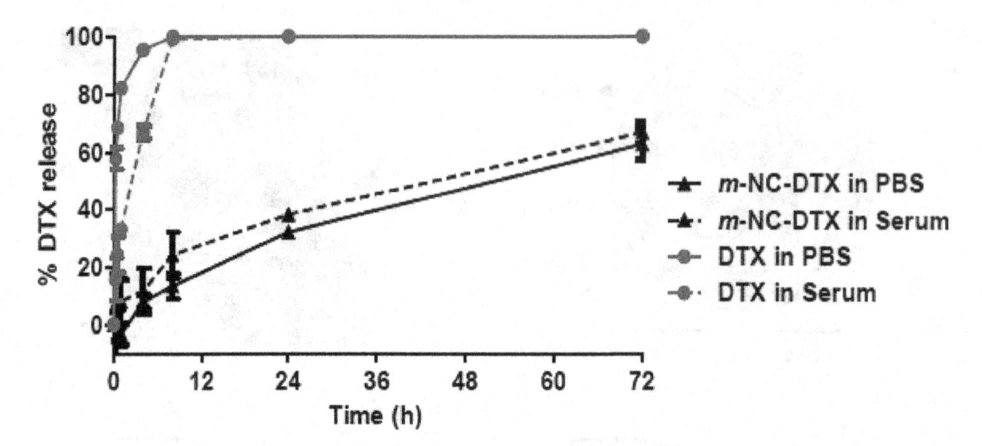

FIGURE 12.18
In vitro drug release profile of DTX loaded magnetic nanocells. (From K. T. Al-Jamal *et al.*, Magnetic Drug Targeting: Preclinical *In Vivo* Studies, Mathematical Modeling, and Extrapolation to Humans, *Nano Lett.*, 16 (2016) 5652–5660, Copyright American Chemical Society, reproduced with permission).

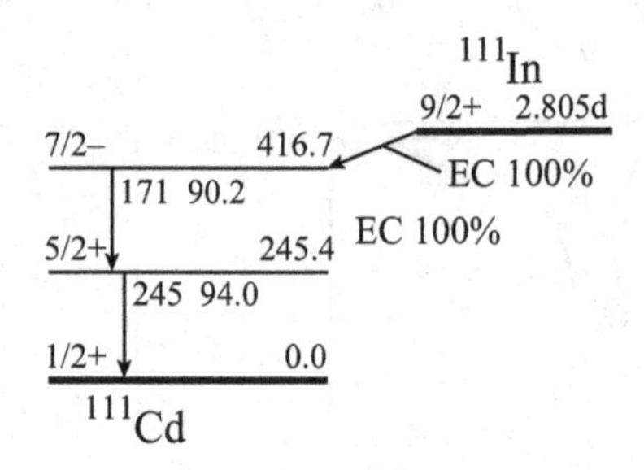

FIGURE 12.19
Decay scheme of In-111 used in gamma scintigraphy imaging.

experimental conditions. The release of the drug from the nanocell is necessary for the drug to act on the tumor. The observed slow release allows for the drug to be effective over a longer period of time, while its greater localization on the tumor site through magnetic guidance targeting, presented below, limits its overall toxicity to healthy organs.

The magnetic nanoparticles within the capsules were labeled with radioactive ^{111}In for radiological detection of their fate after injection into the bloodstream. Indium-111 decays by electron capture to cadmium-111 emitting two gamma rays of 171 and 245 keV energy, respectively, as seen in Figure 12.19. These gamma rays can be detected by gamma cameras enabling the use of gamma-ray scintigraphy for 2-D imaging as well as single-photon emission computed-tomography (SPET-CT) for 3-D imaging of the biodistribution of the radioactively labeled nanoparticles. As seen in Figure 12.19, Indium-111 has a relatively short radioactive half-life, $t_{1/2} = 2.8$ days, which makes it a good candidate as a radioactive tracer.

This technique allows the study of the biodistribution and retention characteristics of the nanoparticles *in vivo*, at different times after injection and/or after organs or tumors were removed following animal sacrifice.

The therapeutical efficacy of magnetic targeting was assessed by injecting mice with cultured CT26 colon carcinoma cancerous cells suspended in PBS solution at pH 7.4. A volume of 20 μL of this solution containing 1×10^6 cancerous cells was injected subcutaneously and bifocally in the two hind feet of BALB/c mice aged four to six weeks. The tumors were allowed to increase in size for at least eight days before magnetic nanocapsules loaded with the anticancer drug were administered. Specifically, mice were injected intravenously *via* a lateral tail vein with 150 μL PBS solution containing ~0.7 MBq (or 0.019mCi) of ^{111}In radioactive nanocells. On one of the tumors of the mouse, a magnet was placed non-invasively attached to the leg by surgical tape immediately above the tumor and left there one hour after drug administration. The magnet was an 8-mm diameter 5-mm height disk-shaped nickel-coated neodymium iron boron ($Nd_2Fe_{14}B$) permanent magnet with a magnetic field strength of 0.43 T and magnetic field gradient of ~70 T/m, evaluated at z = 5 mm above the center, in the xy-plane, of the circular face of the disk.

Figure 12.20 presents the results of the study that included blood clearance profiles, excretion profiles and organ biodistribution profiles, shown in Figure 12.20(a–c), of the radioactive magnetic nanoparticles as assessed by gamma scintigraphy. Tumor drug

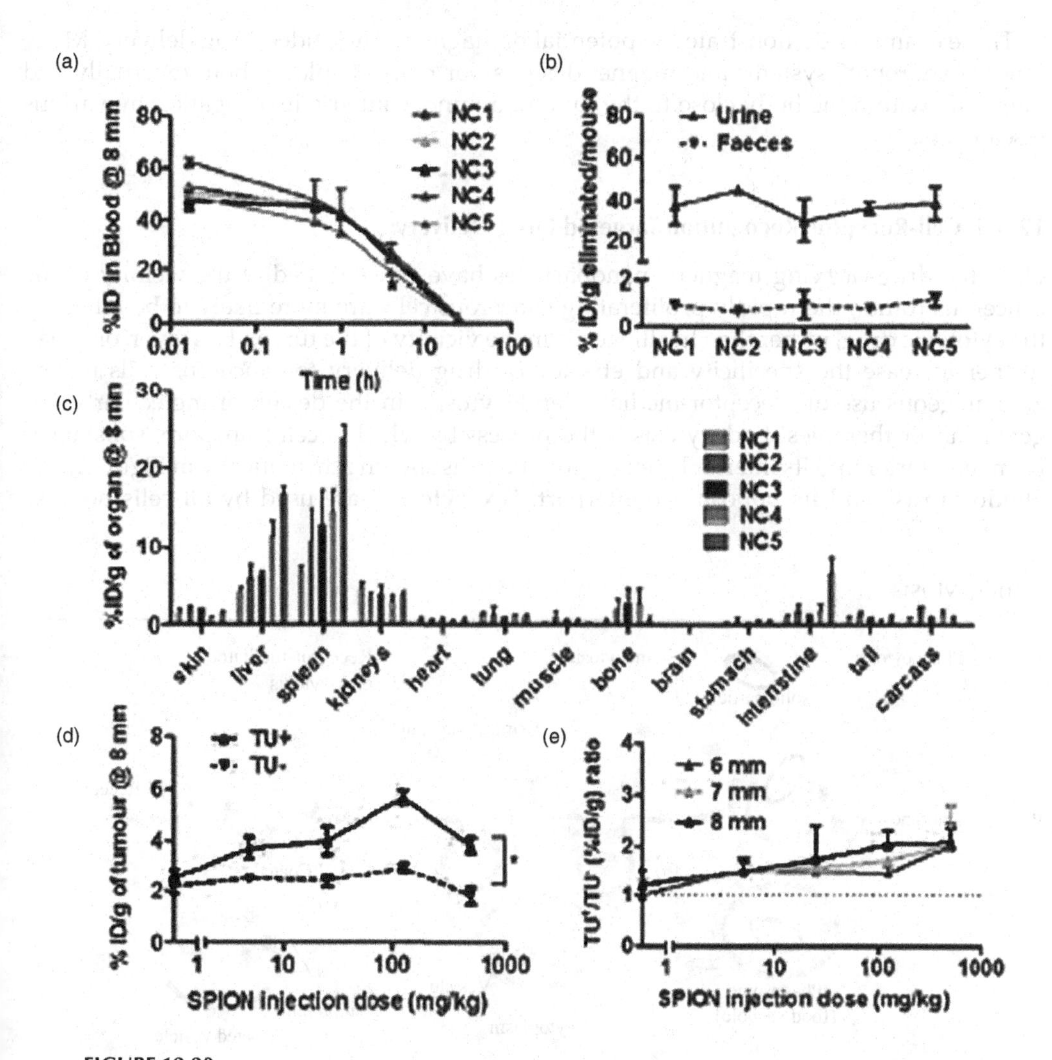

FIGURE 12.20
Blood clearance (a), excretion (b), and organ biodistribution profiles (c). Drug accumulation for magnetically targeted (TU+) and non-magnetically targeted (TU-) tumors (d and e). (Adapted from K. T. Al-Jamal *et al.*, Magnetic Drug Targeting: Preclinical *In Vivo* Studies, Mathematical Modeling, and Extrapolation to Humans, *Nano Lett.*, 16 (2016) 5652–5660 Copyright American Chemical Society, reproduced with permission).

accumulation profiles are given in Figures 12.20(d) and (e), where the results are expressed in terms of percentage of injected dose per organ (%ID/organ) or percentage of injected dose per gram (%ID/g) of the tumor.

There is clearly a difference in drug accumulation in the tumor area for the tumor that experiences no magnetic field (TU-) relative to the one that was exposed to the field (TU+). The targeted tumor TU+ shows increased drug accumulation relative to the non-targeted one. Data are presented as radioactive % injection dose (%ID) per gram of tumor *vs.* SPION injection dose. To examine the effect SPION loading on magnetic targeting efficiency, mice were injected with comparable nanocell doses, but increased loadings of SPIONs from 0% to *ca.* 7% w/w SPION/NC, labeled in the figure NC 1-5. Organs were excised 24 hours post injection.

These examples demonstrate the potential of magnetically guided drug delivery. Many other "nanorobot" systems and magnet designs, for magnets placed both externally and internally within the body close to the tumor, are under intense investigation by various researchers.

12.3.3 Cell-Receptor-Recognition Targeted Drug Delivery

Once the drug-carrying magnetic nanoparticles have concentrated at the vicinity of the cancerous tumor, the rapidly proliferating cancerous cells are more likely to be killed by the cytotoxic drug, rather than healthy cells in the vicinity of the tumor. However, one may further increase the specificity and efficacy of drug delivery to cancerous cells by the advantageous use of "receptor-mediated endocytosis" in the design of molecularly targeted cancer therapies. Endocytosis is the process by which a cell transports substances from its exterior into its interior by engulfing them, as shown schematically in Figure 12.21. "Endocytosis" and its opposite counterpart, "exocytosis", are used by all cells because

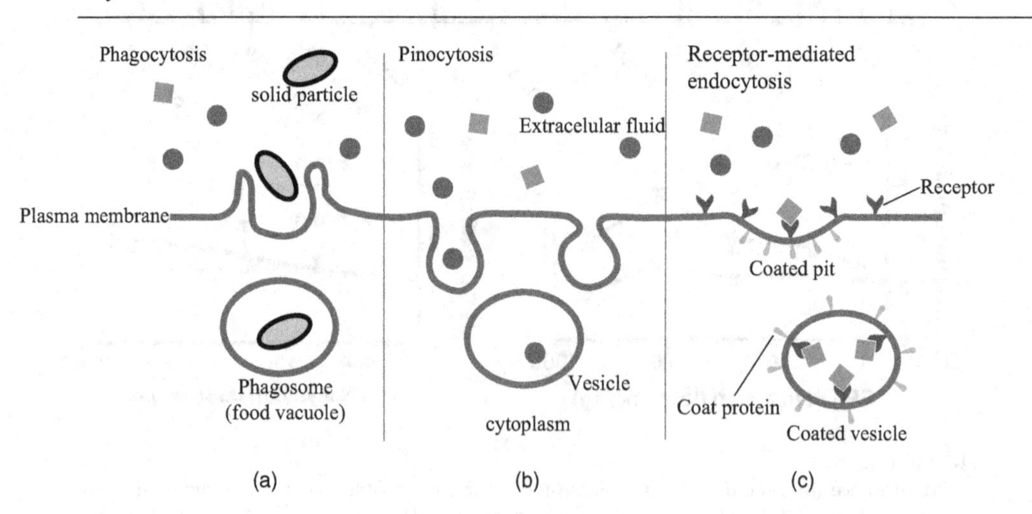

FIGURE 12.21
Schematic depiction of the process of endocytosis (a, b), which is taken advantage of in cell-receptor-mediated targeted drug delivery to cancerous cells (c).

most nutrients important to them are large chemical substances that cannot pass through the cell membrane by passive means. Endocytosis includes phagocytosis (cell eating), Figure 12.21(a), and pinocytosis (cell drinking), Figure 12.21(b). Receptor-mediated endocytosis, also called "clathrin-mediated endocytosis", Figure 12.21(c), is a process by which cells absorb metabolites, hormones, proteins or even viruses, by the inward budding of plasma membrane vesicles containing proteins with receptor sites specific to the molecules being absorbed. The ligand and receptor are then captured by a preformed or forming clathrin-coated pit. A mature pit will pinch off the plasma membrane forming a clathrin-coated vesicle. Incorporation of the foreign substance into the cell is then mediated by the production of clathrin-coated small vesicles (~100 nm diameter) that typically fuse to an early endosome for further sorting by lysosomal recycling, or other trafficking pathways.

A rapidly growing tumor requires various nutrients and vitamins. Therefore, tumor cells overexpress many tumor-specific receptors on their surface, which can be used as targets to deliver cytotoxic agents into tumors. For example, monoclonal antibodies, poly-unsaturated fatty acids, folic acid, aptamers, oligopeptides and hyaluronic acid have been applied as tumor-specific moieties to construct cell-receptor-recognition drug-targeting vehicles.

In addition to cytotoxic drugs for localized chemotherapy, magnetic nanoparticles can be linked to radioactive materials for localized radiation therapy, as well as to DNA for gene therapy.

12.4 Magnetofection

Ever since the advent of genetic engineering in the 1970's, scientists have considered its application to medicine. Genetic engineering is a process that alters the genetic make-up of an organism by either removing or introducing nucleic acids, the building blocks of DNA that carry the building plans of living systems. Thus, nucleic acids can be exploited to make cells repair defective genes, produce a desired protein, or shut down the expression of endogenous disease genes. The result is what is known today as gene therapy, which is the therapeutic delivery of nucleic acid polymers into a patient's cells to treat disease. In this process, the therapeutic DNA must be administered, reach the damaged cells and enter the cells. Therapies usually involve removing cells from patients, modifying their genetic make-up *in vitro* and returning the transformed cells to patients.

The process of introducing foreign DNA into a cell *via* a "gene delivery vector", such as a virus, is known as "transfection". "Magnetofection" refers to magnetic-field-assisted transfection through the use of magnetic nanoparticles and microspheres. To form "magnetic gene delivery vectors", the magnetic nanoparticles must possess functionalities that enable their association with a gene delivery vector. Virus-mediated gene delivery utilizes the ability of a virus to inject its DNA inside a host cell. The gene that is intended for delivery is packaged into a replication-deficient viral particle. Some viruses used to date include retrovirus, lentivirus, adenovirus, adeno-associated virus and herpes simplex virus. However, viruses can only deliver very small pieces of DNA into the cells, the process is labor-intensive and there are risks of random insertion sites resulting in mutagenesis. For this reason, non-viral vectors have also been developed. In magnetofection, the vector could be a virus or naked DNA, producing "viral" *vs.* "non-viral" magnetic gene delivery systems, respectively.

The magnetic properties of the vectors must be strong enough so that they can be concentrated at the target cells under the action of a magnetic field gradient, *in vitro* or *in vivo*. The magnetic particles usually have a magnetic core of magnetite or maghemite encapsulated by a polymer, silica or metal shell, such as gold, functionalized to bind the nucleic acid of interest or the virus containing the therapeutic DNA. The particles may also be precipitated within a porous polymer to form a magnetic microsphere. The magnetic gene vectors are injected into the bloodstream *via* a catheter at a site near the target. Just as in the case of magnetically guided drug delivery, the technique is most useful for target sites near the surface, such as the brain, as magnetic nanoparticles have been demonstrated to cross the blood–brain barrier. For deeper sites, implantable magnets must be considered. For the case of *in vitro* gene transfection, a high-field, high-gradient magnet is positioned under the culture flask or petri dish in which the cells are growing, as shown in Figure 12.22. The colloidal magnetic particles attached to the gene vectors are then concentrated on the target cells, lying on the bottom of the petri dish, by the influence of the magnetic field gradient generated by the magnet.

The cellular uptake of the genetic material is accomplished by receptor-mediated endocytosis through clathrin-dependent pits, which is, as noted earlier, a natural biological process that does not disrupt the cellular membrane architecture. Thus, the cell's membrane structure stays intact, in contrast to other physical transfection methods that may damage the cell membrane. The coupling of the magnetic nanoparticles to gene vectors of any kind, viral or non-viral, results in a dramatic increase of the uptake of these vectors and, thus, in high transfection efficiency. The principal advantage of magnetofection for *in vitro* applications derives from the rapid sedimentation of the colloidal gene-carrying magnetic nanoparticles or microspheres onto the target area, which reduces both the time and amount of vector needed to achieve efficient transfection. It must be that the diffusion barrier, present for vector particles too small to sediment under gravitational forces alone, is overcome by the more rapid kinetics introduced by magnetophoretic forces. It has also been demonstrated that the overall efficiency of transfection can be further improved by using dynamic magnetic fields produced from oscillating arrays of permanent rare-earth magnets. Studies suggest that dynamic magnetic fields can further improve the level of

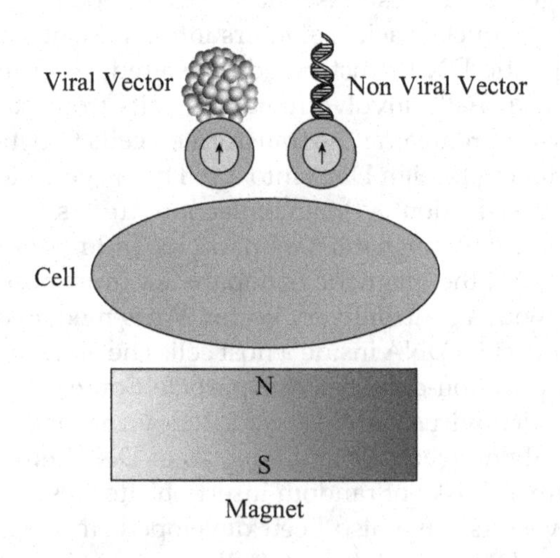

FIGURE 12.22
In vitro magnetofection by viral or non-viral vectors.

transfection more than 10-fold compared to static magnetic fields. It appears that the lateral movement of the magnetic field transmits mechanical forces to cellular membranes bound to magnetic vectors. Mechanical stimuli can affect cellular membrane traffic including endocytosis. In general, magnetic actuation of cellular processes is an active field of research in many areas of biology not only magnetofection. The application of pulsed and/or alternating magnetic fields in combination with static fields is also studied.

In vitro magnetofection plays an important role in viral synchronization used in vaccine development and immune response studies as well as in emerging tissue regeneration therapies with genetically modified cells. Magnetofection is very effective in transferring plasmid DNA into a variety of primary cells including cells hard to transfect such as neurons, which are also very sensitive to toxicity. Magnetofection allows for efficient transfection with low vector doses, minimizing unwanted toxicity side effects.

Both, specific linkers and general electrostatic interactions have been utilized to attach DNA onto magnetic nanoparticles. The negatively charged phosphate backbone of DNA can profitably be utilized for strong electrostatic interactions between DNA and positively charged molecules linked to the surface of the magnetic particles. A popular choice is to coat the magnetic nanoparticles with Polyethyleneimine (PEI) due to its polycationic character. In addition, PEI facilitates lysosomal release of the complex following cell internalization by its ability to buffer the intralysosomal pH, causing the lysosomes to rapture and release their content.

Summarizing, current strategies in systemic drug- and gene delivery seek to exploit all three modes of delivery: (a) the EPR effect, (b) active targeting by the use of external magnets and (c) biological receptor–ligand interaction. The true potential of magnetic guidance lies in its amenability to innovation and integrated technologies, such as multimodal theranostic processes and imaging.

Example 12.5: Effectiveness of Therapy with Magnetofection

As an example of the antitumor effectiveness of non-viral magnetofection, we present the study by Prosen *et al.*, who used SPIONs coated with polyacrylic acid (PPA) and functionalized with polyethylenimine (PEI) for further binding of plasmid DNA (pDNA) encoding against the melanoma cell adhesion molecule (MCAM) ($pDNA^{antiMCAM}$) as magnetofection agents for therapy (L. Prosen *et al.*, Magnetic field contributes to the cellular uptake for effective therapy with magnetofection using plasmid DNA encoding against *Mcam* in B16F10 melanoma *in vivo*, *Nanomedicine (Lond)*, 11 (2016) 627).

In this study, monodisperse SPIONs with a hydrodynamic diameter of about 40 nm were used, while the functionalized nanoparticles (SPIONs-PPA-PEI-pDNA) measured approximately 200–400 nm in diameter and had a zeta potential of about 8 mV at pH 8.0. Figure 12.23 presents an outline of the experimental protocol used.

Mice were injected with B16F10 melanoma cells. Tumors were allowed to grow until they reached a size of 40 mm^3, before mice were given therapeutic injections of the functionalized nanoparticles. On one group of infected mice, an Nd-Fe-B permanent magnet, with a magnetic field of 403 mT and magnetic field gradient of 38 T/m, was placed above the tumor immediately after injection and kept there for 30 minutes; this group is labeled MF (for magnetofection). A second group of infected mice was similarly injected, but no magnet was placed above the tumor after injection; this group is labeled NF (for nanofection). Tumors were excised after a single or three therapeutic injections, as indicated in the figure. Tumor measurements were taken for a period of time up to 6 days after the first injection. Excised tumors were characterized by immunohistochemical (ICH) staining, inductively coupled plasma mass spectrometry (ICP-MS) and transmission electron microscopy (TEM).

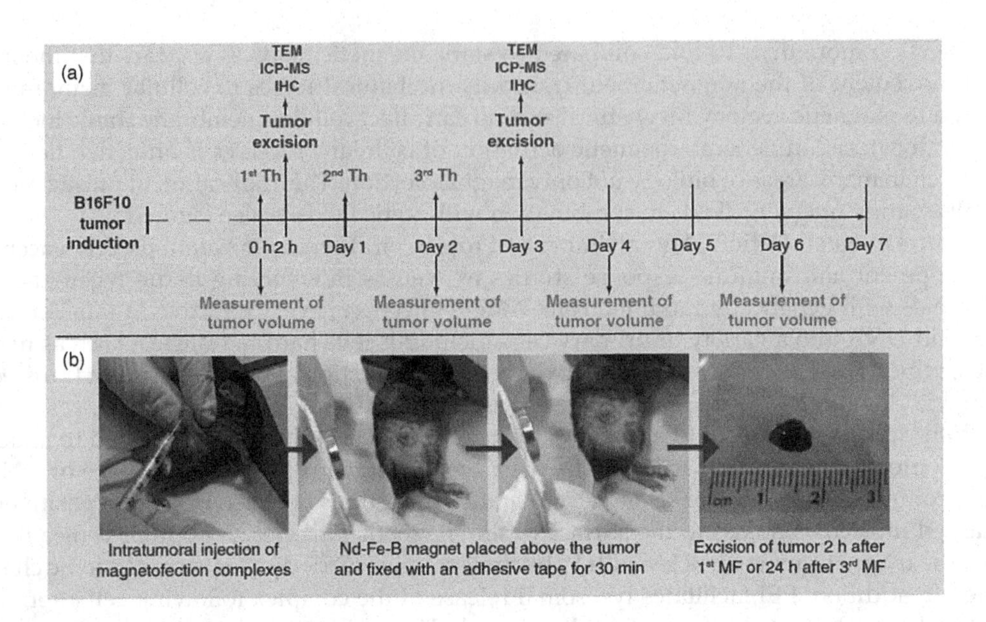

FIGURE 12.23
(a) and (b) give an outline of the experimental protocol of the magnetofection study by Prosen *et al.* (From L. Prosen *et al.*, Magnetic field contributes to the cellular uptake for effective therapy with magnetofection using plasmid DNA encoding against *Mcam* in B16F10 melanoma *in vivo*, *Nanomedicine (Lond)*, 11 (2016) 627, Copyright Future Medicine, reproduced with permission).

Figure 12.24 gives the most representative TEM micrographs of magnetofection complexes present in the tumor cells and the extracellular space after the third injection. Figure 12.24(a) indicates that the magnetofection agents, that is, DNA carrying magnetic nanoparticles, were accumulated within the B16F10 tumor cells, which appear as dark spots, either isolated or aggregated, indicated by arrows. These observed dark spots result from the electron-rich iron ions of the magnetic nanoparticles, which strongly scatter the electron beam.

However, the TEM of the extracellular matrix also indicates the presence of magnetic nanoparticles, Figure 12.24(b). While iron accumulation within the melanoma tumors showed no statistically meaningful difference under NF or MF, as indicated in Figure 12.24(c), the iron accumulation into malignant cells is much higher compared to the extracellular space under MF, that is, the application of an external magnetic field, as indicated in Figure 12.24(d). These results support the fact that the magnetic field contributes to effective magnetofection, that is, the delivery of pDNA within the melanoma cells. The exposure of the melanoma tumors to an external magnetic field immediately after injection contributes to an increased uptake of magnetofection complexes from the extracellular matrix into the cells. That is, the magnetic field retains the magnetofection agents in the vicinity of the tumor longer, increasing the probability for the magnetofection agents to enter the cells. This should lead to better therapeutic effectiveness. The authors' studies on tumor growth under NF and MF treatments indicated that tumors of the MF-treated mice exhibited much slower growth as compared to control (no therapeutics) or NF-treated mice. This study demonstrates the importance of the magnetic field in *in vivo* magnetofection with therapeutic plasmid DNA.

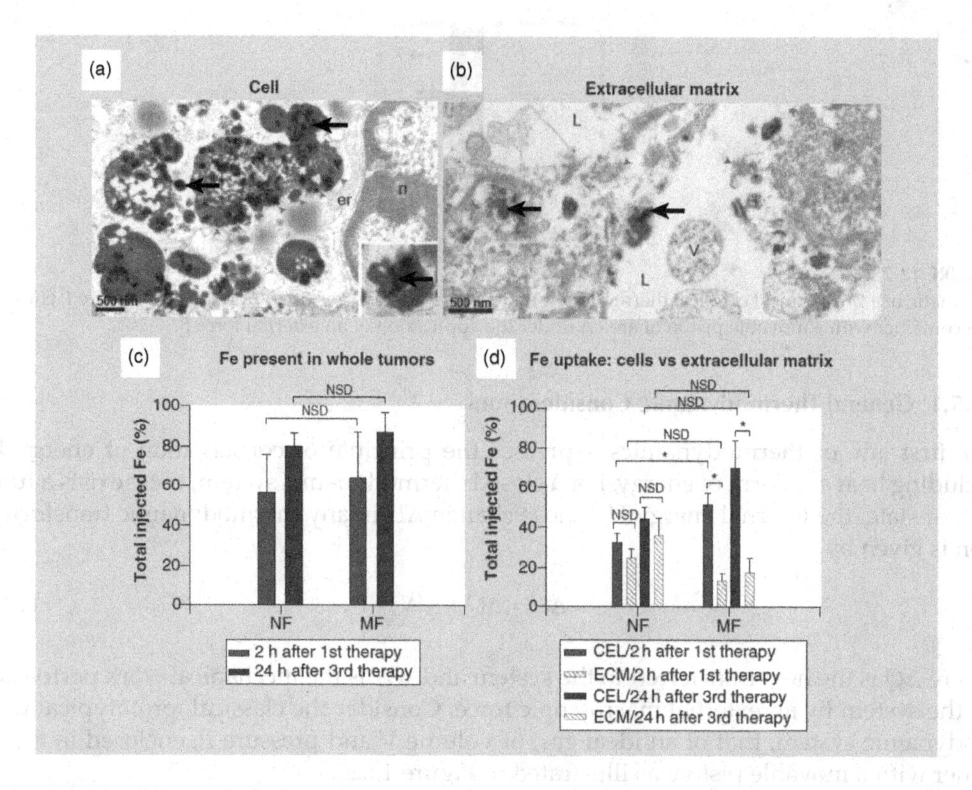

FIGURE 12.24
TEM micrographs of excised tumor cells (a) and extracellular matrix (b). The accumulation of magnetofection complexes in the melanoma tumor (B16F10) after intratumoral injection under the absence (nanofection, NF) (c) and presence (magnetofection, MF) (d) of an external magnetic field. (From L. Prosen *et al.*, Magnetic field contributes to the cellular uptake for effective therapy with magnetofection using plasmid DNA encoding against *Mcam* in B16F10 melanoma *in vivo*, *Nanomedicine (Lond)*, 11 (2016) 627, Copyright Future Medicine, reproduced with permission).

The non-viral gene delivery method described in the example above is currently in a preclinical stage of development with few *in vivo* studies performed so far. However, as a non-viral, non-invasive and painless gene delivery method, it holds great promise and further studies are being anticipated.

12.5 Magnetic Fluid Hyperthermia

Another area of nanomagnetism under intense investigation is the examination of the thermal transfer characteristics of ferrofluids when heated by the application of a time-varying external magnetic field in order to achieve hyperthermia for medical treatment. Here, some basic thermodynamic parameters relevant to hyperthermia are reviewed followed by applications.

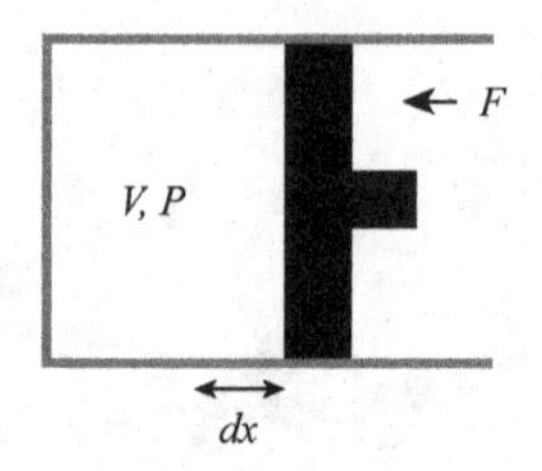

FIGURE 12.25
Schematic of a prototypical classical thermodynamic system of an ideal gas of volume V and pressure P enclosed in a container with a movable piston of area A under the application of an external force F.

12.5.1 General Thermodynamic Considerations

The first law of thermodynamics expresses the principle of conservation of energy by including heat as a form of energy. For a closed thermodynamic system, there exists a function of state, the internal energy U, whose change ΔU in any thermodynamic transformation is given by

$$\Delta U = \Delta Q + \Delta W \tag{12.9}$$

where ΔQ is the heat transferred to the system and ΔW is the mechanical work performed on the system by an external macroscopic force. Consider the classical, prototypical thermodynamic system, that of an ideal gas, of volume V and pressure P, enclosed in a container with a movable piston, as illustrated in Figure 12.25.

Under equilibrium conditions the strength of the external force, F, is related to the pressure, P, by

$$F = PA \tag{12.10}$$

where A is the area of the piston. For an infinitesimal process, the change of the position of the piston by dx results in the performance of work

$$\delta W = Fdx = PAdx = -P\Delta V \tag{12.11}$$

where ΔV is the change (decrease) in the gas volume. When a cycling process (compression followed by expansion) is considered, the work per cycle performed onto the system would be the integral of this quantity over a cycle

$$W = -\oint PdV \tag{12.12}$$

12.5.2 Magnetic Thermodynamic Parameters

For a magnetic substance, rather than an ideal gas, the corresponding thermodynamic variables are the magnetic field, $B = \mu_0 H$, generated by external currents, and the magnetization density, M, of the material. The work done on a system can be obtained under the correspondence of $B \leftrightarrow -P$ and $M \leftrightarrow V$, given by

$$\delta W = \mu_0 HdM \tag{12.13}$$

The first law of thermodynamics for a magnetic system then reads

$$dU = \delta Q + \mu_0 H dM \tag{12.14}$$

As we have discussed, bulk ferromagnets are composed of ferromagnetic domains. The configuration of the domains is such as to minimize the magnetic energy by reducing the surface fields. Upon application of an external field, B, at a fixed temperature less than the Curie temperature, $T < T_C$, the domain walls move in response to the change in B. This is a dissipative process that leads to our familiar magnetic hysteresis loop. By analogy to Eq. (12.12) when a cyclic process is considered, such as the application of an alternating magnetic field, the work performed on the magnetic substance over one complete cycle is

$$W = \mu_0 \oint H dM \tag{12.15}$$

Thus, in this case, the area enclosed within the hysteresis loop represents the amount of work performed on the magnetic substance per cycle. Under adiabatic conditions, that is, $\delta Q = 0$, this work represents the change in internal energy of the magnetic substance per cycle

$$\Delta U = \mu_0 \oint H dM \tag{12.16}$$

As the motion of magnetic walls is a dissipative (frictional) process, some of this internal energy results in heating. Looking at the microscopic processes associated with tracing the hysteresis loop, rotational work is done in orienting individual atomic moments in the direction of the external magnetic field. Upon removal of the magnetizing field, not all atomic moments move completely back to their original orientation; the specimen retains some magnetism, known as remnant magnetization, indicated by M_r in a hysteresis loop (Figure 3.23). In other words, the energy spent during the magnetization process is not completely recovered upon demagnetization, and there is always a definite loss of energy for each hysteresis loop traced. This energy loss is called "hysteresis loss", which is dissipated into heat.

Thus, from a microscopic point of view, the hysteresis loss can be calculated by considering the rotational work performed in orienting the magnetization in the direction of the magnetic field (see Section 1.6.2). According to Eq. (1.18), the strength of the magnetic torque acting on a magnetic moment, $\vec{m}$, when it is at an angle θ with respect to the magnetic field direction, is given by

$$|\vec{\tau}| = |\vec{m} \times \vec{B}| = mB \sin\theta = \mu_0 Hm \sin\theta \tag{12.17}$$

The infinitesimal amount of rotational work done, when all magnetic dipoles within a unit volume of the specimen are rotated by $d\theta$, would therefore be

$$dW = -\mu_0 H (\Sigma m) \sin\theta d\theta \tag{12.18}$$

The negative sign indicates that negative work is done on the system, since the orientational potential energy decreases. Components of the magnetic moment $\vec{m}$ along and perpendicular to the direction of H would be $(m \cdot \cos\theta)$ and $(m \cdot \sin\theta)$, respectively. As the magnetization measured is the projection of the sum of the atomic moments along the direction of H, the magnetization density M is given by

$$M = (\Sigma m)\cos\theta$$

$$dM = -(\Sigma m)\sin\theta d\theta$$

Substituting into Eq. 12.18, we obtain $dW = \mu_0 H dM$ and over a cycle, $W = \mu_0 \oint H dM$, recovering Eq. 12.15. Furthermore, as we argued earlier, under adiabatic conditions, this corresponds to the total change in internal energy, ΔU. Finally, using integration by parts over a cycle, Eq. 12.16 can be recast as

$$\Delta U = -\mu_0 \oint M dH \tag{12.19}$$

Colloidal suspensions of magnetic nanoparticles under the influence of an increasing magnetic field can undergo two distinct physical processes: (a) each nanoparticle can physically rotate or (b) have its magnetization flip along opposite directions of the particle's magnetic anisotropy axis. Both of these processes are dissipative. Only multi-domain magnetic particles, typically larger than ~100 nm, would then exhibit a hysteresis loop due to the motion of magnetic walls. In the absence of magnetic walls, as is the case for single-domain-magnetic nanoparticles, it is the dissipative processes of particle rotation and spin-flipping that contribute to heating. Particle rotation is known as "Brownian relaxation", while the process of spin reversal is known as "Néel relaxation". We want to describe how these processes contribute to heating, as they are crucial to the use of superparamagnetic nanoparticles in the application of magnetic hyperthermia in medicine.

12.5.3 Response of a Ferrofluid to an Alternating Magnetic Field

Consider a ferrofluid, consisting of single-domain- or sub-domain-magnetic nanoparticles. Under the application of an external, alternating magnetic field, the magnetization of the ferrofluid, cannot follow in step with the applied, time-varying magnetic field if the relaxation time of the magnetization is longer than the field reversal time. The phase lag between the applied magnetic field and the magnetization of the ferrofluid results in the conversion of magnetic work to internal energy. Since the ferrofluid consists of superparamagnetic particles, its magnetic properties are characterized by its magnetic susceptibility, as discussed in Section 2.5, for the case of paramagnetic materials. In the presence of an alternating magnetic field, one needs to consider the complex ferrofluid susceptibility $\chi = \chi' - i\chi''$. Then, with an alternating magnetic field of angular frequency ω, expressible in complex representation

$$H(t) = H_0 \cos\omega t = \mathrm{Re}\left[H_0 e^{i\omega t}\right] \tag{12.20}$$

the magnetization can also be given in its complex representation as

$$M(t) = \mathrm{Re}\left[\chi H_0 e^{i\omega t}\right] = H_0\left(\chi'\cos\,\omega t + \chi''\sin\,\omega t\right) \tag{12.21}$$

Thus, χ', the real part of the complex susceptibility is in-phase, while the imaginary part, χ'', is out-of-phase with the applied field. Substituting these expressions for M and H in Eq. (12.19), we obtain

$$\Delta U = \mu_0 \omega H_0^2\left[\chi'\oint \cos\,\omega t \sin \omega t dt + \chi''\oint \sin^2 \omega t dt\right] \tag{12.22}$$

The integration is taken over one cycle, that is the integration limits are from 0 to T, where T is the period of the alternating magnetic field, $T = \frac{1}{f} = \frac{2\pi}{\omega}$. Here, f stands for the cyclic frequency measured in Hz. In Eq. (12.22), the first integral within the bracket vanishes, as the average value of $\cos\omega t \sin \omega t$ over a cycle is zero, while the second integral over $\sin^2\omega t$ has an average value of one half

$$\frac{1}{T}\int_0^T \sin^2 \omega t dt = \frac{1}{2} \tag{12.23}$$

Thus, the change in internal energy per cycle is given by

$$\Delta U = \mu_0 \pi \chi'' H_0^2$$

Finally, multiplying both sides by the cyclic frequency, f, we get the following expression for the power dissipated per cycle, per unit volume of the material

$$P = f\Delta U = \mu_0 \pi \chi'' f H_0^2 \tag{12.24}$$

Equation (12.24) indicates that the power dissipated is a function of the complex susceptibility χ'' of the ferrofluid. This, therefore, is the material parameter that must be optimized in a colloidal suspension of magnetic nanoparticles in order to maximize the efficacy of the ferrofluid as a hyperthermia agent.

When a motionless ferrofluid is exposed to a static magnetic field, H_0, the magnetic moments of the nanoparticles tend to align in the direction of a magnetic field. The ferrofluid attains a magnetization $M = \chi_0 H_0$, where χ_0 is the equilibrium susceptibility. When the applied field is suddenly changed, the magnetization of the sample cannot respond instantaneously. It responds to the new value of the applied field with a characteristic time constant, τ, referred to as the characteristic relaxation time. In the presence of an alternating magnetic field, amplitude H_0 and angular frequency ω, the magnetization becomes dynamic and its time dependence is governed by Eq. (12.25)

$$\frac{\partial M(t)}{\partial t} = \frac{1}{\tau}\left(M_0(t) - M(t)\right) \tag{12.25}$$

where τ is the characteristic time for the ferrofluid to return to its equilibrium position, $M(t)$ is the instantaneous magnetization, and $M_0(t) = \chi_0 H_0 \cos \omega t = \mathrm{Re}\,[\chi_0 H_0 e^{i\omega t}]$ is the equilibrium magnetization and χ_0 is the initial magnetization at zero frequency, that is, under a static magnetic field H_0. Substituting the complex representations for $M_0(t)$ and $M(t)$ in Eq. (12.25), we obtain

$$\chi = \frac{\chi_0}{1+i\omega\tau} \tag{12.26}$$

Equation (12.26) gives the dependence of the complex susceptibility on the angular frequency. This then yields the following expressions for the real and complex components of the susceptibility

$$\chi' = \frac{\chi_0}{1+(\omega\tau)^2} \tag{12.27}$$

and

$$\chi'' = \frac{\chi_0\omega\tau}{1+(\omega\tau)^2} \tag{12.28}$$

Substituting the complex susceptibility given by Eq. (12.28) into Eq. (12.24) and using $\omega = 2\pi f$, we obtain for the power dissipated per unit volume

$$P = \mu_0\pi\chi_0 f H_0^2 \frac{2\pi f\tau}{1+(2\pi f\tau)^2} \tag{12.29}$$

Thus, for a monodisperse ferrofluid, the power density dissipated depends on the equilibrium susceptibility of the material, the relaxation time, the amplitude and the frequency of the oscillating magnetic field. The above expression assumes that χ_0 remains constant throughout the cycling process, which is strictly not correct in the presence of an oscillating field. Actual susceptibility is magnetic-field-dependent. Better approximations involve the use of the Langevin function (see Chapter 2), which gives the magnetization in thermal equilibrium as a function of the applied field.

The temperature rise for a monodispersed ferrofluid is then given by

$$\Delta T = \frac{P\Delta t}{c} \tag{12.30}$$

where c is the specific heat of the ferrofluid and Δt is the duration of time the alternating field is applied. The specific heat is calculated as the volume average of nanoparticle and carrier liquid constituents.

12.5.4 Relaxation Times

The magnetization of an ensemble of colloidal magnetic nanoparticles within a carrier fluid can relax back to its equilibrium position through the two distinct processes or mechanisms mentioned above. In the Brownian mechanism, the magnetic moment of the particle remains locked within the crystal structure, pointing along a specific direction of the anisotropy axis, throughout the process; alignment of the particle moment along the direction of the magnetic field is achieved by the physical rotation of the particle as a whole, and heat is dissipated through friction at the interface of the particle and carrier liquid.

The characteristic relaxation time for the Brownian process is given by

$$\tau_B = \frac{3\eta V_H}{kT} \tag{12.31}$$

As expected, the Brownian relaxation time depends on η, the kinematic viscosity of the medium (or carrier fluid) in which the nanoparticles are suspended. In the above equation, k is Boltzmann's constant (1.38×10^{-23} J/K) and T is the absolute temperature in Kelvin. V_H is the thermodynamic volume of the particle, which is larger than the magnetic volume, $V_M = 4\pi R^3/3$, for a particle of radius R. The hydrodynamic volume is usually defined by the equation

$$V_H = \left(1 + \frac{\delta}{R}\right)^3 V_M \tag{12.32}$$

where δ is the thickness of the surfactant layer that is either adsorbed or functionalized on the particle's surface.

In the Néel relaxation process, the particle as a whole does not rotate, rather the magnetic moment flips to the opposite direction of the uniaxial magnetic anisotropy axis. Néel's theory of spin reversal in monodisperse, single-domain-magnetic nanoparticles was discussed in Chapter 4. The characteristic relaxation time *via* this process is given by

$$\tau_N = \tau_0 \exp\left(\frac{K_u V_M}{kT}\right) \tag{12.33}$$

which is the same as Eq. (4.45). Here, τ_0 is the characteristic attempt time for moment reversal, of the order of 10^{-9} to 10^{-12} s depending on the material, K_u is the uniaxial anisotropy density, V_M is the volume of the magnetic particle, k is Boltzmann's constant and T is the absolute temperature.

As there is always a particle size distribution in a physical ferrofluid system, the Brownian and Néel relaxation processes take place in parallel, resulting in an effective relaxation time, τ, for a typical ferrofluid

$$\frac{1}{\tau} = \frac{1}{\tau_B} + \frac{1}{\tau_N} \tag{12.34}$$

Clearly, the dynamics of magnetic relaxation for a ferrofluid are determined by the competition of these two alternative relaxation processes of very different physical characteristics. The Néel relaxation is dominated by the magnetic properties of the nanoparticles; it is the only relaxation process present in the case of nanoparticles embedded in a solid matrix and therefore unable to rotate. On the other hand, Brownian relaxation is dominated by the physical properties of the carrier fluid, that is, its viscosity, as well as, the molecules functionalizing the surface of the nanoparticles, and thus its hydrodynamic volume; thus, unlike Néel relaxation, Brownian relaxation is possible only in the case of nanoparticles suspended in a fluid.

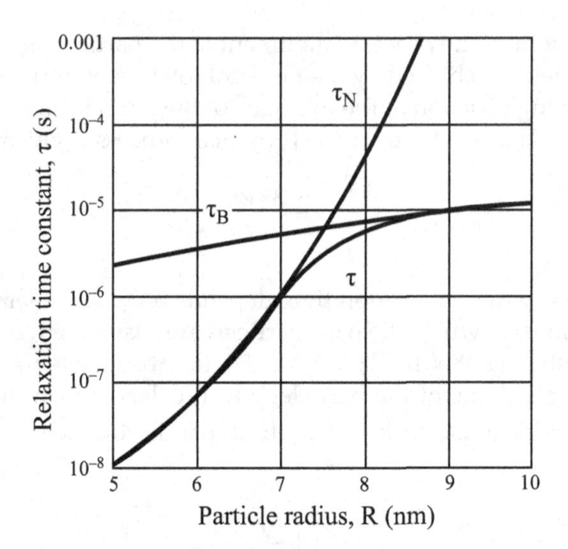

FIGURE 12.26
Relaxation times *vs.* particle radius for magnetite nanoparticles. (From R. E. Rosensweig, Heating Magnetic Fluid with alternating Magnetic Field, *Journal of Magnetism and Magnetic Materials*, 252 (2002) 370, Copyright North Holland (Elsevier), reproduced with permission).

Example 12.6: Heating Magnetic Fluid with Alternating Magnetic Field

Rosensweig was the first to undertake a comprehensive study of the heating rate achieved by a number of ferrofluids (R. E. Rosensweig, Heating Magnetic Fluid with Alternating Magnetic Field, *Journal of Magnetism and Magnetic Materials*, 252 (2002) 370) based on different ferrimagnetic particles, consisting of maghemite (γ-Fe_2O_3), magnetite (Fe_3O_4), cobalt ferrite ($CoFe_2O_4$) and barium ferrite ($BaFe_{12}O_{19}$).

Accounting for both relaxation processes, Eq. (12.34) yields a value for the effective relaxation time τ of the form

$$\tau = \frac{\tau_B \tau_N}{\tau_B + \tau_N} \tag{12.35}$$

For the case of magnetite nanoparticles, this relaxation time is plotted as a function of particle radius in Figure 12.26, along with the individual functional dependence of τ_N and τ_B on particle radius.

The plot shown in Figure 12.26 illustrates that the shorter relaxation time tends to dominate in determining the effective relaxation time for any given size of the particle. Since τ_N increases exponentially (Eq. (12.33)) with the magnetic particle volume V_M, while τ_B increases only linearly (Eq. (12.30)) with the hydrodynamic particle volume V_H, Brownian relaxation processes will dominate in larger nanoparticles, while Néel relaxation processes will dominate in the smaller nanoparticles.

Most nanoparticle synthesis techniques yield log-normal particle size distributions as discussed in Chapter 5. Thus, most ferrofluids are not strictly monodispersed but contain a particle size distribution $g(R)$, where

$$g(R) = \frac{1}{\sqrt{2\pi}\sigma R} \exp\left[\frac{-(\ln R / R_0)^2}{2\sigma^2}\right] \tag{12.36}$$

with

$$\int_0^{\infty} g(R)dR = 1 \tag{12.37}$$

In Eq. (12.36), $\ln R_0$ is the median and σ is the standard deviation of $\ln R$. To obtain the volumetric heat-release rate for a polydispersed ferrofluid, one has to carry out a weighted average over the particle size distribution

$$\langle P \rangle = \int_0^{\infty} P g(R)dR \tag{12.38}$$

Thus, the corresponding temperature rise of an adiabatic ferrofluid sample containing isolated, non-interacting magnetic nanoparticles is given by

$$\langle \Delta T \rangle = \frac{\langle P \rangle \Delta t}{c} \tag{12.39}$$

The analytical integration of Eq. (12.38) is prohibitive. The integrand becomes huge when the expressions for $g(R)$, P, χ and τ are substituted in it. Instead, numerical values of temperature rise were calculated by Rosensweig as examples for various ferrofluids of ferrimagnetic particles consisting of maghemite, magnetite, cobalt ferrite and barium ferrite. These calculations are based on the physical property values listed in Table 12.1, where M_d is the domain magnetization, K is the magnetic anisotropy density of the nanoparticles, c is their specific heat and ρ is the mass density. A value of $\tau_0 = 10^{-9}$ s is employed throughout the Néel relaxation process (Eq. (12.33)).

Figure 12.27(a) gives the calculated heating rate for the case of a monodispersed magnetite sample in a hydrocarbon carrier solution, with magnetic fraction $\varphi = 0.071$, as a function of particle radius and field intensity amplitude ($B_0 = \mu_0 H_0$) at a frequency of 300 kHz. Very substantial heating rates are achievable. An optimum particle size yielding maximum heating exists that is nearly independent of the applied field intensity. Figure 12.27(b) shows the influence of carrier viscosity on the heating rate with the frequency of the alternating magnetic field as a parameter for a sample with a fixed particle size of $R = 7$ nm and magnetic field intensity $B_0 = 0.06$ T.

The influence of polydispersity is illustrated in Figure 12.28 for magnetite particles in a hydrocarbon carrier. It is clearly observed that the presence of polydispersity (increasing σ values) degrades the usefulness of the ferrofluid, showing the incentive to utilize highly monodispersed samples. The calculations are done for a tetradecane carrier fluid having specific heat of 2,080 J kg^{-1} K^{-1}, mass density 765 kg m^{-3} and viscosity 0.00235 kg m^{-1} s^{-1}. Surfactant layer thickness is set at $\delta = 2.0$ nm (Eq. (12.30)).

TABLE 12.1

Physical Properties of the Ferrofluids Considered by Rosensweig

Magnetic Solid	Chemical Formula	M_d (kA m^{-1})	K (kJ m^{-3})	c (J kg^{-1} K^{-1})	ρ (kg m^{-3})
Maghemite	$\gamma \cdot Fe_2O_3$	414	−4.6	~746	4,600
Magnetite	$FeO \cdot Fe_2O_3$	446	23–41	670	5,180
Cobalt ferrite	$CoO \cdot Fe_2O_3$	425	180–200	700	4,907
Barium ferrite	$BaO \cdot 6Fe_2O_3$	380	300–330	~650	5,280

M_d, domain magnetization; K, magnetic anisotropy constant; c, specific heat; ρ, density

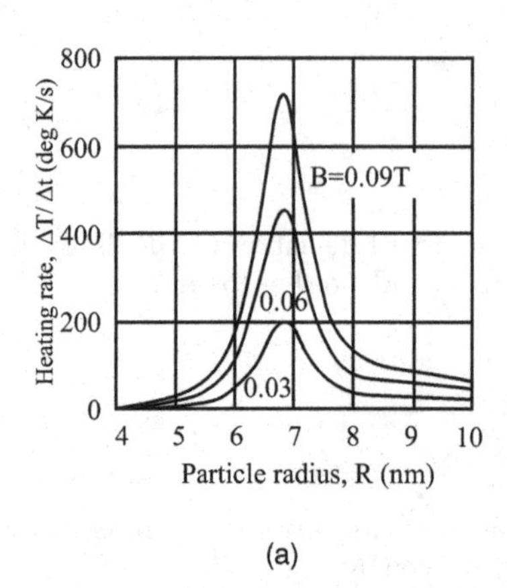

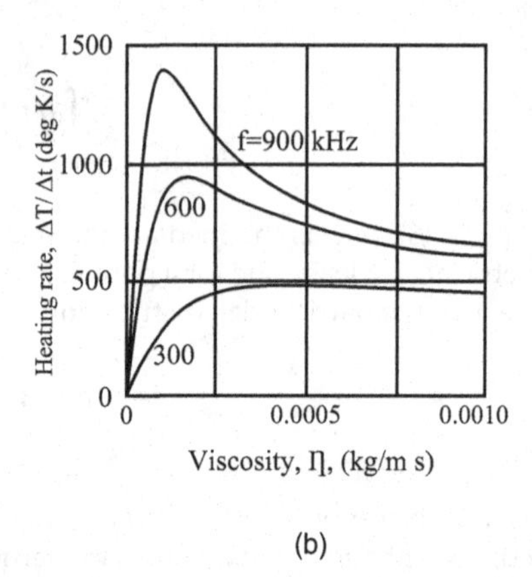

FIGURE 12.27
Temperature rise rate in monodispersed ferrofluid based on magnetite with magnetic fraction φ = 0.071: (a) Heating rate as a function of particle radius for various field strengths (tetradecane carrier, frequency 300 kHz); (b) heating rate as a function of viscosity at various frequencies (magnetic induction B_0 = 0.06 T, R = 7 nm). (From R. E. Rosensweig, Heating Magnetic Fluid with alternating Magnetic Field, *Journal of Magnetism and Magnetic Materials*, 252 (2002) 370, Copyright North Holland (Elsevier), reproduced with permission).

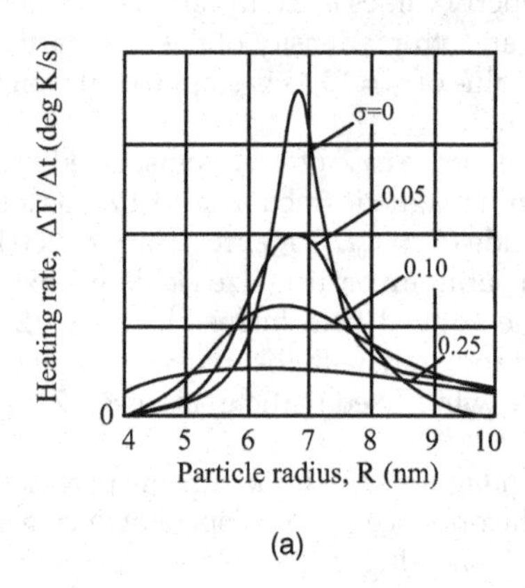

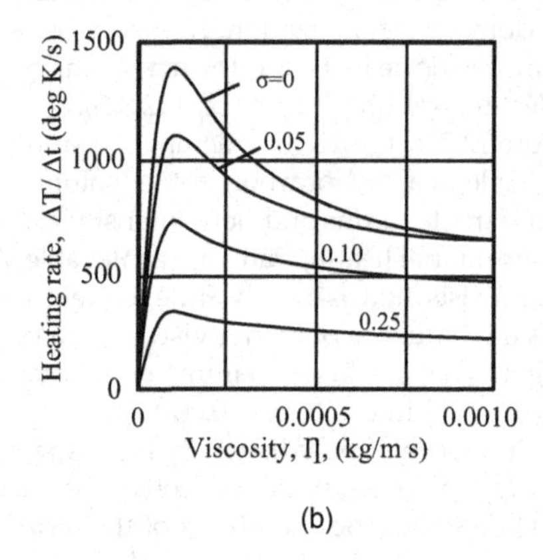

FIGURE 12.28
Temperature rise rate in polydispersed ferrofluid based on magnetite with magnetic fraction φ = 0.071: (a) Tetradecane carrier, f = 300 kHz; (b) Magnetic induction B_0 = 0.06 T, f = 900 kHz and R = 7 nm. (From R. E. Rosensweig, Heating Magnetic Fluid with alternating Magnetic Field, *Journal of Magnetism and Magnetic Materials*, 252 (2002) 370, Copyright North Holland (Elsevier), reproduced with permission).

Figure 12.29 compares heating rates for monodispersions of the various magnetic solids listed in Table 12.1, assuming $\tau_0 = 10^{-9}$ s (Eq. (12.33)). The carrier liquid is tetradecane in all cases. The main differences between the performances in heating rates of the resultant ferrofluids are due to the anisotropy constants, K, and the domain magnetizations, M_d, of the material (Table 12.1).

From the figure, it can be seen that barium ferrite and cobalt ferrite yield the largest heating rates in the size range of typical ferrofluids (R ~ 4–5 nm). The highest heating rates are provided by magnetite and maghemite; however, the size range where this occurs (R ~ 7–11 nm) is larger than that of typical ferrofluids.

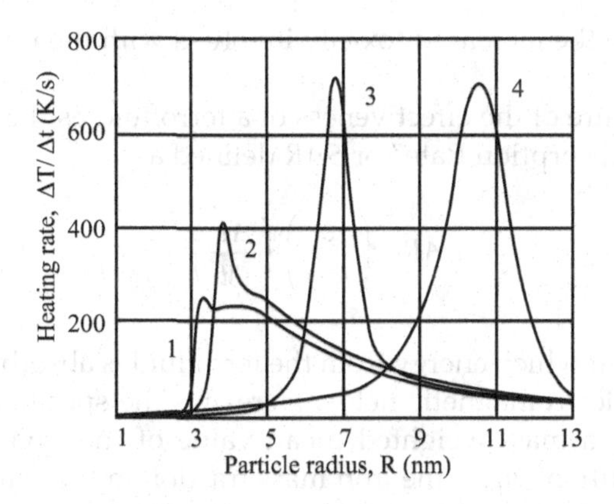

FIGURE 12.29
Comparative heating rates as a function of particle radius for various magnetic solids. 1—Barium ferrite, 2—Cobalt ferrite, 3—Magnetite, 4—Maghemite ($\varphi = 0.071$; $\eta = 0{:}00235$ kg m^{-1} s^{-1}, $f = 300$ kHz, $B_0 = 0.09$ T, $\delta = 2$ nm). (From R. E. Rosensweig, Heating Magnetic Fluid with alternating Magnetic Field, *Journal of Magnetism and Magnetic Materials*, 252 (2002) 370, Copyright North Holland (Elsevier), reproduced with permission).

12.6 Magnetic Fluid Hyperthermia for Cancer Therapy

The remarkable property of ferrofluids to produce considerable heating when subjected to an alternating magnetic field has been utilized for applications in cancer therapy. Ferrofluids can be selectively deposited in tumors, and subsequently subjected to an alternating magnetic field in the radio-frequency (RF) range to produce extremely localized magnetic heating. In this way, the energy absorption and the respective heat delivery are focused on the tumor without influencing the surrounding healthy tissue.

The resulting therapeutic procedure is known as magnetic fluid "hyperthermia". It promotes the increase of temperature in cancerous cells in order to change the functionality of their cellular structures. Achieving an increase in temperature to 41°C or 42°C can induce tumor cell death by altering cell growth and differentiation, which can induce apoptosis. By increasing the temperature further to 44°C or 46°C, one can induce cell ablation. Hyperthermia has many effects on the structure and physiology of the cell. One of the main mechanisms by which hyperthermia induces cell death involves the denaturation and aggregation of nuclear proteins, which are the first steps toward nuclear thermal damage. But the therapeutic effects of hyperthermia treatment are not merely caused by alterations in cell morphology and function. They are also caused by an immune response to thermal stress. Temperatures of about 42°C are already enough to stimulate killer cells, a mechanism of antitumor activity by the immune system with the release of heat shock proteins.

Ferrofluids for hyperthermia applications are usually based on polymer-encapsulated maghemite or magnetite nanoparticles, due to their biocompatibility and low toxicity. Nevertheless, one aims to raise the temperature of a tumor to the desired temperature by employing the minimum amount of ferrofluid possible and in the shortest irradiation time possible. For this reason, research in nanoferrites for hyperthermia applications, because they show the highest heating rates at typical ferrofluid radii (Figure 12.29), is an active area of research despite the increased toxicity involved with the introduction of transition metals other than iron.

A quantitative measure of the effectiveness of a ferrofluid as the hyperthermia agent is given by its "Specific Absorption Rate" or *SAR* defined as

$$SAR = \left(\frac{c}{m_{Fe}} \right) \cdot \left(\frac{\Delta T}{\Delta t} \right)_{initial} \tag{12.40}$$

SAR measures the rate at which energy from the ferrofluid is absorbed by the human body when exposed to RF electromagnetic fields. Here, c is the specific heat of the ferrofluid, which is calculated as a mass-weighted mean value of the iron-oxide core/polymeric shell/carrier water solution, m_{Fe} is the iron mass fraction in the sample, T is the temperature and t is the time the alternating field is on. The larger the initial heating rate after application of the alternating magnetic field, the more effective the ferrofluid. *SAR* is measured in watts per kilogram but is usually reported in watts per gram (W/g).

The principle of using magnetic materials to promote hyperthermia was first introduced in the late 1950s, by R. K. Gilchrist and co-workers. After their seminal work, many empirical studies were carried out to evaluate the therapeutic expression of hyperthermia in different tumor types. However, the results of these early studies did not receive widespread use among academic circles until the 1990s. Today, there have been many *in vitro* and *in vivo* studies including clinical trials. The efficacy of magnetic fluid hyperthermia has been demonstrated through experiments *in vitro* where the application of the technique to tumor cell cultures was observed to promote cell death by necrosis and/or apoptosis, or even *via* thermal ablation. This effectiveness has further been confirmed in small animal model studies, where an increase in survival and, in some cases, a decrease or total regression of tumor mass was observed. In addition, in the first clinical trials, magnetic fluid hyperthermia seems to be promising as an adjuvant therapy modality, combined with other conventional therapies, such as radiation therapy.

The application of magnetic fluid hyperthermia is especially important in the case of tumors that cannot easily be removed by surgery, such as those of the brain or the central nervous system, known as gliomas. Gliomas are a group of heterogeneous primary tumors arising from the glial cells. Malignant gliomas are associated with high morbidity and mortality. Glioblastoma is the most frequent and malignant glioma, and despite advances in diagnosis and treatment, its prognosis remains dismal. Using hyperthermia, new opportunities for the development of effective therapies for malignant gliomas are being pursued.

Example 12.7: Rapid Magnetic Heating Treatment by Highly Charged Maghemite Nanoparticles on Wistar Rats' Exocranial Glioma Tumors at Microliter Volume

As an example of research efforts to develop better therapies against gliomas using hyperthermia, we present the work of Rabias and co-workers (I. Rabias *et al.* Rapid magnetic heating treatment by highly charged maghemite nanoparticles on Wistar rats' exocranial glioma tumors at microliter volume, *Biomicrofluidics*, 4, 024111 (2010)). In their research, these authors have addressed the challenge of treating very small tumors, at the limit of the medical diagnosis resolution, without affecting the peripheral healthy tissue. In the case of small tumors, the efficiency of hyperthermia decreases dramatically due to the extremely small quantity of ferrofluid that can be deposited in such small volumes. To increase the number of magnetic particles deposited within such small-size tumors, the authors needed to use much denser ferrofluids than colloidal stability would dictate. In dense ferrofluids, interparticle interactions can result in particle aggregation and flocculation. To counteract the overall attractive magnetic interactions, the authors investigated the possibility of using a highly charged dextran coat on the magnetic nanoparticles in order to introduce strong electrostatic repulsion among particles. The strong electrostatic repulsion would also allow the particles to endure constant heating for several minutes at high temperatures without agglomeration. In a colloidal suspension during magnetically induced RF heating, the induced rising of temperature increases the average numbers of particle collisions per unit time, which could lead to particle agglomeration. Hence, keeping nanoparticles as far as possible, by increasing electrostatic repulsion, makes it possible to minimize the number of particle collisions and thus prevent agglomeration at high temperatures.

Their nanoparticle synthetic efforts were focused on the optimization of the surface charge of the dextran coating, while keeping the pH close to the physiological acidity value in the living organisms (pH = 7), which is a prerequisite for biological applications. For comparison and optimization reasons, three ferrofluids based on highly monodisperse ~10 nm diameter maghemite nanoparticles carrying different surface charges were investigated: An ionic one, with no coating and zeta potential value of 8 mV (S1), and two dextran coated with zeta potential values of 70 mV (S2) and 350 mV (S3), respectively, with a hydrodynamic radius of ~70 nm.

Figure 12.30 illustrates the magnetic heating effect and stability of aqueous dispersions of the dextran-coated maghemite particles. Samples were heated from room temperature to a maximum temperature value and then left to cool down back to the initial conditions. Heating was performed by placing a sample of the ferrofluids within a heating induction coil, which produced a moderate alternating magnetic field with an amplitude of 11 kA/m and frequency of 150 kHz. The temperature of the ferrofluid sample was monitored using a fiber optic thermometry system that can operate in extremely harsh electromagnetic environments.

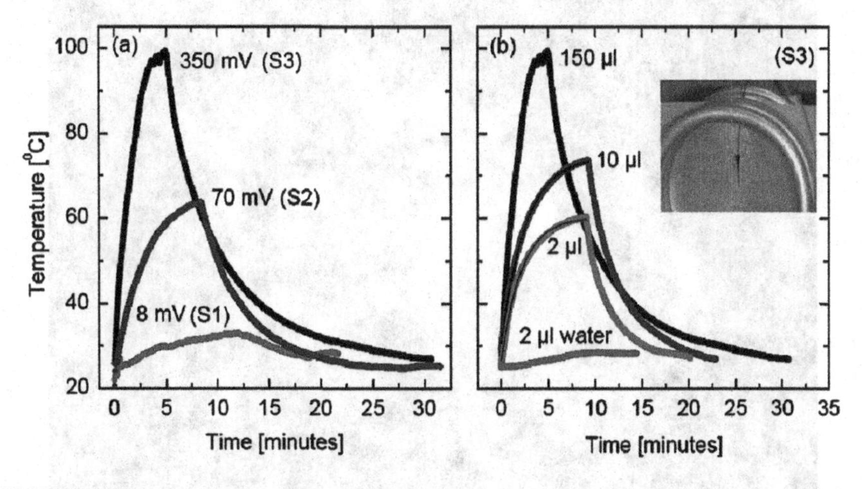

FIGURE 12.30
Magnetic heating effect on aqueous dispersions of dextran-coated maghemite nanoparticles. (a) The influence of the zeta potential on 150 μL of ferrofluids S1, S2, and S3 is indicated. (b) The influence of the volume of ferrofluid S3 is indicated. The curve for 2 μL water was used as the control sample. In the inset, the copper coil of the magnetic heating apparatus with the 10 μL sample and the optical fiber used for the temperature monitoring are shown, after removing the heat insulating cover. (From I. Rabias *et al.* Rapid magnetic heating treatment by highly charged maghemite nanoparticles on Wistar rats' exocranial glioma tumors at microliter volume, *Biomicrofluidics*, 4, 024111 (2010), Copyright American Institute of Physics, reproduced with permission).

Figure 12.30(a) demonstrates the strong effect of the zeta potential upon heating: S1 with the lowest zeta potential value of 8 mV does not produce any significant heating effect and started to agglomerate or flocculate right after heating. S2 exhibits a respectable temperature rise, reaching 60°C within 10 min. Similar to the S1 sample, S2 showed problems with stability and endurance, and it was instantly flocculated after the heating treatment. On the contrary, S3 exhibits an impressive heating effect, reaching 98°C within 3 min, with no traces of flocculation and a remarkable long-term stability, despite the fact that it reached almost the water boiling temperature. Thus, the strong electrostatic repulsion among nanoparticles in S3, due to the high surface charge, guaranteed the colloidal stability of the ferrofluid, even after heating. The *SAR* was calculated at 286 W/g.

Figure 12.30(b) demonstrates the heating effect of S3 for various amounts of ferrofluid 2, 10 and 150 μL, as indicated. Specifically, using 2 μL of the fluid (containing 40 μg of γ-Fe_2O_3) exhibited a temperature rise of 33°C within 10 min, while 150 μL containing 3 mg of γ-Fe_2O_3 reached to temperatures up to 99°C within 3 min with no indication of flocculation, when exposed to a moderate magnetic field with amplitude of 11 kA/m and frequency of 150 kHz. The inset shows an image of a sample within the induction coil and fiber optic probe. Note that the curve labeled 2 μL water, corresponding to a sample of water containing no ferrofluid, does not heat up upon exposure to the alternating electromagnetic field.

In order to demonstrate the ability of the S3 ferrofluid to function *in vivo* and assess its medical potential, the authors used it to perform hyperthermia treatment on small rat tumors. The tumors had been induced by inoculating Wistar male rats exocranially with glioma cells. After the rats had been anesthetized, the tissue covering the top of the skull was gently elevated and the bregma region of the skull was identified. Approximately 8×10^6 glioma cells were inoculated in the area anterior of the bregma. After two weeks, the development of a tumor was palpable and visible. Figure 12.31 shows a magnetic resonance image of the head of an inoculated rat.

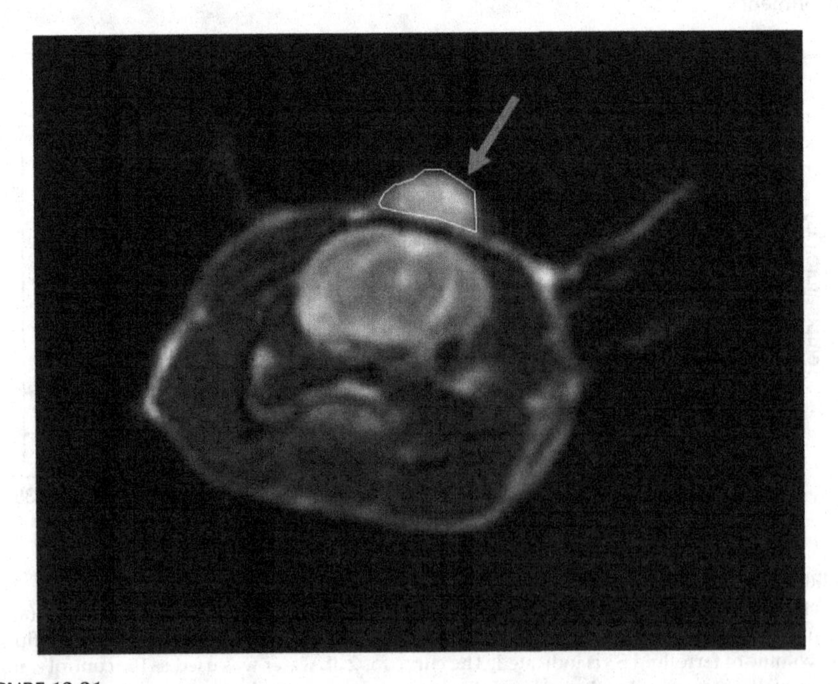

FIGURE 12.31
Magnetic resonance image of the head of an inoculated rat. The glioma tumor area is marked with an arrow. (From I. Rabias *et al.* Rapid magnetic heating treatment by highly charged maghemite nanoparticles on Wistar rats' exocranial glioma tumors at microliter volume, *Biomicrofluidics*, 4, 024111 (2010), Copyright American Institute of Physics, reproduced with permission).

When the tumor reached an average size of 5–10 mm, it was infused with 150 μL of S3 ferrofluid. The rat's head was then placed within the induction-heating coil and an alternating magnetic field of amplitude 11 kA/m and frequency 150 kHz was applied for 20 min, subjecting the tumor to magnetic hyperthermia treatment. Following the treatment, the rats were sacrificed and tumors were excised and fixed in formaldehyde. Paraffin-embedded sections were stained to reveal the nuclei and cytoplasm of cells using hematoxylin and eosin, a standard staining protocol for examining tissue integrity. The stained paraffin-embedded sections were examined under a Nikon Eclipse E800 microscope using 2× and 20× lenses (Figure 12.32).

Figure 12.32 shows the low- and high-resolution images of tumor tissue after treatment with and without ferrofluids. Hyperthermia treatment led to extensive tumor tissue damage and dissolution, shown in panels Figure 12.32(b) and (d), indicating that the obtained temperatures were close to boiling temperatures. Control experiments included the infusion of S3 ferrofluid without subsequent treatment as well as exposure to alternating magnetic fields in the absence of ferrofluid. No tumor tissue damage was observed, as indicated in panels Figure 12.32(a) and (c). Thus, the surface charge coated ferrofluid S3 performed

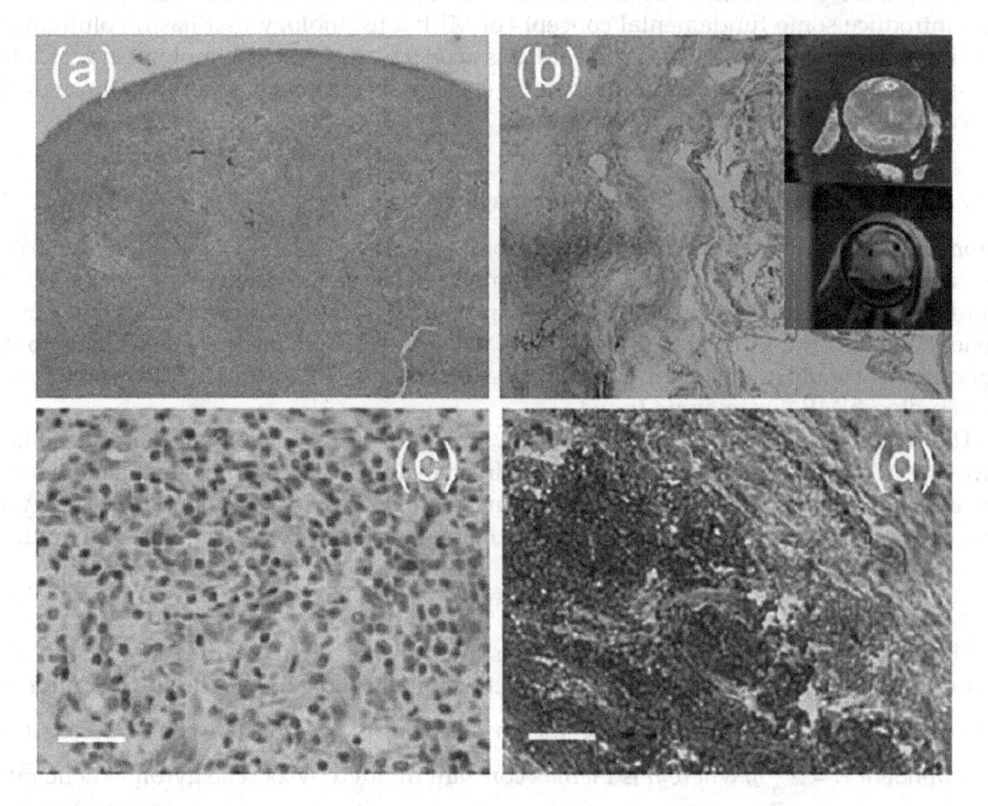

FIGURE 12.32
Panels (a) and (c) showing low- and high magnifications, respectively, represent sections of control tumor tissue. Panels (b) and (d), represent low- and high magnifications, respectively, of ferrofluid-treated tissue. They show extensive damage to the tumor tissue after treatment with ferrofluid. White bars = 200 μm. The lower inset in panel (b) shows the experimental setup for *in vivo* magnetic heating. The upper inset is an infrared image, which demonstrates the ability of the ferrofluid to produce strong localized heating at the tumor position. (From I. Rabias *et al.* Rapid magnetic heating treatment by highly charged maghemite nanoparticles on Wistar rats' exocranial glioma tumors at microliter volume, *Biomicrofluidics*, 4, 024111 (2010), Copyright American Institute of Physics, reproduced with permission). A color version of this image is available at https://www.routledge.com/9781439818466

excellently in the *in vivo* studies due to the preservation of the good dispersion of the ferrofluid within the tumor during the hyperthermia heating. The lower inset in panel (b) shows the experimental setup for *in vivo* magnetic heating. The upper inset is an infrared image of the rat's head during irradiation, which demonstrates the ability of the ferrofluid to produce strong localized heating at the tumor position, while peripheral tissue remains unaffected.

This example amply demonstrates the importance of developing stable, colloidal suspensions in the practice of hyperthermia.

12.7 Magnetic Resonance Imaging (MRI) Contrast Agents

Magnetic nanoparticles are making, yet, another important contribution to nanomedicine as intravenous, MRI contrast agents. In order to describe their mode of action, we must first introduce some fundamental concepts of MRI: a technology that has revolutionized the diagnosis and treatment of disease. MRI is an electromagnetic technology in which the composition of a sample can be probed by sensing its magnetic properties through RF waves. It is based on NMR spectroscopy.

12.7.1 Principles of Magnetic Resonance Imaging

Atomic nuclei with an odd number of nucleons generate a magnetic moment of their own as they spin around their axis. Several common nuclei, including hydrogen (^{1}H), the ^{13}C isotope of carbon, the ^{19}F isotope of fluorine and the ^{31}P isotope of phosphorous, have magnetic moments and, thus, are amenable to NMR spectroscopic measurements. Due to the abundance of water molecules in biological tissue, MRI makes use of ^{1}H-NMR, also referred to as proton-NMR, for imaging tissue.

The magnetic moment of the proton, $\vec{\mu}_p$, is due to its intrinsic quantum mechanical property of spin angular momentum, $\vec{I}$, in direct analogy to the spin magnetic moment of the electron, $\vec{\mu}_s$, due to its spin angular momentum $\vec{S}$, discussed in Chapter 2. The gyromagnetic ratio of the proton, γ_p, connects its magnetic moment to its angular momentum

$$\vec{\mu}_p = \gamma_p \vec{I} \tag{12.41}$$

The concept of the gyromagnetic ratio was first introduced in Chapter 2, for the case of an orbiting electron and its associated magnetic moment (Eq. (2.12)). The angular momentum of a proton, just as is the case for an electron, is quantized with $I = \frac{1}{2}\hbar$ and its z-component $I_z = \pm\frac{1}{2}\hbar$, where $\hbar$ is Plank's constant divided by 2π. The gyromagnetic ratio of the electron is given by

$$\gamma_e = g_e \frac{-e}{2m_e} \tag{12.42}$$

while that for the proton is given by

$$\gamma_p = g_p \frac{e}{2m_p} \tag{12.43}$$

where e is the elementary charge, m_e and m_p are the mass of the electron and proton, respectively, and g_e and g_p are their corresponding g-factors (see Chapter 2). Due to its lighter mass, 1,836 times less massive than the proton, the magnetic moment of the electron is much larger than that of the proton. Electronic magnetic moments are measured in Bohr magnetons, μ_B, while proton magnetic moments are measured in nuclear magnetons, μ_N, defined earlier (Eqs. (2.19) and (2.20)). In the presence of an applied magnetic field, electron-spin resonance occurs at the microwave regime, while proton-spin resonance occurs at much lower frequencies corresponding to the RF region.

In the absence of an applied magnetic field, water proton moments in a biological sample acquire random orientations. When an external magnetic field $\vec{B}$ is applied to the sample, protons experience a torque, which tends to orient their magnetic moment, and thus their spin angular momentum, parallel to the field. The quantization of the angular momentum of the proton results in parallel and antiparallel spin alignment with respect to the applied field, assumed in the z-direction. For a classical magnetic dipole, Eq. (1.22) indicates that the magnetic moment orientation parallel to the magnetic field corresponds to the lower energy state, while the antiparallel orientation is the higher energy state. In analogy to the classical case, each proton at equilibrium will acquire one of these two possible orientations; spin-up or spin-down, as depicted in Figure 12.33(a). These two states differ in orientational potential energy of the proton magnetic moment in the presence of the applied field, with the spin-up $I_z = +\frac{1}{2}\hbar$ state having lower energy compared to the spin-down $I_z = -\frac{1}{2}\hbar$ state. The Heisenberg uncertainty principle prevents the total nuclear magnetic moment, $\vec{I}$, to be perfectly aligned with the applied magnetic field; only its z-component can be aligned or antialigned with the magnetic field. This is the nuclear Zeeman Effect.

Thus, the proton magnetic moment also has x- and y-components, in the plane perpendicular to the field. This results in a torque exerted on the moment by the magnetic field, which causes the proton's magnetic moment to precess about the magnetic field, B, with a well-defined angular frequency, called the angular Larmor precession frequency,

$$\omega_L = \gamma_p B \tag{12.44}$$

Figure 12.33(b) depicts the precessing proton moment about the applied magnetic field. The frequency of precession ($f_L = \omega_L / 2\pi$) is simply the number of complete revolutions the

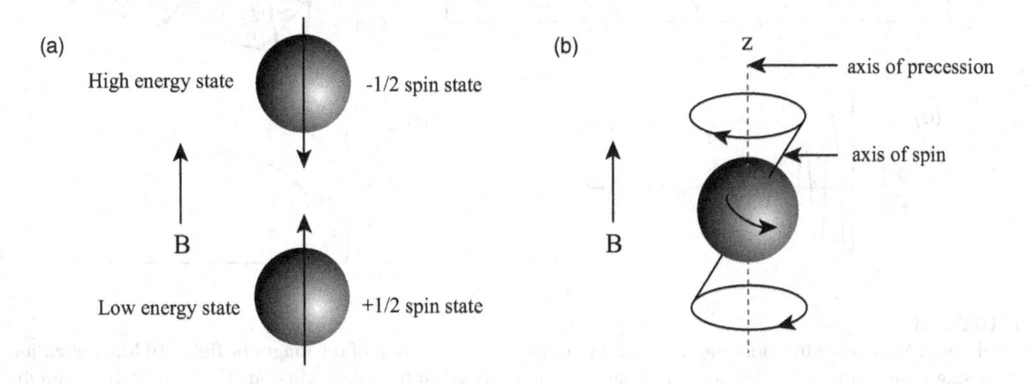

FIGURE 12.33
(a) Spin-up and spin-down states. (b) Proton moment precession in a magnetic field.

proton performs per second, as it precesses around the magnetic field. Its value is directly proportional to the strength of the magnetic field B.

At thermal equilibrium, the relative population of protons with spin-up and spin-down states is given by

$$\frac{N_h}{N_l} = e^{-U/kT} \tag{12.45}$$

Here, N represents the population number, l the lower energy state, h the higher energy state, U their energy difference, k Boltzmann's constant, and T the temperature in Kelvin. In a unit volume of material, containing a large number of protons, the induced macroscopic magnetization is given by

$$\vec{M} = (N_l - N_h)\vec{\mu}_p \tag{12.46}$$

It is this induced proton magnetization density and its variation within the body that MRI relies upon in imaging biological tissue through proton-NMR measurements.

Even though the moment on a proton is exceedingly small, ~1.41×10^{-26} J T^{-1}, the exceedingly large number of protons due to water molecules present in biological tissue leads to a measurable effect in the presence of large magnetic fields. For example, under conditions of thermal equilibrium at room temperature, an externally applied field of 1 T on an ensemble of protons will result in only three of every million proton moments being aligned parallel to the field. However, there are about 6.6×10^{19} protons available in every mm^3 of water, so the effective signal arising from 2×10^{14} proton moments per mm^3 is observable.

Figure 12.34 gives an illustration of the process of magnetic resonance for a large ensemble of protons, a fundamental process used in MRI imaging. The net proton magnetization density $\vec{M}$ is a macroscopic quantity, and thus, can be treated classically. It can point in any

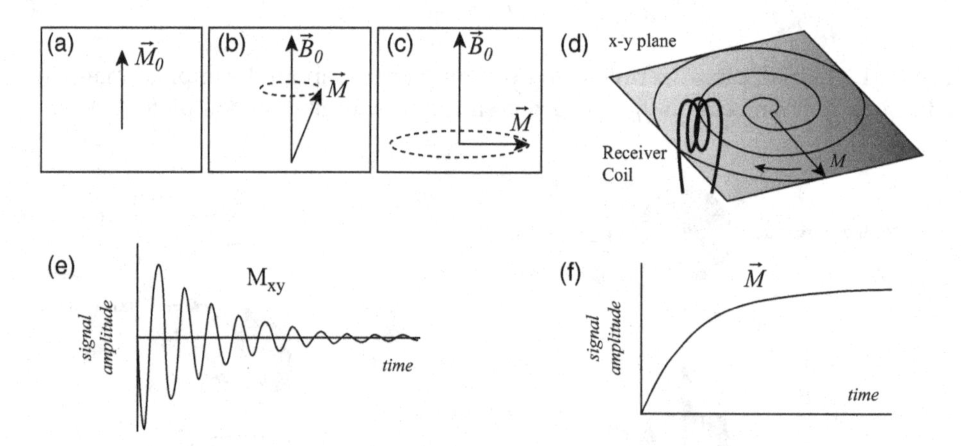

FIGURE 12.34
Principles of MRI. (a) Saturation magnetization pointing in the direction of the magnetic field. (b) Magnetization precessing around the magnetic field. (c) Magnetization flipped in the perpendicular direction to the magnetic field after the application of a pulse. (d) Depiction of receiver coils. (e) Decrease of the transverse magnetization with time after a pulse. (f) Growth of the longitudinal magnetization along the applied field with time.

direction, including along the direction of the applied magnetic field, Figure 12.34(a), acquiring a maximum value or saturation magnetization, M_0, along the field direction, assuming the z-direction (pointing up). When the magnetization is tipped away from $\vec{B}_0$, so that it has x- and y-components, $\vec{M}$ will precess about the applied magnetic field, as shown in Figure 12.34(b). The frequency of precession is the characteristic proton angular Larmor precession frequency in the given field,

$$\omega_0 = \gamma_p B_0 \tag{12.47}$$

The proton gyromagnetic ratio $\gamma_p = 2.67 \times 10^8$ rad s^{-1} T^{-1}, so that in a field of $B_0 = 1$ T, the Larmor precession frequency, $f_0 = \omega_0/2\pi = 42.57$ MHz, falls in the RF range. As discussed in detail below, f_0 determines the frequency of the electromagnetic radiation required for MRI imaging. The fact that it falls in the RF range, which is low-energy non-ionizing radiation, makes MRI a safe medical procedure compared to other imaging modalities, such as computerized tomography that depend on high-frequency ionizing radiation, like X-rays.

A pulse of a second, alternating external field $\vec{B}_1$ perpendicular to $\vec{B}_0$ is applied oscillating with angular frequency ω_0, the proton Larmor precession frequency in the static magnetic field $\vec{B}_0$, Figure 12.34(c). The strength of $\vec{B}_1$ is much weaker than that of $\vec{B}_0$. Nevertheless, the fact that $\vec{B}_1$ oscillates with the Larmor precession frequency of the magnetization vector has the effect of resonantly exciting the proton magnetic moments. As energy is absorbed from the RF-field pulse, the magnetization vector rotates away from the longitudinal direction. The amount of rotation, referred to as the flip angle, depends on the strength and duration of the RF pulse. A 90° flip angle results in the precession of the magnetization $\vec{M}$ within the *xy*-plane, the plane perpendicular to $\vec{B}_0$. This results in a diminution of M_z and the acquisition of a transverse oscillatory component, $M_{x,y}$ in the *xy*-plane. After a short time, $\vec{B}_1$ is removed and the system, over time, returns back to its original equilibrium state. One records the ensuing relaxation process, that is, the loss of the transverse magnetization. In practice, the RF transverse field $\vec{B}_1$ is applied in a pulsed sequence of a duration sufficient to derive a coherent response from the net magnetization of the protons. From the instant that the RF pulse is turned off, the relaxation of the coherent response is measured *via* induced currents in pick-up coils, or receivers, in the MRI scanner. The detection is based on Faraday's Law of induction (Eq. (1.26)). These resonantly tuned detection coils are placed within the *xy*-plane as depicted in Figure 12.34(d) in order to detect the decay of $M_{x,y}$, Figure 12.34(e). The coils are designed not simply to detect the signal but to also enhance the signal intensity by a quality factor of *ca.* 50–100. The pulsed nature of $\vec{B}_1$ is critical to the technology of MRI.

The return to equilibrium is exponential in time, governed by the following equations:

$$M_z(t) = M_0\left(1 - e^{-t/T_1}\right) \tag{12.48}$$

and

$$M_{x,y}(t) = M_0 \sin(\omega_0 t + \varphi) e^{-t/T_2} \tag{12.49}$$

Equation (12.48) gives the exponential growth of the component of the magnetization along the field direction, depicted in Figure 12.34 (f), with time constant T_1, which is known as the longitudinal or spin–lattice relaxation (*T1*-recovery) time constant. Equation (12.49) gives the exponential decay of the transverse magnetization, perpendicular to the field direction, the exponentially dumped oscillation of Figure 12.34(e), where T_2 is the time

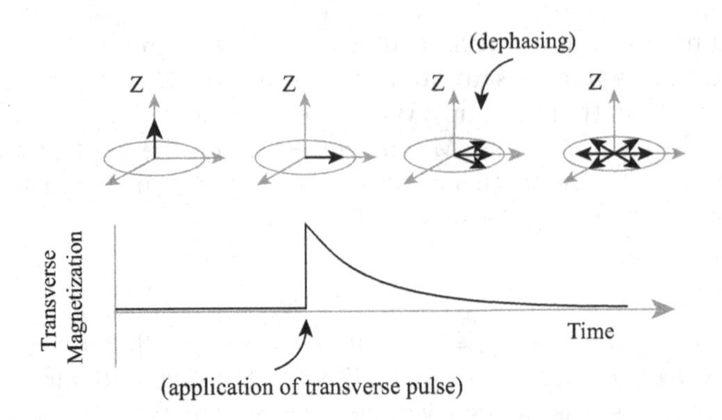

FIGURE 12.35
Depiction of proton-spin dephasing and loss of transverse magnetization.

constant of the exponential decay of the signal's amplitude and is known as the transverse or spin–spin relaxation (*T2*-decay) time constant, and φ is a phase constant.

The longitudinal relaxation is slow and reflects a loss of energy, as heat, from the proton-spin system to its surrounding "lattice"; it is primarily a measure of the dipolar coupling of the proton moments to their surroundings. The transverse relaxation within the *xy*-plane is relatively rapid, and is driven by the loss of phase coherence, or "dephasing", in the precessing protons due to magnetic interactions with each other and with other fluctuating magnetic moments in their environment. Figure 12.35 depicts the process of dephasing. While right after the application of the transverse pulse, all the excited protons rotate within the *xy*-plane coherently, that is with the same phase φ, over time, the protons dephase. The onset of dephasing, whereupon different protons start to acquire different phases, results in a decrease in transverse magnetization.

Figure 12.36 depicts these differences between White Matter, Gray Matter and the Cerebro-Spinal Fluid (CSF) of the brain. The rate at which the longitudinal magnetization grows back (*T1*-recovery) is different for the protons associated with different tissues, due to differences in tissue density, magnetic susceptibility, water content, *etc*. Similarly, the rate at which the transverse magnetization decays (*T2*-decay) is tissue-dependent because protons in different tissues dephase at different rates. This difference is utilized in obtaining the MRI image.

T1-weighted imaging is depicted in Figure 12.36(a), while *T2*-weighted imaging is shown in Figure 12.36(b). For best contrast, the images are obtained at a time when the relaxation curves are widely separated. Thus, the spin–lattice (*T1*-recovery) and spin–spin (*T2*-decay) relaxation processes are utilized to generate a bright and a dark MRI image, respectively. That is, T_1 relaxation leads to an increase in signal intensity and thus causing positive contrast, while T_2 processes result in signal loss and negative contrast. As T_2 relaxation is much faster, it is the preferred mode in MRI imaging.

12.7.2 Mode of Action of Superparamagnetic Contrast Agents

Dephasing, and therefore T_2 relaxation, is affected by local magnetic inhomogeneity in the tissue or in the applied longitudinal field, leading to the replacement of the relaxation time constant T_2 by the shorter relaxation time, T_2^*, where

$$\frac{1}{T_2^*} = \frac{1}{T_2} + \gamma_p \frac{\Delta B_0}{2} \tag{12.50}$$

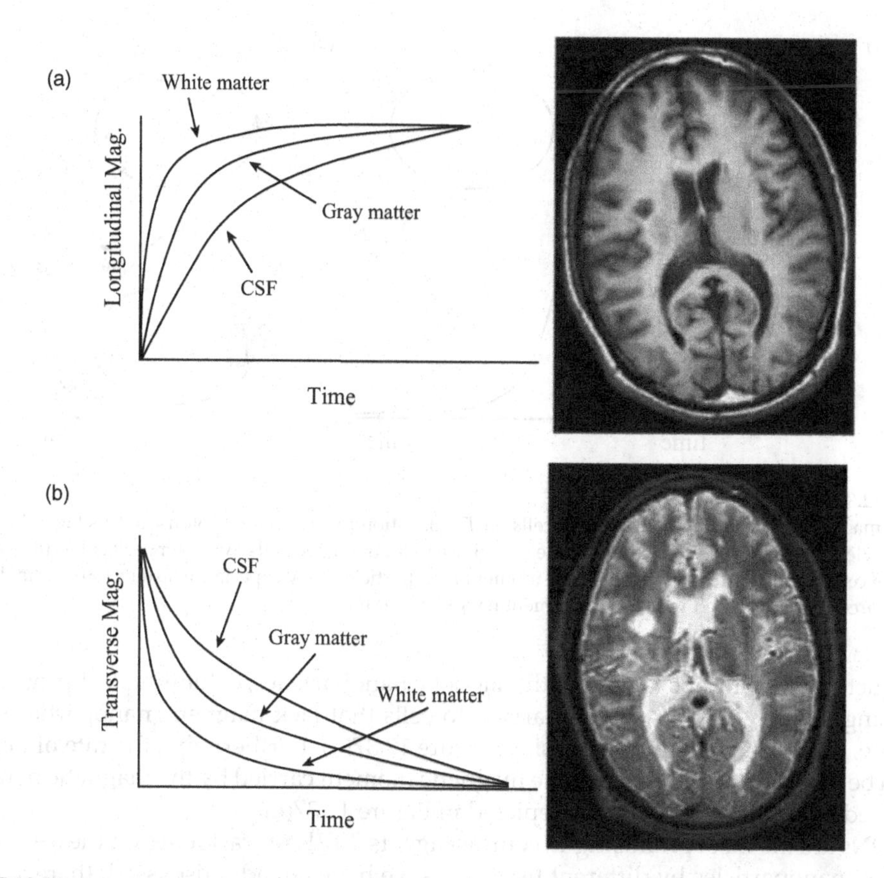

FIGURE 12.36
Brain images with corresponding exponential magnetization graphs for (a) longitudinal, *T1*-weighted imaging, and transverse, *T2*-weighted imaging. (From Robert A. Pooley, *Fundamental Physics of MR imaging, Radiographics*, Volume 25, Number 4, pp. 1087 – 1099, Copyright Radiological Society of North America, reproduced with permission).

Here, ΔB_0 is the variation in the field brought about either through distortions in the homogeneity of the applied field itself, or by local variations in the magnetic susceptibility of the tissue.

Both T_1 and T_2^* can be shortened by using magnetic contrast agents, such as paramagnetic gadolinium compounds, or biocompatible SPIONs. When the contrast agents are in the form of superparamagnetic particles, they act primarily as T_2^* contrast agents. The superparamagnetic particles used are magnetically saturated in the normal range of magnetic field strengths, 0.5–3 T, used in MRI scanners, thereby establishing a substantial locally perturbing dipolar field, contributing to ΔB_0 in Eq. (12.50). Thus, this dipolar field can introduce a significant field perturbation, ΔB_0, which results in fast proton-spin dephasing. This leads to a marked shortening of T_2^* with a less marked reduction of T_1. This mode of action of superparamagnetic nanoparticles as MRI enhancement agents is given by Eq. (12.50). The enhanced image is processed by recording $M_{x,y}$, the magnetization transverse to the applied field, that is, in *T2*-weighted imaging, for best contrast.

Thus, superparamagnetic contrast agents disrupt the local magnetic field by contributing a large ΔB_0 that enhances proton dephasing of water molecules diffusing in their vicinity. Their mode of action as T_2^* contrast agents is depicted in Figure 12.37. The figure depicts

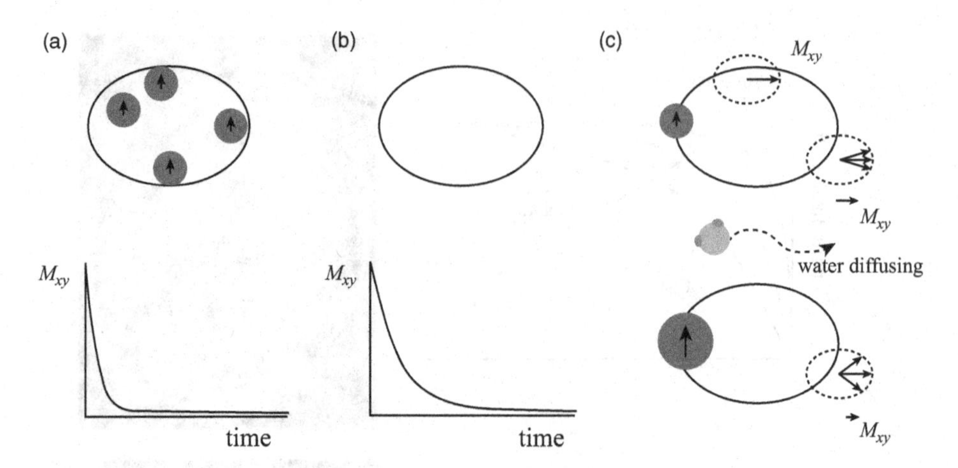

FIGURE 12.37
Effect of magnetic particle internalization in cells on T_2 relaxation times: (a) the protons in cells tagged by magnetic particles have a shorter T_2^* relaxation time than those in (b) untagged cells. As water molecules diffuse in the vicinity of cells tagged with a higher magnetic moment nanoparticles, they experience faster proton-spin dephasing compared to those tagged with lower moment nanoparticles (c).

the reduction of T_2 in cells tagged with magnetic nanoparticles, exhibiting fast proton-spin dephasing, Figure 12.37(a), in comparison to cells that lack magnetic nanoparticles, and therefore, exhibit slower spin dephasing, Figure 12.37(b). Furthermore, the rate of dephasing can be tuned by the strength of the magnetic moment carried by the magnetic nanoparticles used to tag different cells, as depicted in Figure 12.37(c).

SPIONs are the most promising T_2 contrast agents. MRI contrast relies on the differential uptake of nanoparticles by different tissues. As we have already discussed, there is also a size effect: nanoparticles with diameters of ~30 nm or more are rapidly collected by the liver and spleen, while particles with sizes of ~10 nm or less are not so easily recognized. The smaller particles, therefore, have a longer half-life in the bloodstream and are collected by reticuloendothelial cells throughout the body, including those in the lymph nodes and bone marrow, allowing for MRI dark image enhancement of these organs. Similarly, such agents can be used to visualize the vascular system, and to image the central nervous system. The technique is based on the fact that tumor cells lack the effective reticuloendothelial system of healthy cells, and thus do not collect nanoparticulate agents. As a result, tumor cell relaxation times are not altered by the contrast agents. Thus, the enhancement of the tumor cell detection is achieved by darkening the image of the surrounding healthy tissue, that is, negative contrast enhancement. This has been used, for example, to assist the identification of malignant lymph nodes, liver tumors and brain tumors in small animal studies.

12.7.3 Relaxivity of Contrast Agents

The efficacy of MRI enhancement agents is generally presented in terms of their "relaxivity", which is the degree to which the agent can enhance the water proton relaxation rate constant, and thus enhance the sensitivity of detection. Since the contrast agent may affect the T_1 and T_2 relaxation times individually, there are two corresponding relaxivities. The relaxation rate constants are defined as the inverse of the relaxation times ($R_1 = 1/T_1$ and $R_2 = 1/T_2$), normalized to the concentration of the contrast agent. That is, the relaxivity reflects how the

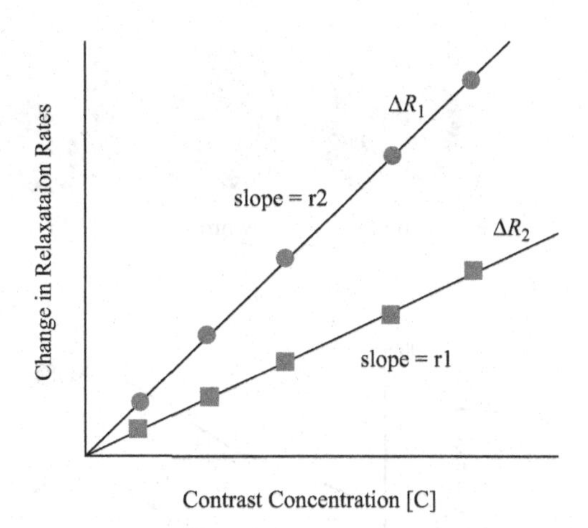

FIGURE 12.38
Measurement of relaxivities of MRI enhancement agents. The change in relaxation rate is plotted as a function of magnetic nanoparticle contrast agent concentration.

relaxation rates of a water solution change as a function of contrast agent concentration [C]. The relaxivities, denoted by $r1$ and $r2$, are then defined by Eqs. (12.51) and (12.52).

$$\Delta R_1 = R_1 - R_1^0 = \frac{1}{T_1} - \frac{1}{T_1^0} = r1 \cdot [C] \tag{12.51}$$

$$\Delta R_2 = R_2 - R_2^0 = \frac{1}{T_2} - \frac{1}{T_2^0} = r2 \cdot [C] \tag{12.52}$$

where the superscript zero denotes the absence of contrast agents in the solution. Thus, the relaxivities in solution are obtained by graphing changes in relaxation rates as a function of contrast concentration, measured in mmol/L, and thus, the relaxivities are given in L/(mmol · s). The slopes of the lines in Figure 12.38 represent $r1$ and $r2$.

Relaxivity depends on the field strength, temperature, and type of solution used. For clinical use, it is typical to report relaxivities at an applied field of 1.5 T, in plasma for a body temperature of 37°C.

Magnetic nanoparticle imaging enhancement agents contribute to ongoing research aiming at the development of ultra-sensitive molecular imaging nanoprobes with enhanced relaxivities for the detection of targeted biological entities at high resolution. Systematic evaluation of the magnetic moment, size and type of spinel metal ferrites, for example, and their relaxivities have been investigated, as presented in the following example.

Example 12.8: Artificially Engineered Magnetic Nanoparticles for Ultra-Sensitive Molecular Imaging

As an example of spinel ferrite nanoparticle enhancement agents, we present a study by Lee *et al.*, published in Nature Medicine (J.-H. Lee *et al.*, Artificially engineered magnetic nanoparticles for ultra-sensitive molecular imaging, *Nature Medicine*, 13 (2007) 95). The authors carried out a comprehensive study on ferrite nanoparticles, $MnFe_2O_4$, Fe_3O_4, $CoFe_2O_4$ and $NiFe_2O_4$, for MRI applications. In particular, they investigated their relaxivities as a function of their size and magnetic anisotropy properties. Figure 12.39 shows the relaxivities of magnetite (Fe_3O_4) and manganese ferrite ($MnFe_2O_4$) nanoparticles as a function of particle size.

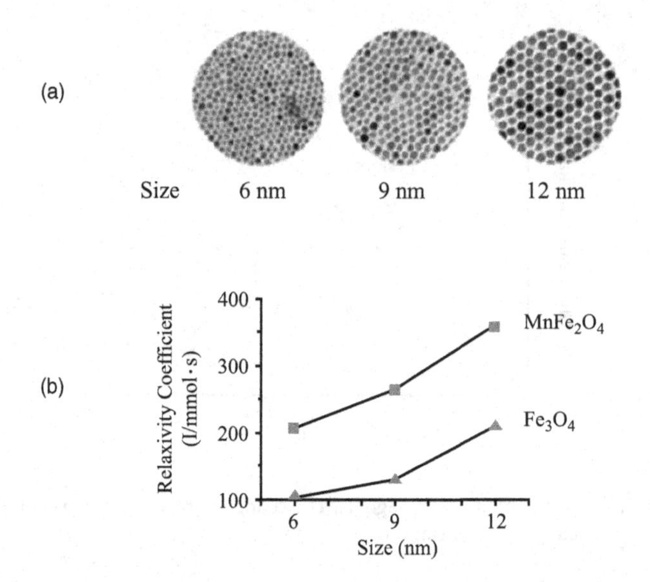

FIGURE 12.39
Size dependence of relaxivity of $MnFe_2O_4$ and Fe_3O_4 nanoparticle contrast agents as a function of particle size. (Adapted from J.-H. Lee *et al.*, Artificially engineered magnetic nanoparticles for ultrasensitive molecular imaging, *Nature Medicine*, 13 (2007) 95, Copyright Nature Research with permission.

It is observed that the relaxivity increases with particle size for both spinel metal oxides, but $MnFe_2O_4$ nanoparticles show consistently higher relaxivities compared to similar size Fe_3O_4 nanoparticles. This is due to the fact that the magnetization value of a superparamagnetic nanoparticle at a certain magnetic field is dependent, not only on the size, but also on the magnetocrystalline anisotropy of the nanoparticle. Manganese ferrite nanoparticles have low magnetocrystalline anisotropy compared to magnetite, resulting in easy magnetization and higher local magnetic susceptibility values, leading to a larger field perturbation, ΔB_0, and therefore, shorter relaxation time and higher relaxivity. The relaxivity values reported by the authors for 12 nm nanoparticles at an applied field of 1.5 T, are 358, 218, 172, and 152 L/(mmol · s) for $MnFe_2O_4$, Fe_3O_4, $CoFe_2O_4$, and $NiFe_2O_4$, respectively. The lower the magnetic anisotropy, the higher the relaxivity value. This allows for fine-tuning and tailoring the relaxivity properties of the enhancement agents for specific applications.

Research on ferrite nanoparticles as MRI contrast agents is a very active area of research. In addition to size and magnetocrystalline anisotropy, other aspects that affect the nanomagnetism of these systems, such as surface anisotropy altered by the type of coating used to render the nanoparticles biocompatible, are being investigated.

Exercises

1 Describe the process of functionalization of magnetic nanoparticles.
2 Define the magnetophoretic mobility of a magnetic particle in a colloidal solution and compare it with that of an immunomagnetically labeled cell.

3. What is the fundamental principle of operation of magnetic separators? Give examples of various separator designs.
4. What force is responsible for magnetically guided drug delivery? What are the advantages and limitations of the technique?
5. Define the concept of magnetofection. Give examples of various magnetofection vectors.
6. Write the first law of thermodynamics for a classical thermodynamic system as opposed to a magnetic system.
7. For a ferrofluid subjected to an alternating magnetic field, what parameters affect the power dissipated per unit volume?
8. Define Néel and Brownian relaxation in ferrofluids.
9. In MRI imaging, what do the relaxation times T_1 and T_2 refer to? How are they measured?
10. MRI contrast agents are characterized by their relaxivities. Define the concept of relaxivity. How is it related to relaxation time?

Afterword

Environmental and Ethical Issues in Nanomagnetism

Nanomagnetism is making important contributions to nanotechnology integrating physics with engineering, biology, chemistry, and medicine. As an important player in the nanotechnology revolution, nanomagnetism is advancing many areas of science and engineering and provides a platform for current and future technological innovations. Magnetic nanoparticles exhibit unique properties due to their size and shape as they lie at the interface between the molecular and solid state, at the intersection of classical and quantum-mechanical behavior, that is, they exhibit quantum-size effects leading to size-dependent physical properties different from those of their bulk counterparts.

Given the expected increased rates of magnetic nanomaterial production, nanomagnetism shares the same ethical and environmental issues of weighing benefit *vs.* risk associated with nanotechnology at large. The potential that magnetic nanoparticles could be released in the environment and possibly adversely affect our ecosystem is an area of scientific inquiry. Due to their high reactivity, it is necessary to investigate the fate and behavior of engineered magnetic nanoparticles in the environment. In particular, their biodegradability must be thoroughly examined when treated as waste products. Due to their small size, magnetic nanoparticles can become airborne, they can migrate into bodies of surface and groundwater and they can be deposited into soils. Therefore, they could potentially enter the human body by inhalation or *via* the food chain, where, as we have seen, they can interact with the biological cell exhibiting nanostructure-dependent biological activity, which could present a risk to public health. Magnetic nanoparticles can be deposited in the respiratory system and have nanostructure-induced toxicity due to their large surface-to-volume ratio, high surface reactivity and unusual morphology, agglomeration into larger particles or fiber-like chains or degradation into smaller particles after deposition. Regulation of the manufacturing and means of disposal of engineered magnetic nanoparticles is needed in order to allow nanomagnetism to fulfill its promise for technological innovation in the miniaturization of devices and in nanomedicine applications, while minimizing the risk.

Magnetic nanoparticles enable new technological innovations in nanomedicine due to their novel properties, such as superparamagnetism, promising great positive impact on human health by contributing to the diagnosis (cell sorting, MRI contrast enhancement) and treatment (magnetically guided drug delivery, hyperthermia) of disease. Molecular MRI can be realized only through the accumulation of contrast agents at the nanoscale to reach the necessary contrast levels. However, before such nanomedicine products acquire widespread use in disease diagnosis, prevention, and treatment, they must undergo thorough testing for toxicity and evaluation of their benefit to risk ratio for *in vivo* applications in pre-clinical and clinical trials. It is generally agreed that safety and risk issues must be thoroughly understood before nanomagnetism can have its full beneficial impact on society and human health.

Once introduced into the circulatory system, magnetic nanoparticles may enter the liver, lymph nodes, spleen, bone marrow, *etc*. Translocation of nanoparticles along nerve fibers could enable entry into the central nervous system, circumventing the blood–brain barrier, where they may cause or trigger neurodegenerative diseases. The mechanisms of interaction between nanoparticles and the immune system need further elucidation; these may vary according to the chemical composition, size range and degree of functionalization of the nanoparticles. Furthermore, the risk associated with magnetic nanoparticles may vary based on the route of entrance. For example, nanoparticles that are benign when injected may be toxic when inhaled. Thus, their toxicity must be further examined according to the nature of exposure: dermal, oral, respiratory, or intravenous. Nanotoxicologists are conducting *in vitro* studies using animal and human cells as well as *in vivo* experiments using small animal model systems to gather information about the cause of toxicity and the distribution, metabolism and excretion of magnetic nanoparticles. Ethical guidelines dictate that associated risks be reasonable relative to potential benefits to patients and society and that risks be minimized at all costs.

Ethics in medical applications of nanomagnetism also raise the issue of social justice related to the equitable accessibility of novel products and procedures once nanomagnetism moves from the research and development (R&D) stage to the market. Due to manufacturer's patents, new medical products are usually very expensive. Thus, in the short term at least, intellectual property rights may exacerbate socio-economic inequalities, as the poor may not afford to avail themselves of the benefits derived from expensive medical innovations. This may widen the gap within and between both individuals and societies. Governmental and industrial policy guidelines are necessary to ensure nanomedicine's affordability for all.

It is, however, certain that no matter how long it takes for nanomedicine to become common practice, nanomedicine will usher a new era in health care where medical procedures will be highly accurate and precise, less painful or toxic, with smaller collateral damage from undesirable side effects compared to those currently practiced. It is anticipated that invasive surgical procedures that require cutting will become a thing of the past.

Appendix: Differential Vector Calculus

A.1 The Ordinary Derivative

The derivative $\frac{df}{dx}$ of a scalar function of one variable, $f(x)$, indicates how rapidly the value of the function changes if the argument x varies by an infinitesimal amount, dx, or

$$df = \left(\frac{df}{dx}\right)dx \tag{A.1}$$

A.2 The Gradient

In the case of a function of three variables, $f(x,y,z)$, the derivative indicates how the value of f changes if each of the variables in the argument varies by an infinitesimal amount, dx, dy, dz.

$$df = \left(\frac{\partial f}{\partial x}\right)dx + \left(\frac{\partial f}{\partial y}\right)dy + \left(\frac{\partial f}{\partial z}\right)dz \tag{A.2}$$

Here, we make use of the partial derivatives of f along each of the three Cartesian coordinate directions. In vector form, this can be expressed as

$$df = \left(\frac{\partial f}{\partial x}\hat{x} + \frac{\partial f}{\partial y}\hat{y} + \frac{\partial f}{\partial z}\hat{z}\right)\cdot\left(dx\,\hat{x} + dy\,\hat{y} + dz\,\hat{z}\right) = \left(\vec{\nabla}f\right)\cdot\left(d\vec{l}\right) \tag{A.3}$$

Here, $\hat{x}$, $\hat{y}$ and $\hat{z}$ indicate unit vectors along the axes of a right-handed orthogonal Cartesian coordinate axis system,

$$d\vec{l} = dx\,\hat{x} + dy\,\hat{y} + dz\,\hat{z} \tag{A.4}$$

gives the generalized increment $d\vec{l}$ along an arbitrary direction and the "del operator" or "gradient", a vector differential operator, is defined by

$$\vec{\nabla} = \frac{\partial}{\partial x}\hat{x} + \frac{\partial}{\partial y}\hat{y} + \frac{\partial}{\partial z}\hat{z} \tag{A.5}$$

Thus, the change in the value of f due to a generalized increment $d\vec{l}$ can be expressed as

$$df = \vec{\nabla} f \cdot d\vec{l} = \left|\vec{\nabla} f\right|\left|d\vec{l}\right|\cos\theta \tag{A.6}$$

where θ is the angle between $\vec{\nabla} f$ and $d\vec{l}$. The maximum change in f occurs when $\cos\theta$ is one, that is, when the generalized increment is along the gradient of f. Thus, the gradient of a scalar function is a vector pointing in the direction of the function's steepest ascent. The gradient operator $\vec{\nabla}$ is a vector operator, as seen from Eq. (A.5), and thus, has all the properties of a vector. In the case of vectors, in addition to the multiplication with a constant, $a\vec{A}$, there are two other vector products defined, the dot product $(\vec{A}\cdot\vec{B})$ and the cross product $(\vec{A}\times\vec{B})$, where $\vec{A}$ and $\vec{B}$ are vectors. One can, therefore, define the product of del with constants, scalar and vector functions of position.

Thus, besides its multiplication with a constant, $a\vec{\nabla}$, there are three other ways $\vec{\nabla}$ can act.

On a scalar function f: $\vec{\nabla} f$ (the gradient)

On a vector function $\vec{v}$, *via* the dot product: $\vec{\nabla}\cdot\vec{v}$ (the divergence)

On a vector function $\vec{v}$, *via* the cross product: $\vec{\nabla}\times\vec{v}$ (the curl).

A.3 The Fundamental Theorem of Calculus

In calculus, the operation of integration is the inverse of differentiation. Thus, for a function of one variable $f(x)$

$$\int_a^b \frac{df(x)}{dx}dx = f(b) - f(a) \tag{A.7}$$

Equation (A.7) is known as the "Fundamental Theorem of Calculus". To evaluate the integral of the derivative of a scalar function of one variable, you only need to evaluate the function at the endpoints of the interval over which you are integrating. The endpoints are the boundaries of the region on the x-axis over which the integration takes place.

In vector calculus, we define three fundamental theorems, corresponding to the three derivatives of gradient, divergence, and curl, as described below.

A.4 The Fundamental Theorem for Gradients

Consider a scalar function of three variables $g(x,y,z)$ or $g(\vec{r})$, where $\vec{r} = x\hat{x} + y\hat{y} + z\hat{z}$ is the position vector. Then,

$$\int_{\vec{a}}^{\vec{b}} \left(\vec{\nabla} g(\vec{r})\right)\cdot d\vec{l} = g(\vec{a}) - g(\vec{b}) \tag{A.8}$$

Equation (A.8) is known as the "Fundamental Theorem of Gradients". In words, this reads in an analogous way to the fundamental theorem of calculus: in order to evaluate the line integral of the gradient of a scalar function in three-dimensional space over a generalized path, you need to evaluate the function at the beginning $(\vec{a})$ and the end $(\vec{b})$ points of the path; that is the boundaries of the line path.

This theorem has two very important corollaries. First, it is seen that the line integral of the gradient of a scalar function is independent of the path taken from $(\vec{a})$ to $(\vec{b})$, and second, the line integral over a closed path must be zero, since the beginning and ending points are the same.

A.5 The Fundamental Theorem for Divergences

Given a vector function $\vec{v}(x,y,z)$ in three-dimensional space, the integral of its divergence, over a volume V, can be expressed in terms of the surface integral of the vector function over a closed surface S bounding the volume V.

$$\int_V (\vec{\nabla}\cdot\vec{v})d\tau = \oint_S \vec{v}\cdot d\vec{a} \tag{A.9}$$

Equation (A.9) is known as the "Fundamental Theorem for Divergences". It is also known as Gauss's Theorem or Green's Theorem. Again, the integral of a derivative of a vector function, this time a divergence, over a volume is determined by evaluating the vector function at a surface bounding the volume and integrating over the surface, that is, the boundary of the volume.

A.6 The Fundamental Theorem for Curls

Given a vector function $\vec{v}(x,y,z)$ in three-dimensional space, the surface integral of its curl over a surface S can be expressed in terms of the line integral of the vector function over a closed path P, bounding S.

$$\int_S (\vec{\nabla}\times\vec{v})\cdot d\vec{a} = \oint_P \vec{v}\cdot d\vec{l} \tag{A.10}$$

Equation (A.10) is known as the "Fundamental Theorem for Curls". Again, the surface integral of the derivative of a vector function, this time its curl, can be determined by evaluating the vector function at the boundary of the surface and taking its line integral over this boundary, which constitutes a closed path.

Since an infinite number of surfaces can be bounded by a specific closed path, or loop, it follows that the surface integral of a curl depends only on the boundary line, not the particular surface considered. In addition, if you consider the process of making the loop smaller and smaller, till you shrink it to a point, it follows that the surface integral of a curl over a closed surface vanishes, as the integration loop degenerates to a point.

The above fundamental theorems of vector calculus may be used to go from the differential to the integral forms of Maxwell's equations and *vice versa*.

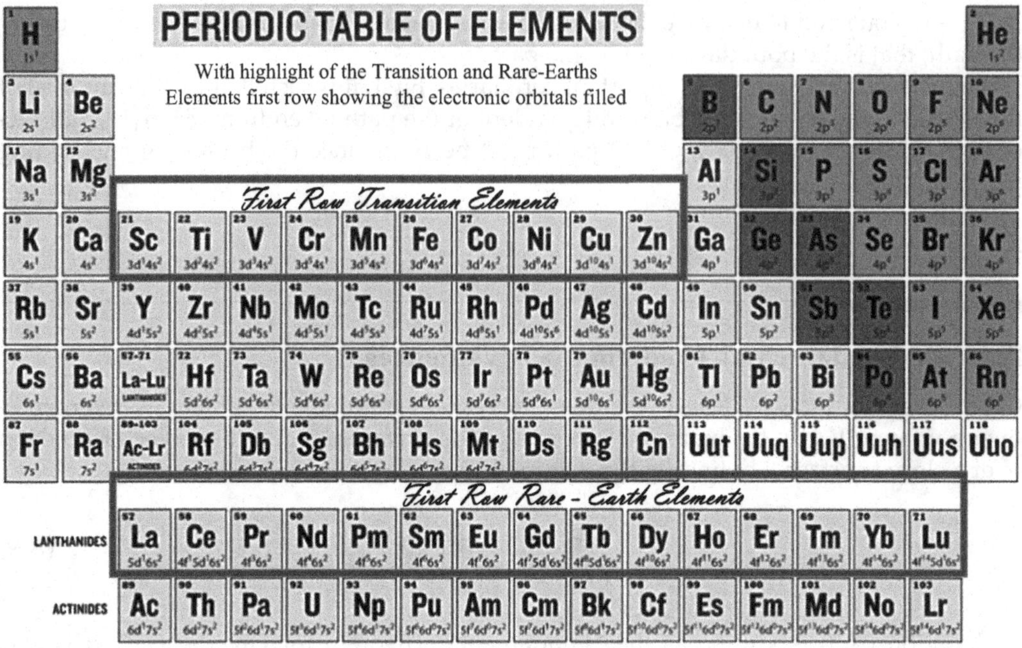

PERIODIC TABLE OF ELEMENTS

With highlight of the Transition and Rare-Earths Elements first row showing the electronic orbitals filled

1	2	3	4	5	6	7	8	9	10	11	12	13	14	15	16	17	18
1 H $1s^1$																	2 He $1s^2$
3 Li $2s^1$	4 Be $2s^2$											5 B $2p^1$	6 C $2p^2$	7 N $2p^3$	8 O $2p^4$	9 F $2p^5$	10 Ne $2p^6$
11 Na $3s^1$	12 Mg $3s^2$											13 Al $3p^1$	14 Si $3p^2$	15 P $3p^3$	16 S $3p^4$	17 Cl $3p^5$	18 Ar $3p^6$
19 K $4s^1$	20 Ca $4s^2$	21 Sc $3d^14s^2$	22 Ti $3d^24s^2$	23 V $3d^34s^2$	24 Cr $3d^54s^1$	25 Mn $3d^54s^2$	26 Fe $3d^64s^2$	27 Co $3d^74s^2$	28 Ni $3d^84s^2$	29 Cu $3d^{10}4s^1$	30 Zn $3d^{10}4s^2$	31 Ga $4p^1$	32 Ge $4p^2$	33 As $4p^3$	34 Se $4p^4$	35 Br $4p^5$	36 Kr $4p^6$
37 Rb $5s^1$	38 Sr $5s^2$	39 Y $4d^15s^2$	40 Zr $4d^25s^2$	41 Nb $4d^45s^1$	42 Mo $4d^55s^1$	43 Tc $4d^55s^2$	44 Ru $4d^75s^1$	45 Rh $4d^85s^1$	46 Pd $4d^{10}5s^6$	47 Ag $4d^{10}5s^1$	48 Cd $4d^{10}5s^2$	49 In $5p^1$	50 Sn $5p^2$	51 Sb $5p^3$	52 Te $5p^4$	53 I $5p^5$	54 Xe $5p^6$
55 Cs $6s^1$	56 Ba $6s^2$	57-71 La-Lu LANTHANIDES	72 Hf $5d^26s^2$	73 Ta $5d^36s^2$	74 W $5d^46s^2$	75 Re $5d^56s^2$	76 Os $5d^66s^2$	77 Ir $5d^76s^2$	78 Pt $5d^96s^1$	79 Au $5d^{10}6s^1$	80 Hg $5d^{10}6s^2$	81 Tl $6p^1$	82 Pb $6p^2$	83 Bi $6p^3$	84 Po $6p^4$	85 At $6p^5$	86 Rn $6p^6$
87 Fr $7s^1$	88 Ra $7s^2$	89-103 Ac-Lr ACTINIDES	104 Rf $6d^27s^2$	105 Db $6d^37s^2$	106 Sg $6d^47s^2$	107 Bh $6d^57s^2$	108 Hs $6d^67s^2$	109 Mt $6d^77s^2$	110 Ds	111 Rg	112 Cn	113 Uut	114 Uuq	115 Uup	116 Uuh	117 Uus	118 Uuo

First Row Transition Elements: Sc, Ti, V, Cr, Mn, Fe, Co, Ni, Cu, Zn

First Row Rare - Earth Elements

LANTHANIDES	57 La $5d^16s^2$	58 Ce $4f^15d^16s^2$	59 Pr $4f^36s^2$	60 Nd $4f^46s^2$	61 Pm $4f^56s^2$	62 Sm $4f^66s^2$	63 Eu $4f^76s^2$	64 Gd $4f^75d^16s^2$	65 Tb $4f^85d^16s^2$	66 Dy $4f^{10}6s^2$	67 Ho $4f^{11}6s^2$	68 Er $4f^{12}6s^2$	69 Tm $4f^{13}6s^2$	70 Yb $4f^{14}6s^2$	71 Lu $4f^{14}5d^16s^2$
ACTINIDES	89 Ac $6d^17s^2$	90 Th $6d^27s^2$	91 Pa $5f^26d^17s^2$	92 U $5f^36d^17s^2$	93 Np $5f^46d^17s^2$	94 Pu $5f^66d^07s^2$	95 Am $5f^76d^07s^2$	96 Cm $5f^76d^17s^2$	97 Bk $5f^86d^17s^2$	98 Cf $5f^{10}6d^07s^2$	99 Es $5f^{11}6d^07s^2$	100 Fm $5f^{12}6d^07s^2$	101 Md $5f^{13}6d^07s^2$	102 No $5f^{14}6d^07s^2$	103 Lr $5f^{14}6d^17s^2$

Index

Page numbers in **bold** indicate tables and page numbers in *italic* indicate figures.